Optical Wireless Communication Theory and Technology

Series Editor

Xizheng Ke, School of Automation and Information Engineering, Xi'an University of Technology, Xi'an, Shaanxi, China

The book series Optical Wireless Communication Theory and Technology aims to introduce the key technologies and applications adopted in optical wireless communication to researchers of communication engineering, optical engineering, and other related majors. The individual book volumes in the series are thematic. The goal of each volume is to give readers a comprehensive overview of how the theory and technology in a certain optical wireless communication area can be known. As a collection, the series provides valuable resources to a wide audience in academia, the communication engineering research community, and anyone else who are looking to expand their knowledge of optical communication.

Xizheng Ke

Principles and Applications of Optical Wireless Orthogonal Frequency-Division Multiplexing

Science Press
Beijing

Springer

Xizheng Ke
School of Automation and Information Engineering
Xi'an University of Technology
Xi'an, Shaanxi, China

ISSN 2731-5967 ISSN 2731-5975 (electronic)
Optical Wireless Communication Theory and Technology
ISBN 978-981-97-7972-7 ISBN 978-981-97-7973-4 (eBook)
https://doi.org/10.1007/978-981-97-7973-4

Jointly published with Science Press
The print edition is not for sale in China mainland. Customers from China mainland please order the print book from: Science Press.

Translation from the Chinese Simplified language edition: "无线光正交频分复用原理及应用" by Xizheng Ke, © Science Press 2018. Published by Science Press. All Rights Reserved.

This Springer imprint is published by the registered company Springer Nature Singapore Pte Ltd.
The registered company address is: 152 Beach Road, #21-01/04 Gateway East, Singapore 189721, Singapore

Preface

Transportation, energy, and communication are the three infrastructures of modern social development. Radio-frequency communication has important applications in modern society, and the application fields involving optical communication will expand.

Optical communication includes optical-fiber and optical wireless communication. Optical-fiber communication corresponds to traditional cable communication and is a type of wired communication. Optical wireless communication is similar to microwave and radio-frequency communication. Optical wireless communication has the bandwidth of optical-fiber communication, and the flexibility of microwave communication. More importantly, optical wireless communication has no spectrum control, good confidentiality, large communication capacity, low cost, easy installation, flexible networking, strong anti-electromagnetic interference ability, and other advantages. Optical wireless communication includes wireless laser communication, ultraviolet communication, and visible-light communication, which can be used for visible-light communication between satellites, between satellites and Earth, and indoors.

Atmospheric turbulence and atmospheric channel interference are problems that optical wireless communications face, and inhibiting atmospheric turbulence and the impact of the atmospheric environment on optical wireless communications is an urgent problem to be solved. To maximize the approximate capacity of optical wireless channels, optical orthogonal frequency-division multiplexing (OFDM) transmission is introduced in optical wireless communications to improve the spectral efficiency of optical transmissions and suppress atmospheric turbulence and atmospheric environmental interference. This book expounds on the principles and applications of optical wireless orthogonal frequency-division multiplexing and verifies the relevant theory through numerical analyses and language-communication experiments.

Since 2002, the author has undertaken two National Natural Science Foundation projects (60977054 and 61377080), two major scientific and technological projects in Shaanxi Province, a project for the Natural Science Foundation of the Shaanxi Provincial Science and Technology Department, a project for the Xi'an Science and Technology Bureau, and a National Science and Technology Innovation Fund project. He and his associates have conducted systematic and in-depth research on the problems involved in the field of optical wireless communication, obtained more than 20 invention patents in the field of optical wireless communication, and won 10 provincial and ministerial science and technology awards, including one first prize and four second prizes.

This book is divided into 10 chapters. It first conducts a systematic and in-depth analysis of the research progress of optical wireless communication and clarifies the importance and benefits of optical wireless OFDM transmission. Then, source coding is discussed, optical wireless OFDM systems are clarified, and the characteristics of optical OFDM are explained through numerical simulations. Theoretical analyses and numerical simulations of the peak-to-average ratio, time synchronization, channel estimation, and channel allocation of OFDM are conducted. Numerical simulations and communication experiments are used to verify the performance of optical OFDM systems and the feasibility of related algorithms.

This book is the crystallization of more than ten years of research by researchers at the Optoelectronic Engineering Technology Research Center of Xi'an University of Technology. Associate Professors Zhao Li and Chen Dan participated in the research work of the project; doctoral students Xie Mengqi, Wang Jiao, and Chen Jinni participated in some of the research work, and master's students Zhang Hongwei, Wu Rui, Wang Wu, Duan Zhongxiong, Li Mengfan, Li Beilei, Liu Juan, and Zhang Tuo participated in the corresponding research work. The author is grateful for the youth and enthusiasm of many teachers and students for over ten years on this topic, and deeply respects their tireless spirit of inquiry in scientific research.

He thanks the National Natural Science Foundation of China (60977054, 61377080), Shaanxi Province "13115" Science and Technology Overall Plan (2011KTCQ01-31), Shaanxi Provincial Department of Education Industrialization Incubation Fund (2010JC17), Xi'an Science and Technology Achievements Conversion Fund (CX12165), Shaanxi Provincial Key Industrial Innovation Chain "Research on Key Technologies of Ultra-long-distance Wireless Laser Communication in Atmospheric Environment" (2017 ZDCXL-GY-06-01), and other funding sources.

The author is grateful to the authors of the numerous references in this book. It is their pioneering research that has enabled him and others to clarify the direction of the development of this topic. The information contained in this book is relatively complex and limited by the author's level of knowledge. Inappropriate or even incorrect points will inevitably appear. Corrections and comments are welcome.

Xi'an, China
Midsummer 2017

Xizheng Ke

Content Summary

Orthogonal frequency-division multiplexing (OFDM) is a type of multi-carrier communication that can improve the spectral efficiency of optical communication and suppress channel interference. This book discusses the theory and application of OFDM in optical wireless communication.

The OFDM transmission system in optical wireless communication is systematically introduced, and the peak-to-average ratio problem, time synchronization, channel estimation, and channel allocation involved in optical OFDM are analyzed in depth. The effects of atmospheric turbulence and the atmospheric environment on OFDM signals are analyzed in detail, and the characteristics of optical OFDM that inhibit atmospheric turbulence are compared and analyzed. Through numerical simulations and experiments, the relevant theories have been verified, and the latest progress in this field has been touched upon.

This book can be used as a reference for undergraduates, graduate students, and scientific and technological workers in fields related to electronic information, computers, and communication engineering.

Contents

Chapter 1
Introduction

Optical wireless communication uses light waves as carriers to transmit information. Corresponding to optical-fiber communication, it is called optical wireless communication, because it does not require transmission cables or other media. This chapter describes the history and existing problems in optical wireless communication.

1.1 Development of Optical Wireless Communication Technology

In 1880, Alexander Graham Bell (1847–1922) successfully conducted an optical telephone experiment using sunlight as the light source, which marked the beginning of modern optical wireless communication [1]. Later, the progress of optical wireless communication technology was limited by the lack of many technical developments, such as the levels of light sources, which slowed its development.

The concept of stimulated radiation, proposed by Albert Einstein (1879–1955), provided the theoretical basis for laser technology. This theory states that a laser is a phenomenon in which excited atoms radiate photons when they transition to a low-energy state or a ground state under the action of an external radiation field. At this point, the energy from the external radiation must be exactly equal to the energy difference between the two levels of the atom. The frequency, phase, propagation direction, and polarization state of the photons emitted by the stimulated radiation and foreign photons are the same.

In 1964, Qian Xue Sen (1911–2009) proposed changing "light amplification by stimulated emission of radiation" to "laser". This radiation output amplifies the light and makes it coherent; thus, the emission direction, frequency, phase, and polarization of multiple photons are exactly the same. The establishment and development of quantum mechanics have enabled a deeper understanding of the microstructure and motion laws of matter. The energy-level distribution, transition, and photon radiation

X. Ke, *Principles and Applications of Optical Wireless Orthogonal Frequency-Division Multiplexing*, Optical Wireless Communication Theory and Technology,
https://doi.org/10.1007/978-981-97-7973-4_1

of microscopic particles have also been more effectively proven, which objectively improved Einstein's stimulated-radiation theory and laid a theoretical foundation for the production of lasers.

In 1951, Edward Mills Purcell (1912–1997) and others successfully realized particle-number inversion in an experiment and obtained 50 kHz stimulated radiation. Charles Townes (1915–2015) and Soviet physicists Nikolay G. Basov (1922–2001) and Aleksandr M. Prokhorov (1916–2002) successfully proposed a design for generating and amplifying microwaves using the principle of stimulated radiation of atoms and molecules.

In 1960, Theodore Maiman (1927–2007) used a high-strength flash tube to stimulate the chromium atoms in a ruby crystal to produce a fairly concentrated thin red-light column. The appearance of ruby lasers provided an opportunity to improve the transmission stability of optical wireless communication systems. With the progress in optoelectronic technology and the urgent requirements of users, optical wireless communication has once again become the focus of attention.

In 1981, the United States conducted a optical wireless communication demonstration experiment between aircraft and ground stations at a missile range in New Mexico. A frequency-doubled yttrium–aluminum–garnet (YAG) laser (532-nm wavelength) was used, with a maximum communication rate of 1 Gbit/s and a maximum communication distance of 100 km [2]. From the mid-1980s to 1994, the U.S. Air Force supported a high-speed inter-satellite optical wireless communication experimental device built by Lincoln Laboratory of the Massachusetts Institute of Technology, which adopted a heterodyne receiving mode to verify the feasibility of coherent wireless laser communication. The experiment used a 30 mW semiconductor laser and an eight-inch telescope system, with a 220 Mbit/s data rate and a 40 km communication distance [3].

In 1998, the Jet Propulsion Laboratory (JPL) undertook an optical communication demonstrator (OCD) and carried out terminal system-level assembly and performance tests. The assembly and experimental results of the terminal were measured by a wireless laser-communication terminal test and evaluation station (LCT-TES) [4]. In February 1998, Lucent Company manufactured a 10 Gbit/s atmospheric laser-communication experimental system, which is equivalent to a ground-to-ground link in space laser communication. The laser signal of the experimental system was transmitted through the atmosphere, and the communication performance was limited by the communication distance, baud rate, climatic conditions, and other factors.

Astro Terra incorporated an automatic-tracking system into its optical wireless communication experimental system to correct the impact of building shaking on beam collimation. The automatic-tracking system used a built-in camera to detect the change of the laser-beam transmission direction, which was fed back to the electronic execution unit to maintain the beam collimation. In 1998, Lucent and Astro Terra successfully tested the prototype of a optical wireless communication system, with a communication-link distance of 2.5 km and a data rate of 2.5 Gbit/s [5].

In 2013, the United States National Aeronautics and Space Administration (NASA) successfully completed an experiment of a lunar laser-communication demonstration (LLCD). The LLCD experiment verified the practicability of deep-space broadband communication. The downlink data sent back to Earth reached 622 Mbit/s, and the uplink data reached 20 Mbit/s. In 2014, NASA used a laser beam to transmit a high-definition video from an international space station to the ground. It took only 3.5 s to successfully transmit it back, which is equivalent to a transmission rate of 50 Mbit/s, which is 10–1000 times higher than the existing RF communication mode. In the test, the ground station first aimed at the space station and sent a laser cursor, which was then locked by the optical parallel array logic system (OPALS) on the space station. It then started to transmit the video.

Since the 1970s, the European Space Agency (ESA) has developed a series of research plans in the field of space wireless laser communication and has conducted research on relevant technologies for optical wireless communication [6]. Examples include the Telecommunications Research Program (TRP) for basic technology research and the Association of European Science and Technology Transfer Professionals (ASTP) for inter-satellite laser communication system and technology research. The semiconductor laser inter-satellite link experiment (SILEX) program has been implemented since 1989 [7].

The plan also included cooperation with Japan to conduct space laser-communication experiments. Japan used the LUCE wireless laser-communication terminal on the optical inter-orbit communications-engineering test satellite (OICETS) to conduct communication experiments with the SILEX terminal. In 1994, the ESA passed a scheme review of the SILEX equipment level, and in 1996, a subsystem performance test was conducted. The purpose of this plan was to develop all unit technologies and components of an inter-satellite wireless laser-communication system and to establish and test it.

Japan has been researching inter-satellite wireless laser communications since the mid-1980s, mainly including the Communication Research Laboratory (CRL), the National Space Development Agency (NASDA), and the Optical and Radio Research Laboratory of the Advanced Telecommunications Research Institute International (ATR) of the Ministry of Posts and Telecommunications. The ATR mainly studied and demonstrated key technologies, such as beam control and modulation, and conducted ground simulation experiments with optical wireless [8–10].

1.2 Research Progress on Light-Wave Propagation in Atmospheric Turbulence

The establishment of a theoretical model of light-wave propagation in atmospheric turbulence directly depends on the refractive-index fluctuation spectrum of atmospheric turbulence. In 1948, von Karman developed the von Karman spectrum used to describe the turbulence in a circular tube [11]; it is widely used to describe models of a turbulent-energy input region. To facilitate theoretical analyses, Russian scientist

Tatarskii introduced the turbulence theory developed by Kolmogorov and Obukhov, and proposed an exponential spectrum model [12–17].

In 1974, Greenwood and Tarazano measured and researched the turbulence spectrum in a terrain environment with a rough underlying surface [18] and assembled the Greenwood–Tarazano spectrum from the measured data. The exponential spectrum is very close to the Greenwood–Tarazano spectrum. The exponential spectrum and Greenwood–Tarazano spectrum are commonly used in many theoretical analyses. However, some experimental studies have shown that the von Karman spectrum is closer to the experimental data than the exponential spectrum.

To study the turbulence effect, more research has focused on the characteristics of the beam spread and light-intensity scintillation. Currently, several theoretical and experimental results have been presented [19–21]. In 1970, Whitman et al. [22] derived the propagation formula of a laser in random media according to the wave-scalar equation and provided a specific analysis using the Kolmogorov spectrum. The results showed that the larger the propagation distance, the larger the beam propagation radius. In 1972, Poitier et al. [23] applied the Born approximation method to discuss the broadening effect of a focused beam in turbulence. In 1975, Fante [24] obtained a long-term beam-propagation formula by using the modified von Karman spectrum and derived a short-term beam-propagation formula using the mutual interference function.

In 1999, Fan [25] studied the intensity distribution and beam spread of Gaussian collimated beams obtained using the Kolmogorov and Mellin transform and obtained the average intensity and long-term beam radius under the von Karman spectrum. The beam spread at various target heights was calculated using the measured height distribution of the near-earth atmospheric-turbulence intensity in Kunming, China. In 1999, Andrews et al. [26–30] studied the influence of internal and external turbulence scales on the long-term beam radius, and obtained long-term beam calibration by numerical simulation. Wei et al. [31] provided the spread radius of a laser beam in an oblique transmission path.

In 2002, Yi et al. [32, 33] extended the classical Rytov method to medium and strong turbulent regions, and extended horizontal propagation to oblique propagation, according to the turbulence atmospheric-structure constant model of the International Telecommunications Union Radio Communications Sector (ITU-R), and obtained zero internal- and external-scale plane waves. They presented the variation of the spherical-wave scintillation exponent with a modified Rytov variance.

In 2004, Yang and Wu [34] and Wei [35] further discussed the scintillation exponential characteristics of oblique Gaussian beams, based on the scintillation model of spherical and plane waves on an oblique path. In 2007, Jinyong et al. [36] determined the average intensity relationship of Gaussian beams propagating in a turbulent atmosphere, turbulent scale factors, short-term spread factor of beams, including turbulent external-scale effects, and the relationship between the Gaussian-beam equivalent radius and the propagation distance, initial beam radius, and optical wavelength under a modified von Karman turbulence spectrum.

When a laser propagates in the near-ground layer, the spot is "broken" into several pieces, owing to the influence of atmospheric turbulence. To study the influence of the spot on beam expansion, Zhang Feizhou et al. [37] calculated the transmission of the emitted beam with different shapes and the beam quality under different turbulence intensities, and statistically analyzed the characteristics of the spot. They found that the total spot expanded as the beam quality deteriorated and the turbulence intensity enhanced. Moreover, the number of speckles increased and the duty cycle decreased; however, the radius of the speckles changed little, which is approximately the beam radius of a vacuum diffraction containing 63.2% of the total energy.

In 2008, Qian Xian Mei et al. simulated a multi-layer phase screen to analyze the changes of statistical parameters, such as the effective radius of the beam, the root mean-square of the spot-centroid drift, power density with the zenith angle, laser wavelength, and initial radius when the laser was transmitted through the atmosphere at an oblique path [38]. In the same year, Chu Xiu Xiang et al. studied the average intensity and spread of multiple Gaussian beams during oblique propagation in atmospheric turbulence, and obtained the relationship between the intensity and spread with altitude, beam-coherence length, beam-waist width, and other parameters [39].

In 2009, Qian Xian Mei carried out numerical simulations of beam propagation along a non-uniform turbulent path and compared three different phase-screen distribution schemes: uniformly distributed, equal Rytov-exponent-spaced, and equal Fried-parameter-spaced phase screens. The results showed that when the beam propagated along a non-uniform turbulent path, the equal Fried-parameter-spaced phase screen was the best [40].

In 2010, Chu Xiu Xiang et al. deduced the expression of the intensity distribution of partially coherent Gaussian Schell-model (GSM) beams during oblique propagation in atmospheric turbulence, studied the cases of relay propagation and direct propagation, and concluded that the intensity distribution of relay propagation was better than that of direct propagation. In addition, they studied the variation of the peak intensity of beams with a target height [41]. In 2012, Zhang Xiaoxin et al. analyzed the influence of the beam order, spatial coherence, zenith angle, and other parameters on the beam width, when the beam propagated in an oblique path, by taking a partially coherent flat-topped beam as an example [42].

In 2013, Duan Meiling et al. studied the average intensity and mean-square beam width of a GSM beam when it was transmitted in three different paths: uplink, downlink, and horizontal through atmospheric turbulence. They found that the beam was less affected when it was transmitted in the downward link through atmospheric turbulence [43].

In 2014, Zhang Yongtao et al. studied the change in the intensity distribution of nonparaxial polychromatic partially coherent Laguerre–Gaussian beams when they propagated in optical wireless. The intensity-distribution characteristics of the beam were determined using its initial parameters. Moreover, it was found that when the beam propagated in optical wireless, its spectral shift was related to the initial parameters and propagation distance of the beam [44].

In 1995, the Phillips Laboratory of the U.S. Air Force conducted an airborne-laser atmospheric-transmission characteristic-measurement experiment. The laser transceiver was installed on two aircraft flying in the stratosphere. The laser-wavefront slope structure function was measured experimentally, and the measurement results were consistent with the theoretical results [45]. In 2000, the Jet Propulsion Laboratory (JPL) carried out an optical-transmission experiment with a distance of 48.5 km. Multiple beams were simultaneously used in the experiment. It was found that, with an increase in the number of beams, the received light scintillation index also decreased [46].

In October 2005, the German Aerospace Center (DLR) conducted an atmospheric laser-transmission measurement experiment over a transmission distance of 142 km. The laser wavelength was 1064 nm. The experimental results showed that the dynamic range of the atmospheric-turbulence conditions was very large. The measured Fried parameters (atmospheric coherent length) ranged from 1.2 to 28 cm, but were typically <6.5 cm [47].

In 2005, Phillips et al. of Central Florida University conducted experimental measurements on beam spread and drift caused by atmospheric turbulence using two 1.06-μm lasers (a single-mode laser and a multimode laser), which were transmitted on horizontal paths of 1, 2, and 5 km. The experimental results showed that when the emission-aperture size was the same, the drift mean-square deviation of the emitted beams of the multimode laser and single-mode laser were basically the same [48].

Currently, developed countries are attempting to develop their own wireless laser-communication systems. Under certain conditions, wireless laser communication can be used as a supplement to radio-frequency (RF) communication. However, with the emergence of high-power integrated laser devices, the problem of laser-transmission attenuation in the atmosphere will be overcome and optical wireless communication will become one of the main means of communication in the future.

1.3 Pulse-Position Modulation, Subcarrier Modulation, and Orthogonal Frequency-Division Multiplexing

The detection technologies of wireless laser-communication systems mainly include direct and coherent detection. Considering the system structure and cost, existing atmospheric laser-communication systems generally adopt intensity modulation. Intensity modulation includes pulse modulation (PM) and subcarrier modulation.

1.3.1 Research Status of Domestic and Foreign Pulse Modulation

Pulse modulation uses the position, width, or amplitude of an optical pulse in a data frame to represent different information. It can be divided into analog and digital modulation [49]. The most common pulse-modulation methods are pulse-like position modulations; that is, using different positions and amplitudes of pulses in the data frame to represent different information. For example, on–off keying (OOK), pulse-position modulation (PPM), multiple pulse-position modulation (MPPM), differential pulse-position modulation (DPPM), differential pulse-interval modulation (DPIM), dual-header PIM (DHPIM), dual-amplitude PIM (DAPIM), and overlapping PPM (OPPM).

Two main types of modulation and demodulation systems exist in wireless laser communication: coherent modulation/optical heterodyne detection and intensity modulation/direct detection (IM/DD). The former has better sensitivity than the latter, but its implementation is more complex. Therefore, intensity modulation/direct detection systems are typically used.

Many modulation methods have been developed for IM/DD systems. The most widely used method is on–off keying (OOK), as mentioned by Kahn et al. in [50]. The coding structure of this modulation method is simple; however, it is vulnerable to external interference and cannot guarantee the reliability of the communication system over long distances.

In 1978, Pierce proposed pulse-position modulation (PPM) and applied it to space optical communication. Compared with OOK modulation, PPM has strong anti-interference ability and high power efficiency [51]. Because PPM modulation uses the position of pulses in a time period composed of multiple equal time intervals to transmit information, frame and slot synchronization are required during demodulation [52]. In 2005, Ding Deqiang designed a PPM modulation and demodulation system using digital signal processing (DSP). Time-slot synchronization and frame synchronization during demodulation were realized using phase-locked loop technology and Gold codes, respectively. Software simulations showed that the system had a strong anti-interference ability [53].

In 2007, Ke Xizheng et al. designed a hardware circuit to realize frame and slot synchronization in atmospheric laser communication using a field-programmable gate array (FPGA) and DSP. This method was simple and could ensure the correct demodulation of information [54]. In 2009, Yin Zhiyun and Ke Xizheng et al. combined PPM modulation with a Bose–Chaudhuri–Hocquenghem (BCH) error-correction code based on a Galois field (GF(3)), discussed the encoding and decoding methods of Hamming and BCH codes based on the GF(3) field, and proved that the BCH error-correction code based on the GF(3) field could correct two random errors [55].

In recent years, with an in-depth study of atmospheric laser-communication theory, the application of PPM in atmospheric laser communication has achieved a series of results [56]. In 1989, Sugiyama and Nosu proposed multi-pulse position modulation (MPPM), and observed in [57] that this modulation method reduced the transmission bandwidth required in PPM and further improved the bandwidth efficiency. Multi-pulse position modulation transmits information through the position of these pulses in the time slot, which improves the channel capacity compared with PPM, but also increases the complexity of the system implementation.

In 1993, Sato and Ohtsuki et al. analyzed the performance of (m, 2) MPPM in [58], including the symbol structure, symbol-error probability, and bit-error probability, and pointed out that under (8, 2) MPPM modulation, the symbol-error probability of pulse continuity was smaller than that of pulse separation, and the bit-error probability performance of (m, 2) MPPM was better than PPM under the same transmission bandwidth and transmission rate.

In 2007, Qin Ling proposed a new (m, 2) MPPM code mapping method and compared it with the list and constellation methods. This method changes the form of the increase in the number of (m, 2) MPPM symbols, reduces the number of symbols that increase with the increase in the number of bits of the input bitstream, and improves the coding efficiency compared with the first two methods [59]. In the same year, Qin Ling analyzed the channel capacity of MPPM without background noise. A Q-ary discrete memoryless erasure channel was used to simulate the MPPM channel, and the channel capacity of the MPPM was derived according to this model.

On this basis, the expressions of the information-transmission rate (channel capacity per second) and capacity energy efficiency (channel capacity per photon received) of MPPM were derived and compared with those of PPM. When the symbol time and slot number were given, the channel capacity of MPPM was obviously better than that of PPM. With an increase in the slot width, the information-transmission rate of MPPM was higher than that of PPM, and when the photon number of a single optical pulse was greater than five, the capacity energy efficiency of MPPM was better than that of PPM [60].

In 1988, Zwillinger [61] proposed differential pulse-position modulation (DPPM) and pointed out that DPPM has a higher throughput than PPM when the bandwidth and average power are constant. In 1999, Shiu and Kahn introduced and analyzed the codeword group structure, packet-error rate, and power spectral density of DPPM modulation [62]. In 2007, Zhao Li and Ke Xizheng proposed a design scheme for DPPM modulation and demodulation systems based on an FPGA in optical wireless communication. The results showed that the scheme could correctly demodulate the signal and that the system performance was stable [63].

In 2008, Sethakaset and Gulliver et al. proposed a new concatenated code, using a marked codeword, and Reed–Solomon (RS) codes as the inner and outer codes, respectively, for error correction and detection, proposed a hard-decision decoding algorithm, and analyzed the performance of the diffusion and non-diffusion channels [64].

In 1967, Das and Sharma proposed pulse-interval modulation (PIM). This modulation method uses the number of time slots between adjacent pulses to represent the transmitted information [65]; therefore, its symbol length is not fixed, which causes demodulation difficulties. Compared with PPM, it does not require symbol synchronization.

In 2000, Cariolaro et al. analyzed the spectrum of PIM and deduced a closed expression of the spectrum [66]. In 1996, Kaluarachi et al. proposed digital pulse-interval modulation (DPIM) and analyzed its symbol structure, transmission capacity, coding characteristics, and spectrum model through simulation [67]. In 1998, Ghassemlooy et al. analyzed the error performance of DPIM. In 2000, Zhang compared

DPIM with OOK and PPM in terms of power efficiency, bandwidth demand, and channel capacity. In 2001, Cariolaro and Erseghe analyzed the spectral characteristics of DPIM [68–70].

In 1999, Aldibbiat et al. proposed dual-header PIM (DHPIM). Compared with PPM, DPIM, and other modulation methods, the symbol structure of this modulation method is more complex and difficult to implement; however, it has a high data-transmission rate, small bandwidth requirements, and built-in symbol synchronization capability [71].

In 2001, Aldibbiat and Ghassemlooy analyzed the time-slot error rate, average transmission power, bandwidth demand, and power spectral density of the DHPIM modulation mode and compared them with those of the OOK, PIM, and PPM modulation modes. It was concluded that the symbol length of DHPIM was less than that of PPM, the bandwidth demand was higher than that of PIM but lower than that of PPM, the time-slot error rate was lower than that of OOK but higher than that of PPM, and the performance of DHPIM with a pulse-width parameter of 2 was better than that with a pulse-width parameter of 1 [72–74].

In 2009, Ghassemlooy and Rajbhandari analyzed the performance of convolutional-coded dual-header pulse-interval modulation (CC-DH-PIM) and compared the slot-error probability between different modulation methods. Compared with DHPIM, the bit-error performance of this modulation method was improved; however, its information-transmission rate is reduced [75]. In 2003, Zhang Kai et al. discussed dual-amplitude pulse-interval modulation (DAPIM), provided the coding structure of this modulation mode, analyzed the performance of the bandwidth, power spectrum, and time-slot error rate, and compared it with the OOK, PPM, and PIM modes. They found that DAPIM had better comprehensive bandwidth and power characteristics [76].

In 2004, Sethakaset and Gulliver proposed differential-amplitude pulse-position modulation (DAPPM) for indoor optical wireless communications. The symbol length and amplitude of this modulation method change. Symbol synchronization is not required during demodulation. The bandwidth requirements were lower than those of PPM and DPPM. For the same symbol length, the transmission capacity was higher than that of PPM and DPPM [77].

In 2007, Ke Xizheng et al. proposed an optical PPM polarization-modulation method, based on the influence of rain, snow, and fog on the communication quality and reliability of atmospheric laser communication, and introduced the modulation principle and key technical problems in modulation/ demodulation [78]. In 2008, Yang Lihong et al. studied the depolarization characteristics of polarized laser propagation under different fog-particle radii in dense fog and mist, and compared the changes in the polarization angle of the polarized laser at different transmission distances [79].

In 2010, Yang Lihong and Ke Xizheng analyzed the bit-error rate of polarized PPM and the influence of the threshold-decision method and angle demodulation on the bit-error rate. The calculation results showed that the bit-error rate of polarized PPM first decreased and then increased with an increase in the threshold coefficient, and the bit-error rate was the lowest when the threshold coefficient was 1/2. Under the same polarization angle, the smaller the noise power after photoelectric detection, the smaller the bit-error rate of the system and, under the same detector noise power, the smaller the polarization angle, the greater the bit-error rate. The bit-error rate was the highest when the polarization angle was 0° and the lowest when the polarization angle was 90° [80].

In 2007, Zhang Tie Ying et al. proposed fixed-length digital pulse-interval modulation (FDPIM) to solve the problems of PPM requiring symbol synchronization, and the DPIM symbol length not being fixed. The coding structure was provided, the average transmit power and bandwidth requirements were analyzed, and the packet-error rate under the atmospheric-turbulence model was derived and compared with OOK, PPM, and DPIM. The results showed that the packet error rate of FDPIM was inferior to that of PPM and DPIM. Owing to OOK, the bandwidth requirements were high, symbol synchronization was not required, and there was no buffer-overflow problem [81].

In the same year, Zhang Tie Ying et al. proposed dual-amplitude pulse-position modulation (DAPPM) to balance the bandwidth demand and power utilization of existing modulation methods. The symbol length of this modulation method was fixed at half that of PPM, the bandwidth demand was less than PPM and DPIM, and the error performance was better than that of OOK and DAPIM [82].

In 2008, Cheng Gang et al. proposed dual-duration pulse-position modulation (DDPPM) to solve the problems of PPM and DPIM, analyzed its bandwidth requirements, average transmission power, transmission capacity, and packet-error rate in weak turbulent channels, and compared it with OOK, PPM, and DPIM. The results showed that DDPPM had a higher power utilization and error performance than OOK. It had a higher bandwidth utilization and transmission capacity than PPM. Compared with DPIM, the symbol length was fixed, and the receiving end did not have problems, such as buffer overflow. In addition, the implementation is easier [83].

In 2009, Mei Hong Sui et al. of Qingdao Ocean University proposed an improved PPM called short PPM (SPPM), based on traditional PPM. Compared with PPM, this modulation method has better bandwidth utilization, more reliable information transmission, and reduced implementation complexity [84]. In the same year, Huan Gai Ping et al. proposed fixed-length dual-amplitude pulse-interval modulation (FDAPIM) to solve the problems of PPM and DPIM. Its packet error rate is inferior to that of PPM and DPIM, similar to that of FDPIM, and significantly better than that of OOK [85].

In the same year, Xu Zhi Yong et al. proposed separated double pulse-position modulation (SDPPM) on the basis of MPPM with two pulses. The modulation method removes the continuous-pulse combinations from the pulse combination of (m, 2) MPPM to reduce the impact of intersymbol crosstalk and the difficulty of receiving a decision. The simulation results showed that the bit-error rate of SDPPM is lower than that of MPPM with two pulses and that the bandwidth requirements are higher than those of MPPM [86].

In 2010, Cheng Gang et al. proposed dual-pulse pulse-interval modulation (DPPIM). Its symbol structure contains starting and marking pulses. The number of time slots between the two pulses represents the transmitted information. It not only has built-in symbol synchronization but also has a fixed symbol length without cache redundancy and overflow. Compared to PPM and FDPIM, it significantly shortens the symbol length [87].

In 2002, Pang Zhi Yong et al. made a comprehensive analysis and comparison of OOK, PPM, DPPM, and DPIM, including power efficiency, bandwidth requirements, transmission capacity, and anti-intersymbol interference. The results showed that when the symbol length was greater than four, the power utilization of PPM, DPIM, and DPPM was higher than that of OOK. The bandwidth demand was the highest in PPM, the lowest in OOK, and that of DPIM was higher than that of DPPM. OOK and PPM have the same transmission capacity, and the transmission capacity of DPPM is higher than that of DPIM [88].

Based on the work of Pang Zhi Yong et al., Chen Junhong and Yang Xiaoli added DHPIM to the types of modulation methods and added an analysis of the time-slot error rate to the content. The results showed that these modulation methods have their own advantages, and DPIM and DHPIM have more advantages in terms of bandwidth utilization and time slot error rate [89].

In 2006, Mahdiraji and Zahedi did work similar to that of Pang Zhi Yong et al. DPPM was missing in the modulation modes, and its power spectral density was analyzed. In terms of packet error-rate performance, OOK had the largest and PPM had the smallest [90]. In 2009, Azzam et al. analyzed the bandwidth requirements and power-efficiency performance of the OOK, PPM, MPPM, OPPM, and DAPPM (differential) modulation modes [91].

In 2012, Ke Xizheng and Chen Jinni compared the performances of OOK, PPM, DPIM, DAPPM, DHPIM, and other modulation modes. The average symbol length, average transmit power, bandwidth requirements, unit transmission rate, power spectral density, channel capacity, peak power, slot error rate, and packet error rate were analyzed and compared, and the concept of pulse-like position modulation was proposed.

The results showed that among these modulation methods, OOK had the advantages of a simple structure and small bandwidth, while PPM had disadvantages of needing symbol and frame synchronization. DPIM had the advantages of no symbol synchronization, high bandwidth efficiency, and power efficiency; however, it had the disadvantages of a relatively complex structure and variable symbol length. Similar to DPIM, DHPIM did not require symbol synchronization. The former had a higher bandwidth efficiency and transmission capacity than the latter, and its symbol structure was more complex [92].

These modulation methods are interrelated. Each modulation method is based on the shortcomings of existing modulation methods. However, each modulation method has some shortcomings. Thus far, nearly 20 types of pulse-like position modulation have been proposed, and they all still have some problems that need to be solved.

1.3.2 Research Status of Domestic and Foreign Subcarrier-Intensity Modulation

In 1993, Huang and Takayanagi [93] proposed subcarrier phase-shift keying (PSK) modulation for wireless laser communication. Through experimental verification, they found that the signal-to-noise ratio (SNR) required by the OOK system to achieve the same bit-error rate performance was 6 dB higher than that of the subcarrier PSK intensity-modulation system.

Subsequently, Qi Lu and Qing Chong Liu [94] simulated and analyzed the bit-error rate of subcarrier binary phase-shift keying (BPSK) intensity modulation and OOK under atmospheric turbulence. The decision threshold of the former is "0," and the latter should select an appropriate threshold, according to the turbulence intensity and change in additive noise. Even if the signal-to-noise ratio is extremely high, the bit-error rate of OOK mode cannot be infinitely small. Under the same scintillation level, to achieve the same bit-error rate, the signal-to-noise ratio required by OOK mode is 3 dB higher than that required by subcarrier BPSK intensity modulation.

Error-control coding is used in subcarrier intensity-modulation systems to reduce the high bit-error rate caused by light-intensity flickering. In 2006, Sharma and Chadha [95] studied the performance of a low-density parity check (LDPC) code multi-channel subcarrier BPSK and quaternary phase-shift keying (QPSK) intensity-modulation system. They obtained the upper bound of the bit-error rate of the LDPC coding system using maximum-likelihood demodulation in an atmospheric optical channel with light-intensity flicker. They verified that the bit-error rate (BER) performance of the atmospheric optical-communication system using LDPC coding was higher than that of a turbo-code system, and the coding gain of the LDPC code system was 5 dB higher than that of a turbo-code system under the same information length and coding rate.

In 2008, Popoola and Ghassemlooy [96] combined subcarrier differential phase-shift keying (DPSK) intensity modulation with spatial-diversity technology to study the bit-error rate performance of the system in a negative-exponential atmospheric-turbulence environment. When four photodetectors were used at the receiving end, the gain of the received optical power of the system at the bit-error rate could theoretically reach 38 dB.

In 2010, Ben Naila and Bekkali [97] studied the performance of a BPSK subcarrier intensity-modulation system based on average aperture technology. When the turbulence in the atmospheric channel changed from weak to strong, the performance of the average-aperture-based modulation system improved significantly, compared with the BPSK subcarrier intensity-modulation system. A tradeoff exists between the receive aperture, carrier noise ratio, and bit-error rate.

1.3.3 Proposal of Optical Wireless Communication Orthogonal Frequency-Division Multiplexing (OWC-OFDM)

To enhance the ability of wireless laser-communication systems to suppress atmospheric turbulence, Wang et al. [98] proposed an asymmetric amplitude-limiting optical orthogonal frequency-division multiplexing modulation technology in 2011. This modulation scheme has a high power-utilization ratio and is suitable for wireless laser-communication systems with limited optical power. Its performance is superior to that of traditional OOK and DC-offset optical OFDM modulation schemes. However, the complete recovery of asymmetric cut signals depends on the selection of the carrier frequency. If a frequency offset occurs, it will be detrimental to the signal recovery.

In 2011, Chatzidiamantis and Lioumpas [99] applied adaptive-transmission technology to a subcarrier PSK intensity-modulated optical wireless communication system. This method uses the constant envelope of the subcarrier PSK, adaptively changes the modulation order of the subcarrier PSK, according to the instantaneous turbulence and fading states and the predetermined bit-error rate, and improves the channel capacity of optical wireless communication without increasing the transmission power or sacrificing the bit-error rate.

To improve the utilization of the frequency band, an atmospheric laser-communication system can adopt higher-order modulation. Quadrature amplitude modulation (QAM) is a type of joint keying of the amplitude and phase. When the minimum distances between constellation points are equal, more constellation points can be accommodated in the constellation diagram; that is, higher-order modulation can transmit higher-rate data within the limited frequency band.

In 1994, Le Goff and Glavieux [100] combined turbo codes with QAM modulation to improve the spectral efficiency of transmission, and the performance of this system was better than that of a trellis-coded modulation system in the Gaussian and Rayleigh channels. In 2005, Dong In Kim and Ekram Hossain [101] applied high-order modulation and multi-code format to a cellular wideband code-division multiple-access network, solving the problems of forward-link data transmission at a dynamic rate, and power adaptation in the network. In 2011, Lei Zhang and Kschischang [102] studied the channel capacity of an LDPC-code high-order modulation system with multiple edge parameters. This nested coding structure can provide a flexible rate selection for a single decoder.

In 1966, Chang proposed the concept of orthogonal frequency-division multiplexing [103]. With the application requirements of broadband digitalization and the maturity of very-large-scale integration (VLSI) CMOS chips in the 1990s, OFDM technology has received widespread attention. In 1999, OFDM was defined as the European digital-broadcasting standard, which established its status as an important modulation technology and indicated the arrival of the era when OFDM would be widely used [104]. OFDM modulation technology was adopted among the following standards: European digital-video broadcasting, wireless local-area network (WiFi;

IEEE 802.11a/g), wireless metropolitan-area network (WiMAX; 802.16e), asymmetric digital subscriber line (ADSL; ITU G.992.1), and fourth-generation (4G) mobile-communication technology [105].

In 1971, Weinstein of Bell Laboratories proposed that OFDM systems could modulate and demodulate using inverse-discrete and discrete Fourier transforms (IDFT and DFT, respectively) [106]. In 1981, Hirosaki was the first to successfully complete the implementation of a fast Fourier transform (FFT) in an OFDM system [107]. In 1990, Bingham made a detailed analysis and introduction of an OFDM system [108], and predicted that OFDM would become more vigorous and develop rapidly. In 2001, Cisco was the first company to design a commercial communication protocol based on OFDM: vector OFDM (VOFDM) [109]. OFDM has become an important development direction for communication standards.

OFDM was applied to optical communication later than it was to the RF field. In 1996, Pan and Green published a paper on optical OFDM [110] and then successively researched optical OFDM [111–113]. In 2001, Dixon et al. [114] observed that OFDM technology could be used to suppress mode dispersion in multimode fibers because multimode-fiber channels exhibit multipath-fading characteristics similar to those of wireless channels. Therefore, early research on optical OFDM mainly focused on its application to multimode fiber [115–118].

Optical OFDM has attracted more attention because three distinct technical proposal standards [120, 121] have emerged for long-distance communication, including direct-detection optical OFDM (DDO-OFDM) and coherent optical OFDM (CO-OFDM) [119].

In 2001, Yi Sun designed an OFDM system model and applied it to a optical wireless communication (OWC) system, which is the first report of this type of application [122]. In the same year, Roy You proposed two methods to reduce the peak-to-average power ratio (PAPR) in OFDM systems [123] and applied them to OWC-OFDM systems. In 2003, Ohtsuki proposed the concept of applying multicarrier modulation to wireless laser communication and introduced the principle and characteristics of OWC-OFDM [124].

In 2006, Gonzalez et al. proposed a scheme for applying optical wireless adaptive OFDM systems to indoor environments. The scheme had a high data rate and did not require channel equalization [125]. In the same year, Shieh et al. proposed an OFDM system based on coherent detection [126]. In 2007, Yang Tang et al. proposed two models of OWC-OFDM systems [127]. In 2008, Cvijetic first conducted an indoor OWC communication experiment with a 10-Gbit/s communication rate and a 2.5 km transmission distance [128].

In 2009, Cvijetic introduced a short-range OWC transmission test [129]. She adopted polarization multiplexing (polmux)-QPSK modulation and coherent-detection technology, as well as a multiple-input multiple-output (MIMO) mode to improve the OWC transmission rate. The transmission rate of a single channel reached 112 Gbit/s at a distance of 1.5 km.

In 2009, Stotts described an optical/RF hybrid wireless network [130]. This hybrid communication network considered both the reliability of the RF link and the cheapness of the optical link, and provided users with high-speed and long-distance mobile network services. The Air Force Research Laboratory submitted an integrated radio-frequency/optical tactical targeting-network technology plan (IRON-T2) for future implementation.

OWC systems could be be used in large-scale telecommunication networks. Unfortunately, owing to their low rate, large intersymbol interference, unreliable links, etc., further applications of OWC systems were limited. However, as OFDM modulation technology in OWC has matured, the above disadvantages have been gradually overcome. In 2010, Bekkali described the status and role of the OWC-OFDM system in next-generation communications [131].

Today, many research institutions have verified through optical-OFDM experiments that the transmission rate can reach 100 Gbit/s or even higher, and a hyperspectral efficiency of 2 bit/s/Hz can transmit 1000-km coherent waves on a single-mode fiber.

Since 2008, Xi'an University of Technology has conducted extensive theoretical and experimental research on OWC-OFDM systems. In 2008, Zhang Hongwei proposed a probability algorithm, based on a combination of a Golay complementary sequence and Reed–Muller codes, to reduce the PAPR and implemented it on an FPGA [9]. In 2009, Zhao Li combined OFDM technology with OWC technology in his doctoral dissertation and studied some theoretical and technical problems of OWC-OFDM systems that urgently needed to be solved [132].

In 2009, owing to the defects of the traditional baseband OWC-OFDM system model, Zhao Li proposed an improved ground-band model that encoded the signal to reduce the module complexity of optical up/down conversions. This coding method could also be used for system synchronization [133]. In 2012, Xie Mengqi studied the impact of atmospheric dispersion on an OWC-OFDM system [134]. In 2014, Zhang Tuo built a optical wireless communication system based on subcarrier 16-PSK modulation, and experimentally verified the feasibility of the designed system for realizing point-to-point communication [135].

In 2015, Li Beilei built a system model for OWC-OFDM. Through an experimental comparison of a 100-m communication distance and a 1-km communication distance, it was concluded that atmospheric attenuation and atmospheric turbulence in the wireless laser channel would have an impact on the transmission of optical OFDM signals. This impact becomes increasingly obvious with the increase in communication distance [136].

In 2016, Liu Juan conducted an experimental study on PAPR in optical wireless communication and verified theoretical problems [137]. In 2007, Duan Zhongxiong studied channel-allocation technology in optical wireless OFDM [138]. In 2008, Wang Wu studied the use of a particle filter for optical wireless channel estimation [139]. In recent years, high-speed communication systems have been rapidly developed. It can be seen that many different communication systems adopt the technical scheme of multicarrier modulation.

1.4 Characteristics of OWC-OFDM

1.4.1 Advantages of OFDM

The basic idea of OFDM is to transmit orthogonal multichannel parallel signals in overlapping sub-bands to improve bandwidth utilization. Multicarrier transmission divides the data stream into several sub-bit streams so that each data element will have a much lower bit rate. The low-rate multistate symbols formed by such a low bit rate are used to demodulate the corresponding subcarriers to form a transmission system in which multiple low-rate symbols are sent in parallel.

Therefore, using multicarrier technology, high-speed data streams can be converted through serial–parallel conversions, so that the duration length of the data symbols on each subcarrier is relatively increased, and the intersymbol interference (ISI) caused by the time dispersion of the channel can be reduced. In addition, in multicarrier systems, equalization is generally not required.

In a general fading environment, equalization is not an effective method for improving the system performance in multicarrier systems. Because the essence of equalization is to compensate for the intersymbol interference caused by multipath channels, and OFDM technology itself uses the diversity characteristics of multipath channels, equalization is generally unnecessary.

In a single-carrier system, a single instance of fading or interference can lead to the failure of the entire link. However, in a multicarrier system, only a few subchannels are affected by deep fading at a given time. Multicarrier systems can not only suppress interference effectively, but also improve the spectrum efficiency.

One of the most significant characteristics of optical communication is its large communication capacity and high code rate. Introducing multicarrier modulation technology into laser-communication technology can improve the code rate or reduce the device-speed requirements under the same bandwidth conditions. This is conducive to reducing the relative cost and system complexity of the communication system.

The integration of these two technologies, giving full play to their respective advantages and avoiding their disadvantages, is a bright spot of this technology. It has become a popular technology in wireless communication research. OFDM technology has the following advantages over other modulation methods.

1.4.1.1 High Frequency-Band Utilization

Because FFT processing causes each subcarrier to partially overlap, the band utilization can theoretically approach the Nyquist limit. Overlapping orthogonal subcarriers effectively avoid interchannel interference, improve the traditional method of using guard bands to divide ion channels, and improve the frequency utilization.

1.4.1.2 Strong Impulse-Noise-Interference Resistance

OFDM systems convert a high-speed serial-transmission digital stream into multiple low-speed data streams transmitted in parallel on the subchannel. This increases the symbol period by many times; that is, it disperses the impact of impulse interference on multiple symbols transmitted in parallel, which weakens their relative strength; hence, it has a strong ability to suppress impulse noise.

1.4.1.3 Strong Narrowband-Interference Resistance

Because narrowband interference affects only a few subchannels, the vast majority of subchannels can still carry out normal information transmissions. The impact of narrowband interference on the entire system can be greatly reduced through adaptive modulation or joint subchannel coding.

1.4.1.4 Strong Multipath-fading Resistance

Because the OFDM system divides the entire available bandwidth into many narrowband channels, the symbol period becomes much longer for each subchannel, and the frequency response on a single subchannel becomes much flatter, flattening the frequency-selective fading and greatly reducing the intersymbol crosstalk caused by the multipath effect.

1.4.1.5 Suitability for High-Speed Data Transmissions

The OFDM adaptive-modulation mechanism can adaptively select different modulation modes for each subcarrier, according to changes in the channel and background noise. Thus, the system can concentrate more data on a channel under good conditions for high-speed transmission. Therefore, OFDM technology is suitable for high-speed data transmissions.

1.4.2 Disadvantages of OFDM

Although OFDM has many advantages and has become the most promising modulation technology, it also has its own shortcomings, mainly in the following aspects.

1.4.2.1 High Peak-to-Average Power Ratio

Because OFDM symbols are superimposed by multiple independently modulated subcarrier signals, when each subcarrier phase is the same or similar, the superimposed signal is modulated by the same initial phase signal. This results in a larger instantaneous power peak, which further leads to a higher peak-to-average power ratio (PAPR).

When multiple subcarriers have a peak value of the same polarity at a certain time, that is, when the information to be transmitted has a long phase consistency, its instantaneous peak power will be very high, resulting in a high peak-to-average ratio. This requires some components in the system to have a large linear dynamic range. If the dynamic range cannot meet the signal-change requirements and there is nonlinearity, serious multicarrier-intermodulation noise interference will occur.

Because the dynamic range of general power amplifiers is limited, a signal with a large peak-to-average ratio can easily enter the nonlinear region of the power amplifier, resulting in nonlinear distortion of the signal, obvious spectrum-spread interference, and in-band signal distortion. This results in serious degradation of the performance of the entire system. The peak-to-average ratio (PAPR) is a major technical obstacle in OFDM.

1.4.2.2 Sensitivity to Frequency Offsets

When an OFDM signal is demodulated, the carrier frequency of each channel signal must be consistent with the nominal value to ensure the orthogonality of each channel signal. In principle, as long as there is a 1% relative-frequency offset, there will be approximately −40 dB of signal leakage between the subchannels. Therefore, measures must be taken to effectively eliminate or reduce the frequency deviation to avoid mutual crosstalk between the subchannels.

1.4.3 OWC-OFDM Characteristics

Multicarrier technology can effectively suppress atmospheric turbulence and atmospheric scattering-effect interference and improve spectral efficiency. Therefore, introducing multicarrier-modulation technology into optical wireless communication systems can improve the code rate and reduce the device-speed requirements under the same bandwidth. This is conducive to reducing the relative cost and complexity of the communication system, and breaks through the bottleneck of optical communication. OWC and OFDM are effectively combined to form an OWC-OFDM system, which has the following advantages over traditional single-carrier OWC systems.

(1) Most of the signal processing in the system is performed in the RF domain. RF equipment is much more mature than the corresponding optical equipment. The frequency selectivity of microwave filters and the frequency stability of microwave oscillators are also significantly better than those of the corresponding optical devices. The phase noise of a microwave oscillator is much smaller than that of a partial-feedback laser diode, and RF detection is easier to realize than optical-domain detection.
(2) Forward error-correction coding is easy to load. if a single carrier modulates the signal bandwidth to 2.5 Gbit/s, it is easily affected by the atmospheric channel, owing to the narrowing of the optical pulse. At this time, it is difficult to add forward error corrections. If multicarrier modulation is adopted, the code rate of each subcarrier is relatively low, and forward error correction can be easily realized.
(3) An OWC-OFDM system decomposes the high-speed data stream into multiple low-speed data streams, which are used to modulate n channel subcarriers, and are then transmitted in parallel. It can realize broadband and high-capacity communications in a relatively simple manner. It constitutes a flexible and convenient multichannel optical-signal transmission system, which can provide multiple users with voice, data, image, and other services.
(4) The OFDM system has a strong multipath-interference resistance and spectrum-selective fading ability. Because the OFDM system disperses the data to multiple subcarriers, the symbol rate of each subcarrier is significantly reduced, thus reducing the impact of multipath transmissions. If a cyclic prefix is added as the protection interval, the time-delay spread caused by atmospheric scattering can be eliminated.

1.4.4 Differences Between RF OFDM and OWC-OFDM

Although RF and wireless-laser communications have some similarities, they also have significant differences, as follows.

(1) Different signal formats: The OFDM symbol is a complex signal. In an RF communication system, the receiver can simultaneously detect the strength and phase of the signal, while in an intensity modulation / direct detection (IM/DD) optical communication system, the receiver can only determine the strength of the optical signal, not its phase information.
(2) Different system structures. In an optical communication system, the transmitter and receiver require electro-optical conversion and photoelectric conversion modules, respectively; RF communication does not.
(3) Different channels: The atmospheric laser-communication channel is an atmospheric channel, and the attenuation of the atmospheric laser signal is much greater than that for RF.
(4) Higher requirements for peak-to-average ratio suppression: The OWC-OFDM system uses a Mach-Zehnder modulator (MZM) for electro-optical conversions.

The transfer characteristic curve of the modulator is a cosine function, which presents a significant nonlinear problem. Therefore, an OWC-OFDM system is more sensitive to PAPR than an RF OFDM system.

1.4.5 Problems to Be Solved in OWC-OFDM

Although OWC-OFDM has many advantages over traditional single-carrier systems, owing to the differences between RF OFDM and OWC-OFDM, as well as the shortcomings of OFDM and OWC technologies, many problems must be solved when combining them, including the following.

(1) System modeling: Although the principles of OFDM modulation technology in RF and wireless-laser communication systems are the same, because of the significant differences between the two systems, the RF OFDM system model cannot be simply applied to the OWC system. Rather, the OWC-OFDM system model must be reasonably constructed, according to the characteristics and principles of an OWC communication system.
(2) Source coding and channel coding: Owing to the characteristics of atmospheric scattering and turbulence, corresponding measures must be taken during system design to ensure normal communication under various weather conditions. Channel and source coding can significantly improve the anti-interference ability of communication systems. Traditional channel-coding methods, such as RS, convolutional, and turbo codes, can be used in OFDM systems.
(3) Synchronization technology: Because an OFDM signal is composed of multiple mutually orthogonal subcarrier signals, ensuring this orthogonality is very important for the OFDM system; therefore, the synchronization requirements are relatively strict. Three types of synchronization are required in an OWC-OFDM system: carrier, symbol, and sample.
(4) Channel estimation: Channel estimation refers to the method and process used by the receiver to obtain the channel-state information. As the channel of an OWC-OFDM system is an atmospheric channel, it is affected by the scattering of dust, raindrops, fog, and other particles. Although a cyclic protection interval is introduced into OFDM, simple frequency-domain equalization can be used at the receiver to eliminate the multipath interference caused by atmospheric scattering. However, an accurate channel-frequency response on each subcarrier must be known before frequency-domain equalization. Therefore, channel estimation must be performed before OFDM equalization. The accuracy of the channel estimation directly determines the performance of the receiver.

1.5 Key Technologies in OWC-OFDM

At present, the key challenges of OWC-OFDM technology include synchronization, channel estimation, peak-to-average power ratio, source and channel coding, adaptive modulation technology, and laser and detector selection.

1.5.1 Synchronization

Synchronization technology is a practical problem that should be solved in any communication system. Reliable data transmission is impossible without an accurate synchronization algorithm. This is the premise of information transmission.

The synchronization performance has a significant impact on the performance of OFDM systems. OFDM divides the available communication bandwidth into several orthogonal subbands that overlap each other but can still maintain their orthogonality. The OFDM signal waveform is composed of multiple superimposed subcarrier signals, and each subcarrier is distinguished by its orthogonality. Ensuring this orthogonality is very important for OFDM systems; therefore, the synchronization requirements are relatively strict.

OFDM-system synchronization mainly includes carrier, sample, and symbol synchronization. Carrier synchronization is required for signal demodulation, and the signal frequency at the receiving point must be in the same frequency and phase as the carrier at the sending end. Symbol synchronization distinguishes the boundaries of each symbol so that the start and end times of the IFFT and FFT are consistent. Sample synchronization ensures that the sampling time at the receiving end is completely consistent with that at the sending end.

1.5.2 Channel Estimation

If channel interference is not considered, the received signal on the n subchannels is equal to the product of the transmitted signal on each channel and the instantaneous value of the channel spectral characteristics. When atmospheric conditions change, a pilot signal can be sent at a fixed time interval to estimate the spectral characteristics of the channel, and the received signal on each subchannel can be divided by the spectral characteristics of the channel to correctly demodulate the signal.

Macroscopically, channel estimation can be divided into non-blind, blind, and semi-blind. Non-blind estimation means that in the estimation phase, a pilot is first used to obtain the channel information of the pilot position, and then preparations are made for obtaining the channel information of the entire data-transmission phase.

Blind channel estimation means that the channel estimation value is obtained using the corresponding information-processing technology without using pilot information. Compared with traditional non-blind channel-estimation technology, blind channel-estimation technology greatly improves the transmission efficiency of the system. However, blind channel-estimation algorithms generally have a slow convergence speed, which hinders their application in an actual system.

Consequently, semi-blind channel estimation was developed. It compromises between data-transmission efficiency and convergence speed; that is, it uses fewer training sequences to obtain channel information. Making a reasonable trade-off between the estimation effect and implementation complexity is a problem worth studying. This trade-off is typically achieved through the selection of iteration times.

1.5.3 Peak-to-Average Power Ratio [9]

Because OFDM technology can not only improve the spectrum efficiency but also effectively suppress the multipath effect, it has been adopted by a variety of wired and wireless access standards. However, OFDM signals are composed of multiple independent modulated subcarrier signals. According to the central limit theorem, the time-domain signal approximately follows a Gaussian distribution and the synthetic signal may produce a relatively large peak power, which leads to a large peak-to-average power ratio (PAPR).

A high PAPR naturally requires a large linear range of analog-to-digital (A/D) and D/A conversions, and a high-power amplifier in the system; otherwise, it causes spectrum diffusion and band distortion of the transmission signal, resulting in an increase in the bit-error rate and an increase in system complexity. Therefore, reducing the PAPR has always been key in OFDM systems.

1.5.4 Source and Channel Coding

Owing to the characteristics of atmospheric fading and turbulence, it is necessary to take appropriate measures during system design to ensure normal communication in various weather conditions and reduce the bit-error rate of the system through channel-coding technology to improve the system performance.

The basic principle of OFDM is to decompose a high-speed data stream into multiple low-speed data streams and then simultaneously transmit the data on multiple subcarriers. QPSK, orthogonal QPSK (OQPSK), QAM, minimum-shift keying (MSK), etc. can be used to modulate parallel subcarriers using the data signal in OFDM. Because the data are decomposed into multiple low-speed subcarriers for parallel transmission, the symbol duration increases and the intersymbol interference (ISI) is reduced.

The insertion of a cyclic prefix in an OFDM system can avoid ISI and reduce intercarrier interference (ICI); however, some subchannels may be completely submerged after channel fading. Therefore, even if error-free detection can be achieved on most subcarriers, the bit-error rate (BER) of the entire system will increase because of the influence of individual subchannels with a small received signal amplitude. To avoid this phenomenon, it is necessary to introduce forward error-correction coding (FEC).

1.5.5 Adaptive Modulation Technology

Owing to the attenuation, multipath, and time-varying nature of an atmospheric channel, it is necessary to monitor the channel status in real time to make more effective use of resources. The advantage of OFDM technology is that the number of information bits and transmission power of each subcarrier can be dynamically adjusted according to the frequency-selective fading of the channel to optimize system performance.

1.5.6 Laser Selection

Lasers are the key component of laser communication, among which semiconductor lasers are the most widely used. They have the advantages of a simple structure, anti-vibration, small volume, long service life, and convenient modulation. The selection of semiconductor lasers for optical wireless communication should consider the following standards:

(1) When light is transmitted in the atmosphere, some wavelengths are easily absorbed. These atmospheric absorption windows must be avoided.
(2) The inheritance and interoperability of an optical-fiber communication system must be considered.
(3) Owing to atmospheric and geometric losses, a large laser emission power is required. However, eye-safety standards should also be considered.
(4) A laser with good stability and beam characteristics should be used.

1.5.7 Detector Selection

Optical reception is the process of converging the received modulated optical signals through an optical receiving lens, followed by filtering and photoelectric conversion by a photoelectric detector. The light-sensitive range of the detector must be consistent with that of the laser.

Chapter Summary

Optical wireless communication (OWC) has become a hot topic in the communication-technology field in recent years because of its low cost, flexible networking, and lack of frequency licensing. However, owing to the limitations of optical dispersion and the speed of electronic devices, the transmission rate of the system is limited, and the bit-error performance is not ideal. To fully exploit the potential of OWC, support a higher information-transmission rate, and improve the bit-error characteristics of the system, a new transmission technology with high spectral efficiency and a strong ability to resist atmospheric effects must be adopted.

Among the various wireless solutions that can provide high-speed transmission, orthogonal frequency-division multiplexing (OFDM) technology can best represent the development direction of optical wireless communication. This book combines two mature technologies, OFDM and OWC, and studies the theoretical and technical problems that urgently need to be solved in OWC-OFDM systems.

References

1. Wang HX (2001) Technology of laser communication in atmosphere. Infrared Laser Eng 30(2):123–127
2. Fenner WR (1866) Future trends in crosslink communications. SPIE 1993:1–8
3. Jeganathan M et al (1999) Lessons learnt from the optical communications demonstrator (OCD). SPIE 3615:23–30
4. Pease R (1999) Optical laser-communications system carve niche in metro markets. Lightwave (9):23–27
5. Wen P, Yu H (1998) Development and status of Laser Communication in America. J Univ Electron Sci Technol China 27(5):541–545
6. Yu H, Hua L (1998) Development Status of European Inter-satellite Laser Communications. J Univ Electron Sci Technol China 27(5):552–556
7. Lutz HE (1997) Optical communications in space-twenty years of ESA effort. ESA Bull 01.1(91):25–31
8. Yin D, Dezhong P (1998) Development of free-space Laser Communication of Japan. J Univ Electron Sci Technol China 27(5):546–551
9. Zhang H (2008) The Search and Implement of reducing peak-average-power in free space optical communication multicarrers modulation system. Xi'an University of Science and Technology, Xi'an
10. Zhao Y (2007) Research on simulation of wireless laser communication. Jiangsu University of Science and Technology, Nanjing
11. von Karman. Progress in the statistical theory of turbulence. Proc Natl Acad Sci 34:530–539
12. Tatarskii VI (1961) Wave propagation in turbulent medium. McGraw-Hill, New York
13. Tatarskii VI (1971) The effects of the turbulent atmosphere on wave propagation. US Department of Commerce, Springfield
14. ChernoV LA (1960) Wave propagation in random medium. McGraw-Hill, New York
15. Tatarskii VI (1978) Wave propagation in a turbulent medium. McGraw-Hill, New York
16. Rao RZ (2005) Propagation of light in a turbulent atmosphere. Anhui Science and Technology Press, Hefei
17. Lukin VP, Pokasov VV (1981) Optical wave phase fluctuations. App Opt 20(1):421–135

18. Greenwood DP, Tarazano DO (1974) A proposed form for the atmospheric micro-temperature spatial spectrum in the input range. USAF Rome Air Development Center. Technical Report RADC-TR
19. Li X-Q, Ji X-L, Zhu J-H (2013) Higher-order intensity moments of optical beams in atmospheric turbulence. Acta Phys Sin 62(4):044217
20. Wang SCH, Plonus MA (1979) Optical beam propagation for a partially coherent source in the turbulent atmosphere. Opt Soc Am 69(9):1297–1304
21. Yi XX (2005) Propagation characteristics of Gaussian beams in turbulent atmosphere. University of Electronic Science and Technology of China, Xian
22. Whitman AM, Beran MJ (1970) Beam spread of laser light propagating in a random medium. J Opt Soc Am A 60(12):1595–1602
23. Poirier JL (1972) Beam spreading in a turbulent medium. J Opt Soc Am A 62(7):893–898
24. Fante RL (1975) Electromagnetic beam propagation in turbulent media. Proc IEEE 63(12):1669–1688
25. Fan C (1999) Effects of atmospheric turbulence on Gaussian Beam propagation at oblique angles. J Quantum Electron 16(6):519–525
26. Andrews LC (1995) Single-pass and double-pass propagation through complex paraxial optical systems. J Opt Soc Am A 12(1):137–150
27. Andrews LC, Miller WB (1995) The mutual coherence function and the backscatter amplification effect for a reflected Gaussian-beam wave in atmospheric turbulence. Wave Random Media 5(2):167–182
28. Andrews LC, Philips RL, Al-Halash MA (1999) Theory of optical scintillation. J Opt Soc Am A 16(6):1417–1429
29. Andrews LC, Phillips RL, Hopen C et al (2001) Theory of optical scintillation: Gaussian-beam wave model. Wave Random Media 11(2):271–291
30. Andrews LC, Phillips RL, Miller WB (1997) Mutual coherence function for a double-passage retroreflected atmospheric turbulence. App Opt 36(3):698–708
31. Wei H-Y, Wu Z-S (2008) Spreading and wander of laser beam propagation on slant path through atmospheric turbulence. Chinese J Radio Sci 23(4):611–615
32. Yi XX, Guo L, Wu Z (2005) Study on the optical scintillation for Gaussian beam propagation in the slant path through the atmospheric turbulence. Acta Optica Sinica 25(4):433
33. Luo ZM, Wu ZS, Guo LX (2002) Study of the scintillation for optical wave propagation in the slant path through the atmospheric turbulence by considering the inner scale effect. J Xidian Univ 29(4):455–459
34. Yang RK, Wu ZS (2004) Study of scintillation for a Gaussian beam propagating in atmospheric turbulence on earth-space paths. In: Asia-pacific radio science conference proceedings
35. Wei H (2009) Propagation of laser beams and characteristics of target echoes in turbulent atmosphere at oblique angles. Master's Thesis, Xi'an University of Electronic Science and Technology
36. Jinyong C, Xiwen Q, Yuehong H, Fei Z, Zhichao L, Shuanglian F (2015) Numerical simulation of jittering effect of laser beams propagation. Infrared Laser Eng 44(S1):46–49
37. Zhang FZ, Yang YK (2007) The characteristic of the cracked beam spot. Acta Optica Sinica 27(4):567
38. Wenyue Z, Xianmei Q, Ruizhong R, Huihua W (2019) Evaluation technology of high energy laser atmospheric propagation performance. Infrared Laser Eng 48(12):1203002–1203002
39. Chu X, Liu Z, Wu Y (2008) Propagation of a general multi-Gaussian beam in turbulent atmosphere in a slant path. J Opt Soc Am A 25(1):74–79
40. Qian X-M, Zhu W-Y, Rao R-Z (2009) Acta Phys Sin 58(9):6633–6639
41. Chu XX, Liu ZJ, Wu Y (2010) Comparison between relay propagation and direct propagation of Gaussian-Schell-model beam in turbulent atmosphere along a slant path. Chin Phys B 19(9):094201-1-6
42. Zhang XX, Dan YQ, Zhang B (2012) Spreading of partially coherent flat-topped beams propagating along a slant path in turbulent atmosphere. Acta Optica Sinica 32(12):1201001

43. Duan ML, Li JH, Wei JL (2013) Influence of different propagation paths on the propagation of laser in atmospheric turbulence. Optoelectron Lett 9(6):0477–0480
44. Zhang Y, Liu L, Wang F, Cai Y, Ding C, Pan L (2014) Average intensity and spectral shifts of a partially coherent standard or elegant Laguerre-Gaussian beam beyond paraxial approximation. Opt Quant Electron 46:365–379
45. Silbaugh EE, Welsh BM, Roggemann MC (1996) Characterization of atmospheric turbulence phase statistics using wave front slope measurements III. J Opt Soc Am A 13(12):2453–2460
46. Biswas A, Wright MW (2002) Mountain-top-to-mountain-top optical link demonstration: part I. IPN Progr Rep 42–149
47. Perlot N, Giggenbach D, Henniger H, Horwath J, Knapek M (2006) Measurements of the beam-wave fluctuations over a 142 km atmospheric path. SPIE v01.6304
48. Phillips RL, Andrews LC, Stryjewski J et a1 (2006) Beam wander experiments: terrestrial path. SPIE, v01.6303.
49. Mahdiraji GA, Zahedi E (2006) Comparison of selected digital modulation schemes (OOK, PPM and DPIM) for optical wireless communications. In: 4th student conference on research and development, vol 4, pp 5–10
50. Kahn JM, Barry JR (1997) Wireless infrared communication. IEEE Proc 85(2):265–298
51. Pierce JR (1978) Optical channels: practical limits with photon counting. IEEE Trans Commun 26(12):1819–1821
52. Vilnrotter VA, Rodemich ER (1986) A synchronization technique for optical PPM signals. IEEE 45(2):1–6
53. Ding D, Ke X (2005) Design and simulation study of a PPM modulation and demodulation system for atmospheric laser communication. Opt Commun Technol 01:50–52
54. Ke X-Z, Zhao L, Ding D (2007) The achievement of time-slot and frame synchronization in atmosphere laser communication. Semicond Optoelectron 25(5):721–724
55. Yin Z, Ke X, Zhang B (2009) Research on error correction coding over GF field in wireless laser communication. J Electron Meas Instrum 23(7):23–28
56. Li Y, Lei H, Yin Q (2004) Design of PPM modulation and demodulation system based on CPLD. Inf Technol 28(1):67
57. Sugiyama H, Nosu K (1989) MPPM: a method for improving the band utilization efficiency in optical PPM. IEEE Lightw Technol 7(3):465–472
58. Sato K, Ohtsuki T, Sasase I (1993) Performance analysis of (m,2) MPPM with imperfect slot synchronization. IEEE Pacific Rim, pp 765–768
59. Qin L, Ke X-Z (2007) A study of mapping scheme for dual-pulse MPPM. J Xi'an Univ Technol 23(3):269–272
60. Qin L, Ke X-Z (2007) Analysis of optical MPPM channel capacity without background noise. Opto-Electron Eng 34(7):107–110
61. Zwillinger D (1988) Differential PPM has a higher throughput than PPM for the band-limited and average-power-limited optical channel. IEEE Trans Inf Theory 34(5):1269–1273
62. Shiu D, Kahn JM (1999) Differential pulse-position modulation for power-efficient optical communication. IEEE Trans Commun 47(8):1201–1210
63. Zhao L, Ke X, Liu J (2007) Research on differential pulse-position modulation in optical wireless communication. Laser J 28(2):63–64
64. Sethakaset U, Gulliver TA (2008) Performance of differential pulse-position modulation with concatenated coding over optical wireless communications. IET Opt Wirel Commun 2(1):45–52
65. Das J, Sharma PD (1967) Pulse-interval modulation. Electron Lett 3(6):288–289
66. Cariolaro G, Erseghe T, Vangelista L. Stationary model of pulse interval modulation and exact spectral evaluation. IEEE 660–664
67. Kaluarachi ED, Ghassemlooy Z, Wilson B (1996) Digital pulse interval modulation for optical free space communication links. IEEE Electr Eng 996–1000
68. Ghassemlooy Z, Hayes AR, Seed NL (1998) Digital pulse interval modulation for optical communication. IEEE Commun Mag 12:95–99

69. Ghassemlooy Z, Hashemi SK, Amiri M (2008) Spectral analysis of convolutional coded DPIM for indoor optical wireless communications. IEEE Proc 102–105
70. Cariolaro G, Erseghe T, Vangelista L (2001) Exact spectral evaluation of the family of digital pulse interval modulated signals. IEEE Trans Inf Theory 47(7):2983–2992
71. Aldibbiat N, Ghassemlooy Z, Saatchi R (1999) Pulse interval modulation-dual header (PIM-DH). In: 2nd International conference on information, communications & signal processing (ICICS'99) Singapore, pp 7–10
72. Aldibbiat NM, Ghassemlooy Z, Laughlin RMC (2001) Error performance of dual header pulse interval modulation (DH-PIM) in optical wireless communications. IEE Proc 148(2):91–96
73. Aldibbiat NM, Ghassemlooy Z, McLaughlin R (2002) Dual header pulse interval modulation for dispersive in door optical wire less communication systems. IEE Proc Circ Dev Syst 149(3):187–192
74. Dibbiat NMA, Ghassemlooy Z, Mcbughlin R (2001) Spectral characteristics of dual header pulse interval modulation (DH-PIM). IEE Proc Commun 148(5):280–286
75. Ghassemlooy Z, Rajbhandari S (2009) Convolutional coded dual header pulse interval modulation for line of sight photonic wireless links. IET Optoelectron 3(3):142–148
76. Zhang K, Zhang H, Gong M (2003) Performance of dual-amplitude pulse interval modulation for wireless infrared communications. J Infrared Millim Waves 22(6):411–414
77. Sethakaset U, Gulliver TA (2004) Differential amplitude pulse-position modulation for indoor wireless optical channels. In: IEEE global telecommunications conference, 2004. Globecom'04, pp 1867–1871
78. Ke XZ, Yin ZY, Yang LH (2007) Polarization modulation with PPM and its key technique in atmospheric laser communication. Semicond Optoelectron 28(4):553–555+560
79. Lihong Y, Xizheng K, Dongdong M (2008) Depolarization characteristics of the polarized laser in atmosphere. Opto-Electron Eng 35(11):62–67
80. Yang LH, Ke XZ (2010) BER of polarization PPM based on atmospheric optical communication. Chin J Sci Instrum 31(7):1664–1668
81. Wang H, Zhang T, Zhu Y et al (2007) 无A novel fixed length digital pulse interval modulation for optical wireless communications. Chin J Lasers 34(12):1655
82. Zhang T, Wang H, Zhu Y, Sun X (2007) Dual-amplitude pulse position modulation for optical wireless communication. Laser J 28(6):71–73
83. Gang C, Hongxing W, Sun X (2008) A new modulation scheme of optical wireless communications. Chin J Lasers 35(12):1914
84. Sui MH, Yu X, Zhou Z (2009) The modified PPM modulation for underwater wireless optical communication. In: IEEE international conference on communication software and networks, pp 173–177
85. Aiping H, Yangyu F, Wei L, Bai B (2009) Fixed length dual-amplitude pulse interval modulation for optical wireless communications. Chin J Lasers 36(3):602
86. Xu Z, Shen L, Wang J et al (2009) Study on separated double pulse position modulation and its performance. J Commun 30(11):113–119
87. Cheng G, Wang HX, Sun XM, Zhang T (2010) Dual pulse-pulse interval modulation for optical wireless communications. Chin J Lasers 37(7):1750
88. Pang ZY, Piao DZ, Zou CY (2002) Performance comparison of several modulation schemes in optical communication. J Guilin Inst Electron Technol 22(5):1–4
89. Chen J, Yang X. Comparative study on several modulation modes of wireless laser communication. In: Jiangxi optical society, annual meeting of 13 provinces and cities, pp 31–34
90. Mahdiraji GA, Zahedi E (2006) Comparison of selected digital modulation schemes (OOK,PPM and DPIM) for optical wireless communications. In: IEEE 4th student conference on research and development, pp 5–10
91. Nazmy Azzam MH, AbouiSeoud AK (2009) Bandwidth and power efficiency of various PPM schemes for indoor optical wireless communications. In: 26th national radio science conference, pp 1–11

92. Ke X, Chen J (2012) Performance comparison of various pulse position modulation in wireless laser communication. Laser Technol 36(1):67–76
93. Huang W, Takayanagi J, Sakanaka T (1993) Atmospheric optical communication system using subcarrier PSK modulation. In: IEEE International Conference on Communications, vol 3, pp 1597–1601
94. Lu Q, Liu Q, Mitchell GS (2004) Performance analysis for optical wireless communication systems using subcarrier PSK intensity modulation through turbulent atmospheric channel. In: Global Telecommunications Conference, vol 3, pp 1872–1875
95. Sharma M, Chadha D (2006) Maximum likelihood decoding of low density parity check coded subcarrier modulation in optical wireless systems. In: India conference, pp 1–6
96. Popoola WO, Ghassemlooy Z, Fellow IET (2008) BER performance of DPSK subcarrier modulated free space optics in fully developed speckle. In: 6th international symposium on communication systems, networks and digital signal processing, pp 273–277
97. Naila CB, Bekkali A, Kazaura K (2010) BPSK intensity modulated free-space optical communications using aperture averaging. In: International conference on photonics, pp 1–5
98. Wang Y, Cao J (2011) Performance analysis of atmospheric laser communication system based on asymmetrically clipped optical orthogonal frequency division multiplexing intensity modulation 40(1):36–40
99. Chatzidiamantis ND, Lioumpas AS (2011) Adaptive subcarrier PSK intensity modulation in free space optical systems. IEEE Trans Commun 59(5):1368–1377
100. Le Goff S, Glavieux A (1994) Turbo-codes and high spectral efficiency modulation. In: International conference on communication, vol 2, pp 645–649
101. Kim DI, Hossain E, Bhargava VK (2005) Dynamic rate and power adaptation for forward link transmission using high-order modulation and multicode formats in cellular WCDMA networks. IEEE Trans Wirel Commun 4(5):2361–2372
102. Zhang L, Kschischang FR (2011) Design of multi-edge-type LDPC codes for high-order coded modulation. In: 12th Canadian workshop on information theory, pp 5–8
103. Chang RW (1966) Synthesis of band-limited orthogonal signals for multichannel date transmission. Bell Syst Technol J 45(6):96–1775
104. Li Z (2006) Research and simulation of key technologies in orthogonal frequency division multiplexing (OFDM). Beijing Jiaotong University, Beijing
105. Shieh W, Djordjevic I (2001) OFDM for optical communication. Publ House Electron Ind 32(3):21–22
106. Weinstein SB, Ebert PM (1971) Data transmission by frequency division multiplexing using the discrete fourier transform. IEEE Trans Commun 19(5):628–634
107. Hirosaki B (1981) An orthogonally multiplexed QAM system using the discrete fourier transform. IEEE Trans Commun 29(7):982–989
108. Bingham JAC (1990) Multicarrier modulation for data transmission: an idea whose time has come. IEEE Commun Mag 5–14
109. Ayanoglu E et al (2001) OFDM broadband wireless transmission and its advantages over single carrier modulation. In: Proceedings of the ICC 2001 conference, vol 42, no 6, Helsinki, pp 1660–1664
110. Chang RW (1996) Synthesis of band-limited orthogonal signals for multichannel date transmission. Bell Syst Technol J 45(3):1775–1796
111. Pan Q, Green RJ (1996) Bit-error-rate performance of lightwave hybrid AM/OFDM systems with comparison with AM/QAM systems in the presence of clipping impulse noise. IEEE Photo Technol Lett 32(8):278–280
112. Shi Q (1997) Error performance of OFDM-QAM in subcarrier multiplexed fiber-optical transmission. IEEE Photon Technol Lett 32(9):845–847
113. You R, Kahn JM (2001) Acerage power reduction techniques for multiple-subcarrier intensity-modulated optical signals. IEEE Trans Commun 49(6):2164–2171
114. Ma CP, Kuo JW (2004) Orthogonal frequency division multiplex with multi-level technology in optical storage application. Jpn J Appl Physisc Part 1: Regular Pap Short Notes Rev Pap 43(2):4876–4878

115. Dixon BJ, Pollard RD, Iezekeil S (2001) Orthogonal frequency-division multiplexing in wireless communication systems with multimode fiber feeds. IEEE Trans Microw Theory Tech 49(3):1404–1409
116. Jolley NE, Kee H, Rickard R, Tang JM, Cordina K (2005) Generation and propagation of a 1550 nm 10 Gb/s optical orthogonal frequency division multiplexed signal over 1000 m of multimode fibre using a directly modulated DFB. In: Optical Fiber Communications Conference paper no. OFP3. Anaheim, CA
117. Tang JM, Lane PM, Shore KA (2006) High-speed transmission of adaptively modulated optical OFDM signals over multimode fibers using directly modulated DFS. Lightw Technol 24(3):429–441
118. Lowery A, Armstrong J (2005) 10 Gbit/s multimode fiber link using power-efficient orthogonal-frequency-division multiplexing. Opt Express 13(3):3–9
119. Shieh W, Athaudage C (2006) Coherent optical Orthogonal frequency division multiplexing. Electron Lett 587–589
120. Lowery AJ, Du L, Armstrong J (2006) Orthogonal frequency division multiplexing for adaptive dispersion compensation in long haul WDM systems. In: Optical Fiber Communications Conference, paper no. PDP 39. Anaheim, CA
121. Djordjevic I, Vasic B (2006) Orthogonal frequency division multiplexing for high-speed optical transmission. Opt Express 14(5):3767–3775
122. Sun Y (2001) Bandwidth-efficient wireless OFDM. IEEE J Sel Commun 19(11):2267–2278
123. You R, Kahn JM (2001) Average power reduction techniques for multiple subcarrier intensity modulated optical signals. IEEE Trans Commun 49(12):2164–2171
124. Ohtsuki T (2003) Multiple-Subcarrier modulation in optical wireless communications. IEEE Commun Mag 41(3):74–79
125. Gonzalez O et al (2006) Adaptive OFDM system for communication over the indoor wireless optical channel. IEEE Proc Optoelectron 153(4):139–144
126. Shieh W, Athaudage C (2006) Coherent optical orthogonal frequency division multiplexing. Electron Lett 42(10):587–589
127. Djordjevic IB, Vasic B, Neifeld MA (2007) LDPC-Coded OFDM for optical communication system with direct detection. IEEE J Sel Top Quant Electron 13(5):1446–1454
128. Cvijectic N, Qian D, Wang T (2008) 10 Gb/s free-space optical transmission using OFDM. In: Conference on optical fiber communication/national fiber optic engineers conference, vol 12, no 5, San Diego, CA, United states, pp 561–572
129. Cvijetic N, Qian D, Yu J et al (2009) 100 Gb/s per-channel free-space optical transmission with coherent detection and MIMO processing. In: ECOC'09, 35th European conference optical communication 2009, Vienna, Austria, p 1
130. Stotts LB, Andrews LC, Cherry PC et al (2009) Hybrid optical RF airborne communications. Proc IEEE 97(6):1109
131. Bekkali A et al (2010) Transmission analysis of OFDM-based wireless services over turbulent radio-on-FSO links modeled by gamma-gamma distribution. Photon J IEEE 2(3):510
132. Zhao L (2009) FSO-OFDM research on key technologies of the system. Xi'an University of Technology, Xian
133. Zhao L, Ke X (2009) An improved FSO-OFDM baseband model. Semiconductor optoelectronics 30(2):277–280
134. Xie M (2012) Influence of atmospheric dispersion on FSO-OFDM system. Xi'an University of Technology, Xian
135. Zhang T (2014) Experiment of optical wireless communication system with subcarrier 16PSK modulation. Xi'an University of Technology, Xian
136. Li B (2015) Experimental study of FSO-OFDM system. Xi'an University of Technology, Xian

137. Liu J (2016) Experimental study on PAPR control method in FSO-OFDM system. Xi'an University of Technology, Xi'an
138. Duan, Xiong Z (2007) Research on channel allocation technology of FSO-OFDM system. Xi'an University of Technology, Xi'an
139. Wang W (2008) Particle filtering based semi-blind estimation for OFDM time-varying channels. Xi'an University of Technology, Xi'an

Chapter 2
Subcarrier Modulation Techniques

A carrier is a radio wave of a specific frequency that carries data. A subcarrier is an electronic communication-signal carrier that is carried on top of another carrier so that both the signals can efficiently propagate simultaneously. To distinguish the optical carrier, the electrical carrier premodulated by the analog baseband signal is called the subcarrier.

There are two main modulation schemes in optical wireless communication systems: one is intensity modulation/direct detection (IM/DD) and the other is coherent modulation/heterodyne detection. Three main modulation methods are suitable for IM/DD atmospheric optical wireless communication systems: on–off keying (OOK) [1–5] modulation, pulse modulation [6–12], and subcarrier intensity modulation. This chapter introduces optical wireless subcarrier modulation techniques.

2.1 Subcarrier Intensity Modulation

Subcarrier modulation involves modulating binary-bit information onto a certain carrier first and then using this modulated subcarrier to modulate the optical carrier, so that a certain parameter of the optical carrier corresponds to the modulated signal. The optically modulated signal is sent by the transmitting optical system and transmitted by the atmospheric channel. The photodetector uses the photoelectric effect to detect the light energy collected on the photosensitive surface and converts it into a current signal proportional to the number of incident photons or incident light power. The information data are demodulated by the subcarrier electromodulator.

X. Ke, *Principles and Applications of Optical Wireless Orthogonal Frequency-Division Multiplexing*, Optical Wireless Communication Theory and Technology,
https://doi.org/10.1007/978-981-97-7973-4_2

2.1.1 Generation and Detection of the Subcarrier Intensity Modulation Signal

Figure 2.1 shows a optical wireless communication (OWC) system with N subcarrier signals. In the subcarrier modulation system, m(t) is the radio-frequency subcarrier signal after pre-modulating the signal source d(t), and is used to modulate the intensity of the optical carrier emitted by the laser. After a serial-to-parallel conversion, M-ary phase-shift keying (MPSK) subcarrier modulation converts one symbol into in-phase branch data I and quadrature branch data Q at a given time; its magnitude is $\{a_{ic}, a_{is}\}_{i=1}^{N}$. According to the I and Q channel data, it can be mapped to the corresponding phase.

Because the subcarrier signal $m(t)$ is a sinusoidal signal, positive and negative elements exist; thus, it is necessary to add a DC bias b_0 to $m(t)$ and inject it into the laser as a driving current. In an N-way subcarrier intensity-modulation (SIM)-OWC system,

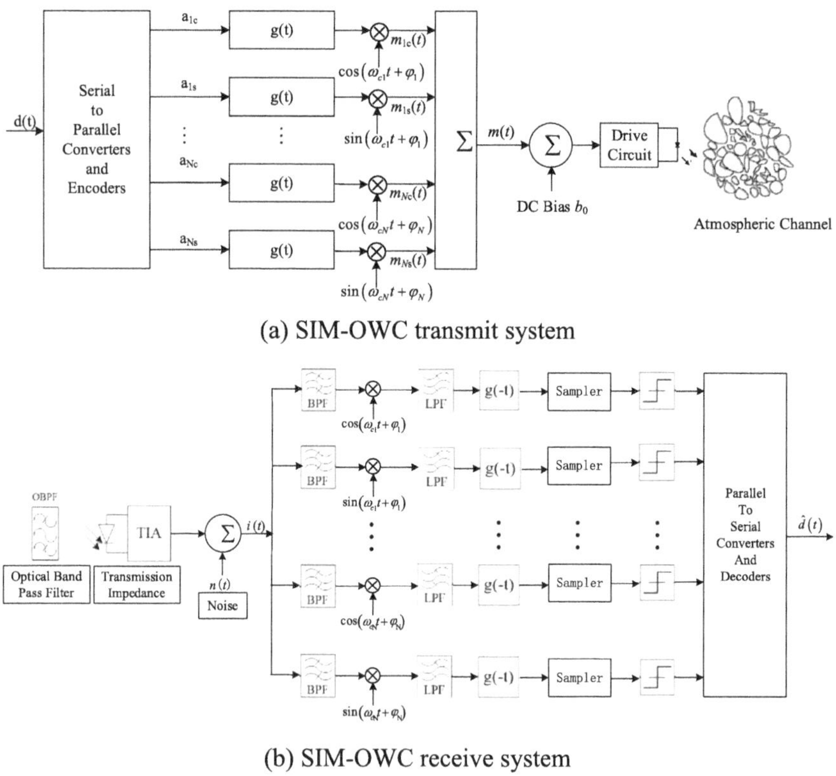

(a) SIM-OWC transmit system

(b) SIM-OWC receive system

Fig. 2.1 N-channel subcarrier intensity-modulation optical wireless communication system

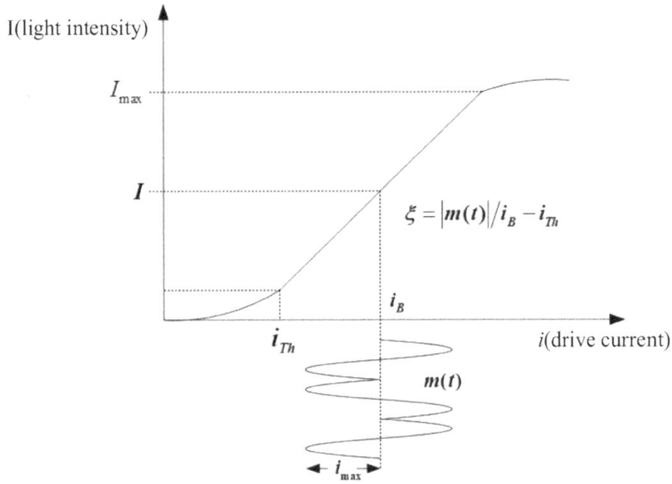

Fig. 2.2 Subcarrier laser-analog modulation characteristic curve

$$m(t) = \sum_{i=1}^{N} m_i(t). \tag{2.1}$$

During a given symbol duration, the radio-frequency (RF) subcarrier-modulated signal can generally be expressed as

$$m_i(t) = g(t)a_{ic}\cos(\omega_{ci}t + \varphi_i) + g(t)a_{is}\sin(\omega_{ci}t + \varphi_i), \tag{2.2}$$

where g(t) is the pulse-shaping function, and the carrier frequency and phase are $\{\omega_{ci}, \varphi_i\}_{i=1}^{N}$. When the receiver adopts direct detection, the light-intensity signal is photoelectrically converted into a current signal $I(t)$:

$$I(t) = RA(t)[1 + \xi m(t)] + n(t), \tag{2.3}$$

where R is the photoelectric-conversion constant, $n(t)$ is white noise with a Gaussian distribution but a non-zero mean value, light-modulation index is $\xi = \left|\frac{m(t)}{i_B - i_{th}}\right|$, and the laser-modulation characteristic curve is shown in Fig. 2.2.

2.1.2 Error Rate of BPSK Subcarrier Modulation

Figure 2.3 shows the principle of a optical wireless communication system using binary phase-shift keying (BPSK) subcarrier modulation. For light-intensity modulation/direct detection (IM/DD) communication systems, the light intensity received

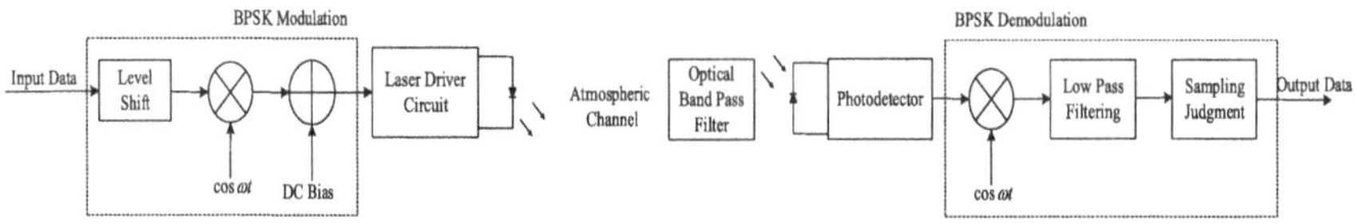

Fig. 2.3 BPSK subcarrier-modulation optical wireless communication system

by the receiver can be expressed as

$$P(t) = A(t)P_s(t) + n(t). \tag{2.4}$$

For the subcarrier modulation system, the light intensity emitted by the optical transmitter is

$$s(t) = 1 + \xi[s_i(t)\cos\omega_c t - s_q(t)\sin\omega_c t], \tag{2.5}$$

where $s_i(t) = \sum_j g(t - jT_s)\cos\Phi_j$ is the in-phase signal, $s_q(t) = \sum_j g(t - jT_s)\sin\Phi_j$ is the quadrature signal, ξ is the modulation index, and $0 < m \leq 1$. Φ_j is the jth bit phase, $g(t)$ is the gate pulse, and T_s is the symbol time. The light intensity received by the receiver is

$$P(t) = \frac{P_{\max}}{2}A(t)\{1 + \xi[s_i(t)\cos\omega_c t - s_q(t)\sin\omega_c t]\}. \tag{2.6}$$

After passing through the photodetector, the output electrical signal is

$$\begin{aligned} I(t) = {} & \frac{P_{\max}R}{2}A(t)\{1 + \xi[s_i(t)\cos\omega_c t - s_q(t)\sin\omega_c t]\} \\ & + n_i(t)\cos\omega_c t - n_q(t)\sin\omega_c t, \end{aligned} \tag{2.7}$$

where R is the photoelectric conversion constant. $n_i(t)$ and $n_q(t)$ are non-zero mean Gaussian white noise with variance σ_g^2.

If a subcarrier modulation system is required to transmit the same optical power as an OOK system, the power-spectral density of the signal received by the receiver is

$$I(f) = A(f) + \frac{B(f - f_c) + B(f + f_c)}{2} + \frac{N(f - f_c) + N(f + f_c)}{2}. \tag{2.8}$$

In $B(f) = A(f) * Z(f)$, the slow fading of the channel depends on the DC component $A(f)$. If the carrier frequency f_c is sufficiently high, it is assumed that $f_c > B_A + B_B$ and f_c is the intermediate frequency (IF); B_A is the bandwidth of the $A(f)$ single sidebands, and B_B is the bandwidth of the $B(f)$ single sidebands.

Unnecessary terms can be filtered out by a receiver band-pass filter. The filtered signal is subjected to coherent demodulation for carrier-phase recovery, and after filtering out high-frequency components through a low-pass filter, the in-phase signal of the output signal is obtained:

$$r_i(t) = \frac{P_{\max}R}{2}\xi A(t)s_i(t) + n_i(t), \tag{2.9}$$

as well as quadrature signals:

$$r_i(t) = \frac{P_{\max}R}{2}\xi A(t)s_q(t) + n_q(t). \tag{2.10}$$

When the BPSK subcarrier-modulation method is used, if the atmospheric-fading effect is not considered and the channel characteristics follow a Gaussian distribution, the system bit-error rate can be expressed as

$$P_e = Q\left(\sqrt{2SNR}\right), \tag{2.11}$$

where $SNR = \frac{(P_{\max}/2)^2(R)^2\xi^2}{2\sigma_g^2}$. Considering the effect of atmospheric fading, the demodulated signal is

$$r(t) = \frac{P_{\max}R}{2}[\xi A(t)s(t) + n(t)]/2. \tag{2.12}$$

Assuming that the "0" and "1" codes are sent with equal probability, that is, $p(1) = p(0) = 0.5$, the bit-error rate of the BPSK optical wireless communication system is

$$P_e = p(1)p(r|1) + p(0)p(r|0). \tag{2.13}$$

(1) In the case of weak turbulence, the light-intensity fluctuation A(t) follows a lognormal distribution. For BPSK subcarrier modulation, the conditional probability-density function of the received signal is

$$p(r|x) = \begin{cases} \frac{\exp(-\sigma_l^2/2)}{2\pi\sigma_l\sigma_g}\int\limits_0^{\infty}\frac{1}{t^2}\exp\left\{-\left[\frac{\ln^2 x}{2\sigma_l^2}+\frac{(\xi r-t)^2}{2\sigma_g^2}\right]\right\}dt\ x=\pm 1 \\ \frac{\exp(-\sigma_l^2/2)}{2\pi\sigma_l\sigma_g}\int\limits_{-\infty}^{0}\frac{1}{t^2}\exp\left\{-\left[\frac{\ln^2 x}{2\sigma_l^2}+\frac{(\xi r+t)^2}{2\sigma_g^2}\right]\right\}dt\ x=0 \end{cases}. \tag{2.14}$$

For BPSK modulation, the decision threshold is set to 0, and Eq. (2.14) is inserted into Eq. (2.13) to obtain

$$P_e = \frac{\exp(-\sigma_l^2/2)}{\sqrt{2\pi}\sigma_l}\int\limits_0^{\infty}\frac{1}{x^2}\exp\left(-\frac{\ln^2 x}{2\sigma_l^2}\right)Q\left(\frac{x}{\sigma_g}\right)dx, \tag{2.15}$$

where

$$SNR = \frac{(P_{\max}/2)^2 (R)^2 \xi^2}{2\sigma_g^2}.$$

(2) When the light-intensity fluctuation $A(t)$ follows the gamma–gamma distribution, for BPSK subcarrier modulation, the conditional probability-density function of the received signal is [13]

$$p(r|x) = \begin{cases} \frac{2}{\sqrt{2\pi}\sigma_g \Gamma(\alpha)\Gamma(\beta)} \left(\frac{\alpha\beta}{\xi}\right) \int\limits_0^\infty t^{\frac{\alpha+\beta}{2}} K_{\alpha-\beta}\left(2\sqrt{\frac{\alpha\beta t}{\xi}}\right) \exp\left\{-\left[\frac{(r-t)^2}{2\sigma_g^2}\right]\right\} dt & x = \pm 1 \\ \frac{2}{\sqrt{2\pi}\sigma_g \Gamma(\alpha)\Gamma(\beta)} \left(\frac{\alpha\beta}{\xi}\right) \int\limits_0^\infty t^{\frac{\alpha+\beta}{2}} K_{\alpha-\beta}\left(2\sqrt{\frac{\alpha\beta t}{\xi}}\right) \exp\left\{-\left[\frac{(r+t)^2}{2\sigma_g^2}\right]\right\} dt & x = 0 \end{cases}. \tag{2.16}$$

Substituting Eq. (2.16) into Eq. (2.15), we obtain

$$P_e = \frac{(\alpha\beta)^{\frac{\alpha+\beta}{2}}}{\Gamma(\alpha)\Gamma(\beta)} \int\limits_0^\infty x^{\frac{\alpha+\beta}{2}-1} K_{\alpha-\beta}\left(2\sqrt{\alpha\beta t}\right) erfc\left(-\frac{\xi x}{\sqrt{2}\sigma_g}\right) dx. \tag{2.17}$$

2.1.3 Error Rate of FSK Subcarrier Modulation

The binary frequency-shift keying (FSK) modulation time-domain expression is [14]

$$e_{2FSK}(t) = b(t)\cos(\omega_1 t + \varphi_1) + \overline{b(t)}\cos(\omega_2 t + \varphi_2), \tag{2.18}$$

where $b(t)$ is the baseband signal, which can be expressed as

$$b(t) = \sum_{n=-\infty}^{\infty} a_n g(t - nT_s),$$

where

$$a_n = \begin{cases} 0, & \text{probability P} \\ 1, & \text{probability } 1 - \text{P} \end{cases}. \tag{2.19}$$

The intensity of the emitted light is

$$s(t) = 1 + \sum_{n=-\infty}^{\infty} a_n g(t - nT_s)\cos(\omega_1 t + \varphi_1)$$

$$+\sum_{n=-\infty}^{\infty} \overline{a_n} g(t - nT_s)\cos(\omega_2 t + \varphi_2). \tag{2.20}$$

Without loss of generality, the emitted light can be expressed as

$$s(t) = 1 + \sum_{n=-\infty}^{\infty} a_n g(t - nT_s)\cos(\omega_1 t) + \sum_{n=-\infty}^{\infty} \overline{a_n} g(t - nT_s)\cos(\omega_2 t). \tag{2.21}$$

The received signal is expressed as

$$r(t) = A(t) + \sum_{n=-\infty}^{\infty} a_n g(t - nT_s) A(u, t)\cos(\omega_1 t)$$

$$+ \sum_{n=-\infty}^{\infty} \overline{a_n} g(t - nT_s) A(u, t)\cos(\omega_2 t) + n(t). \tag{2.22}$$

The first part of Eq. (2.22) can be filtered out by a band-pass filter, and the resulting received signal is

$$r(t) = \sum_{n=-\infty}^{\infty} a_n g(t - nT_s) A(u, t)\cos(\omega_1 t)$$

$$+ \sum_{n=-\infty}^{\infty} \overline{a_n} g(t - nT_s) A(u, t)\cos(\omega_2 t) + n(t). \tag{2.23}$$

Using the synchronous detection method, assuming that the symbol sent at time $T_s = 0$ is "1," the waveforms of the two signals sent to the sampling decider for comparison are

$$\begin{cases} x_1(t) = A(t) + n_1(t) \\ x_2(t) = n_2(t). \end{cases} \tag{2.24}$$

$n_1(t)$ and $n_2(t)$ are normal random variables with variance σ_g^2, whereas sampling value $x_1(t) = A(t) + n_1(t)$ is a normal random variable with mean $A(t)$ and variance σ_g^2; sampling value $x_2(t) = n_2(t)$ is also a normal random variable with non-zero mean and variance σ_g^2. Because $x_1 < x_2$ will cause the "1" code to be incorrectly judged as a "0" code at this time, the error probability at this time is P_{e1} (here, $A(t)$ is replaced by a):

$$P_e = p(x_1 < x_2) = p[(a + n_1) < n_2] = p(a + n_1 - n_2 < 0). \tag{2.25}$$

Let $z = a + n_1 + n_2$; then, z is also a normal random variable, and the mean is a and the variance is σ_z^2, where $\sigma_z^2 = 2\sigma_g^2$; therefore, the probability-density function

$p(z)$ of z is

$$p(z)=\frac{1}{\sqrt{2\pi}}\exp\left[-\frac{(z-a)^2}{2\sigma_z^2}\right]=\frac{1}{2\sqrt{\pi}\sigma_g}\exp\left[-\frac{(z-a)^2}{4\sigma_g^2}\right]. \tag{2.26}$$

The previous known probability-density function $A(t)$ is

$$p(A)=\frac{1}{\sqrt{2\pi}\sigma_l A}e^{-\frac{\left(\ln A+\sigma_l^2/2\right)^2}{2\sigma_l^2}}. \tag{2.27}$$

From Eqs. (2.26) and (2.27), it can be known that the joint probability-density function is

$$p(r|s_1)=\frac{\exp(-\sigma_l^2/2)}{4\pi\sigma_l\sigma_g}\int_0^{\infty}\frac{1}{x^2}\exp\left\{-\left[\frac{\ln^2 x}{2\sigma_l^2}+\frac{(r-x)^2}{4\sigma_g^2}\right]\right\}dx. \tag{2.28}$$

Because the probability of the code being misjudged as "1" when sending "0" is equal to that of it being misjudged as "0" when sending "1," the bit-error rate in both cases is the same; therefore, let $P(0)=0.5$. The total bit-error rate for

$$p_e=\frac{1}{\sqrt{2\pi}\sigma_l}\exp\left(-\frac{\sigma_l^2}{2}\right)\int_0^{\infty}\frac{1}{x^2}\exp\left\{-\frac{\ln^2 x}{2\sigma_l^2}\right\}Q\left(\frac{x}{\sqrt{2}\sigma_g}\right)dx. \tag{2.29}$$

2.1.4 *Error Rates of MPSK and MQAM Subcarrier Modulation*

For a quadrature phase-shift keying (QPSK) subcarrier modulation system, the atmospheric-fading effect is not considered, and when the channel follows a Gaussian distribution, according to Eq. (2.11), where $SNR=\frac{(P_{\max}/2)^2(R)^2\xi^2}{2\sigma_g^2}$, the bit-error rate of the system can be obtained [15]. Considering the effect of atmospheric fading, the demodulated signal can be expressed as

$$r_i(t)=\left[\frac{P_{\max}R}{2}\xi A(t)s_i(t)+n_i(t)\right]/2 \tag{2.30}$$

$$r_q(t)=\left[\frac{P_{\max}R}{2}\xi A(t)s_q(t)+n_q(t)\right]/2. \tag{2.31}$$

The bit-error rate of a QPSK system based on a lognormal channel is [12, 16]

$$p_e = \frac{\exp(-\sigma_l^2/2)}{2\sqrt{\pi}\sigma_l}\int_0^{\infty}\frac{1}{x^2}\exp\left(-\frac{\ln^2 x}{2\sigma_l^2}\right)Q\left(\frac{x}{\sigma_g}\right)dx. \quad (2.32)$$

For multiple phase-shift keying (MPSK) subcarrier modulation, when the channel is assumed to be Gaussian and the atmospheric-fading effect is not considered, the system bit-error rate can be expressed as

$$P_e = \frac{2}{\log_2^M}Q\left(\sqrt{\log_2^M SNR}\sin(\pi/M)\right)M \geq 4. \quad (2.33)$$

The bit-error rate of an MPSK system based on the lognormal atmospheric channel model is [17]

$$P_e = 1 - \int_{-\pi/M}^{\pi/M}\frac{1}{2\pi\sigma_l}\int_0^{\infty}\exp\left\{-\frac{r^2}{2\sigma_l^2} - \frac{\left(\ln r + \sigma_l^2\right)^2}{2\sigma_l^2}\right\}$$
$$\left\{\frac{1}{\sqrt{2\pi}} + \frac{r\cos\theta}{\sigma_g}\left[1 - Q\left(\frac{r\cos\theta}{\sigma_g}\right)\right] \times \exp\left[\frac{r^2\cos^2\theta}{2\sigma_g^2}\right]\right\}drd\theta. \quad (2.34)$$

For square MQAM subcarrier modulation, if atmospheric-fading effects are not considered, the system bit-error rate using coherent demodulation is

$$BER = k_1 f(\sqrt{k_2\gamma}), \quad (2.35)$$

where $k_1 = 2(1 - 1/\sqrt{M})$ and $k_2 = 3/(M - 1)$. Then, the bit-error rate of the weakly turbulent channel modulated optical-communication system is [18]

$$BER = \frac{k_1}{\sqrt{\pi}}\int_{-\infty}^{\infty} f\left\{\sqrt{k_2\gamma/4}\exp\left(\sqrt{2}\sigma_l x - \sigma_l^2/2\right)\right\}\exp(-x^2)dx. \quad (2.36)$$

2.1.5 Analysis of Subcarrier Modulation Performance

Using light-intensity flicker variance values $\sigma_l^2 = 0.01$, $\sigma_l^2 = 0.04$, and $\sigma_l^2 = 0.09$, Eq. (2.14) simulates the bit-error rate performance of an OWC system using OOK and BPSK modulation. The OOK detection uses fixed-threshold detection with threshold $T = e^{-\sigma_l^2/2}$.

Figure 2.4 shows that when the scintillation index is $\sigma_l = 0.1$ and signal-to-noise ratio (SNR) = 10 dB, the bit-error rate of the OOK modulation system is 1.5×10^{-4}, whereas the bit-error rate of the BPSK subcarrier modulation system reaches 1.9 ×

Fig. 2.4 Using OOK and BPSK modulation under different system bit-error rates [17]

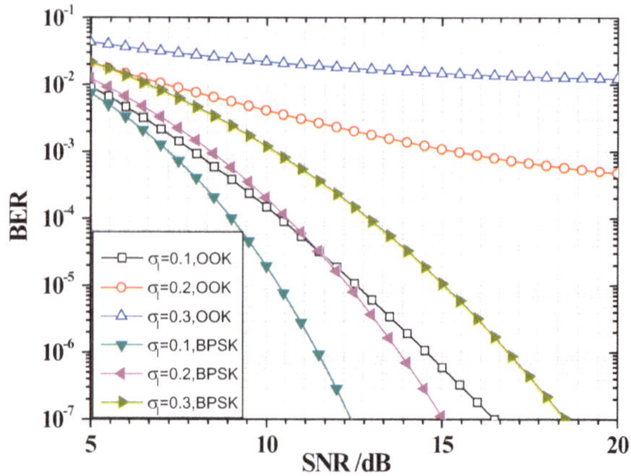

10^{-5}. When $\sigma_l = 0.2$ and SNR = 10 dB, the bit-error rate of OOK modulation is approximately 4×10^{-3}, whereas that of BPSK modulation is approximately 2.1×10^{-4}. At the same scintillation index of $\sigma_l = 0.3$ and SNR = 20 dB, the bit-error rate of the BPSK modulation system reaches 9.9×10^{-9}; however, for OOK modulation, with the continuous increase in the signal-to-noise ratio, the system bit-error rate decreases very slowly, to approximately 1×10^{-2}.

Figure 2.5 shows that when the scintillation index is $\sigma_l = 0.1$ and SNR = 10 dB, the bit-error rate of the 2-FSK modulation system is 1.3×10^{-3}, whereas the bit-error rate of the BPSK subcarrier modulation system reaches 1.9×10^{-5}; at $\sigma_l = 0.2$, when the SNR = 10 dB, the 2-FSK modulation system has a bit-error rate of 3.4×10^{-3}, and the bit-error rate of BPSK modulation reaches 2.1×10^{-4}. At $\sigma_l = 0.3$, when SNR = 20 dB, the bit-error rate of 2-FSK modulation is 8.9×10^{-7}, and the bit-error rate of BPSK modulation reaches 9.9×10^{-9}. The simulation results show that the BPSK intensity-modulation system can suppress the light-intensity scintillation effect caused by atmospheric turbulence more effectively than the 2-FSK intensity-modulation system.

Figure 2.6a, b show the simulation results of the bit-error rate performance under different light-intensity flicker variances. As the flicker variance increases, the error performance of various subcarrier modulations degrades. When the scintillation index is $\sigma_l = 0.1$ and BER = 10^{-3}, BPSK has a signal-to-noise ratio gain margin of 9 dB and 13 dB, respectively, compared with QPSK and 8-PSK. When the scintillation index is $\sigma_l^2 = 0.2$ and SNR = 10 dB, the bit-error rate of BPSK is approximately 6×10^{-3}, and the bit-error rates of QPSK, 8-PSK, 16-PSK, and 16-QAM are approximately 7×10^{-2}, 1.7×10^{-2}, 4×10^{-1}, and 2×10^{-1}, respectively. It can also be seen from Fig. 2.6 that among several subcarrier modulation systems, BPSK has the best error performance, followed by 2-FSK, QPSK, 8-PSK, and 16-QAM. The modulation method with the worst error performance is 16-PSK.

With the increase in the subcarrier modulation order, under the same signal-to-noise ratio and flicker variance, although the bit-error rate is degraded, the

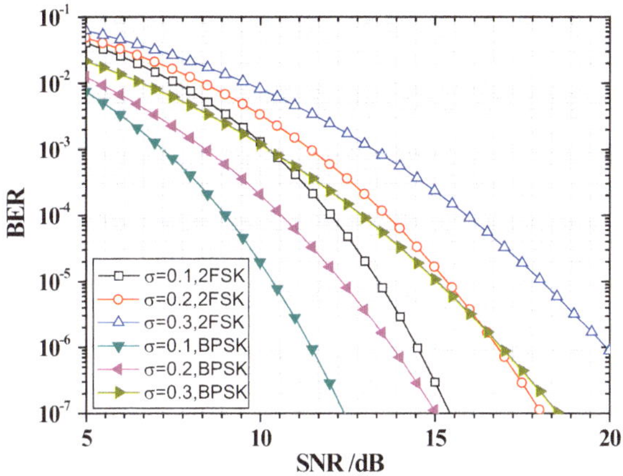

Fig. 2.5 Using OOK and BPSK modulation under different σ_l system bit-error rates [17]

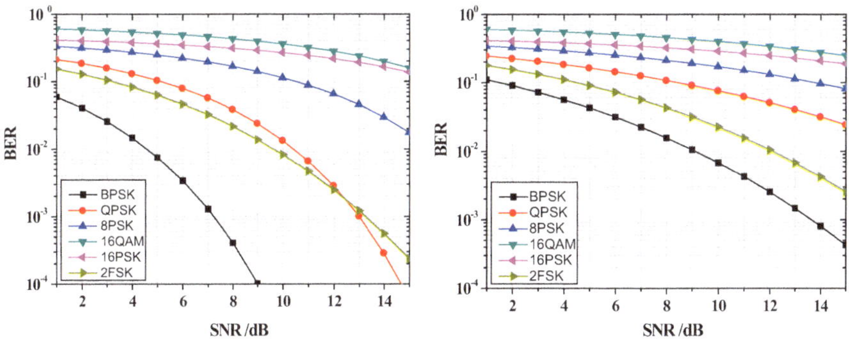

(a) $\sigma_l = 0.01$ (b) $\sigma_l = 0.2$

Fig. 2.6 System bit-error rate using different subcarrier modulation methods [17]

information-transmission rate is increased, and more information can be transmitted in the same frequency band, which improves the frequency-band utilization. Therefore, an appropriate modulation method should be selected, according to the bandwidth requirements of the system, and a reasonable choice should be made between the transmission bandwidth and bit-error rate characteristics.

2.2 Unlimited QPSK-Like Sinusoidal Modulation

Non-clipped quadrature phase-shift keying (NC-QPSK) quasi-sine modulation performs a half-wave rectification transformation on the basis of QPSK modulation and then performs intensity modulation on the laser. Instead of limiting the negative half-axis of the electrical signal, half-wave rectification is performed on the electrical signal, and the transmission of the information loaded on the signal is not affected. The laser intensity can be directly modulated without adding a DC bias. This modulation is called "unclipping."

2.2.1 Principle of NC-QPSK-Like Sinusoidal Modulation

The NC-QPSK sinusoidal modulation process is shown in Fig. 2.7. First, a QPSK signal is generated using the phase-modulation method. The corresponding constellation diagram is shown in Fig. 2.8, and the phase-logic relationship is shown in Table 2.1. The process is as follows: The binary digital baseband signal $S(t)$ is divided into two bipolar signals after a serial–parallel conversion (that is, the in-phase I signal and the quadrature Q signal, where $I^2 + Q^2 = 1$). The QPSK signal is obtained by multiplying the modulated carriers and then combining them; finally, the NC-QPSK signal is obtained after the half-wave rectification and transformation of the modulated QPSK signal.

The demodulation process of the NC-like sinusoidal modulation signal is shown in Fig. 2.9. A method similar to "incoherent demodulation" is used here. Assuming that the input signal at the receiving end is $S'_{\text{NC - QPSK}}(t)$ and the length is M, the band-pass filter has an ideal rectangular-transmission characteristic, which makes the signal pass through without distortion. The signal $S''_{\text{NC - QPSK}}(t)$ output from the

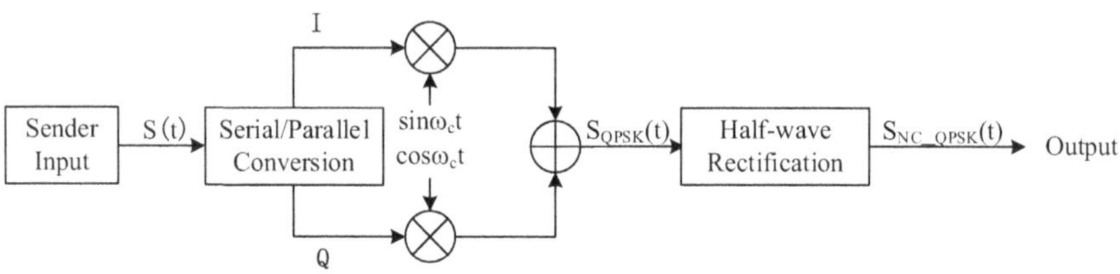

Fig. 2.7 NC-QPSK-like sinusoidal signal-modulation process

Fig. 2.8 NC-QPSK-like sinusoidal modulation constellation

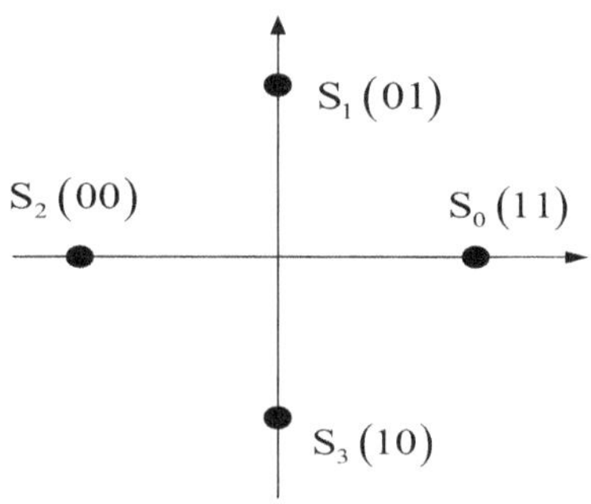

Table 2.1 NC-QPSK-like sinusoidal-modulation phase coding logical relationships

I	0	0	1	1
Q	0	1	0	1
I-way forming	0	1	−1	0
Q-way forming	−1	0	0	1
I-channel modulation	No	0°	180°	No
Q-channel modulation	180°	No	No	0°
Composite phase	180°	90°	270°	0°

band-pass filter must go through a low-pass filter to further filter the noise. The filter selection and parameter settings determine the performance of the system to a large extent. Therefore, the low-pass filter is set to a transfer function *xrc* with a length of N and a low-pass characteristic, which can be expressed as

$$xrc = 0.5 + 0.5 * \cos(\pi f * t). \tag{2.37}$$

Its spectrum is shown in Fig. 2.10, and the amplitude in the figure is the result of normalization.

In the frequency domain, the signal is multiplied by the transfer function *xrc*, which can better filter out high-frequency and sideband noise. In the time domain, the signal $S''_{\text{NC-QPSK}}(t)$ and *xrc* need to be convolved, and the $(N - 1)/2$ group delays at both ends are removed before the output. After filtering, it is necessary to set a reasonable threshold to make a sampling decision.

To better find the optimal value of the decision threshold, several simulation verifications were carried out under a Gaussian channel, weak-turbulence channel (scintillation index $0.1 \le \sigma_I^2 \le 0.5$), and mixed channel, and the average values were obtained. The obtained experimental results are shown in Figs. 2.11 and 2.12. It can be seen that the larger the threshold value is, the better. The probability of misjudgments increases for large and small threshold values.

The optimal decision threshold under a Gaussian channel is 0.5, according to the empirical value [19] combined with the simulation results, and the optimal decision threshold under a weak turbulent channel and mixed channel is approximately 0.3.

As shown in Fig. 2.13, after appropriately adjusting the decision threshold, the sampling decision of the signal amplitude is completed within one signal cycle, and the signal amplitude is determined by one-quarter, one-half, and three-quarter cycles of the signal. The amplitude value is used as the reference value to compare with the best threshold value.

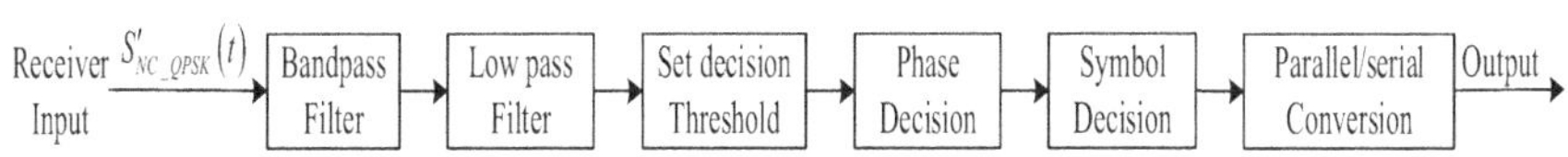

Fig. 2.9 Schematic diagram of NC-QPSK class sinusoidal demodulation

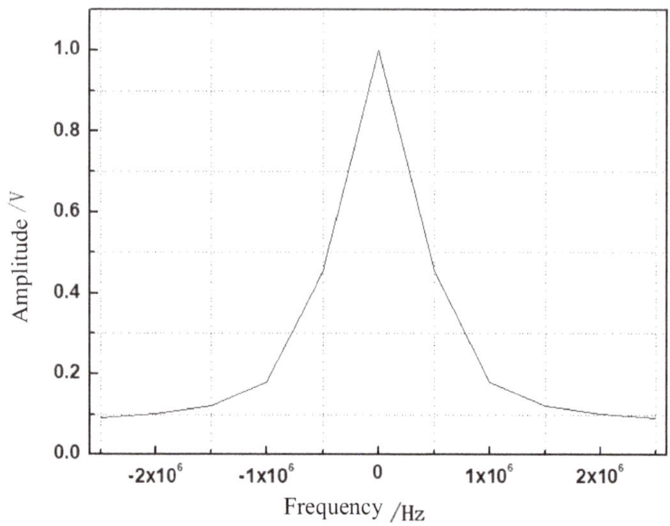

Fig. 2.10 Spectrum of transfer function xrc

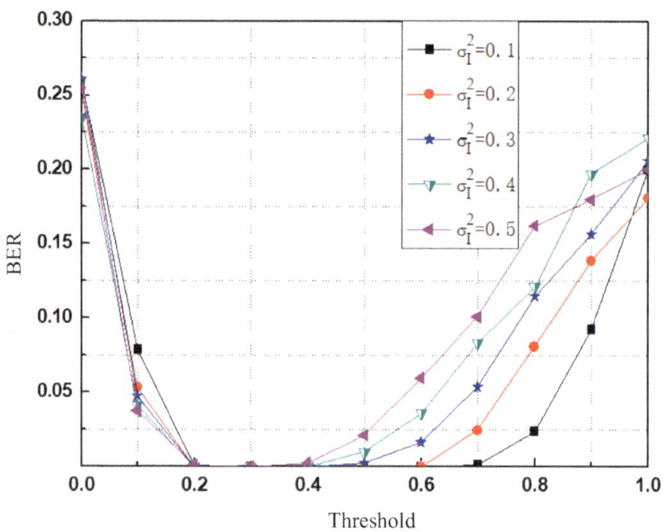

Fig. 2.11 Bit-error rates of different decision thresholds under weak turbulence [20]

If $A_{1/4} \le A_{th}, A_{1/2} \le A_{th}$, and $A_{3/4} \le A_{th}$, the phase corresponding to this waveform is 0°; if $A_{1/4} \ge A_{th}, A_{1/2} \le A_{th}$, and $A_{3/4} \le A_{th}$, the phase corresponding to this waveform is 90°. If $A_{1/4} \le A_{th}$, $A_{1/2} \ge A_{th}$, and $A_{3/4} \le A_{th}$, the phase corresponding to this waveform is 180°. If $A_{1/4} \le A_{th}$, $A_{1/2} \le A_{th}$, and $A_{3/4} \ge A_{th}$, the phase corresponding to this waveform is 270°.

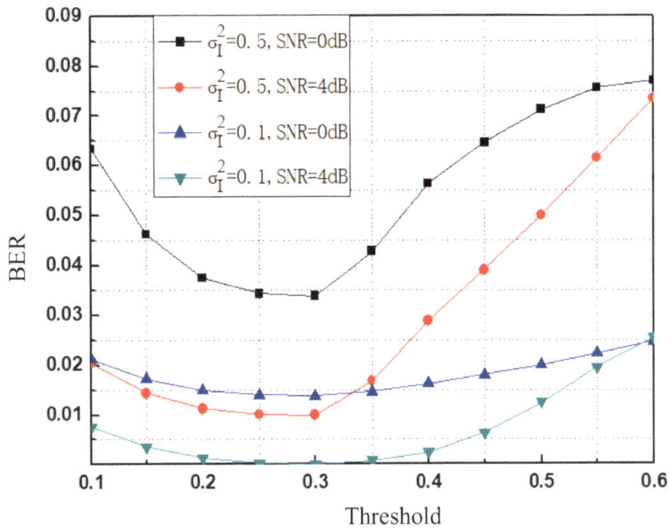

Fig. 2.12 Bit-error rates of different decision thresholds under weak turbulent and Gaussian mixed channels [20]

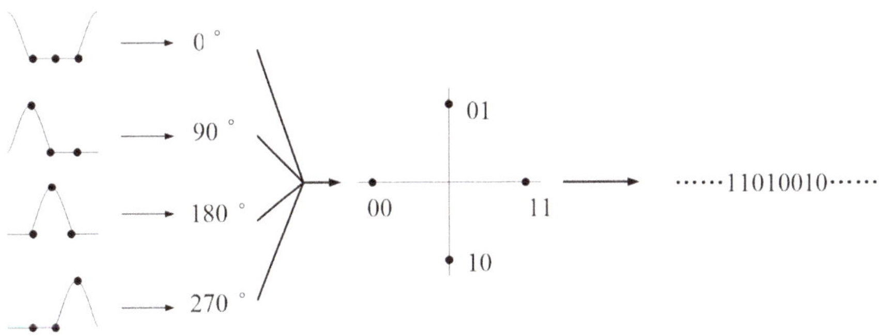

Fig. 2.13 Schematic diagram of phase and symbol decisions

Among them, A_{th} is the optimal threshold value, and $A_{1/4}$, $A_{1/2}$, and $A_{3/4}$ represent the amplitudes at one-quarter, one-half, and three-quarter cycles of the signal, respectively. Inverse mapping is performed on the symbols, according to the determined phase (refer to Table 2.1 for the inverse mapping rules), and the original baseband signal is finally restored after the symbols are output through a parallel–serial conversion.

2.2.2 *Performance Analysis of Non-limiting QPSK Sinusoidal Modulation*

We adopted atmospheric turbulence with a lognormal light-intensity scintillation distribution model to analyze the power efficiency and bit-error rate (BER) of NC-QPSK-like sinusoidal modulation and a DC-bias subcarrier intensity-modulation atmospheric laser-communication system. The outage probability and channel capacity were simulated and analyzed.

2.2.2.1 Power Utilization

The larger the DC offset, the lower the power needed to transmit a useful signal. In addition, the transmission of a laser in the atmospheric channel will be affected by various factors, and a large amount of power is lost. To ensure that the receiving power can meet certain requirements, it is necessary to continuously increase the transmitting power to meet this demand. The transmission equation of the optical wireless communication system can be expressed as [21]

$$P_r = P_t \times \eta_t \times \eta_s \times \eta(L, \beta) \times \eta_A \times \tau(L), \tag{2.38}$$

where P_r is the received optical power at transmission distance L, P_t is the average transmit power, η_t and η_s are the efficiencies of the transmitting and receiving antennas, respectively, η_A is the variation of the receiving efficiency caused by the aiming error, and $\tau(L) = 10^{-\alpha L/10}$ is the atmospheric transmittance. The attenuation coefficient α(dB/km) is related to the transmission distance L (km), and $\eta(L, \beta)$ is the receiving efficiency caused by the beam expansion, which is related to the transmission distance L (km), the beam divergence angle β(mrad), and the laser aperture D (mm). Its expression can be simplified as

$$\eta(L, \beta) = \left(e^{-2D^2/L^2\beta^2} - 1\right)/\left(e^{-2} - 1\right). \tag{2.39}$$

Assuming that the efficiencies of η_t and η_s are each 40%, the aiming error η_A is 50%, and the laser aperture D is 80 mm. According to the expressions of Eq. (2.39) and $\tau(L)$, we can obtain

$$P_r = 8 \times 10^{-2} P_t \cdot \eta(L, \beta) \cdot 10^{-\alpha L/10}. \tag{2.40}$$

According to the expression of $\eta(L, \beta)$, Eq. (2.40) can be simplified to

$$P_r = 0.09 P_t \times \left(1 - e^{-0.0128/L^2\beta^2}\right) \times 10^{-\alpha L/10}. \tag{2.41}$$

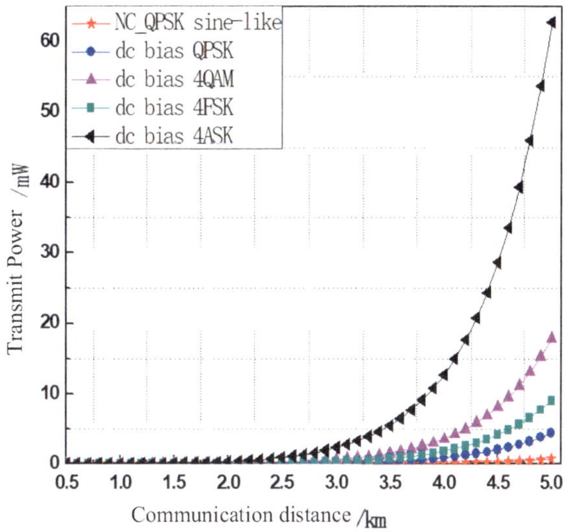

Fig. 2.14 Transmit power required by NC-QPSK-like sinusoidal modulation and DC-bias modulation at different communication distances [20]

Assume that the detection sensitivity is $P_{\min} = P_r = 0.1$ μW, the atmospheric attenuation coefficient α is 5 dB/km, and β is 0.2 mrad. Figure 2.14 shows the relationship between the required transmit power of the laser and the communication distance in the NC-QPSK-like sinusoidal modulation and DC-bias modulation in an atmospheric laser-communication system.

It can be seen from Fig. 2.14 that when the communication distance is the same, the required transmit powers for various methods are DC-bias 4-amplitude-shift keying (ASK), DC-bias square 4-QAM, DC-bias 4-FSK, DC-bias QPSK, and NC-QPSK in descending order; that is, the minimum transmit power is required for NC-QPSK-like sinusoidal modulation.

At $L = 5$ km, the transmit power required for NC-QPSK-like sinusoidal modulation is only 1.2% of DC-biased 4-ASK and 16.8% of DC-biased QPSK. When the communication distance becomes larger, the transmission power of the laser must also increase. However, compared with DC-bias modulation, the transmission power required by NC-QPSK sinusoidal modulation varies slowly with the transmission distance. After reaching a certain value—e.g., the communication distance of DC-bias 4-ASK reaches 3 km and the communication distance of DC-bias square 4-QAM reaches 4 km—the transmit power of the laser rises sharply with the increase of distance; therefore, using NC-QPSK sinusoidal modulation can improve the power utilization of an atmospheric laser-communication system.

2.2.2.2 Bit-Error Rate

Figure 2.15 shows the bit-error rate curves of the NC-QPSK-like sinusoidal modulation system and DC-biased QPSK modulation system. When there is no turbulence

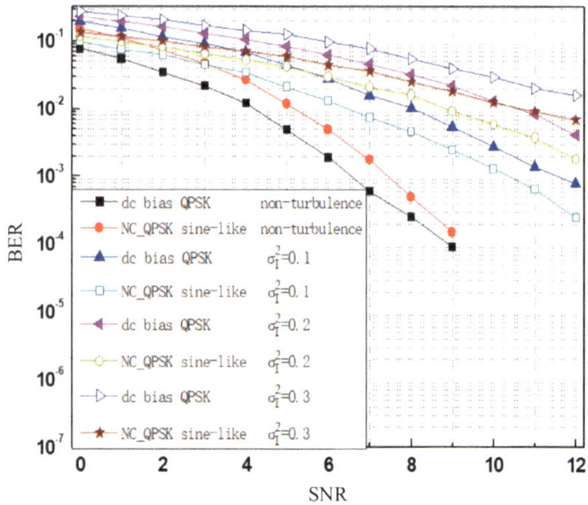

Fig. 2.15 Bit-error rate of NC-QPSK-like sinusoidal modulation and DC-bias QPSK modulation system [20]

in the atmospheric channel, the atmospheric laser-communication system using DC-bias QPSK modulation needs a signal-to-noise ratio of 6 dB to achieve a bit-error rate of 10^{-3}, whereas the NC-QPSK-like sinusoidal modulation system needs a signal-to-noise ratio of 7 dB to achieve the same bit-error rate. The former outperforms the latter by 1 dB.

When the turbulence-free SNR is 9 dB, the bit-error rates of the NC-QPSK-like sinusoidal modulation and DC-biased QPSK modulation are roughly equivalent. When the scintillation index is $\sigma_I^2 = 0.2$, the system needs to achieve a bit-error rate of 10^{-2}. The SNR required by the atmospheric laser-communication system using NC-QPSK-like sinusoidal modulation is 2 dB lower than that required by the DC-biased QPSK modulation system. At scintillation indices of $\sigma_I^2 = 0.1$ and $\sigma_I^2 = 0.3$, the performance of the system is also improved by 1 dB.

Figure 2.16 compares the bit-error rates of the NC-QPSK-like sinusoidal modulation and DC-bias modulation (square 4-QAM, 4-FSK, 4-ASK) systems when the scintillation index is $\sigma_I^2 = 0.2$. For the system to achieve a bit-error rate of 10^{-2}, the SNRs required for atmospheric laser-communication systems using NC-QPSK-like sinusoidal modulation and DC-bias square 4-QAM are 7 dB and 9 dB, respectively, whereas the signal-to-noise ratio required for the DC-bias 4ASK and 4FSK to achieve the same bit-error rate exceeds 12 dB. When the signal-to-noise ratio is >10 dB, the bit-error rate of DC-bias 4-FSK is close to that of DC-bias 4-ASK.

Through the analysis of Figs. 2.15 and 2.16, it can be seen that under a Gaussian channel without turbulence, the bit-error performance of the "quasi-incoherent demodulation method" used in NC-QPSK quasi-sine modulation is lower than that of the "quasi-incoherent demodulation method" used in DC-bias QPSK modulation when the signal-to-noise ratio is small.

The difference is not significant when the signal-to-noise ratio is relatively large and in the weak turbulent channel because the NC-QPSK sinusoidal modulation system does not lose transmission. Given the condition of the information, its zero

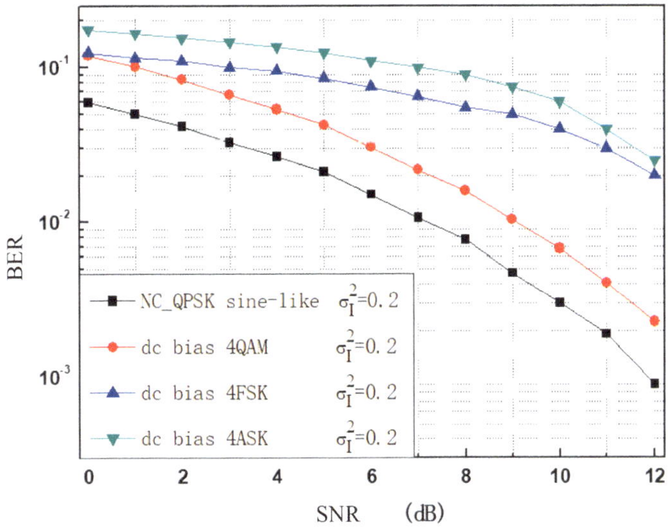

Fig. 2.16 Bit-error rate of NC-QPSK-like sinusoidal modulation and DC-bias modulation system [20]

amplitude is not affected by multiplicative noise, thus greatly improving the bit-error performance.

2.2.2.3 Outage Probability

The interruption probability is another expression of channel capacity. When the channel capacity cannot meet the required user rate, an interruption event will occur. This event presents a certain probability distribution and depends on the average signal-to-noise ratio of the channel and its channel-fading distribution model.

When the instantaneous signal-to-noise ratio is higher than the target SNR threshold, it indicates that the channel quality is good and normal communication can be guaranteed. If the power-utilization rate of the modulation mode is high, the probability that the instantaneous signal-to-noise ratio is lower than the target signal-to-noise ratio threshold when passing through the atmospheric random channel is smaller, and the probability of interruption is lower. Its expression is as follows [22]:

$$
\begin{aligned}
P_{out} &= P_r(\mu \leq \mu_{th}) = P_r\left(\frac{(\eta h)^2}{N_0} \leq \mu_{th}\right) \\
&= P_r\left(h \leq \frac{\sqrt{\mu N_0}}{\eta}\right) = \int_0^{\frac{\sqrt{\mu N_0}}{\eta}} f(h)dh,
\end{aligned}
\tag{2.42}
$$

where μ_{th} is the threshold value of the system signal-to-noise ratio, the instantaneous T/N ratio is $\mu = (\eta h)^2/N_0$, η is the photoelectric-conversion efficiency, h is the light intensity, and N_0 is the variance of the additive white Gaussian noise.

In the case of weak turbulence, the light-intensity fluctuation $f(h)$ follows a lognormal distribution, and the probability-density function of the lognormal distribution is substituted into Eq. (2.42). It can be seen that the interruption probability P_{out} is the integral of the lognormal distribution probability-density function; that is, P_{out} at this time is expressed as a cumulative distribution function:

$$P_{out} = \int_0^{\frac{\sqrt{\mu N_0}}{\eta}} \frac{1}{h\sqrt{2\pi\sigma_0^2}} \exp\left(-\frac{\left(\ln h + \sigma_0^2/2\right)^2}{2\sigma_0^2}\right) dh. \tag{2.43}$$

$\overline{\mu} = (\eta E[h])^2/N_0$ is defined as the average signal-to-noise ratio of a channel that is not affected by atmospheric turbulence; then, the probability-density function of the instantaneous signal-to-noise ratio can be transformed into

$$P\left(\sqrt{\frac{\mu}{\overline{\mu}}}\right) = \frac{1}{\sqrt{2\pi}\sigma_0\sqrt{\frac{\mu}{\overline{\mu}}}} \exp\left(-\frac{\left(\ln\sqrt{\frac{\mu}{\overline{\mu}}} + \sigma_0^2/2\right)^2}{2\sigma_0^2}\right). \tag{2.44}$$

After the transformation, the interruption probability can be expressed as

$$P_{out} = P_r(\mu \leq \mu_{th}) = P_r\left(\frac{\mu}{\overline{\mu}} \leq \frac{\mu_{th}}{\overline{\mu}}\right). \tag{2.45}$$

According to Eqs. (2.43), (2.44), and (2.45), the outage probability can be expressed as

$$\begin{aligned} P_{out} &= 1 - \frac{1}{2} erfc\left[\frac{-0.5 \times \ln\left(\frac{\overline{\mu}}{\mu_{th}}\right) + \sigma_0^2/2}{\sqrt{2\sigma_0^2}}\right] \\ &= 1 - \frac{1}{2} erfc\left[\frac{-\frac{1}{2} \times \left(\ln\left(\frac{\overline{\mu}}{\mu_{th}}\right) - \ln\left(\sigma_I^2 + 1\right)\right)}{\sqrt{\ln\left(\sigma_I^2 + 1\right)^2}}\right], \end{aligned} \tag{2.46}$$

where $erfc(y) = \frac{2}{\sqrt{\pi}} \int_y^\infty \exp(-t^2) dt$ is the error-compensation function. The relationship between the received optical power and the light intensity is

$$P_r = A_r h, \tag{2.47}$$

where A_r is the receiving antenna area. According to Eq. (2.45), combined with Eqs. (2.46) and (2.47), the expression of the interruption probability P_{out} under weak turbulence is

$$\begin{aligned} P_{out} &= 1 - \frac{1}{2}erfc\left[\frac{-\frac{1}{2}\times\left(\ln\left(\frac{\overline{\mu}}{\mu_{th}}\right) - \ln\left(\sigma_I^2+1\right)\right)}{\sqrt{\ln\left(\sigma_I^2+1\right)^2}}\right] \\ &= 1 - \frac{1}{2}erfc\left\{\frac{-\frac{1}{2}\times\left[\ln\left(\frac{0.09P_t\times\left(1-e^{-0.0128/L^2\beta^2}\right)}{A_r^2N_0\mu_{th}}\right) - \ln\left(\sigma_I^2+1\right)\right]}{\sqrt{2\ln\left(\sigma_I^2+1\right)}}\right\}. \end{aligned} \tag{2.48}$$

After normalizing $\left(0.09\times\left(1-e^{-0.0128/L^2\beta^2}\right)\times 10^{-\alpha L/10}\right)^2\eta^2/A_r^2N_0$ in the formula, the outage probability is only related to the transmit power and normalized threshold. Figure 2.17 shows the variation curve of the interruption probability of the NC-QPSK-like sinusoidal modulation system with the SNR threshold under different scintillation indices when the transmit power is the same. The results show that when the scintillation index increases from $\sigma_I^2 = 0.1$ to $\sigma_I^2 = 0.3$, the system performance decreases.

Figure 2.18 shows the curve of the interruption probability of the NC-QPSK-like sinusoidal and DC-bias modulation systems when the scintillation index is $\sigma_I^2 = 0.3$ with the SNR threshold. When the SNR normalization threshold is 0.5, the outage probability of the NC-QPSK sine-like modulation system is 0.3, whereas that of the DC-biased QPSK and DC-biased 4-FSK is approximately 0.5, that of the

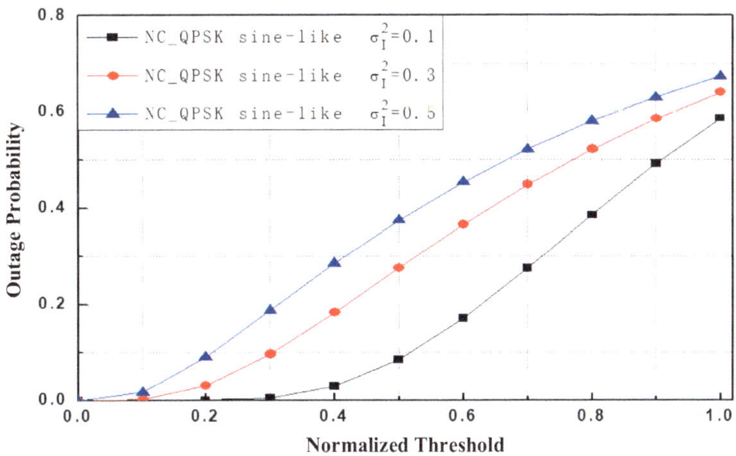

Fig. 2.17 Outage probability of an NC-QPSK-like sinusoidal modulation system under different turbulence intensities [20]

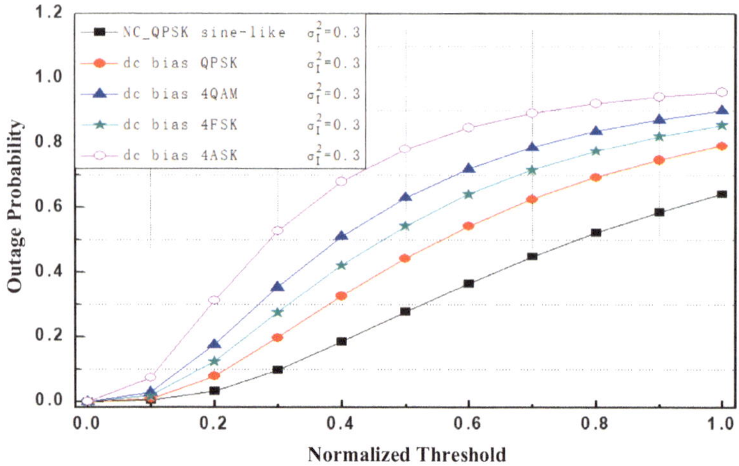

Fig. 2.18 NC-QPSK-like sinusoidal modulation and DC-bias modulation system-outage probability [20]

DC-biased square 4-QAM is approximately 0.63, and that of the DC-biased 4ASK is approximately 0.78. It can be seen that the performance of the NC-QPSK-like sinusoidal modulation system has greatly improved, compared with the DC-bias modulation (square 4-QAM, 4-FSK, 4-ASK) system.

2.2.2.4 Channel Capacity

The channel capacity is the maximum information rate that a channel can transmit without errors. For a single-user channel with only one source and one sink, it is a number in bits per second or bits per symbol. It represents the maximum amount of information that can be transmitted per second or per channel symbol. Alternatively, an information rate less than this number must be transmitted on this channel without errors. The following mainly analyzes and considers the average channel capacity of different modulation methods under turbulent channels.

(1) **Non-turbulent channel**

For discrete input and continuous-output discrete memoryless additive white Gaussian noise (AWGN) channels, the channel capacity can be defined as the maximum value of the average mutual information about the source probability-density function [23]:

$$C = \max_{\{p(x)\}} I(X; Y). \tag{2.49}$$

Consider an M-ary signal, where $\{X_m(t),\ m = 1, 2, \ldots, M\}$ is an equally spaced finite-energy waveform, each of which can be expanded on a set of standard quadrature sets:

$$X_m(t) = x_m\varphi(t), (m = 1, 2, \ldots, M). \tag{2.50}$$

A one-to-one correspondence is established between each signal waveform $X_m(t)$ and its coordinate vector $\overline{x}_m = (x_1, x_2, \ldots x_i, x_m)$. The channel capacity of the signal can be obtained by Eq. (2.49) as

$$C = \max_{P(x_m, m=1,2,\ldots,M)} \sum_{m=1}^{\infty} \int_{-\infty}^{\infty} P(x)P(y/x) \times \left[\log_2 \frac{P(y/x_m)}{P(y)}\right] dy. \tag{2.51}$$

From Eq. (2.51), it can be deduced that the channel capacity of one-dimensional modulation and two-dimensional modulation are respectively [24]

$$C = \log_2(M) - \frac{1}{M\sqrt{\pi}} \sum_{m=1}^{M} \int_{-\infty}^{\infty} \exp(-t^2) \times \log_2\left[\sum_{j=1}^{M} \exp\left(-\frac{2t(y - x_i)}{\sqrt{N_0}} - \frac{(y - x_m)^2}{N_0}\right)\right] dt \tag{2.52}$$

$$C = \log_2(M) - \frac{1}{M\pi} \sum_{m=1}^{M} \int_{-\infty}^{\infty} \exp(-|t|^2) \times \log_2\left[\sum_{j=1}^{M} \exp\left(-\frac{2\mathrm{Re}\{t(y - x_i)^*\}}{\sqrt{N_0}} - \frac{|y - x_m|^2}{N_0}\right)\right] dt, \tag{2.53}$$

where M is the modulation order, and the noise variance per dimension is $N_0/2$.

The average SNR γ of one-dimensional modulation and two-dimensional modulation is expressed by Eqs. (2.54) and (2.55), respectively [24]:

$$\gamma = \frac{2}{MN_0} \sum_{i=1}^{M} x_i^2 \tag{2.54}$$

$$\gamma = \frac{2}{MN_0} \sum_{i=1}^{M} |x_i|^2. \tag{2.55}$$

(2) **Atmospheric-turbulence channel**

In an atmospheric-turbulence channel, assuming that the channel is memoryless, stationary, and ergodic and considering Eqs. (2.4) and (2.24), the following can be obtained [25]:

$$y = sx + n = \eta hx + n, \tag{2.56}$$

where y represents the signal at the receiving end, x represents the modulated signal, h represents the instantaneous light intensity of the received signal, η represents the photoelectric-conversion efficiency of the receiving end, and n is the noise of the receiving end device, which is n $\approx$ N(0.5, N_0/2) Gaussian white noise; η can be expressed as

$$\eta = \frac{ve\lambda}{h_p c}, \tag{2.57}$$

where v is the quantum efficiency, e is the electron charge, λ is the signal wavelength, h_p is Planck's constant, and c is the speed of light.

The optical wireless channel is a random time-varying channel, and the signal-to-noise ratio of the instantaneous electrical signal received by the receiver is also a random variable, expressed as [27]

$$\gamma = \frac{s^2}{N_0} = \frac{\eta^2 h^2}{N_0}. \tag{2.58}$$

The expression of the average channel capacity with respect to the signal-to-noise ratio in a turbulent channel is [27]

$$\langle C \rangle = \int_0^\infty B \log_2(1 + \overline{\gamma}) p_\gamma(\gamma) d\gamma, \tag{2.59}$$

where $p_\gamma(\gamma)$ is the probability-density function of the signal-to-noise ratio of γ, which varies according to the distribution that the light intensity h obeys, and $\overline{\gamma} = \frac{(\eta E[h])^2}{N_0} = \frac{\eta^2}{N_0}$ is the average signal-to-noise ratio after light-intensity normalization.

In a weakly turbulent channel, the light intensity h follows a lognormal distribution, and the probability-density function of the signal-to-noise ratio in the weakly turbulent case can be obtained as

$$p_\gamma(\gamma) = \frac{1}{2\gamma\sigma_0\sqrt{2\pi}} \exp\left(-\frac{\left(\ln(\gamma/\overline{\gamma}) + \sigma_0^2\right)^2}{8\sigma_0^2} \right). \tag{2.60}$$

Substituting Eq. (2.60) into Eq. (2.59), the expression of the average channel capacity with respect to the signal-to-noise ratio is

$$\langle C \rangle = \frac{B}{2\ln(2)\sigma_0\sqrt{2\pi}} \int_0^\infty \frac{\ln(1+\gamma)}{\gamma} \exp\left(-\frac{\left(\ln(\gamma) - \ln(\overline{\gamma}) + \sigma_0^2\right)^2}{8\sigma_0^2} \right) d\gamma. \tag{2.61}$$

Let $u = \left(\ln(\gamma) - \ln(\overline{\gamma}) + \sigma_0^2\right)$; then, Eq. (2.61) can be expressed as

$$\langle C\rangle = \frac{B}{2\ln(2)\sigma_0\sqrt{2\pi}}\left[\int_0^{\infty} \ln\left(1+e^{-\mu}\right)e^{-\frac{(\mu+\xi)^2}{8\sigma_0^2}}du + \int_0^{\infty} \ln(1+e^{\mu})e^{-\frac{(\mu-\xi)^2}{8\sigma_0^2}}du\right]. \tag{2.62}$$

Because $\ln(1 + e^u) = u + \ln(1 + e^{-u})$, substituting into the above formula yields

$$\langle C\rangle = \frac{B}{2\ln(2)\sigma_0\sqrt{2\pi}}\left[\int_0^{\infty} \ln\left(1+e^{-\mu}\right)e^{-\frac{(\mu+\xi)^2}{8\sigma_0^2}}du + \int_0^{\infty} \left[\ln\left(1+e^{\mu}\right)+u\right]e^{-\frac{(\mu-\xi)^2}{8\sigma_0^2}}du\right]. \tag{2.63}$$

Because $d\left(e^{-\frac{(\mu-\xi)^2}{8\sigma_0^2}}\right) = -\frac{1}{4\sigma_0^2}ue^{-\frac{(\mu-\xi)^2}{8\sigma_0^2}} + \frac{\xi}{4\sigma_0^2}e^{-\frac{(\mu-\xi)^2}{8\sigma_0^2}}$, we can obtain

$$\int_0^{\infty} ue^{-\frac{(\mu-\xi)^2}{8\sigma_0^2}}du = 4\sigma_0^2 e^{-\frac{\xi^2}{8\sigma_0^2}} + \xi\sqrt{2\pi}\sigma_0 erfc\left(-\frac{\xi}{2\sqrt{2}\sigma_0}\right). \tag{2.64}$$

For $0 \le x \le 1$,

$$\ln(1+x) = \sum_{k=1}^{\infty}\frac{(-1)^{k+1}}{k}x^k. \tag{2.65}$$

For $0 \le e^{-u} \le 1$, $u \ge 0$ has

$$\ln\left(1+e^{-u}\right) = \sum_{k=1}^{\infty}\frac{(-1)^{k+1}}{k}e^{-ku}. \tag{2.66}$$

Then, the average channel capacity under a weakly turbulent channel can be expressed as [27]

$$\begin{aligned}\langle C\rangle &= \frac{B}{2\ln(2)\sigma_0\sqrt{2\pi}}\sum_{k=1}^{\infty}\frac{(-1)^{k+1}}{k}\left(\int_0^{\infty} e^{-k\mu}e^{-\frac{(\mu+\xi)^2}{8\sigma_0^2}}du + \int_0^{\infty} e^{-k\mu}e^{-\frac{(\mu-\xi)^2}{8\sigma_0^2}}du\right)\\ &+ \frac{2\sigma_0 B}{\ln(2)\sqrt{2\pi}}e^{-\frac{\xi^2}{8\sigma_0^2}} + \frac{\xi B}{2\ln(2)}erfc\left(-\frac{\xi}{2\sqrt{2}\sigma_0}\right).\end{aligned} \tag{2.67}$$

Because $\int_0^{\infty} e^{-k\mu}e^{-\frac{(\mu\pm\xi)^2}{8\sigma_0^2}}du = e^{-\frac{\xi^2}{8\sigma_0^2}}\sqrt{2\pi}\sigma_0 \mathrm{erfcx}\left(\sqrt{2}\sigma_0 k \pm \frac{\xi}{2\sqrt{2}\sigma_0}\right)$, the closed expression of the average channel capacity about the average signal-to-noise ratio is

$$\begin{aligned}\langle C\rangle &= \frac{B}{2\ln(2)}e^{-\frac{\xi^2}{8\sigma_0^2}}\sum_{k=1}^{\infty}\frac{(-1)^{k+1}}{k}\left[\operatorname{erfcx}\left(\sqrt{2}\sigma_0 k+\frac{\xi}{2\sqrt{2}\sigma_0}\right)+\operatorname{erfcx}\left(\sqrt{2}\sigma_0 k-\frac{\xi}{2\sqrt{2}\sigma_{+0}}\right)\right]\\ &+\frac{2\sigma_0 B}{\ln(2)\sqrt{2\pi}}e^{-\frac{\xi^2}{8\sigma_0^2}}+\frac{\xi B}{2\ln(2)}\operatorname{erfc}\left(-\frac{\xi}{2\sqrt{2}\sigma_0}\right).\end{aligned} \tag{2.68}$$

The channel capacities of each modulation method in a optical wireless communication system are not the same. From the mathematical expression of the average channel capacity, it can be seen that the key to distinguishing the average channel capacity of different modulation methods is the signal-to-noise ratio. However, the expression of the signal-to-noise ratio cannot distinguish different modulation methods; thus, it should be changed to an appropriate expression.

We changed the optical intensity in Eq. (2.38) to optical power to represent the change of a laser, and the average optical power of the signal at the receiving end is expressed as [28]

$$P_R(h) = P_T\left(\frac{\eta A}{\lambda L}\right)^2 h, \tag{2.69}$$

where h is the channel state, P_T is the average power of the transmitter signal, η is the photon efficiency of the transmitter and receiver, A is the area of the telescope, expressed as $A = \pi D^2/4$, where D is the diameter of the telescope, λ is the working wavelength, and L is the distance between the transmitter and receiver.

The photocurrent detected on the detector at the receiving end is $I = \eta_z P_R$, and η_z is the photoelectric-conversion efficiency; therefore, the electrical signal power at the receiving end is $P_E = I^2 R = (\eta_z P_R)^2 R$, and the electrical-to-noise ratio at the receiving end is

$$\gamma(h) = \frac{P_E}{2\sigma_n^2} = \frac{\eta_z^2 P_T^2 R}{2\sigma_n^2}\left(\frac{\eta A}{\lambda L}\right)^4 h^2. \tag{2.70}$$

When other parameters are fixed, the average channel capacity at this time is only related to the average power of the transmitting end, and the average channel capacity of different modulation modes can be represented by this relationship. The average channel capacity can be expressed as

$$\langle C\rangle = \int_0^{\infty} B\log_2(1+\gamma(h))f_h(h)dh. \tag{2.71}$$

In the formula, the signal-to-noise ratio is shown in Eq. (2.70).

In the case of weak turbulence, the channel state h follows a lognormal distribution, and the SNR at the receiving end is expressed in relation to the average power at the transmitting end. Combining Eqs. (2.69), (2.70), and (2.71), the expression of the average channel capacity of different modulation methods with respect to the transmit power P_T is as follows:

$$\langle C\rangle = \frac{B}{\ln 2 \times \sqrt{2\pi \ln(\sigma_I^2+1)}} \int_0^\infty \ln\left(1+\frac{\eta_z^2 P_T^2 R}{2\sigma_n^2}\left(\frac{\pi D^2 \eta}{4\lambda L}\right)^4 h^2\right) \times h^{-1} \exp\left\{-\frac{\left[\ln\left(h\sqrt{(\sigma_I^2+1)}\right)\right]^2}{2\ln(\sigma_I^2+1)}\right\} dh. \tag{2.72}$$

Numerical simulations are carried out according to Eq. (2.72), where the photoelectric-conversion efficiency η_z is 0.5, the photon efficiency η is 0.8, the receiving-end aperture D is 8 cm, the wavelength is 850 nm, and the noise variance is 2.5×10^{-5} A.

Figure 2.19a, b show the average channel-capacity comparison curves of the NC-QPSK-like sinusoidal modulation system and the DC-offset modulation (QPSK, square 4-QAM, 4-FSK, 4-ASK) system without turbulence. It can be clearly seen from the figure that as the signal-to-noise ratio increases, the average channel capacity of the system keeps increasing.

When the signal-to-noise ratio is higher than 8 dB, the average channel capacity of the DC-biased square 4-QAM modulation system does not change significantly with the increase of the signal-to-noise ratio, while the average channel capacities of the NC-QPSK-like sinusoidal modulation and the DC-biased QPSK system are still continuously improving. It can also be seen from the figure that the average channel capacities of the NC-QPSK-like sinusoidal modulation and DC-biased QPSK are slightly larger than that of the DC-biased square 4-QAM under the same signal-to-noise ratio. The average channel capacities of the DC-biased 4-FSK and DC-biased 4-ASK modulation systems are close to the one-dimensional Shannon limit [30].

Figure 2.19a, b show the average channel-capacity curves of the NC-QPSK-like sinusoidal modulation system and DC-biased QPSK modulation system when the scintillation indices are $\sigma_I^2 = 0.1$ and $\sigma_I^2 = 0.3$, respectively. It can be seen that the average channel capacity increases with the increase of transmit power, but as the scintillation index increases, the average channel capacity of the system decreases. The average channel capacity of the NC-like sinusoidal modulation system is approximately 2 bits/symbol/Hz higher in the curve with a scintillation index of $\sigma_I^2 = 0.1$ than the curve with a scintillation index of $\sigma_I^2 = 0.3$. When the scintillation index is $\sigma_I^2 = 0.3$, the average channel capacity of the NC-like sinusoidal modulation system is approximately 2–3 bits/symbol/Hz higher than that of the DC-biased QPSK modulation system.

Figure 2.19c shows the average channel-capacity comparison curve between the NC-QPSK-like sinusoidal modulation system and the DC-bias modulation (square 4-QAM, 4-FSK, 4-ASK) system when the scintillation index is $\sigma_I^2 = 0.3$. As can be seen from the figure, with the increase of transmit power, the average channel capacity of the system increases. The average channel capacities of various modulation modes are given in descending order: NC-QPSK-like sinusoidal modulation, DC-bias 4-FSK,

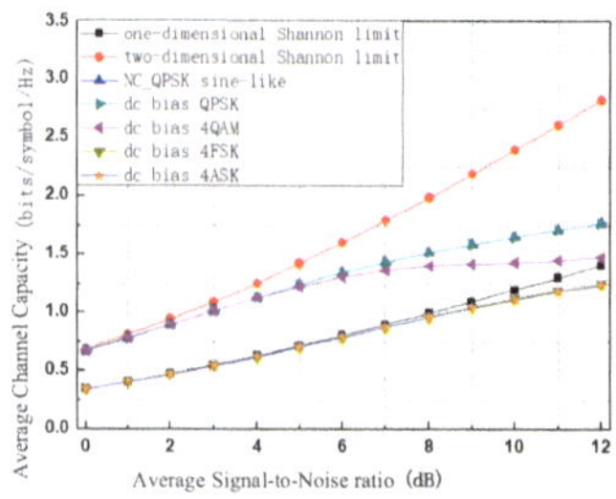

(a) No turbulence (NC_QPSK, QPSK, square 4-QAM, 4-FSK, 4-ASK)

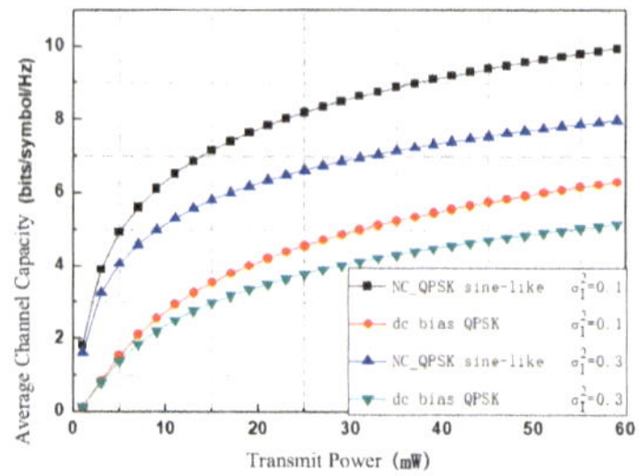

(b) Weak turbulence, $\sigma_I{}^2 = 0.1$ and $\sigma_I{}^2 = 0.3$, $L = 5$ km (NC_QPSK, QPSK)

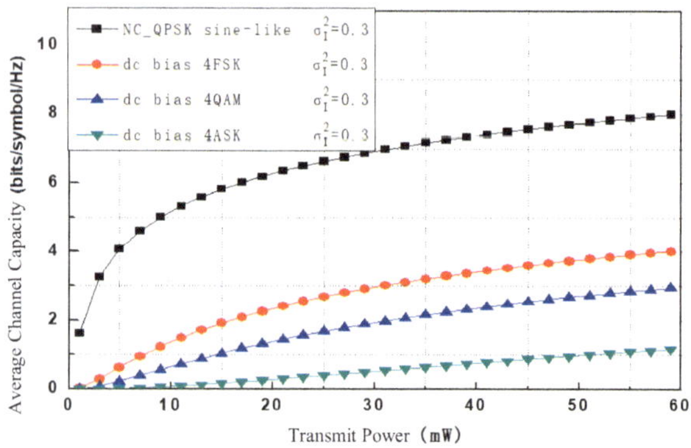

(c) Weak turbulence, $\sigma_{I2} = 0.3$, $L = 5$ km (NC_QPSK, square 4-QAM, 4-FSK, 4-ASK)

Fig. 2.19 Channel capacity of NC-QPSK-like sinusoidal modulation and DC-bias modulation systems [29]

DC-bias square 4-QAM, and DC-bias 4-ASK. Under the same conditions, the average channel capacity of the NC-like sinusoidal modulation system is approximately 2–7 bits/symbol/Hz higher than that of the DC-biased 4-ASK modulation system.

2.3 16-PSK Modulation Principles and Simulation

The multiple phase-shift keying (MPSK) or multi-digit digital phase-modulation modulation method includes M different symbols, each of which is composed of K binary bits ($M = 2K$), at a symbol interval of T_S. The carrier phase corresponds to one of the M discrete phases, and the phase of the sine carrier changes with the changing law of the input symbols. Multi-system digital phase modulation, also known as a polyphase system, is a generalization of the two-phase system. It uses a variety of different carrier phase states to characterize digital information.

2.3.1 Vector Representation of MPSK Signals

2.3.1.1 Expression of an MPSK Signal

In digital phase modulation [12], M signal waveforms with different phases can be expressed as

$$\begin{aligned} s_m(t) &= \left[g_T(t)e^{j2\pi(m-1)/M}e^{j2\pi f_c^t}\right] \qquad (m = 1, 2, 3, \ldots\ldots, M, 0 \le t \le T_S) \\ &= \mathrm{g_T(t)}\cos\left[2\pi f_c\mathrm{t} + \frac{2\pi}{M}(m-1)\right], \end{aligned} \tag{2.73}$$

where $\mathrm{T_S}$ is a symbol period of M-ary phase-shift keying, $\mathrm{T_S} = (\log_2 M)T_b = KT_b$; T_b is the binary symbol interval; $g_T(t)$ is the baseband transmit-filter impulse response; and f_c is the sinusoidal carrier frequency. We expand Eq. (2.73) to obtain

$$\begin{aligned} s_m(t) &= \mathrm{g_T(t)}\left\{\left[\cos\frac{2\pi}{\mathrm{M}}(\mathrm{m}-1)\right]\cos 2\pi f_c\mathrm{t} - \left[\sin\frac{2\pi}{\mathrm{M}}(\mathrm{m}-1)\right]\sin 2\pi f_c\mathrm{t}\right\} \\ &= \mathrm{g_T(t)}[\mathrm{a_{m_c}}\cos 2\pi f_c\mathrm{t} - \mathrm{a_{m_s}}\sin 2\pi f_c\mathrm{t}]. \end{aligned} \tag{2.74}$$

In

$$\mathrm{a_{m_c}} = \cos\frac{2\pi}{\mathrm{M}}(\mathrm{m}-1),\ \mathrm{a_{m_s}} = \sin\frac{2\pi}{\mathrm{M}}(\mathrm{m}-1) \quad (m = 1, 2, 3, \ldots\ldots, M, 0 \le t \le T_S), \tag{2.75}$$

$\{\mathrm{a_{m_c}}\}$ and $\{\mathrm{a_{m_s}}\}$ are groups of multi-level amplitude sequences. Within a symbol period of $\mathrm{T_S}$, it is necessary to ensure that

$$\mathrm{a_{m_c}^2} + \mathrm{a_{m_s}^2} = 1 \qquad 0 \le t \le T_S. \tag{2.76}$$

In each M-ary symbol period $\mathrm{T_S}$, the waveforms of each MPSK signal have the same energy:

$$E_S = \int_0^{T_S} s_m^2(t)dt = \frac{1}{2}\int_0^{T_S} g_T^2(t)dt = \frac{1}{2}E_g \qquad m = 1, 2, 3, \ldots\ldots, M, \tag{2.77}$$

where E_g is the energy of pulse $\mathrm{g_T(t)}$; if $\mathrm{g_T(t)}$ is a rectangular pulse, set

$$\mathrm{g_T(t)} = \sqrt{\frac{2E_S}{T_S}} \qquad 0 \le t \le T_S. \tag{2.78}$$

If we substitute Eq. (2.73) into Eq. (2.74), we obtain

$$s_m(t) = \sqrt{\frac{2E_S}{T_S}}\left[\mathrm{a}_{\mathrm{m_c}}\cos 2\pi f_c \mathrm{t} - \mathrm{a}_{\mathrm{m_s}}\sin 2\pi f_c \mathrm{t}\right] \qquad 0 \le t \le T. \tag{2.79}$$

It can be seen from Eq. (2.79) that the multi-level amplitude-keying signals of the two orthogonal carriers are superimposed on each other to form the MPSK signal.

Two-Dimensional Vector Representation of MPSK Signals

Each MPSK signal waveform can be composed of a complete linear combination of two normalized orthogonal functions, $f_1(t)$ and $f_2(t)$:

$$f_1(t) = \sqrt{\frac{2}{T_S}}\cos 2\pi f_c t \qquad 0 \le t \le T_S \tag{2.80}$$

$$f_2(t) = \sqrt{\frac{2}{T_S}}\sin 2\pi f_c t \qquad 0 \le t \le T_S. \tag{2.81}$$

The orthogonal expansion of MPSK is

$$s_m(t) = s_{m1}f_1(t) + s_{m2}f_2(t) \quad , 0 \le t \le T_S, \tag{2.82}$$

where

$$\begin{aligned} s_{m1} &= \int_0^{T_S} s_m(t)f_1(t)dt \\ &= \sqrt{E_S}\cos\frac{2\pi}{\mathrm{M}}(\mathrm{m}-1) \\ &= \sqrt{E_S}\mathrm{a}_{\mathrm{m_c}} \qquad m = 1, 2, 3, \ldots\ldots, M \end{aligned} \tag{2.83}$$

$$\begin{aligned} s_{m2} &= \int_0^{T_S} s_m(t)f_2(t)dt \\ &= \sqrt{E_S}\sin\frac{2\pi}{\mathrm{M}}(\mathrm{m}-1) \\ &= \sqrt{E_S}\mathrm{a}_{\mathrm{ms}} \qquad m = 1, 2, 3, \ldots\ldots, M \end{aligned} \tag{2.84}$$

$$s_m(t) = \sqrt{E_S}[\mathrm{a}_{\mathrm{m_c}}f_1(t) + \mathrm{a}_{\mathrm{m_s}}f_2(t)] \qquad m = 1, 2, 3, \ldots\ldots, M. \tag{2.85}$$

The MPSK signal can be represented by a two-dimensional vector as

$$s_m = [s_{m1},\ s_{m2}] = \left[\sqrt{E_S}\mathrm{a}_{\mathrm{m_c}},\ \sqrt{E_S}\mathrm{a}_{\mathrm{ms}}\right] \quad m = 1, 2, 3, \ldots\ldots, M. \tag{2.86}$$

2.3.2 16-PSK Signal Modulation Method

Equation (2.86), when $M = 16$, demonstrates 16-PSK modulation. The phase of the 16-PSK sinusoidal carrier has 16 different values, and each value represents a group of symbols. In 16-PSK, the interval between adjacent phases in the values of these 16 phases is equal, and each phase value uniquely corresponds to a group of symbol information. The mathematical expression of the 16-PSK signal is expressed as

$$s_{16\mathrm{PSK}}(t) = \sqrt{\frac{2E_S}{T_S}}\left[\cos\left[\frac{2\pi}{\mathrm{M}}(\mathrm{m}-1)\right]\cos 2\pi f_c\mathrm{t} - \sin\left[\frac{2\pi}{\mathrm{M}}(\mathrm{m}-1)\right]\sin 2\pi f_c\mathrm{t}\right], \tag{2.87}$$

where T_S is the symbol period of 16-PSK, and f_c is the sinusoidal carrier frequency. 16-PSK has 16 different phases: $\pi/16$, $3\pi/16$, $5\pi/16$, $7\pi/16$, $9\pi/16$, $11\pi/16$, $13\pi/16$, $15\pi/16$, $17\pi/16$, $19\pi/16$, $21\pi/16$, $23\pi/16$, $25\pi/16$, $27\pi/16$, $29\pi/16$, and $31\pi/16$. Figure 2.20 shows the constellation diagram of the 16-PSK modulated signal [1].

As can be seen from Fig. 2.20, among the 16 different four-bit symbols of 16-PSK, there is only one difference in binary bit information between adjacent symbols; that is, the four-bit symbol information of 16-PSK is different from the sine. The phase values of the carrier conform to the phase-logic relationship of the Gray code.

Using the phase-logic relationship of the Gray code in the 16-PSK modulation and demodulation system can reduce the bit-error rate of the system. The 16-PSK

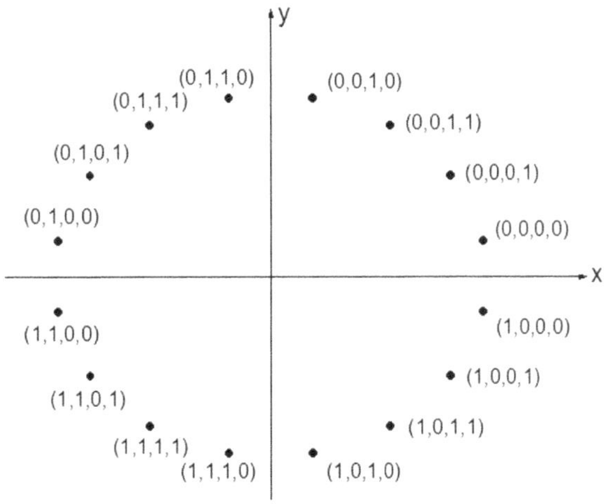

Fig. 2.20 Constellation diagram of a 16-PSK signal

signal is inevitably affected by channel noise during the transmission process. The received 16-PSK sinusoidal carrier-phase signal "polluted" by the noise may be close to the adjacent sinusoidal carrier phase and indistinguishable. The symbols will be misjudged as adjacent symbols.

Two methods exist for generating a 16-PSK modulation signal: phase modulation and phase selection.

(1) **Phase-modulation method**

We expand Eq. (2.87) further:

$$\begin{aligned} s_{16\text{PSK}}(t) &= I(t)\cos w_c t - Q(t)\sin w_c t \\ &= A_0\cos(w_c t + \varphi(t)) \qquad m = 1, 2, 3, \ldots\ldots, 16, \quad 0 \le t \le T_S \end{aligned} \tag{2.88}$$

where $A_0 = \sqrt{I^2(t) + Q^2(t)}$, $I(t) = \sqrt{\frac{2E_S}{T_S}}\cos\left[\frac{2\pi}{\text{M}}(\text{m}-1)\right]$, and $Q(t) = \sqrt{\frac{2E_S}{T_S}}\sin\left[\frac{2\pi}{\text{M}}(\text{m}-1)\right]$ are the in-phase and quadrature components, respectively, and $\frac{2\pi}{\text{M}}(\text{m}-1) = \varphi(t) = tg^{-1}\frac{I(t)}{Q(t)}$.

According to Eq. (2.88), the principle block diagram of the 16-PSK phase-modulation method is obtained as shown in Fig. 2.21.

It can be seen from Fig. 2.21 that the serial–parallel conversion module converts the input binary-code stream into in-phase branch data and quadrature branch data. In the constellation-mapping module, according to the input in-phase and quadrature branch data, they are mapped to the corresponding phase, and the corresponding in-phase branch output $A_0\cos(\varphi_m)$ and quadrature branch output $A_0\sin(\varphi_m)$ are calculated.

Subsequently, baseband shaping is used to reduce the intersymbol interference (ISI) and make the modulated signal suitable for transmission in the channel; the signal is completed in the multiplier module. The filtered data of the in-phase branch and quadrature branch are multiplied by quadrature carriers $\cos(w_c t)$ and $\sin(w_c t)$ of the same frequency, respectively. Finally, the data of the two branches are superimposed in the adder to complete the modulation of the 16-PSK signal.

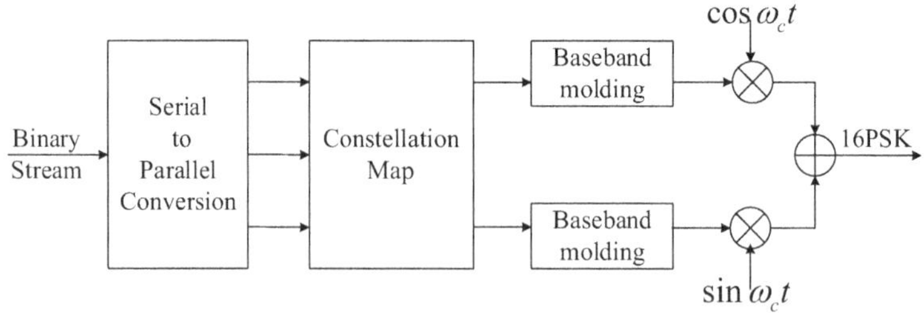

Fig. 2.21 16PSK phase-modulation method modulation-principle block diagram

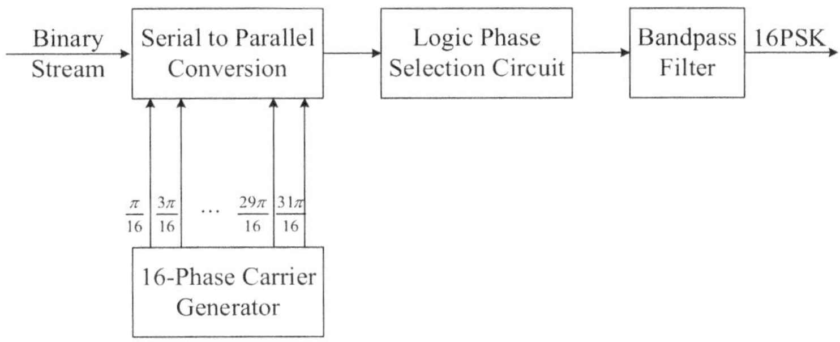

Fig. 2.22 16-PSK phase-selection method modulation-principle block diagram

(2) **Phase-selection method**

The modulation-principle block diagram of the 16-PSK modulation signal is shown in Fig. 2.22. The modulation principle of the phase-selection method is as follows: The serial–parallel conversion module converts the input binary-code stream sequence into a four-bit group of symbols for output; the logic-selection circuit is based on the input. It controls the 16-phase carrier generator to output the corresponding phase carrier from the 16 different-phase carriers, and then passes through the band-pass filter to finally generate the required 16-PSK modulation signal.

For example, when the four-bit symbol is 0000, a carrier with a phase of $\pi/16$ is output; when it is 0001, a carrier with a phase of $3\pi/16$ is output; when it is 0011, a carrier with a phase of $5\pi/16$ is output; when it is 1011, a carrier with a phase of $27\pi/16$ is output; when it is 1001, a carrier with a phase of $29\pi/16$ is output; when it is 1000, a carrier with a phase of $31\pi/16$ is output, etc.

2.3.3 16-PSK Modulation and Demodulation Performance Analysis

The power-spectral density, bandwidth efficiency, bit-error rate, and channel capacity of the 16-PSK modulation and demodulation system are analyzed below.

2.3.3.1 Power-Spectral Density and Bandwidth Efficiency

For the 16 different symbols of the 16-PSK signal, each symbol is composed of four bits $(b_{4i}, b_{4i+1}, b_{4i+2}, b_{4i+3})$; thus, the duration of one symbol is

$$T = T_b \log_2 16 = 4T_b, \tag{2.89}$$

where T_b is the duration of one bit. The baseband power-spectral density of the 16-PSK signal is expressed as

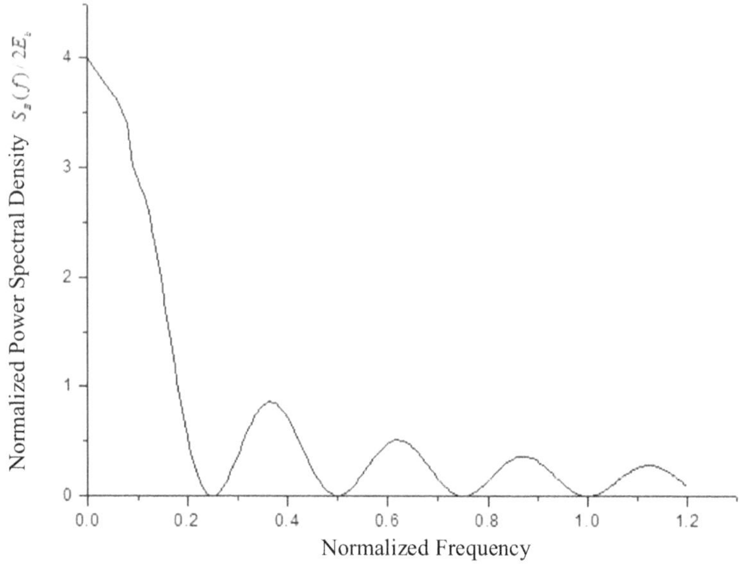

Fig. 2.23 16-PSK power spectrum

$$\begin{aligned} S_B(f) &= 2E \sin c^2(Tf) \\ &= 2E_b \log_2 16 \sin c^2(T_b f \log_2 16) \\ &= 8E_b \sin c^2(4T_b f), \end{aligned} \tag{2.90}$$

where $\sin c^2(4T_b f) = \frac{\sin^2(4\pi T_b f)}{(4\pi T_b f)^2}$. Figure 2.23 shows the power spectrum of 16 PSK and the change trend of the normalized power-spectral density $S_B(f)/2E_b$ relative to the normalized frequency fT_b.

It can be seen from Fig. 2.23 that the main lobe of the power spectrum of the 16-PSK signal is defined by the frequency point where the power-spectral density is zero. For the 16-PSK signal bandwidth, the main-lobe spectrum width is often used for measurement, and it can be clearly seen that most of the 16-PSK signal power is contained in the zero–zero bandwidth.

The main lobe of the power spectrum of a 16-PSK signal is defined by a well-defined spectral null (frequency point where the power-spectral density is zero). When the zero–zero bandwidth contains the main lobe of the 16-PSK signal power spectrum, it also contains most of the signal power. Therefore, a simple and common method to measure the 16-PSK signal bandwidth is using the main lobe spectral width.

The channel bandwidth required to transmit a 16-PSK signal is

$$B = \frac{2}{T}, \tag{2.91}$$

Table 2.2 Bandwidth utilization of an M-ary PSK signal

Base	2	4	8	16
$\rho(bs^{-1}Hz^{-1})$	0.5	1	1.5	2

where T is the duration of one of the 16 different symbols in the 16-PSK signal, which can be calculated using Eq. (2.91), and the bit rate is $R_b = 1/T_b$. Then, the relationship between the channel bandwidth required by the 16-PSK signal and bit rate R_b can be expressed by the following equation:

$$B = \frac{2R_b}{\log_2 16} = \frac{R_b}{2}. \tag{2.92}$$

On this basis, the bandwidth efficiency of the 16-PSK signal can be obtained as

$$\rho = \frac{R_b}{B} = \frac{\log_2 16}{2} = 2. \tag{2.93}$$

Table 2.2 shows the bandwidth efficiency of PSK in 2, 4, 8, and hexadecimal.

It can be seen from Table 2.2 that in the MPSK modulation mode, with the increase in the base number M, the bandwidth efficiency also increases; the distance is also reduced, which increases the bit-error rate. The system bit-error rate increase, caused by the increase of M, can be reduced by increasing the system signal-to-noise ratio.

2.3.3.2 Influence of White Gaussian Noise on the Phase Identification of a 16-PSK System

In an MPSK modulation and demodulation communication system, the frequency of the carrier is much larger than the bandwidth of the signal, and the transmission process of the signal can be equivalent to a narrowband random process. After passing through the system, the white Gaussian noise $n(t)$ becomes narrowband white Gaussian noise $n'(t)$ [26]:

$$n'(t) = n_I'(t)\cos(w_c t) - n_Q'(t)\sin(w_c t), \tag{2.94}$$

where $n_I'(t)$ and $n_Q'(t)$ are the quadrature components. The mathematical expression of the known 16-PSK signal is shown in Eq. (2.87), and the received signal can be obtained from Eqs. (2.87) and (2.94):

$$\begin{aligned} s'_{16\mathrm{PSK}}(t) &= s_{16\mathrm{PSK}}(t) + n'(t) \\ &= [I(t) + n_I'(t)]\cos w_c t - [Q(t) + n_Q'(t)]\sin w_c t \\ &= a(t)\cos w_c t - b(t)\sin w_c t \\ &= A'(t)\cos(w_c t + \theta(t)), \end{aligned} \tag{2.95}$$

where $A'(t) = \sqrt{[I(t) + n_I'(t)]^2 + [Q(t) + n_Q'(t)]^2} = \sqrt{a^2(t) + b^2(t)}$ and $\theta(t) = \arctan \frac{Q(t)+n_Q'(t)}{I(t)+n_I'(t)} = \arctan \frac{a(t)}{b(t)}$.

Therefore, the phase difference $\Delta\theta$ caused by white Gaussian noise is

$$\begin{aligned}\Delta\theta &= \theta(t) - \varphi(t)\\ &= \arctan \frac{Q(t) + n_Q'(t)}{I(t) + n_I'(t)} - \arctan \frac{Q(t)}{I(t)}.\end{aligned} \tag{2.96}$$

Equation (2.96) can be further simplified to obtain

$$\begin{aligned}\tan\Delta\theta &= \tan(\arctan \tfrac{Q(t)+n_Q'(t)}{I(t)+n_I'(t)} - \arctan \tfrac{Q(t)}{I(t)})\\ &= \tfrac{I(t)n_Q'(t)-Q(t)n_I'(t)}{I^2(t)+Q^2(t)+I(t)n_I'(t)+Q(t)n_Q'(t)}.\end{aligned} \tag{2.97}$$

From Eq. (2.96), the phase difference $\Delta\theta$ caused by white Gaussian noise can be converted into

$$\Delta\theta = \arctan\left(\frac{I(t)n_Q'(t) - Q(t)n_I'(t)}{I^2(t) + Q^2(t) + I(t)n_I'(t) + Q(t)n_Q'(t)}\right). \tag{2.98}$$

2.3.3.3 Effects of Noise and Fading on the BER of a 16-PSK System

When the phase-variation range of the signal point caused by noise is $\pm\frac{\pi}{16}$, the signal point can be correctly demodulated at the receiving end; hence, the bit-error rate is

$$P_{e.16PSK} = 1 - \int_{-\frac{\pi}{16}}^{\frac{\pi}{16}} f(\theta)d\theta, \tag{2.99}$$

where $f(\theta)$ is the probability-density function of θ. To find $f(\theta)$, we first find the conditional probability-density function $f_{A\theta/\phi}(A_t, \theta_t/\phi)$, where A_t is the amplitude of the signal under the influence of noise and θ_t is the phase of the input signal. According to the two-dimensional joint normal distribution, we can obtain

$$f_{A\theta/\phi}(A_t, \theta_t/\phi) = \frac{A_t}{2\pi\sigma^2} \exp\left\{-\frac{1}{2\sigma^2}[A_t^2 + A_0^2 - 2A_0A_t\cos(\phi - \theta_t)]\right\}. \tag{2.100}$$

We apply the method for finding the function distribution of random variables, which can be obtained from the above formula:

$$f_{A\theta/\phi}(A_t,\theta_t/\phi)=\frac{A_t}{2\pi\sigma^2}\exp\left\{-\frac{1}{2\sigma^2}[A_t^2+A_0^2-2A_0A_t\cos(\phi-\theta_t)]\right\}. \tag{2.101}$$

From Eq. (2.101) we can obtain

$$\begin{aligned}
f(\theta/\phi)&=\int_0^\infty f(A_t,\theta/\phi)dA_t\\
&=\int_0^\infty \frac{A_t}{2\pi\sigma^2}\exp\left\{-\frac{1}{2\sigma^2}[A_t^2+A_0^2-2A_0A_t\cos(\phi-\theta_t)]\right\}dA_t\\
&=\int_0^\infty \frac{A_t}{2\pi\sigma^2}\exp\left\{-\frac{1}{2\sigma^2}[A_t^2-2A_0A_t\cos(\phi-\theta_t)+A_0^2\cos^2(\phi-\theta_t)-A_0^2\cos^2(\phi-\theta_t)+A_0^2]\right\}dA_t\\
&=\frac{1}{2\pi}\exp\left\{-\frac{A_0^2-A_0^2\cos^2(\phi-\theta_t)}{2\sigma^2}\right\}\int_0^\infty \frac{A_t}{\sigma^2}\exp\left\{-\frac{[A_t-A_0\cos(\phi-\theta_t)]^2}{2\sigma^2}\right\}dA_t\\
&=\frac{1}{2\pi}\exp\left\{-\frac{A_0^2\sin^2(\phi-\theta_t)}{2\sigma^2}\right\}\int_0^\infty \frac{A_t}{\sigma^2}\exp\left\{-\frac{[A_t-A_0\cos(\phi-\theta_t)]^2}{2\sigma^2}\right\}dA_t.
\end{aligned} \tag{2.102}$$

Let $a=A_0\cos(\phi-\theta_t)$; then, (2.102) is transformed into

$$\begin{aligned}
f(\theta/\phi)&=\frac{1}{2\pi}\exp\left[-\frac{A_0^2\sin^2(\phi-\theta_t)}{2\sigma^2}\right]\left[\int_0^\infty \frac{A_t-a+a}{\sigma^2}\exp\left[\frac{(A_t-a)^2}{2\sigma^2}\right]\right]dA_t\\
&=\frac{1}{2\pi}\exp\left[-\frac{A_0^2\sin^2(\phi-\theta_t)}{2\sigma^2}\right]\left[\frac{1}{\sigma^2}\int_0^\infty (A_t-a)\exp\left[\frac{(A_t-a)^2}{2\sigma^2}\right]dA_t+\frac{a}{\sigma^2}\int_0^\infty \exp\left[\frac{(A_t-a)^2}{2\sigma^2}dA_t\right]\right.
\end{aligned} \tag{2.103}$$

where the first integral in (2.103) is

$$\frac{1}{\sigma^2}\int_0^\infty (A_t-a)\exp\left[\frac{(A_t-a)^2}{2\sigma^2}\right]dA_t. \tag{2.104}$$

Let $z=A_t-a$; then, $dz=dAt$, $At:0\to\infty, z:-a\to\infty$. Therefore, Eq. (2.104) is further simplified to

$$\frac{1}{\sigma^2}\int_{-a}^\infty z\exp\left(-\frac{z^2}{2\sigma^2}\right)dz=\exp\left[-\frac{A_0^2\cos^2(\phi-\theta_t)}{2\sigma^2}\right], \tag{2.105}$$

where the second integral in (2.104) is

$$\frac{a}{\sigma^2}\int_0^\infty \exp\left[\frac{(A_t-a)^2}{2\sigma^2}dA_t.\right. \tag{2.106}$$

Let $z = \frac{A_t - a}{\sqrt{2}\sigma}$; then, $dz = \frac{1}{\sqrt{2}\sigma} dA_t, At : 0 \to \infty, z : -\frac{a}{\sqrt{2}\sigma} \to \infty$

$$\frac{a}{\sigma^2} \int_{-\frac{a}{\sqrt{2}\sigma}}^{\infty} \exp(-z^2)\sqrt{2}\sigma\, dz = \frac{a}{\sigma}\frac{\sqrt{\pi}}{\sqrt{2}}\left[2 - erfc\left(\frac{a}{\sqrt{2}\sigma}\right)\right]$$

$$= \frac{\sqrt{\pi} A_0 \cos(\phi - \theta_t)}{\sqrt{2}\sigma}\left[(1 + erf\left(\frac{a}{\sqrt{2}\sigma}\right)\right], \quad (2.107)$$

where $erf(x) = \frac{2}{\sqrt{\pi}} \int_0^x \exp(-z^2)dz$ is the error function.

$erfc(x) = 1 - erf(x) = \frac{2}{\sqrt{\pi}} \int_x^\infty \exp(-z^2)dz$ is the complementary error function;

$$f(\theta/\phi) = \frac{1}{2\pi} \exp\left[-\frac{A_0^2 \sin^2(\phi - \theta_t)}{2\sigma^2}\right]\left\{\exp\left[-\frac{A_0^2 \sin^2(\phi - \theta_t)}{2\sigma^2}\right] + \frac{\sqrt{\pi} A_0 \cos(\phi - \theta_t)}{\sqrt{2}\sigma}\left[\left(1 + erf\left(\frac{a}{\sqrt{2}\sigma}\right)\right]\right\}$$

$$= \frac{1}{2\pi} \exp\left[-\frac{A_0^2}{2\sigma^2}\right] + \frac{A_0 \cos(\phi - \theta_t)}{2\sqrt{2\pi}\sigma} \exp\left[-\frac{A_0^2 \sin^2(\phi - \theta_t)}{2\sigma^2}\right]\left[\left(1 + erf\left(\frac{A_0 \cos(\phi - \theta_t)}{\sqrt{2}\sigma}\right)\right] \quad (2.108)$$

Let $r = \frac{A_0^2}{2\sigma^2}$ be the signal-to-noise ratio; then,

opto. (2.109)

If $\phi = 0$,

$$f(\theta) = \frac{1}{2\pi} \exp[-r] + \frac{\sqrt{r}}{2\sqrt{\pi}} \cos(\theta_t) \exp[-r \sin^2(\theta_t)]\left[(1 + erf(\sqrt{r}\cos(\theta_t))\right]. \quad (2.110)$$

Therefore, under the influence of an additive white Gaussian noise (AWGN) channel, the bit-error rate of the 16-PSK system is

$$P_{16PSK} = 1 - \int_{-\frac{\pi}{16}}^{\frac{\pi}{16}} f(\theta)d\theta$$

$$= 1 - \left[\frac{1}{16}e^{-r} + \frac{1}{2}erf\left(\sqrt{r}\sin\left(\frac{\pi}{16}\right)\right) + \int_0^{\sqrt{r}\sin(\frac{\pi}{16})} \frac{1}{\sqrt{\pi}} e^{-x^2} erf(\sqrt{r - x^2})dx\right]. \quad (2.111)$$

When r >> 1,

$$P_{16PSK} = 1 - \int_{-\frac{\pi}{16}}^{\frac{\pi}{16}} f(\theta)d\theta = erfc(\sqrt{r}\sin(\pi/16)). \quad (2.112)$$

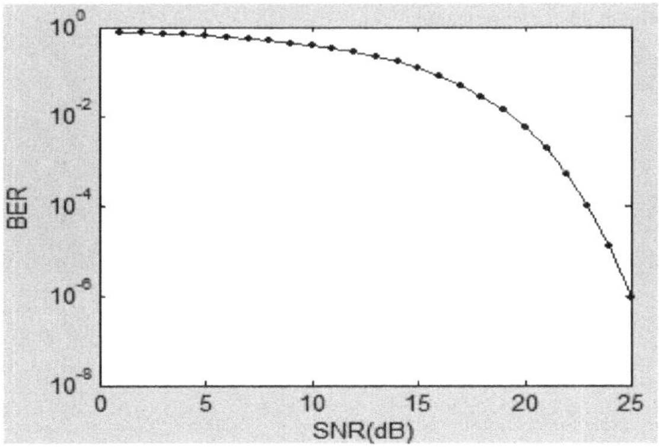

Fig. 2.24 Bit-error rate curve under an AWGN channel

Figure 2.24 shows the bit-error rate curve of the 16-PSK signal after passing through the AWGN channel.

It can be seen from Fig. 2.24 that when the AWGN channel and the effects of Rayleigh fading exist simultaneously, the amplitude and phase of the signal are significantly distorted, resulting in a sharp increase in the bit-error rate. In this case, the bit-error rate of a 16-PSK system is

$$\begin{aligned} P_{16PSK} &= \int_0^\infty P_e(r)f(r)d\gamma \\ &= 1 - \frac{1}{16(1+r_0)} - \frac{1}{2r_0}\int_0^\infty erf\left(\sqrt{r}\times\sin\frac{\pi}{16}\right)e^{-\frac{r}{r_0}}dr \\ &\quad - \frac{1}{r_0\sqrt{\pi}}\int_0^\infty erf\left(\sqrt{r-x^2}\right)\int_0^{\sqrt{r}\sin\frac{\pi}{16}} e^{-(x^2+\frac{r}{r_0})}drdx \end{aligned} \tag{2.113}$$

where $P_e(r)$ can be obtained using Eq. (2.113). $r = \frac{\alpha^2 E_b}{N_0}$ represents the ratio between the signal energy and the noise power-spectral density when the channel attenuation is $\alpha; f(r) = \frac{1}{r_0}e^{-(\frac{r}{r_0})}, r \geq 0; r_0 = \frac{E_b}{N_0}E[\alpha^2]$.

When $r >> 1$, Eq. (2.113) can be transformed into

$$P_{16PSK} = \int_0^\infty P_e(r)f(r)dr = \frac{1}{r_0}\int_0^\infty erfc(\sqrt{r}\sin\frac{\pi}{16})e^{-\frac{r}{r_0}}dr. \tag{2.114}$$

Figure 2.25 shows the bit-error rate curve of the 16-PSK signal after an AWGN channel and Rayleigh attenuation. We compare the bit-error rate curves of 16-PSK through an AWGN channel and a coexisting AWGN and Rayleigh attenuation channel. When SNR = 20 dB, BER = 5.798×10^{-3} under the AWGN channel, and

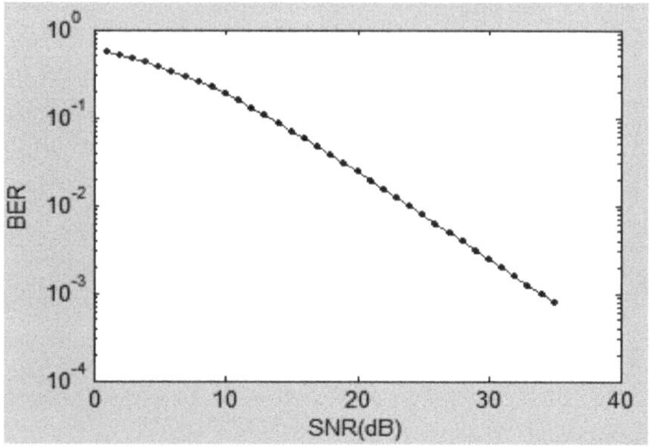

Fig. 2.25 Bit-error rate curve under AWGN and Rayleigh fading channels

BER = 2.39×10^{-2} under the coexisting AWGN and Rayleigh fading channel. With the increase in the SNR, the communication quality under the AWGN channel is evidently better than that under the coexistence of AWGN and Rayleigh fading.

2.3.4 Analysis of Experimental Results of 16-PSK Subcarrier Modulation

Experiments were carried out on the optical wireless communication system, and a large number of experimental data under different typical weather conditions were measured. By analyzing a large amount of data, it was found that different weather conditions have different effects on the performance of wireless communication systems.

2.3.5 16-PSK Modulation-Signal Power-Spectral Density

Welch's method was used to estimate and analyze the power spectrum of the experimental data. The Welch method uses algorithms, such as signal-segment overlap, windowing functions, and fast Fourier transforms (FFTs), to estimate the self-power spectrum of a signal sequence [31]:

$$P(f) = \frac{1}{ML} \sum_{i=1}^{L} \left| \sum_{n=0}^{M-1} x_i(n) d(n) e^{-j2\pi fn} \right|^2. \tag{2.115}$$

The experimental data $x(n)$ are divided into L segments. In the program, the segments are overlapped for half of the segment length; thus, $x(n)$ can be divided into 2L − 1 segments of data in total, and then each data segment is multiplied in the time domain by a window function. An FFT is conducted, and finally the square of the spectral amplitude of each piece of data is added and averaged to obtain the power-spectral density estimate of the experimental data $x(n)$.

Figures 2.26, 2.27, 2.28 and 2.29 show the single-sided power-spectral density estimation diagrams of data collected under different meteorological conditions. It can be seen that the highest components of the power spectrum on cloudy, rainy, and snowy days are all around the 1.5625 MHz carrier frequency, which also characterizes the distribution of light energy. The power spectrum of the signal on snowy days is broadened at the main peak of the power spectrum of the signal on cloudy and rainy days. Moreover, owing to the influence of the Doppler effect, the peak of the spectrum on snowy days is shifted to the right by 20 kHz, which is mainly caused by the movement of the medium (Fig. 2.30).

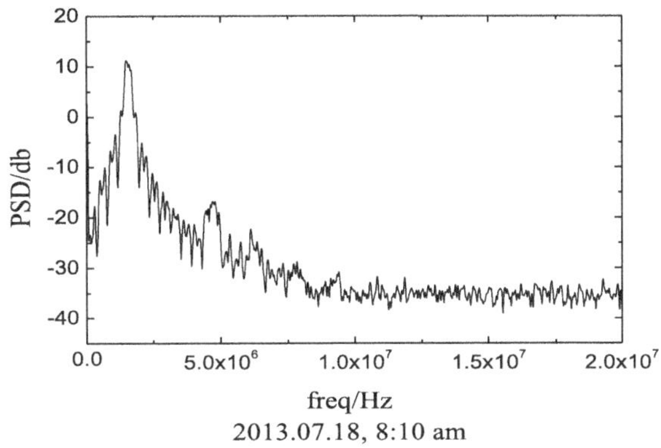

Fig. 2.26 Power-spectrum estimation of cloudy weather data [32–34]

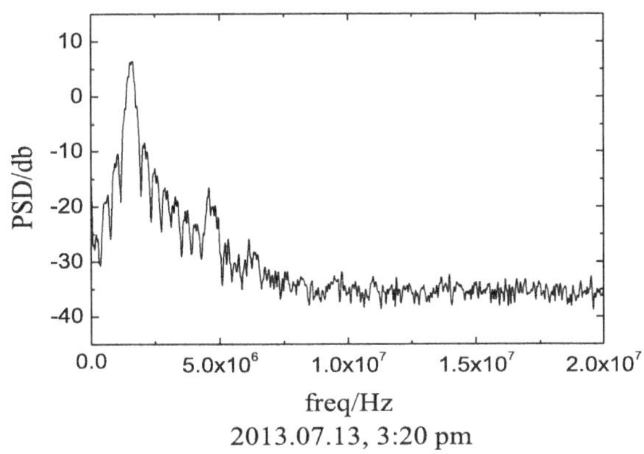

Fig. 2.27 Power-spectrum estimation of light-rain weather data [32–34]

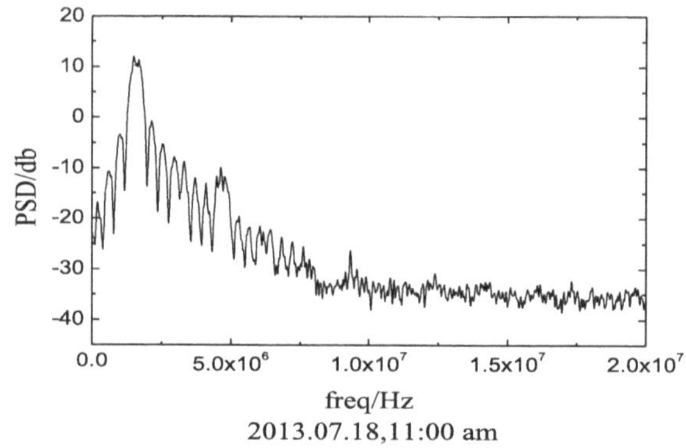

Fig. 2.28 Estimated power spectrum of moderately-rainy weather data [32–34]

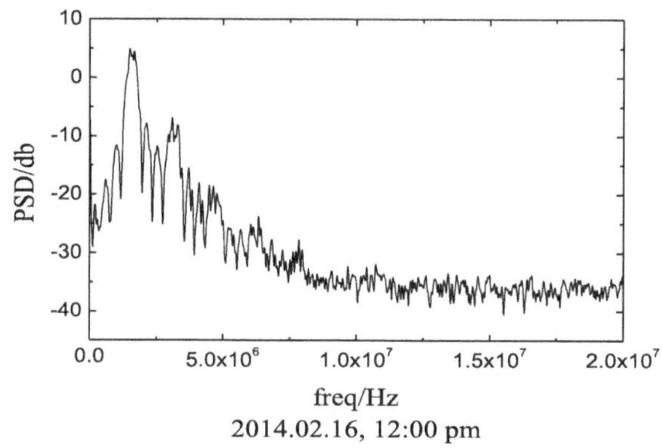

Fig. 2.29 Power-spectrum estimation of snowy weather data [32–34]

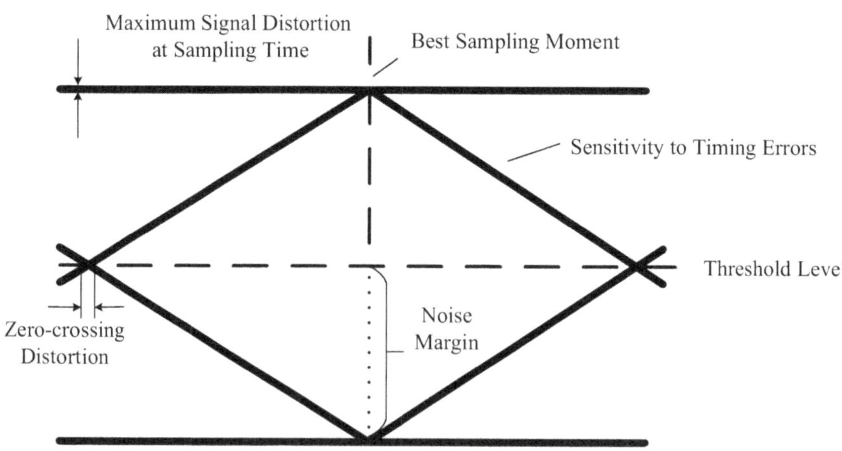

Fig. 2.30 Relationship between the eye diagram and system performance

2.3.5.1 16-PSK Modulation Eye-Diagram Analysis Under Different Weather Conditions

An eye diagram is a graph formed by the accumulation of a series of digital signals on an oscilloscope and contains a wealth of information. The influence of intersymbol crosstalk and noise can be observed from the eye diagram, which reflects the overall characteristics of the digital signal, thereby allowing the pros and cons of the communication system to be estimated.

The size of the "eye" of the eye diagram reflects the strength of intersymbol crosstalk. The larger the "eye" is and the more correct the eye diagram is, the smaller the intersymbol crosstalk; otherwise, the greater the intersymbol crosstalk. When noise is present, the noise will be superimposed on the signal and the traces of the observed eye diagram will become blurred. If there is intersymbol interference at the same time, the "eyes" will be opened less.

Compared with the eye diagram without intersymbol crosstalk, the original clear and correct thin traces have become blurred lines, which are not very correct. The greater the noise, the wider and blurrier the trace, the greater the crosstalk between symbols, and the less correct the eye diagram.

By observing eye diagrams under different weather conditions, the influence of weather factors on the performance of wireless communication systems can be studied and analyzed. Figures 2.31, 2.32, 2.33 and 2.34 present the eye diagrams of the system under different weather conditions.

The "size" of the eye opening can indicate the strength of the intersymbol crosstalk, and the trace can show the strength of the noise. The louder the noise, the wider and blurrier the stitches will be. From Figs. 2.31, 2.32, 2.33 and 2.34, the following can be seen:

(1) The "eye" of the eye diagram is the largest on cloudy days, and the line trace of the eye diagram is relatively clear. Intersymbol interference effects are minimal.

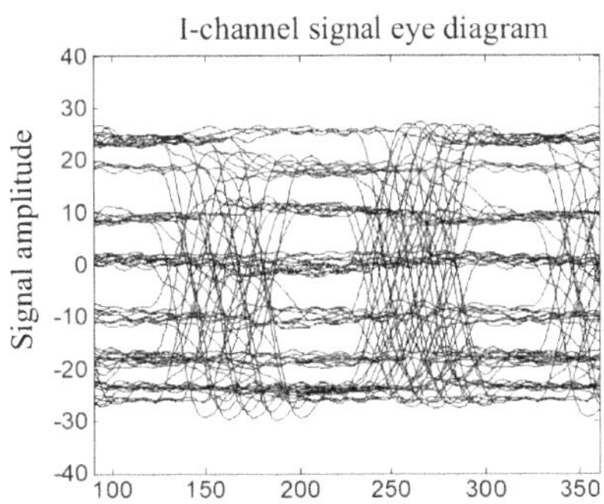

(a) I-channel signal

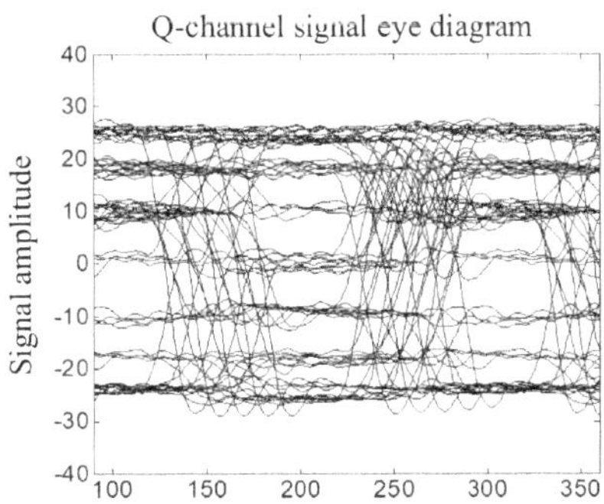

(b) Q-channel signal

Fig. 2.31 Eye diagram of cloudy weather data [32–34]. 2013.07.18, 8:10 am

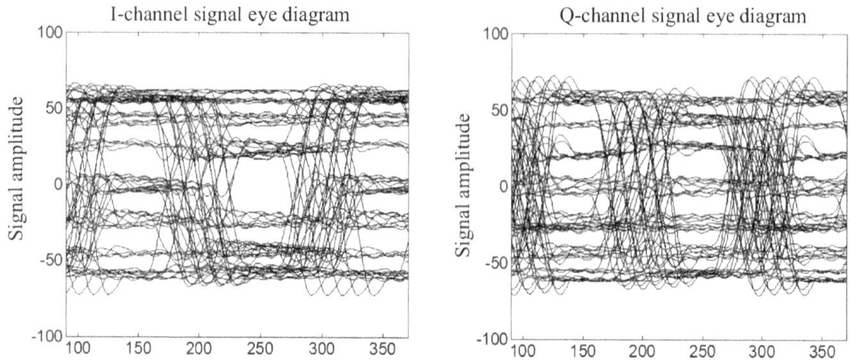

Fig. 2.32 Eye diagram of light-rain weather data [32–34]. 2013.07.13, 3:20 pm

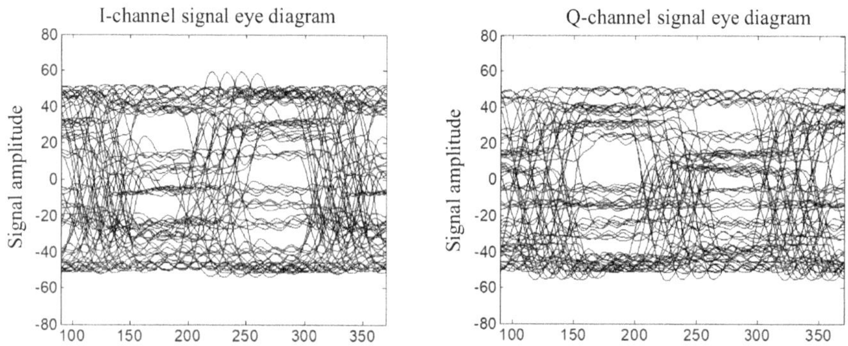

Fig. 2.33 Eye diagram of moderately rainy weather data [32–34]. 2013.07.18, 11:00 am

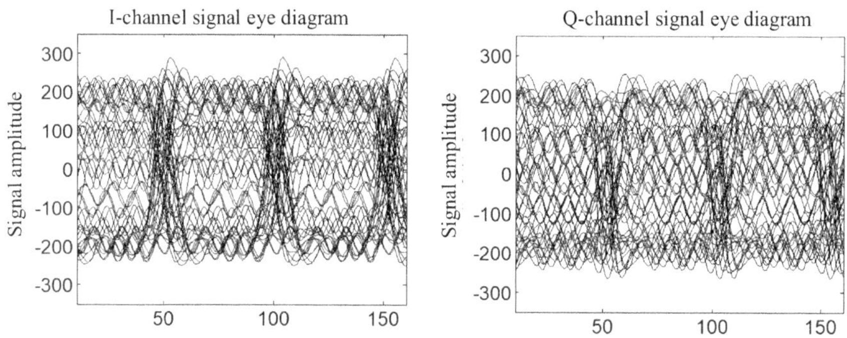

(a) I-channel signal (b) Q-channel signal

Fig. 2.34 Eye diagram of snow weather data [32–34]. 2014.1.16, 12:00 am

(2) The system is affected by the scattering of raindrops, and the scattering intensity in moderate rain is greater than that in light rain, so the "eye" of the eye diagram becomes smaller in light and moderate rain, and the line trace of the eye diagram in moderate rain is smaller than that in light rain. The traces are more blurred.
(3) On snowy days, the "eyes" of the eye diagram are basically closed, and the stitches of the eye diagram are the most blurred.

2.3.5.2 16-PSK Modulation Constellation Analysis Under Different Weather Conditions

In digital modulation, constellation diagrams of different modulation types can be used to observe the change of signal phase, noise interference, phase-transfer trajectory between vector points, etc. The atmospheric channel is a time-varying channel with memory. Atmospheric scattering and atmospheric turbulence have a great influence on the shape of the modulated signal constellation. The resulting phase ambiguity and Doppler frequency shift bring great difficulties to constellation detection. In particular, the influence of different weather conditions on the transmission of subcarrier signals is also different, and the constellation diagram of the signal at the receiving end shows vector-point dispersion, rotation, and deformation.

Figures 2.35, 2.36, 2.37 and 2.38 show the 16-PSK signal constellation diagrams at the receiving end after passing through an atmospheric channel under different weather conditions. It can be seen from the comparison that when the weather conditions are cloudy, or have light or moderate rain, the signal constellation diagram basically reflects the phase and amplitude information of the signal.

Considering the different degrees of signal attenuation under different weather conditions, cloudy days have the least impact on the transmission quality, and light rain and moderate rain have a greater impact on the transmission quality. This is

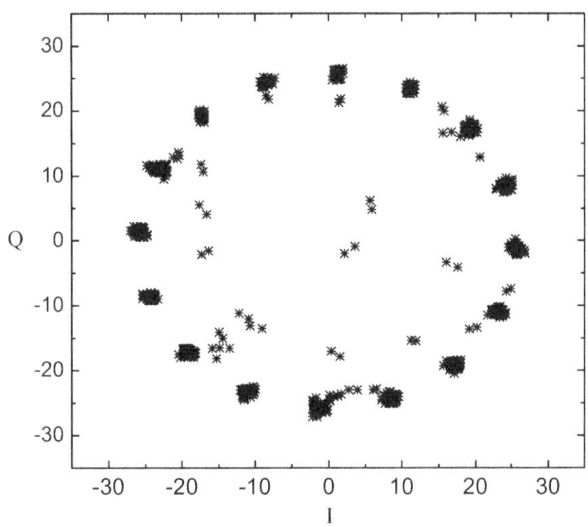

Fig. 2.35 Constellation of snowy weather data [32–34]

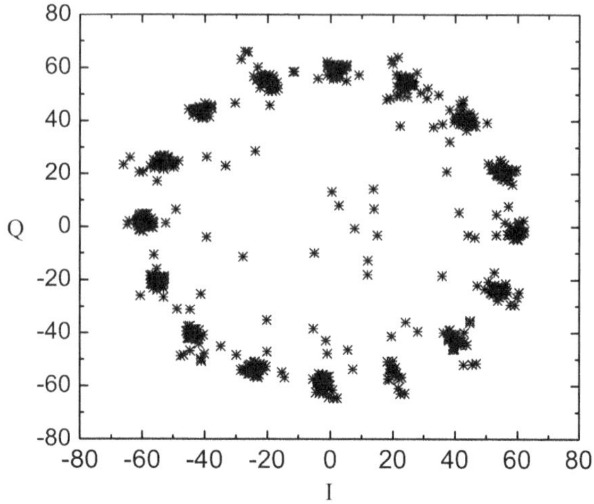

Fig. 2.36 Constellation diagram of light-rain weather data [32–34]

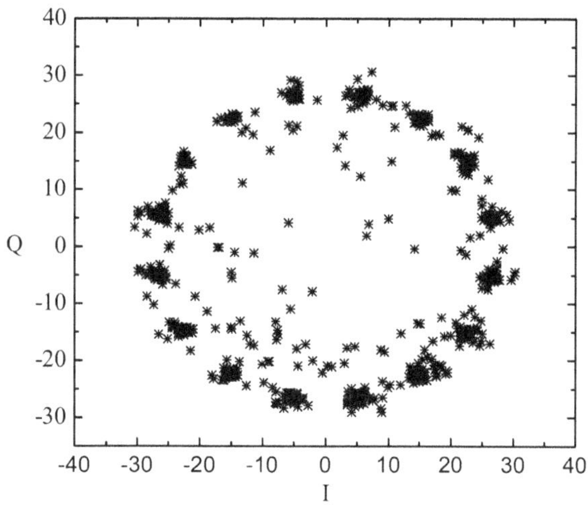

Fig. 2.37 Constellation diagram of moderately rainy weather data [32–34]

because the atmospheric channel is relatively stable on cloudy days, and the optical signal will be scattered by suspended particles in the atmosphere, absorbed by molecules, and affected by atmospheric turbulence, and the signal will be less affected by atmospheric attenuation. The information changes, causing the distribution of some vector points to deviate far from the ideal position.

When it rains, owing to the high density of fog and raindrops in the air, there is almost no scintillation effect on the optical signal. At this time, the optical signal is mainly affected by atmospheric attenuation factors, mainly because raindrops and fog droplets block the beam propagation. A part of the light energy is scattered in all directions, forming a scattering effect. The scattering effect of the atmosphere is related to the number and size of particles. The more particles, the more serious

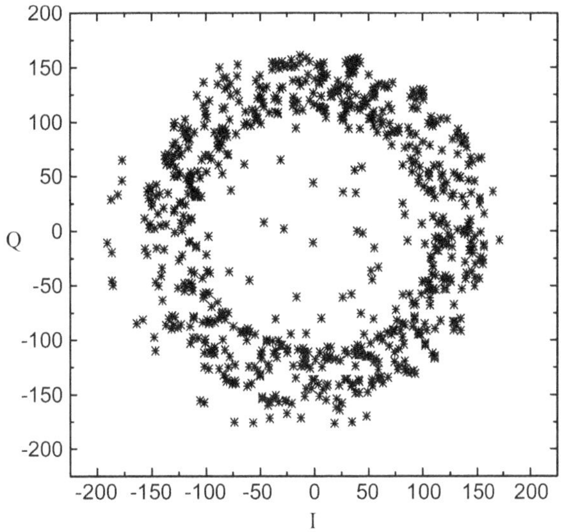

Fig. 2.38 Constellation of snowy weather data [32–34]

the scattering and the greater the attenuation. Therefore, the attenuation effect of the amplitude and phase information of a light-rain light signal is smaller than that of the moderate-rain light-signal amplitude and phase information.

When it snows, the distribution of each vector point on the received signal constellation diagram is scattered and the phase information represented by the constellation points cannot be distinguished. It can be seen that snow has a great influence on the transmission quality, and the optical signal is more affected by attenuation than on cloudy days and the attenuation is greater in rainy days.

Intersymbol interference and channel noise are two main factors that affect the performance of communication systems. The bit-error rate is the largest when the weather is snowy, reaching 2.42×10^{-2}, whereas the bit-error rate is the smallest when it is cloudy, which is 2.9×10^{-3}. The bit-error rates for light rain and moderate rain are 3.5×10^{-3} and 4.8×10^{-3}, respectively.

2.4 64-QAM Modulation and Demodulation

2.4.1 Basic Principles of a 64-QAM System

2.4.1.1 Modulation

The general expression for quadrature amplitude modulation is [31]

$$S(t) = A_m \cos \omega_c t + B_m \sin \omega_c t, \tag{2.116}$$

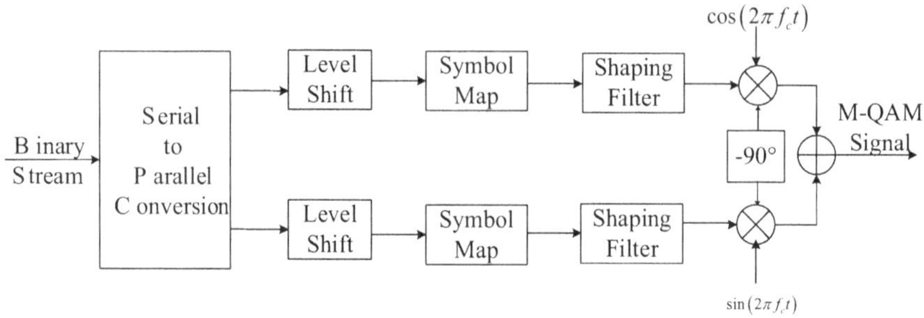

Fig. 2.39 Block diagram of an M-QAM modulator

where $0 \le t \le T_b$ and T_b are the symbol lengths. The modulation signal S(t) is formed by adding two mutually orthogonal carrier waves. $\{A_m\}$ and $\{B_m\}$ are variables that can hold multiple discrete values; therefore, this modulation method is called quadrature amplitude modulation (QAM). In Eq. 2.116, m = 1, 2, ..., M, where M is the number of levels of A_m and B_m. Amplitudes A_m and B_m in QAM can be expressed as

$$\begin{cases} A_m = d_m A \\ B_m = c_m A \end{cases}, \tag{2.117}$$

where A is the fixed amplitude, and (d_m, c_m) is determined by the input signal and is the coordinate point of the modulated signal in the signal space.

Figure 2.39 shows the block diagram of the M-QAM modulator. First, the binary data b_n passes through a serial-to-parallel converter, which divides the binary information with a rate of R_b bits per second into two sequences, consisting of odd-numbered positions and even-numbered positions.

When an M-QAM system is adopted, the symbol-conversion table of the upper branch converts each group of k/2 even bits into transmission symbols a_n^I, where M = 2k. Likewise, each group of k/2 odd bits entering the lower leg is grouped into transmission symbol a_n^Q. a_n^I and a_n^Q are random variables in set $\{\pm 1, \pm 3, \cdots, \pm(\sqrt{M}-1)\}$ whose value is in base $\sqrt{M}$.

For example, with 64-QAM, the character set for a_n^I and a_n^Q is $\{-7, -5, -3, -1, 1, 3, 5, 7\}$. The pulse-shaping output in the upper branch is an octal polar pulse sequence $I(t) = \sum_{n=-\infty} a_n^I v(t - nT)$, and the same carrier $A_c \cos(2\pi f_c t)$ is modulated by I(t). Similarly, the quadrature carrier $A_c \sin(2\pi f_c t)$ is modulated by the octal polar pulse sequence $Q(t) = \sum_{n=-\infty}^{\infty} a_n^Q v(t - nT)$ generated by the pulse-shaping filter of the lower branch. The M-QAM signal can be expressed as [35]

$$x(t) = \sqrt{\frac{2E_s}{T}} [I(t) \cos(2\pi f_c t) - Q(t) \sin(2\pi f_c t)] \tag{2.118}$$

$$\begin{cases} I(t) = C_0 \sum\limits_{n=-\infty}^{\infty} a_n^I v(t - nT) \\ Q(t) = C_0 \sum\limits_{n=-\infty}^{\infty} a_n^Q v(t - nT) \end{cases}. \tag{2.119}$$

In the formula, $\sqrt{E_s/T}$ is the usual unit energy pulse. The constant C_0 is introduced so that the average energy of the signal set equals E_s. The average energy E_s is

$$E_s = \sum_{m=1}^{\sqrt{M}} \sum_{n=1}^{\sqrt{M}} E\{a_m^I, a_n^Q\} \int_0^T |s_{mn}(t)|^2 dt, \tag{2.120}$$

where $s_{mn}(t)$ is the transmit waveform corresponding to the constellation point (a_m^I, a_n^Q):

$$s_{mn}(t) = \sqrt{\frac{2E_s}{T}} C_0 [a_m^I v(t) \cos(2\pi f_c t) - a_m^Q v(t) \sin(2\pi f_c t)]. \tag{2.121}$$

Substituting Eq. (2.121) into Eq. (2.120), we can obtain

$$\begin{aligned} E_s &= \frac{2E_s C_0^2}{M} \sum_{m=1}^{\sqrt{M}} \sum_{n=1}^{\sqrt{M}} [(a_m^I)^2 + (a_n^Q)^2] \frac{1}{2} \int_0^T \left| \frac{v(t)}{\sqrt{T}} \right|^2 dt \\ &= \frac{E_s C_0^2}{M} \sum_{m=1}^{\sqrt{M}} \sum_{n=1}^{\sqrt{M}} [(a_m^I)^2 + (a_n^Q)^2]. \end{aligned} \tag{2.122}$$

From Eq. (2.122), we can obtain

$$C_0^2 = \frac{M}{\sum_{m=1}^{\sqrt{M}} \sum_{n=1}^{\sqrt{M}} [(a_m^I)^2 + (a_n^Q)^2]}. \tag{2.123}$$

For the M-QAM constellation diagram, according to Eq. (2.122), we can obtain

$$C_0 = \sqrt{\frac{3}{2(M-1)}}, \tag{2.124}$$

if the following basis functions are used [36]:

$$\begin{cases} \phi_1(t) = \sqrt{\frac{2}{T}} \cos(2\pi f_c t) \\ \phi_2(t) = \sqrt{\frac{2}{T}} \sin(2\pi f_c t). \end{cases} \tag{2.125}$$

In the $\phi_1 \sim \phi_2$ plane, all M-QAM signals in the nth symbol interval can be expressed in vector form as

$$s = (a_n^I C_0 \sqrt{E_s}, a_n^Q C_0 \sqrt{E_s}). \tag{2.126}$$

According to Eq. (2.126), we can obtain

$$d_{\min} = 2C_0 \sqrt{E_s}. \tag{2.127}$$

We substitute Eq. (2.124) into Eq. (2.127) to obtain

$$d_{\min} = 2\sqrt{\frac{3E_s}{2(M-1)}}. \tag{2.128}$$

Figure 2.40 shows the details of the 64-QAM constellation diagram.

2.4.1.2 Demodulation

As shown in Fig. 2.41, the basic principle of coherent demodulation is as follows: The 64-QAM modulated signal is multiplied by two orthogonal carriers to generate two data channels: in-phase branch I and quadrature branch Q. The I- and Q-channel signals are filtered by low-pass filters to remove the high-frequency components. The judgment module judges the filtered two-way signals, obtains the two-way symbol information, converts the eight-level data into two-level data, and finally combines the two-way signals into one through the serial–parallel conversion module to restore the signal.

At the receiving end of the system, the modulated signal is multiplied by two mutually orthogonal carriers to recover the baseband signal. At this time, the expression of the in-phase component signal is

$$\begin{aligned} I(t) &= Y(t)\cos\omega_0 t \\ &= (A_m \cos\omega_0 t + B_m \sin\omega_0 t)\cos\omega_0 t \\ &= \frac{1}{2}A_m + \frac{1}{2}A_m \cos 2\omega_0 t + \frac{1}{2}B_m \sin 2\omega_0 t. \end{aligned} \tag{2.129}$$

The expression for the quadrature component signal is

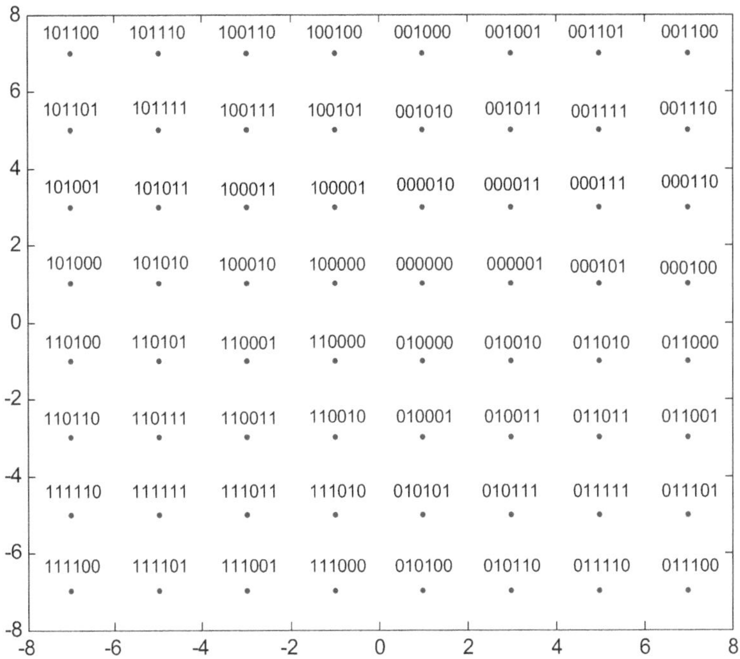

Fig. 2.40 64-QAM constellation details

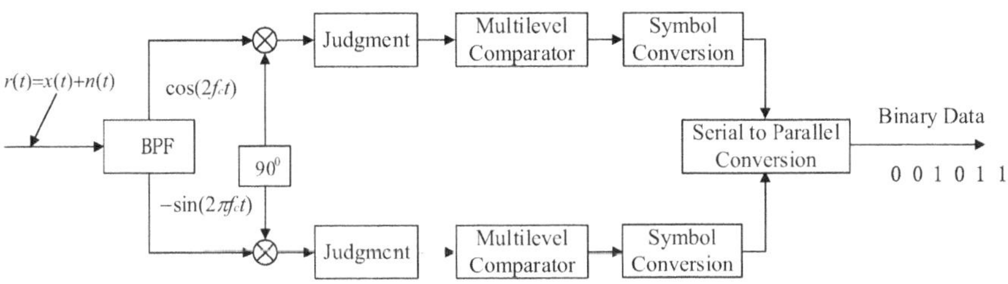

Fig. 2.41 Principle block diagram of 64-QAM demodulation

$$
\begin{aligned}
Q(t) &= Y(t)\sin\omega_0 t \\
&= (A_m\cos\omega_0 t + B_m\sin\omega_0 t)\sin\omega_0 t \\
&= \frac{1}{2}B_m - \frac{1}{2}B_m\cos 2\omega_0 t + \frac{1}{2}A_m\sin 2\omega_0 t.
\end{aligned}
\tag{2.130}
$$

2.4.1.3 QAM Error Performance

In an M-QAM system, the probability of correct symbol reception [37] is

$$P_c = \text{Probability of correct acceptance of way I} \times \text{Probability of correct acceptance of way Q}$$

$$(1 - P_0) \times (1 - P_0). \quad (2.131)$$

In the formula, the probability of a wrong decision $P_0 = I$ path and Q path alone. According to Eq. (2.128), the symbol-error probability P_e [38] in the M-QAM system can be obtained as

$$P_e = 1 - P_c = 1 - (1 - P_0)^2 = 2P_0 - P_0^2 \approx 2P_0, \quad (2.132)$$

where $P_0 << 1$. In a $\sqrt{M}$-ary pulse-amplitude modulation (PAM) system, the symbol-error probability P_0 [29] is

$$P_0 = \frac{2(\sqrt{M} - 1)}{\sqrt{M}} Q\left(\frac{d_{\min}}{\sqrt{2N_0}}\right). \quad (2.133)$$

We substitute Eq. (2.131) into Eq. (2.133) to obtain

$$P_0 = \frac{2(\sqrt{M} - 1)}{\sqrt{M}} Q(\sqrt{\frac{3E_s}{(M - 1)N_0}}). \quad (2.134)$$

Combining Eqs. (2.133) and (2.134), the symbol-error probability of the M-QAM system can be obtained as follows:

$$P_e = \frac{4(\sqrt{M} - 1)}{\sqrt{M}} Q\left(\sqrt{\frac{3E_s}{(M - 1)N_0}}\right), \quad (2.135)$$

because the relationship between the bit-error rate and the symbol-error rate is

$$BER = \frac{P_e}{\log_2 M}. \quad (2.136)$$

According to Eqs. (2.135) and (2.136), substituting $E_s = kE_b = \log_2 M \times E_b$, the bit-error rate of the M-QAM system can be obtained:

$$BER = \frac{P_e}{\log_2 M} = \frac{4(\sqrt{M} - 1)}{(\log_2 M)\sqrt{M}} Q\left(\sqrt{\frac{3E_b(\log_2 M)}{(M - 1)N_0}}\right). \quad (2.137)$$

Therefore, the BER of the 64-QAM system is

$$BER_{64-QAM} = \frac{P_e}{\log_2 64} = \frac{4(\sqrt{64}-1)}{(\log_2 64)\sqrt{64}} Q\left(\sqrt{\frac{3E_b(\log_2 64)}{(64-1)N_0}}\right)$$
$$= \frac{7}{12} Q\left(\sqrt{\frac{2E_b}{7N_0}}\right). \tag{2.138}$$

2.4.2 64-QAM Signal Modulation and Demodulation Simulation Analysis

2.4.2.1 64-QAM Modulation

The modulation part is mainly divided into three parts: serial–parallel conversion, level conversion, and constellation mapping.

(1) **Serial–parallel conversion and level conversion**

The serial–parallel conversion converts one serial sequence into two parallel subsequences, I and Q. The rate of the converted subsequences is half of the original sequence rate, and each symbol is determined by different binary bits h to determine different orders. According to the value of M, a 2-M level conversion is performed on the subsequences, and the conversion rules are $a_k = (2i - 1 - M), i = 1, 2, \ldots M$. Among them, $M = 2^{h/2}$, and i represents the corresponding binary data: 000–1, 001–2, 010–3, 011–4, 100–5, 101–6, 110–7, and 111–8.

Table 2.3 shows the relationship between the serial–parallel conversion and level conversion of 64-QAM. At this time, $h = 6$ and $M = 8$, that is, the six-bit information represents a symbol. First, three consecutive bits are taken from the six-bit symbol sequence to the I and Q sequences. Then, the corresponding i value is found according to the three-bit data; the available level amplitudes are $-7, -5, -3, -1, 1, 3, 5$, and 7. Table 2.3 shows the conversion relationship of the Q sequence between 000 and 111, when the I sequence is 000. The conversion methods of the other 56 states are similar to those in Table 2.3.

(2) **Constellation mapping**

As shown in Fig. 2.42, any combination of the amplitude and phase of I and Q will be mapped to a point in polar coordinates. The state of each constellation point is represented by six-bit data. After combining I and Q, a total of 64 is obtained. In this state, the 64-QAM constellation diagram is generated.

(3) **Up-conversion**

After the constellation mapping is completed, the in-phase mapping value is multiplied by carrier $\sin(\omega t)$, the quadrature mapping is multiplied by carrier $\cos(\omega t)$, and the two data channels are added to complete the 64-QAM signal modulation.

Table 2.3 Level conversion and serial–parallel conversion

Sequence	*I* sequence *I*/*i* value	I subsequence a_k	*Q* subsequence *Q*/*i* value	*Q* sequence a_k
000000	000/1	−7	000/1	−7
000001	000/1	−7	001/2	−5
000010	000/1	−7	010/3	−3
000011	000/1	−7	011/4	−1
000100	000/1	−7	100/5	1
000101	000/1	−7	101/6	3
000110	000/1	−7	110/7	5
000111	000/1	−7	111/8	7

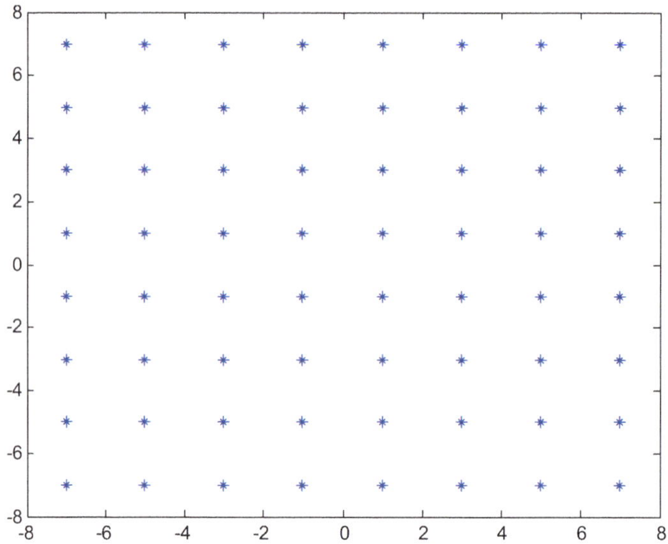

Fig. 2.42 64-QAM constellation diagram

The in-phase-branch and quadrature-branch data waveforms and 64-QAM signal waveforms are shown in Fig. 2.43.

Figure 2.44 shows the final 64-QAM modulated signal generated by the in-phase and quadrature components after passing through the adder. Figure 2.45 shows the power-spectral density of the modulated signal. At this time, the sampling frequency of signal $F_s = 10{,}000$ (Hz) and the carrier frequency $F_c = 1250$ (Hz). It can be seen from the figure that the power spectrum of the output signal is relatively stable, and the highest component of the power spectrum is relatively concentrated, which also shows the energy distribution.

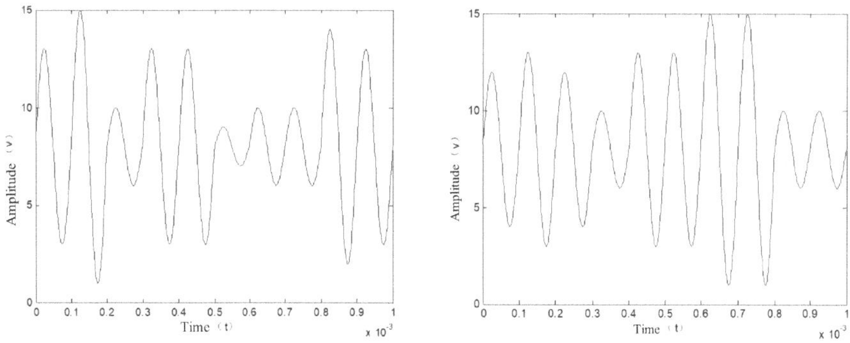

(a) In-phase component (b) Orthogonal component

Fig. 2.43 Up-conversion result

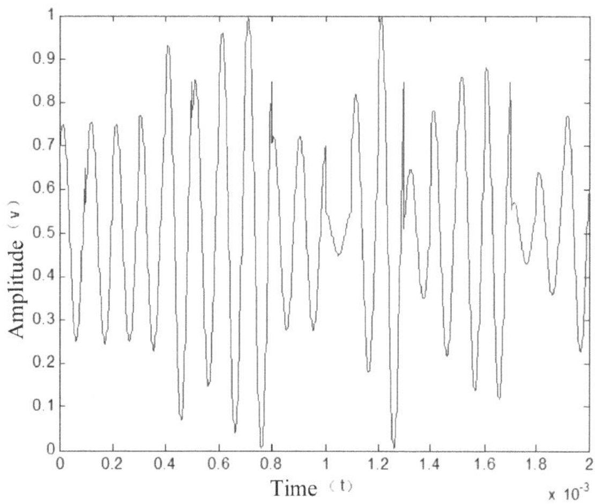

Fig. 2.44 64-QAM modulation signal waveform

2.4.2.2 64-QAM Demodulation Simulation

The demodulation process is equivalent to the inverse modulation process. First, the received signal is multiplied by a carrier with the same frequency as the transmitted signal for down-conversion processing, and then the high-frequency components are filtered out by a low-pass filter, leaving only the low-frequency signal. Next, the signal is inversely mapped, and then i, $i = (a_k + 1 + M)/2$, is calculated according to the amplitude of the electrical signal, where M = 8 is the level amplitude of the signal, and i = 1, 2, 3, 4, 5, 6, 7, 8. Subsequently, the value of *i* is converted into the corresponding binary data.

Next, three-bit consecutive data are taken from the *I* and *Q* sequences and combined into a set of six-bit binary numbers to represent a symbol. Table 2.4 shows the recovered binary sequence when the amplitude of the *I* sequence is −7

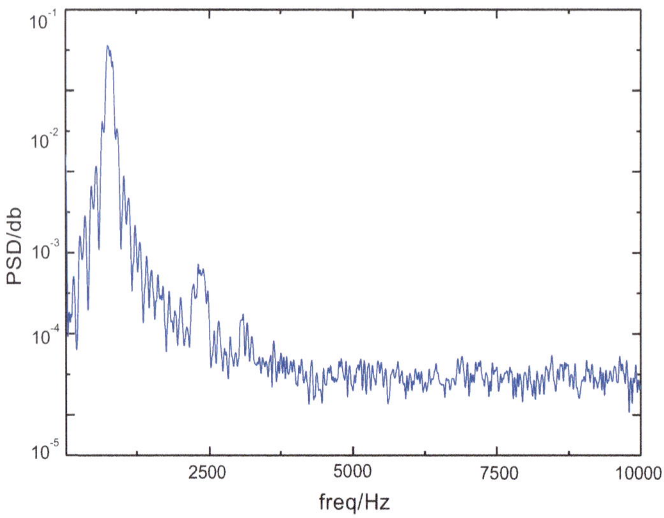

Fig. 2.45 Transmitter-signal power-spectral density

Table 2.4 Level conversion and parallel–serial conversion

I sequence a_k	Subsequence *I* code/*i* value	*Q* sequence a_k	Subsequence *Q* code/*i* value	Output sequence (*I* merged with *Q*)
−7	000/1	−7	000/1	000000
−7	000/1	−5	001/2	000001
−7	000/1	−3	010/3	000010
−7	000/1	−1	011/4	000011
−7	000/1	1	100/5	000100
−7	000/1	3	101/6	000101
−7	000/1	5	110/7	000110
−7	000/1	7	111/8	000111

and amplitude of the *Q* sequence varies from −7 to 7. The decoding methods for the other seven amplitude levels are similar to those in Table 2.4. A total of 64 sets of data are decoded.

2.4.3 Simulation and Experimental Results Analysis

According to the principle of a 64-QAM system, a simulation model was established, and the constellation diagram, bit-error rate, and eye diagram of the system were simulated and analyzed under the conditions of additive white Gaussian noise, gamma–gamma channel submultiplicative noise, and their mixed noise.

2.4.3.1 Constellation-Diagram Analysis

(1) **Influence of white Gaussian noise on the system constellation diagram**

Figure 2.46 shows the constellation diagram of the demodulation end of the system under different SNR cases: SNR = 30, SNR = 20, and SNR = 15. The abscissa represents the amplitude of the I-channel signal, and the ordinate represents the amplitude of the Q-channel signal. Both the abscissa and the ordinate are dimensionless. The constellation diagram has 64 points, and the magnitude of each point is taken from a certain value in (−7, −5, −3, −1, 1, 3, 5, 7).

According to Fig. 2.46a–c, after the system is affected by Gaussian white noise in the channel, the degree of aggregation of each constellation point in the constellation

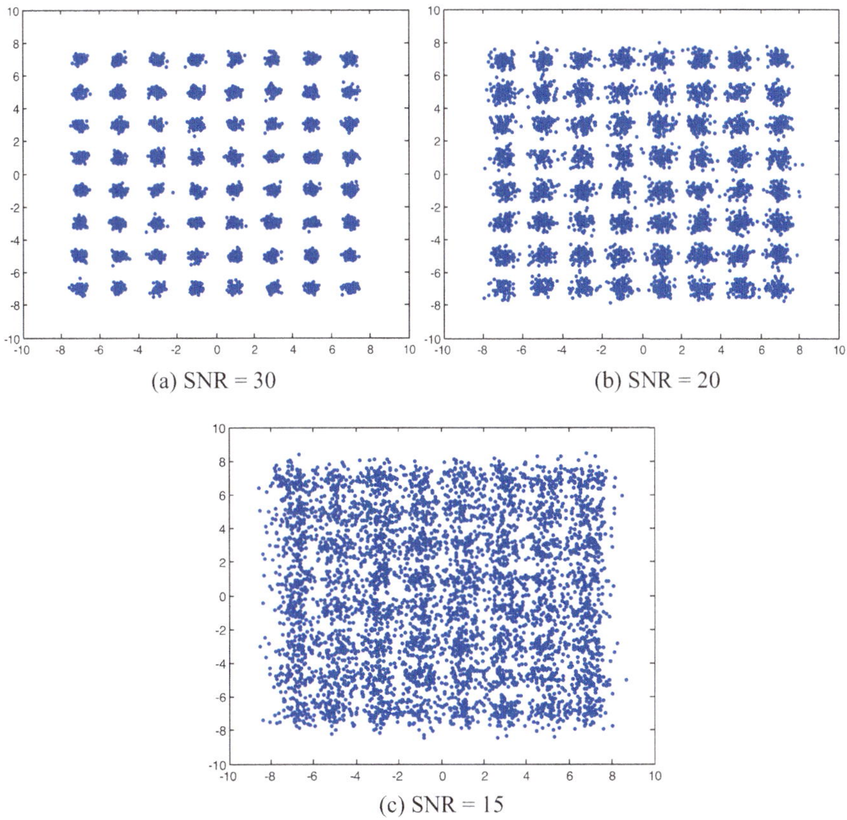

Fig. 2.46 Constellation diagram of a 64-QAM demodulated signal under different SNR [29]

chart will be affected to a certain extent, where the larger the SNR, the smaller the impact on the system, and the higher the degree of aggregation of constellation points.

(2) **Influence of multiplicative noise on the system constellation diagram**

The 64-QAM modulated signal passes through an atmospheric channel following a gamma–gamma distribution, and the signal is affected by multiplicative noise. The influence of 64-QAM modulation on the constellation diagram at the receiving end is analyzed under the weak-turbulence characteristics of the gamma–gamma distribution atmospheric-turbulence channel model. Figure 2.47 represents the constellation diagrams of the demodulation end under different light-intensity fluctuation variances.

Among them, Fig. 2.47a is the constellation diagram of the 64-QAM demodulation terminal under ideal conditions. Figure 2.47b–f are $\sigma_R^2 = 0.02$, $\sigma_R^2 = 0.05$, $\sigma_R^2 = 0.08$, $\sigma_R^2 = 0.1$, and $\sigma_R^2 = 0.15$. 64-QAM modulates the constellation diagram of the receiver signal. Comparing Fig. 2.47a, b, in the case of weak turbulence intensity, such as light-intensity fluctuation variance $\sigma_R^2 = 0.02$, the channel noise has little impact on the transmission system, the constellation diagram as a whole is clear, and the aggregation degree of each constellation point is very high.

It can be seen from Figures (e) and (f) that with the increase of the variance of the light-intensity fluctuation, the channel noise has an increasing influence on the system, which is shown in the constellation diagram as follows: The distribution of the signal points is not concentrated, the constellation points are not concentrated, and the degree of aggregation is low. When $\sigma_R^2 = 0.15$, many of the 64 points in the constellation diagram have aliasing.

Comparing Fig. 2.50a, b with Fig. 2.47a–f, it can be concluded that after the signal passes through an atmospheric channel that follows a gamma–gamma distribution, the constellation diagram will undergo a radial shift; however, when it passes through the Gaussian white noise channel, there will be no radial shift.

(3) **Influence of mixed noise on the system constellation diagram**

The influence of 64-QAM modulation on the demodulated signal constellation at the receiver is analyzed under a mixed-noise channel model considering weak turbulence and white Gaussian noise. Figures 2.48, 2.49 and 2.50 show the constellation diagrams of the demodulator under different SNR and light-intensity fluctuation variances.

Figures 2.48, 2.49, 2.50 and 2.51 show the effect of the mixed noise on the demodulation-end constellation diagram when the SNRs are 30 and 20 and the channel's light-intensity fluctuation variance $\sigma_R^2 = 0.02$. Comparing the two constellation diagrams in Fig. 2.48a, b, it can be seen that when the variance of the light-intensity fluctuation is equal, the smaller the SNR, the higher the point-aggregation degree in the constellation diagram, and the better the system performance.

Comparing Figs. 2.48, 2.49, 2.50 and 2.51, when SNR = 30 and $\sigma_R^2 = 0.02$, the aggregation degree of the constellation points is good, and can be clearly identified, and each constellation point has no radial offset. However, when the SNR is kept constant and the variance of the light-intensity fluctuation is $\sigma_R^2 = 0.08$ and $\sigma_R^2 = 0.1$,

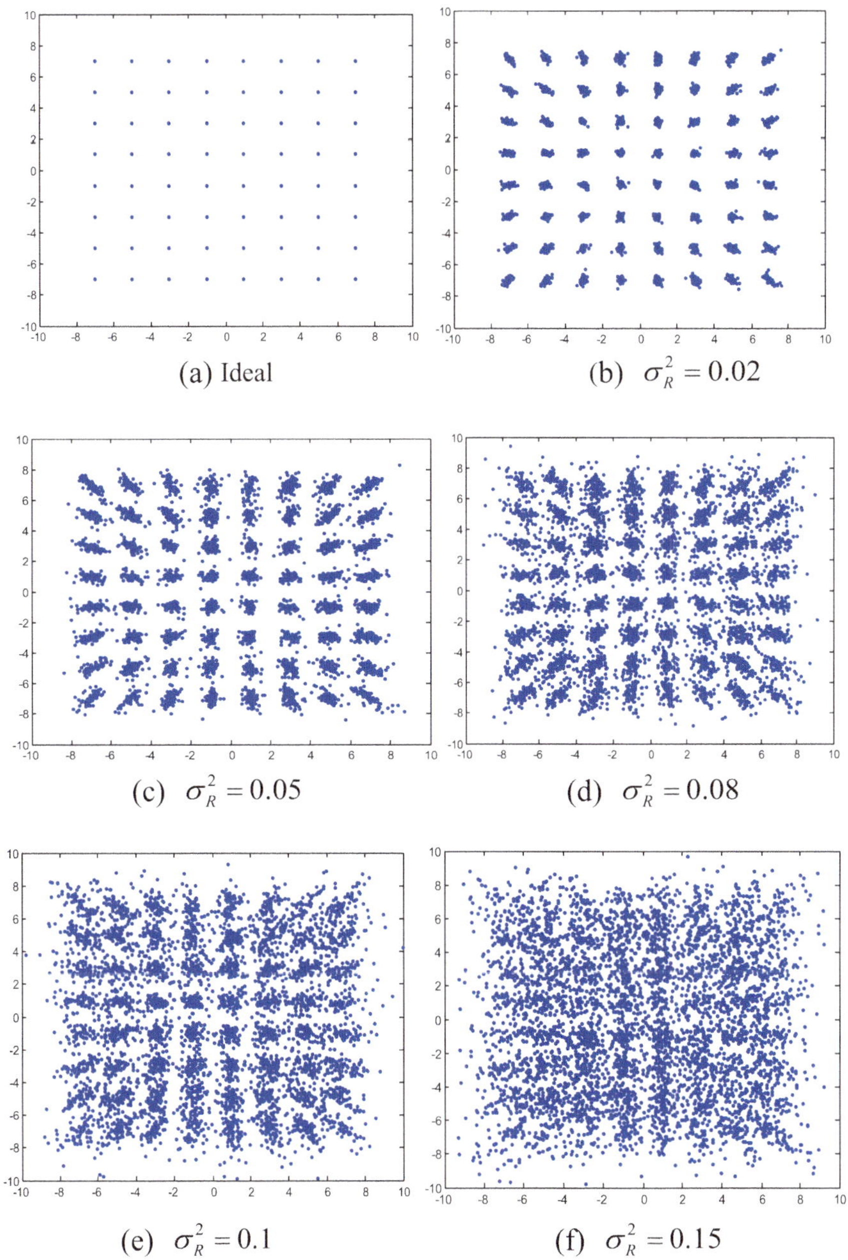

Fig. 2.47 Constellation diagram of the 64-QAM demodulated signal with different light-intensity fluctuation variances [29]

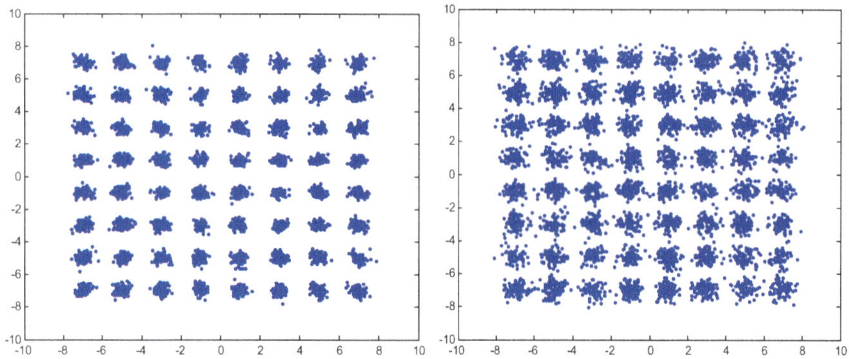

(a) SNR = 30 (b) SNR = 20

Fig. 2.48 Signal constellation diagram of the receiver under 64-QAM modulation when $\sigma_R^2 = 0.02$ [29]

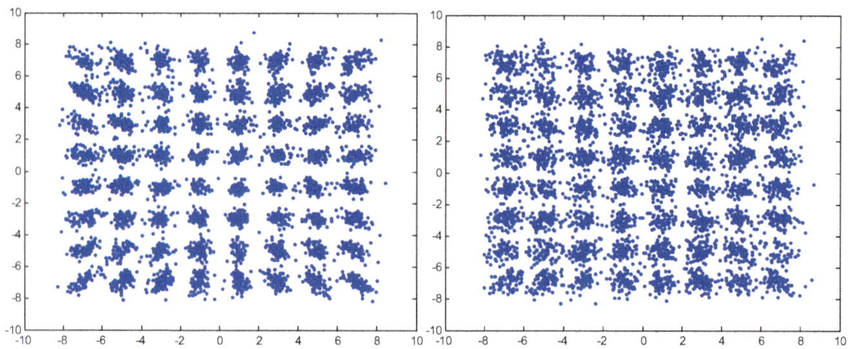

(a) SNR = 30 (b) SNR = 20

Fig. 2.49 Constellation diagram of the receiver signal under 64-QAM modulation when $\sigma_R^2 = 0.05$ [29]

as the variance of the light-intensity fluctuation increases, the aggregation degree of the constellation points steadily decreases, and the system performance worsens. Especially when $\sigma_R^2 = 0.1$, the aliasing between constellation points is very serious, and they are difficult to identify. In addition, as the value of σ_R^2 increases, the radial offset of each constellation point becomes increasingly obvious, and the variance of the light-intensity fluctuation changes. The degree has a greater impact on the system performance.

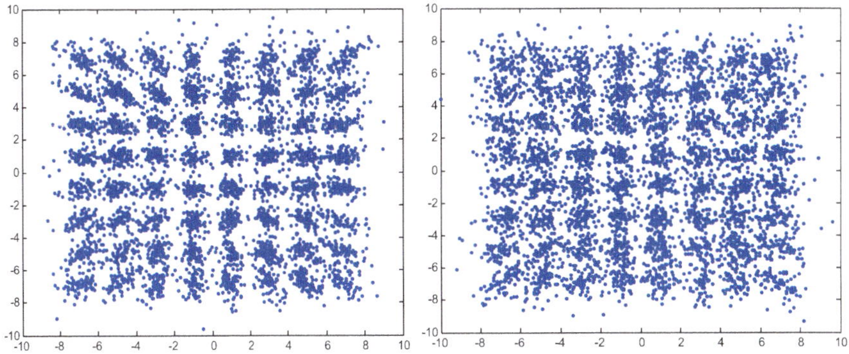

(a) SNR = 30 (b) SNR = 20

Fig. 2.50 Constellation diagram of the receiver signal under 64-QAM modulation when $\sigma_R^2 = 0.08$ [29]

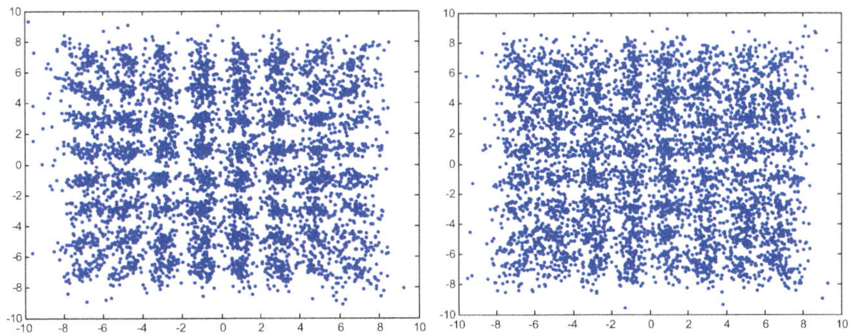

(a) SNR = 30 (b) SNR = 20

Fig. 2.51 Constellation diagram of the receiver signal under 64-QAM modulation when $\sigma_R^2 = 0.01$ [29]

2.4.3.2 Bit-Error Rate Analysis

Figures 2.52 and 2.53 show the system bit-error rates in the cases of multiplicative and mixed noise.

As can be seen from Fig. 2.52, when the value of σ_R^2 is <0.1, the effect of multiplicative noise on the system bit-error rate is small; when the value of σ_R^2 is between 0.1 and 0.4, the slope of the curve is very steep, and the influence of multiplicative noise is relatively high; when the value of σ_R^2 is >0.4, the influence of multiplicative noise on the system gradually decreases.

As can be seen from Fig. 2.53, when the multiplicative noise received by the system is <0.01 and 0.05, the system bit-error rate is mainly affected by additive noise. With the increase of σ_R^2, the influence of multiplicative noise on the system

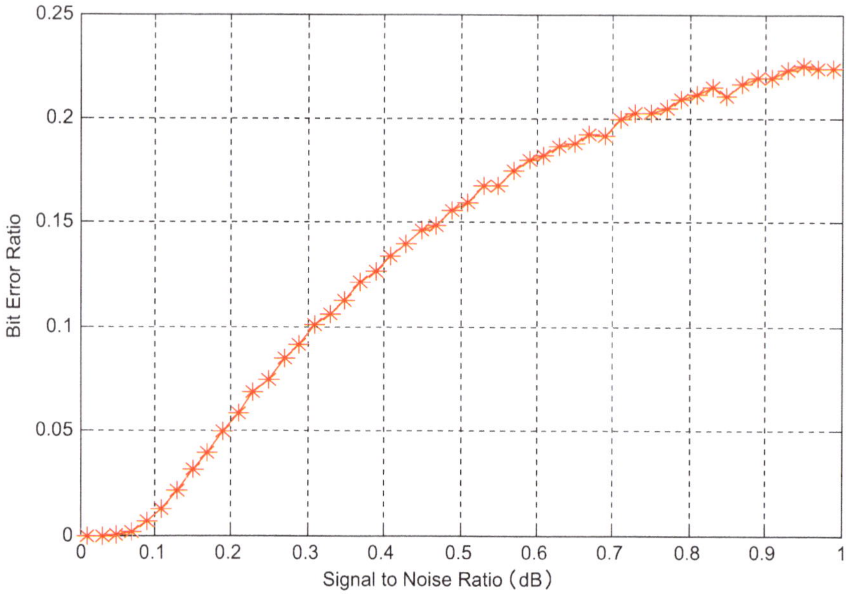

Fig. 2.52 Bit-error rate curve in the case of multiplicative noise [29]

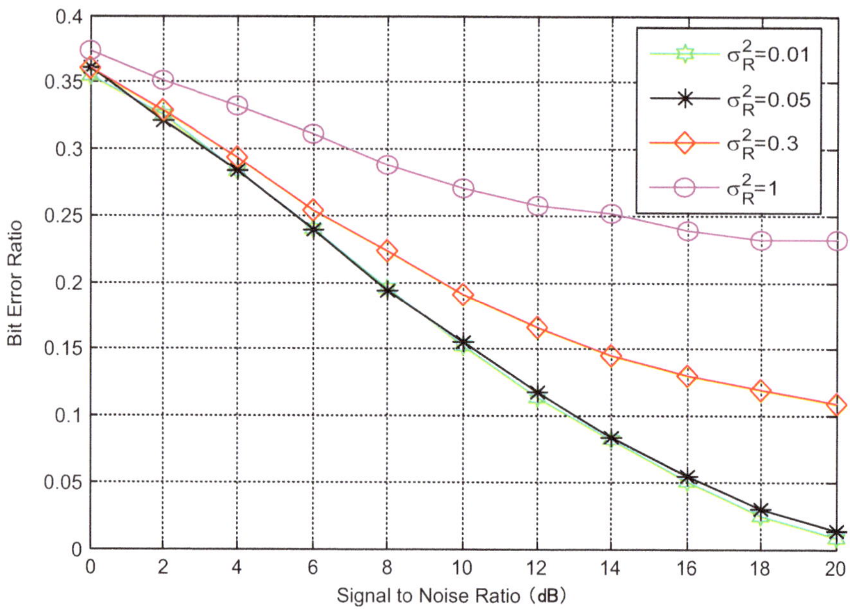

Fig. 2.53 Bit-error rate curve under mixed noise [29]

gradually becomes prominent. When $\sigma_R^2 = 0.3$ *and* $SNR = 20$, the system bit-error rate reaches more than 0.1. It can also be seen that when the value of σ_R^2 is >0.05, the system bit-error rate is mainly affected by multiplicative noise.

2.4.3.3 Eye-Diagram Analysis

By observing the eye diagrams under different noise conditions, the influence of channel noise on the performance of a optical wireless communication system can be analyzed. Figures 2.54 and 2.55 show the eye diagrams of the system under different light-intensity fluctuation variances. It can be seen from the figures that at $\sigma_R^2 = 0.02$, the "eyes" are opened the most, and the eye-diagram traces are relatively clear. At this time, the impact of intersymbol interference on the system is minimal. At $\sigma_R^2 = 0.05$, the "eyes" of the eye diagram are basically closed, and the eye-diagram trace is very blurred; thus, the channel noise has a great impact on the system.

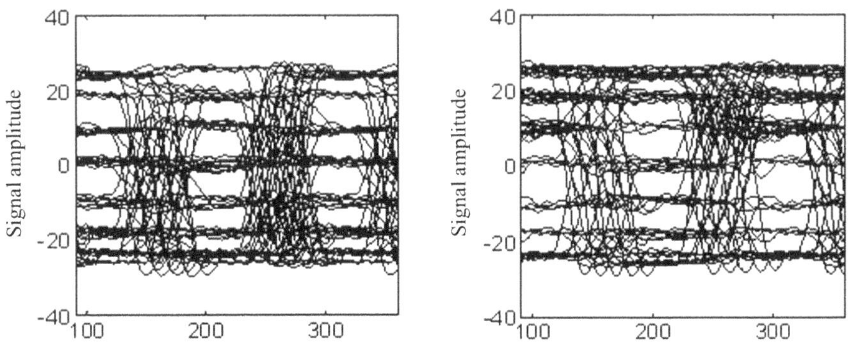

Fig. 2.54 Data eye diagram at $\sigma_R^2 = 0.02$ [29]

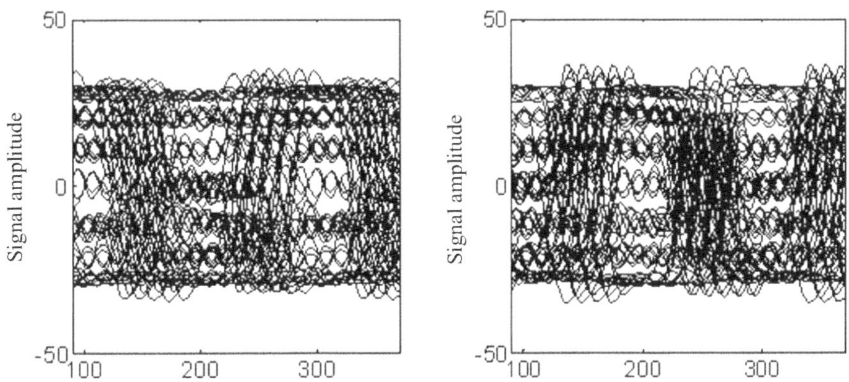

Fig. 2.55 Data eye diagram at $\sigma_R^2 = 0.05$ [29]

2.4.4 Analysis of 64-QAM Subcarrier-Modulation Experimental Results

We conducted experiments at close range, and analyzed the performance of a 64-QAM system through power-spectrum estimation, constellation diagrams, bit-error rates, and eye diagrams.

2.4.4.1 64-QAM Experimental-Data Power-Spectrum Estimation

The Hanning window function is used to overlap signal segments, and algorithms such as windowing functions and fast Fourier transforms (FFTs) are used to estimate the power-spectrum of a signal sequence [29]. The data collected in the experiment are divided into L segments, and then each piece of data is multiplied by a window function in the time domain, and an FFT is performed. Finally, the squares of the spectral amplitudes of each piece of data are added and averaged to obtain the required power-spectral density:

$$P(f) = \frac{1}{MU}\sum_{i=1}^{L}\left|\sum_{n=0}^{M-1} x_i(n)d(n)e^{-j2\pi fn}\right|^2, \tag{2.139}$$

where U is the normalization factor, $d(n)$ is the data window, and M is the length of each segment.

Figure 2.56 shows the power spectrum of the signal at the receiver; the signal sampling frequency $f_s = 10000$ (Hz), carrier frequency $f_c = 1250$ (Hz), and sampling period $T = 1/10000$. It can be seen from the figure that the power spectrum of the signal is relatively stable, and the highest component of the power spectrum is relatively concentrated in the frequency range of 361–441 Hz, with a peak value of 0.04676 dB, which also shows the light-energy distribution.

2.4.4.2 Eye-Diagram Analysis of 64-QAM Experimental Data

Figure 2.57 shows the eye diagram of the received data. The noise margin of the system and the intersymbol interference are analyzed by the height of the eye opening.

Observing Fig. 2.57, it can be clearly seen that the eye diagram of the receiving end has a large opening height, is relatively straight, and the line trace is relatively clear. This indicates that the guard interval in the 64-QAM system can well suppress the impact of intersymbol crosstalk on system performance. Thus, the probability of errors occurring when the receiver performs sampling is reduced.

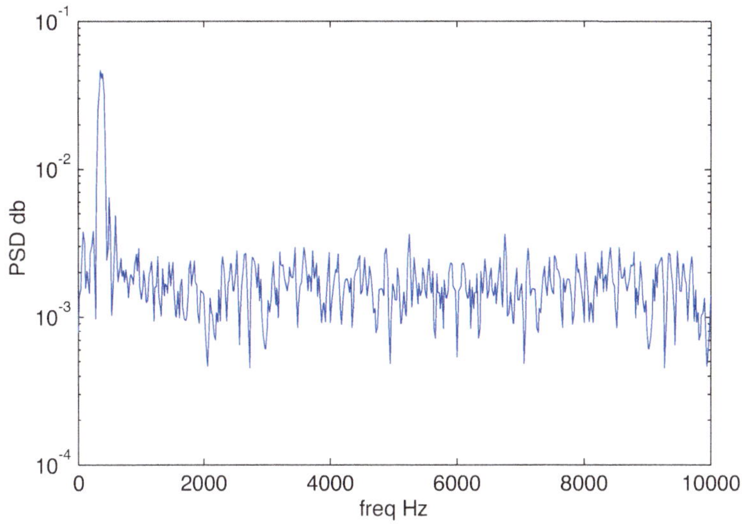

Fig. 2.56 Data power-spectrum estimation [39]

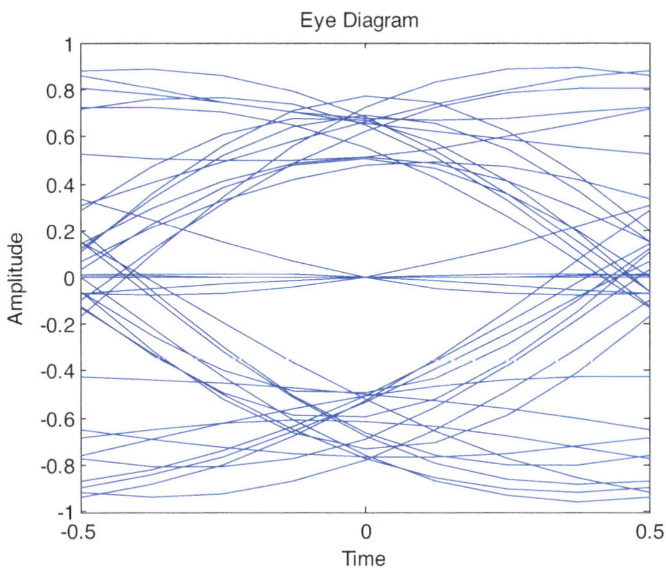

Fig. 2.57 64-QAM demodulation-signal eye diagram [39]

2.4.4.3 64-QAM Experiment Constellation Analysis

Figure 2.58 shows the received-data constellation diagram. It can be seen from the figure that when the distance is relatively close, the constellation points have no aliasing, and the distribution is relatively concentrated and clear. Because each constellation point has a radial offset, the radial offsets of the constellation points closer to the center of the constellation diagram are smaller, and the radial offsets of the constellation points closer to the boundary are larger. It can be seen from this that, although the signal is affected by various noises in the transmission channel, the influence of multiplicative noise on the system is more obvious.

The 64-QAM bit-error rate calculation formula is

$$\begin{aligned} BER_{64-QAM} &= \frac{P_e}{\log_2 64} = \frac{4(\sqrt{64}-1)}{(\log_2 64)\sqrt{64}} Q(\sqrt{\frac{3E_b(\log_2 64)}{(64-1)N_0}}) \\ &= \frac{7}{12} Q\left(\sqrt{\frac{2E_b}{7N_0}}\right), \end{aligned} \tag{2.140}$$

where P_e is the systematic symbol-error probability, E_b/N_0 is the SNR/bit, E_b is the average energy/bit, and $Q(\cdot)$ is defined as

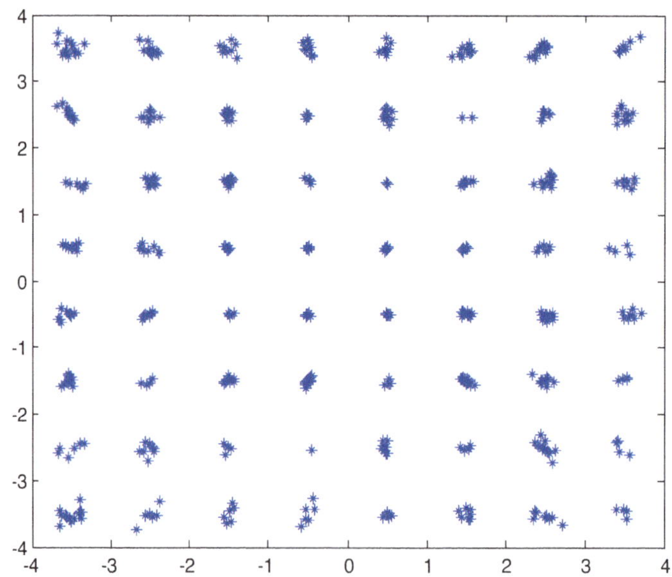

Fig. 2.58 Constellation diagram of a 64-QAM demodulated signal [39]

$$Q(u) = \int_{u}^{\infty} \frac{1}{\sqrt{2\pi}} e^{-\frac{x^2}{2}} dx, u > 0. \tag{2.141}$$

The bit-error rate $BER = 1.6667 \times 10^{-5}$ can be obtained through the above formula. It can be seen that during short-range communication, the influence of noise in the atmospheric channel on the system is relatively weak, the bit-error rate is small, and the system performance is better. For a optical wireless 64-QAM communication system, when the channel noise is only multiplicative [39, 40] and the value of σ_R^2 is <0.1, the multiplicative noise has little effect on the system bit-error rate. When the value of σ_R^2 is between 0.1 and 0.4, the system is highly affected by multiplicative noise; when the value of σ_R^2 is >0.4, the influence of multiplicative noise on the system is gradually reduced.

When the channel noise is a mixture of additive and multiplicative noise, when the multiplicative noise received by the system is <0.01 and 0.05, the bit-error rate and the degree of convergence of constellation points are mainly affected by additive noise. With the increase of σ_R^2, the influence of multiplicative noise on the system gradually increases. When $\sigma_R^2 = 0.3$ *and* $SNR = 20$, the system bit-error rate reaches more than 0.1. It can also be seen that when the value of σ_R^2 is >0.05, the system bit-error rate is mainly affected by multiplicative noise.

These results show that the optical wireless 64-QAM communication system has good transmission efficiency in point-to-point communication when the distance is relatively short. The transmission of QAM signals in the atmosphere is mainly affected by multiplicative noise. The signal-constellation diagram of the optical wireless QAM modulation system after multi-mode blind equalization obviously converges, and the system bit-error rate is reduced. Under weak turbulent conditions, the system has better transmission performance.

Chapter Summary

Through theoretical analyses and numerical simulations, it was found that compared with OOK, BPSK modulation was well able to suppress the light-intensity scintillation caused by atmospheric turbulence. Although the bit-error rate was degraded, the information-transmission rate increased, more information was transmitted in the same frequency band, and frequency-band utilization rate was improved. Therefore, an appropriate modulation method should be selected, according to the system requirements, to optimize the system performance.

Pairs of atmospheric laser-communication systems using NC-QPSK-like sinusoidal modulation and DC-bias subcarrier intensity-modulation (QPSK, square 4-QAM, 4-FSK, and 4-ASK) were analyzed separately in terms of power utilization, bit-error rate, outage probability, and channel capacity.

16-PSK and 64-QAM modulation systems with Gaussian noise and Rayleigh distribution noise were analyzed, the effects of random atmospheric-channel additive noise, multiplicative noise, and mixed noise on optical wireless communication were studied, and their errors were analyzed and calculated, including the bit-error rate performance. The convergence of the constellation diagram at the receiving end was compared and analyzed under different signal-to-noise ratios and light-intensity fluctuation variances.

References

1. Ke X (2011) Coding theory in wireless laser communication systems. Science Press Beijing
2. Ke X, Tian X (2015) Modeling and simulation of two-dimensional correlated K-distributed turbulence channel in optical communication. Acta Optica Sinica 4(10):33–41
3. Ke X, Zhao L, Ding D (2007) Realization of slot synchronization and frame synchronization in atmospheric laser communication. Semicond Optoelectron 28(5):721–724
4. Deqiang D, Xizheng K (2005) Design and simulation of PPM modulation and demodulation system for atmospheric laser communication. Opt Commun Technol 29(1):50–52
5. Ding D (2005) Research on PPM modulation and demodulation system in atmospheric laser communication. Xi'an University of Technology, Xi'an
6. Ling Q, Xizheng K (2007) Capacity analysis of optical MPPM channel without background noise. Opto-Electron Eng 34(7):107–110
7. Qin L, Ke X (2007) A study on two-pulse MPPM coding mapping method. J Xi'an Univ Technol 23(3):269–272
8. Qin L, Du Y, Ke X, Yang L (2008) Realization of three-pulse MPPM encoding and decoding system in atmospheric laser communication. Semicond Optoelectron 29(3):403–406+433.L
9. Qin L (2007) Design of multi-pulse modulation in atmospheric laser communication system. Xi'an University of Technology, Xi'an
10. Zhao L, Ke X, Liu J (2007) Research on DPPM modulation and demodulation technology in OWC. Laser J 28(2):63–64
11. Zhao L (2006) Research on DPPM modulation and demodulation in atmospheric laser communication. Xi'an University of Technology, Xi'an
12. Liu M (2010) Performance analysis of pulse-like position modulation in optical wireless communication. Xi'an University of Technology, Xi'an
13. Madani NM, Hadi J, Fakhraie SM (2005) Design and implementation of a fully digital 4FSK demodulator. IEEE 23
14. Zhang X (2003) Application of multicarrier modulation in communication systems. Opt Commun Technol 27(10):1–2
15. Huang W, Takayanagi J, Sakanaka T, Nakagawa M (1993) Atmospheric optical communication system using subcarrier PSK modulation. IEICE Trans Commun E76–B(9):1169–1177
16. Yuksel H, Davis CC (2006) Aperture averaging analysis and aperture shape invariance of received scintillation in free space optical communication links. SPIE 6304:63041E-1–63041E-11
17. Chen D (2011) Research on SIM-FSO system and its key technologies for suppressing atmospheric effects. Xi'an University of Technology, Xi'an
18. Perez-Jimenez R, Rabadan JA (1996) Improved PPM modulations for high spectral efficiency IR-WLAN. In: Seventh IEEE international symposium on personal, indoor and mobile radio communications, vol 1, pp 262–266
19. Lee DCM, Kahn JM, Audeh MD (1997) Trellis-coded pulse-position modulation for indoor wireless infrared communications. IEEE Trans Commun 45(9):1080–1087
20. Ke X, Deng L (2013) Performance research of limited QPSK-like sine modulation atmospheric laser communication system. Chin J Lasers 2(10):0205001-1–0205001-8

21. Zou C, Ao F, Huang X (2000) Capacity analysis of optical PPM channel without background noise. Acta Electron Sinica 22(4):682–686
22. Wang H, Cheng G, Sun X (2007) Performance analysis of dicode pulse position modulation for optical wireless communications. In: International conference on wireless communications, networking and mobile computing, vol 1, pp 3027–3030
23. Shiu D (1999) Differential pulse-position modulation for power-efficient optical communication. IEEE Trans Commun 47(8):1201–1210
24. Sui M, Yu X, Zhou Z (2009) The modified PPM modulation for underwater optical wireless communication. In: International conference on communication software and networks, vol 1, pp 173–177
25. Hayes AR, Ghassemlooy Z, Seed NL (1998) Optical wireless communication using digital pulse interval modulation. Opt Wirel Commun 61–69
26. Ghassemlooy Z, Hayes AR, Seed NL (1998) Digital pulse interval modulation for optical communications. IEEE Commun Mag 36(12):95–99
27. Ghassemlooy Z, Hashemi SK, Amiri M (2008) Spectral analysis of convolutional coded DPIM for indoor optical wireless communications. IEEE Proceedings, pp 102–105
28. Rong J, Chen Y, Hu Y (2003) Research on transmission characteristics and simulation of laser in turbulent atmosphere. Opt Commun Technol (11):44–46
29. Xia G, Xiong W, Sun H (2008) Channel capacity of one-dimensional and two-dimensional modulation in AWGN channel. Acta Astronautica Sinica 28(2):180–184
30. Hu Z, Tang J (2005) Research on digital pulse interval modulation in atmospheric optical wireless communication system. J Commun 7(3):75–79
31. Wu L, Ying J, Geng B (2008) Influence of atmospheric turbulence on laser transmission. Laser Infrared 10(4):974–977
32. Chen D, Ke X, Zhang T, Lu D (2015) Experimental research on subcarrier optical wireless communication based on 16PSK modulation. Chin J Lasers 42(10):205–212
33. Ke X, Chen D, Pan D (2010) Simulation and bit error rate performance analysis of 16PSK system. Laser J 31(1):41–43
34. Zhang T (2014) Experiment of optical wireless communication system with subcarrier 16PSK modulation. Xi'an University of Technology, Xi'an
35. Xing J, Xu G (2005) Influence of atmospheric turbulence on laser communication system. Acta Photonica Sinica 34(12):1850–1852
36. Chen C, Jiang H, Yang H et al (2009) Research progress on suppression techniques of atmospheric turbulence effects in atmospheric optical communication. Acta Armamentarii 30(6):779–791
37. Han L (2013) Research on performance and compensation methods of space optical communication under atmospheric turbulence. Harbin Institute of Technology, Harbin
38. Jian W, Chunping Y, Jianbin L (2005) Theory of light transmission in the atmosphere. Beijing University of Posts and Telecommunications Press, Beijing, pp 158–160
39. Biyao S (2016) Experimental research on 64-QAM modulation in optical wireless communication system. Xi'an University of Technology, Xi'an
40. Chen D, Ke X (2012) Research on subcarrier modulation recognition in optical wireless communication system with weak atmospheric turbulence channel. Laser Optoelectron Progr 49(7):66–72

Chapter 3
Nonlinear Characteristic of Semiconductor Lasers and Its Correction

Semiconductor lasers are energy-conversion devices in optical communication, and their characteristics have a great impact on signal transmission and detection. When a signal is loaded onto the laser, nonlinear distortion can occur near the threshold. This chapter discusses the structure and working principle of the laser and the causes of nonlinearity, and discusses the use of appropriate compensation methods for correcting nonlinearity.

3.1 Semiconductor Lasers

Semiconductor lasers are mainly composed of three parts: a working substance, pump source, and optical resonator [1]. Among them, the working substance is the cause of the stimulated radiation and the amplification of the light generated by the laser; it is the core of the entire laser. The pump source excites the working substance by providing energy to achieve a population-inversion distribution. Different types of working substances require different pumping methods. The optical resonator provides positive feedback to establish the laser oscillation, and its parameters directly affect the quality of the output laser beam (Fig. 3.1).

Semiconductor lasers have three excitation modes: electric-injection, optical-pump, and high-energy electron-beam. Electric-injection semiconductor lasers are generally junction semiconductor diodes made of materials such as gallium arsenide (GaAs), indium arsenide (InAs), and indium antimonide (InSb), which are excited by injecting a current along the forward-bias voltage. The junction-plane region produces stimulated emission.

Optically pumped semiconductor lasers generally use N-type or P-type semiconductor single crystals (such as GaAs, InAs, InSb, etc.) as the working substance, and use the light emitted by other lasers as the optical-pump excitation. High-energy

X. Ke, *Principles and Applications of Optical Wireless Orthogonal Frequency-Division Multiplexing*, Optical Wireless Communication Theory and Technology,
https://doi.org/10.1007/978-981-97-7973-4_3

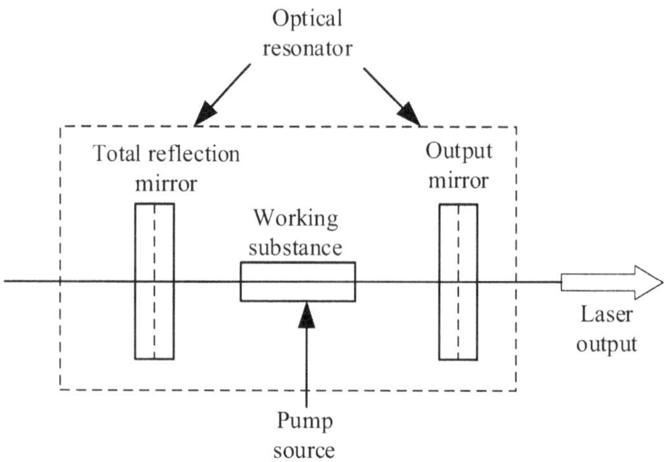

Fig. 3.1 Schematic diagram of the structure of a semiconductor laser

electron beam–excited semiconductor lasers generally use N-type or P-type semiconductor single crystals (such as lead sulfide (PbS), cadmium sulfide (CdS), zinc oxide (ZnO), etc.) as the working material, and are excited by injecting high-energy electron beams from the outside. One widely used semiconductor-laser device is an electric-injection-type GaAs diode laser with a double heterostructure.

The current-injection excitation method stimulates radiation in the junction-plane area by adding a forward-bias voltage to the PN junction, similar to a forward-biased diode. A thin semiconductor layer, called the active layer, is inserted between the N and P regions. It has the same lattice constant as the cladding, but with a smaller band gap and a higher refractive index. The active layer also acts as an optical waveguide.

The natural cleavage surface of the semiconductor crystal is used as a mirror to form an optical resonant cavity. Generally, the end that does not emit light is coated with a high-reflection multilayer dielectric film, and the light-emitting surface is coated with an anti-reflection membrane. For semiconductor lasers with a Fabry-Pérot (F-P) cavity, it is convenient to use the natural cleavage plane of the crystal perpendicular to the plane of the PN junction to form the F-P cavity surface, generate self-excited oscillation in the cavity, and emit the laser beam from one edge of the resonator (Fig. 3.2).

3.1.1 Working Principle of a Semiconductor Laser Diode

Semiconductor-laser diodes consist of a forward-biased P-N structure with P-type and N-type semiconductors formed by doping with impurity atoms. As the doping concentration increases, when doping N-type impurities, the quasi-Fermi level in

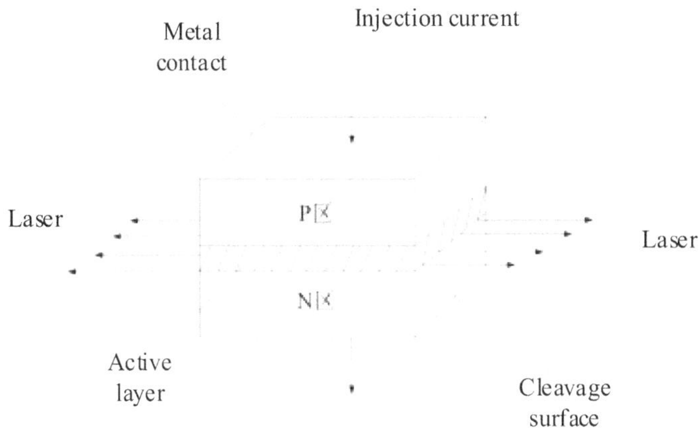

Fig. 3.2 Schematic diagram of a working model of a semiconductor laser

the semiconductor band gap will move to the conduction band; when doping P-type impurities, it will move to the valence band (Fig. 3.3).

Because the Fermi level is continuous in thermal equilibrium, electrons and holes must diffuse through the P-N junction, and the ionized impurities create an internal potential difference. When a forward-bias voltage is applied to the P-N junction externally, the internal potential difference becomes smaller, thereby generating a current. When the current reaches a certain value, a population inversion (i.e., the formation of a carrier distribution opposite the thermal equilibrium state) will occur. A population inversion is a prerequisite for laser generation.

The probability of stimulated emission between two energy levels is related to the difference in the number of particles between the two energy levels. Under normal circumstances, the number of atoms on low-energy level E_1 is greater than the number

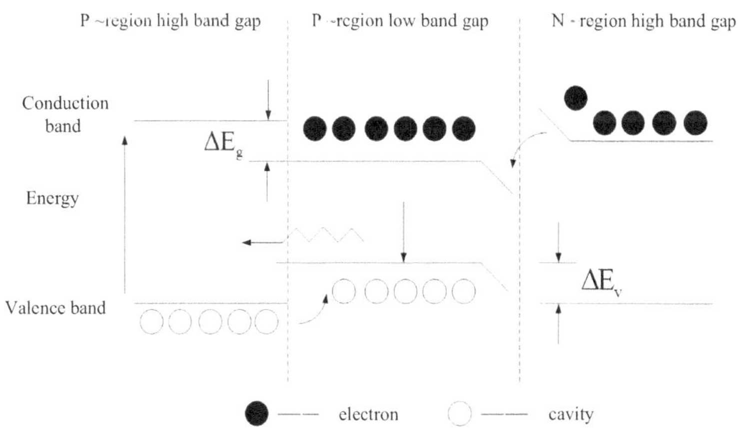

Fig. 3.3 Schematic diagram of the working principle of a semiconductor laser

of atoms on high-energy level E_2; no laser light is generated in this case. To produce a laser, the number of atoms on high-energy level E_2 must be greater than the number of atoms on low-energy level E_1; when many atoms are on E_2, stimulated radiation occurs, and the light is enhanced (also called light amplification).

To achieve this goal, we must try to excite a large number of atoms from the ground state to the metastable E_2 level, and the number of atoms in high-energy level E_2 can greatly exceed the number of atoms in low-energy level E_1. This inverts the population between energy levels E_2 and E_1. In the presence of a large number of inversions, the P-N junction emits highly coherent light: lasing.

The phenomenon in which light is generated from the spontaneous recombination of electrons and holes is called spontaneous emission. When the photons generated by spontaneous radiation pass near the emitted electron–hole pairs, the pairs are stimulated to recombine to generate new photons, thereby producing stimulated radiation [2].

3.1.2 *Classification of Semiconductor Lasers*

According to the optical-waveguide structure of the active layer, lasers can be divided into refractive index–guided lasers and gain-guided lasers. The former can be divided into strong refractive-index guided and weak refractive-index guided lasers. According to the optical-feedback mechanism in the resonator, they can be divided into Fabry-Pérot (F-P) lasers, distributed feedback (DFB) lasers, distributed Bragg reflection (DBR) lasers, and quantum well (QW) lasers.

According to the output optical power, they can be divided into low-power (<100 mW), medium-power (<1 W), and high-power (a few W or more) semiconductor lasers. According to the material, they can be divided into GaAlAs, GaAs, InGaAsP/InP, and InGaAs/GaAs semiconductor lasers. According to the wavelength, they can be divided into visible-light lasers, infrared long-wavelength lasers, and far-infrared long-wavelength lasers.

3.1.3 *Characteristics of Semiconductor Lasers*

3.1.3.1 Volt-Ampere Characteristics

As shown in Fig. 3.4, the core part of a semiconductor laser is a P-N junction, which has a V–I characteristic curve similar to that of an ordinary diode. When the forward voltage is less than a certain value, the current is extremely small and no light is emitted. When the voltage exceeds a certain value, the forward current increases rapidly with the voltage and emits light; this is called the threshold voltage or turn-on voltage. The current through the diode is exponentially related to the voltage until the threshold current is reached.

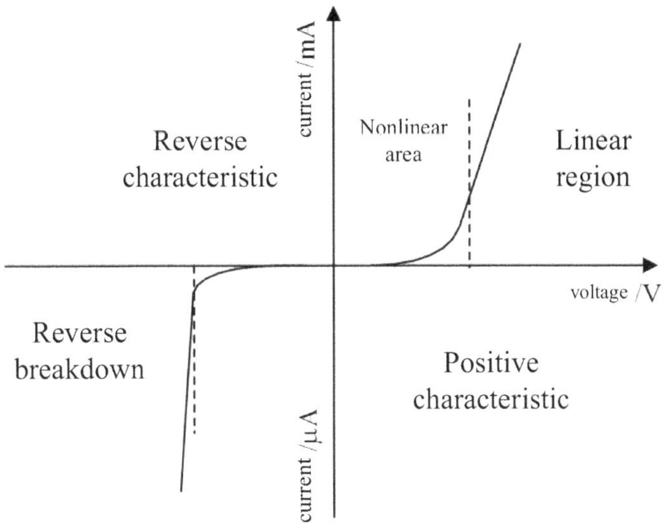

Fig. 3.4 V–I characteristic curve of a semiconductor laser diode

$$I = I_0\left(e^{\alpha_j(V-IR)} - 1\right), \tag{3.1}$$

where I_0 is the saturation current and α_j is the diode parameter. When the threshold current is reached, the diode current is linearly related to the voltage.

$$V \approx \frac{E_g}{e} + IR, \tag{3.2}$$

where E_g is the forbidden-band width, e is the unit charge, and the slope is the series resistance R.

3.1.3.2 Photoelectric Characteristics

As shown in Fig. 3.5, the I–L characteristic refers to the relationship between the output optical power (light) of a semiconductor laser and the injection current. When the injection current is small, the number of particles cannot be reversed in the active area, and spontaneous emission is dominant. At this time, the laser emits fluorescence over a very wide spectrum; with the increase of the injection current, the particles are emitted in the active area.

With the population inversion, the stimulated emission begins to increase; however, when the injection current is less than the threshold current, the gain in the resonant cavity is insufficient to overcome the loss, and a certain oscillation mode cannot be established in the cavity. The output light at this time is relatively small.

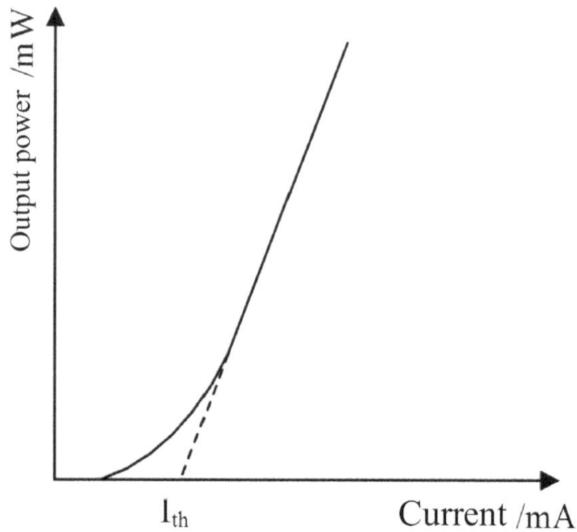

Fig. 3.5 I–L characteristic curve of a semiconductor laser diode

Intense fluorescence, known as a "superradiant" state, can only be produced when the injection current reaches a threshold value. Then, a well-defined lasing pattern is produced, with a sudden narrowing of the spectrum and the appearance of a single peak. The injection current cannot be increased indefinitely. Generally, the maximum operating current should not exceed four times the threshold current; otherwise, the device will be easily damaged.

As shown in Fig. 3.5, a DC offset should be added to the subcarrier modulation signal, which ensures that the signal is non-negative when the light source is intensity modulated. Moreover, to obtain linear modulation, the operating point should be in the linear region of the I–L characteristic curve to ensure that the optical signal is not distorted. Usually, the bias current is chosen near the midpoint of the linear region.

3.1.3.3 Electro-Optical Conversion Efficiency

Semiconductor lasers directly convert electrical power into optical power. One of the important parameters is the conversion efficiency, which is usually measured using the power efficiency and quantum efficiency. Power efficiency refers to the ratio of the electrical power in the semiconductor laser converted to the output optical power.

$$\eta_P = \frac{P_{\text{light}}}{IV + I^2R}, \tag{3.3}$$

where P_{light} is the emitted laser power (i.e., the output optical power of the laser), I is the injection current, V is the forward voltage drop of the laser, and R is the series resistance of the laser.

Quantum efficiency includes internal quantum efficiency, external quantum efficiency, and external-differential quantum efficiency. The internal quantum efficiency is defined as

$$\eta_i = \frac{\tilde{S}}{\tilde{N}}, \tag{3.4}$$

where $\tilde{S}$ is the number of photons generated per second in the active region, and $\tilde{N}$ is the number of electron–hole pairs injected into the active region. Semiconductor lasers are highly efficient electro-optical conversion devices; generally $\eta_i > 70\%$. The external quantum efficiency is defined as

$$\eta_{ex} = \frac{P_{\text{light}}/hf}{I/q} = \frac{P_{\text{light}}}{I} \cdot \frac{q}{hf}, \tag{3.5}$$

where q is the electron charge and hf is the photon energy.

The external quantum efficiency mainly reflects the fact that the light cannot be fully emitted, owing to scattering, diffraction, and absorption losses in the cavity. It can be seen from the above formula that there is a nonlinear relationship between η_{ex} and I. This cannot be used to directly describe the characteristics of the device; therefore, the concept of external-differential quantum efficiency is introduced. The external-differential quantum efficiency is defined as

$$\eta_D = \frac{\left(P_{\text{light}} - P_{th}\right)/hf}{(I - I_{th})/q}, \tag{3.6}$$

where P_{th} is the output optical power at the threshold. $P_{\text{light}} \gg P_{th}$; therefore, the above formula can be written as

$$\eta_D = \frac{P_{\text{light}}/hf}{(I - I_{th})/q} = \frac{P_{\text{light}}}{I - I_{th}} \cdot \frac{q}{hf}. \tag{3.7}$$

The external-differential quantum efficiency η_D is the slope of the linear part above the threshold value in the I–L characteristic curve. It can be used to intuitively compare the efficiency difference between different lasers. In general, η_D can reach more than 50%.

3.1.3.4 Spectrum Characteristics

Light waves are produced by electrons in the motion of atoms. The movement of electrons inside the atoms of various substances is different; thus, the light waves they emit are also different. The spectrum is the pattern in which the dispersed monochromatic light is arranged in order according to the wavelength (or frequency), after the polychromatic light is split by a dispersion system (such as a prism or grating); this is called the optical spectrum.

The largest part of the spectrum, the visible spectrum, is the portion of the electromagnetic spectrum that is visible to the human eye, and electromagnetic radiation in this wavelength range is called visible light. Spectral characteristics are the structure of the laser spectrum output when the laser is working.

$$\Delta\lambda_0 = \frac{\lambda_0^2}{2nL}, \tag{3.8}$$

where $\Delta\lambda_0$ is the longitudinal-mode interval, n is the refractive index of the active region of the laser, L is the cavity length of the laser, and λ_0 is the peak wavelength of the laser.

3.1.3.5 Temperature Characteristics

Optoelectronic devices are made of semiconductor materials, and the temperature has a great influence on their optoelectronic properties. As shown in Fig. 3.6, with the increase of temperature, the threshold of a laser diode (LD) gradually increases, the output optical power gradually decreases, and the external-differential quantum efficiency gradually decreases.

The stability of semiconductor lasers is closely related to the temperature. The approximate relationship between the threshold and temperature can be expressed as

$$J_{th}(T) = J_0 \frac{T - T_0}{T_0}, \tag{3.9}$$

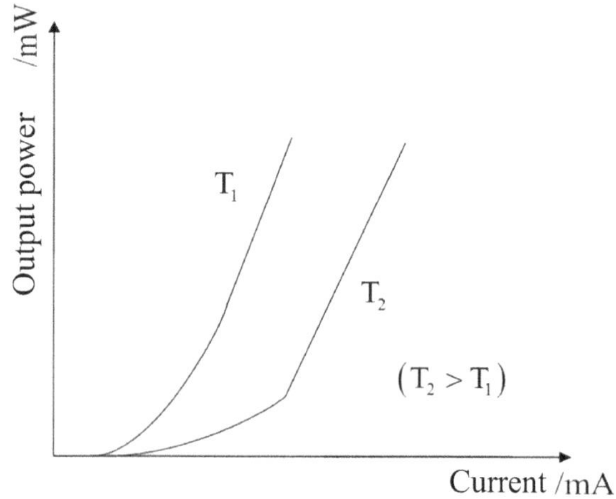

Fig. 3.6. Temperature characteristic curve of a semiconductor laser diode

where J_0 is the threshold current density, J_{th} at room temperature T. T_0 is a physical parameter that characterizes the temperature characteristics of the semiconductor laser; it is called the characteristic temperature. It can be seen from Eq. (3.9) that the larger T_0 is, the more the threshold current density J_{th} changes with the temperature T. The smaller the current density, the more stable the laser is.

3.2 Nonlinear Characteristics of Semiconductor Lasers

The nonlinear distortion of a laser comes from three aspects. The first is frequency-dependent nonlinear distortion, which belongs to the inherent nonlinearity of the laser; the second is frequency-independent nonlinear distortion, which can be measured by the I–L characteristic curve. The third is clipping distortion, owing to the presence of the threshold current. They are described in Table 3.1.

3.2.1 *Dynamic Nonlinearity*

Dynamic nonlinearity is the nonlinearity formed by the internal electro-optical coupling of the laser. It can be described by the large-signal dynamic single-mode rate equation of the laser [3]:

$$\frac{dN(t)}{dt} = \frac{I(t)}{qV_a} - g_0\frac{[N(t) - N_0]S(t)}{1 + \varepsilon S(t)} - \frac{N(t)}{\tau_n} \tag{3.10}$$

Table 3.1 Nonlinearity of lasers

Nonlinear type	Source	Reason
Frequency dependent (dynamic)	Relaxation oscillation	When the modulation frequency is close to the relaxation-oscillation frequency of the laser, the nonlinearity is caused by the interaction of the carriers and photons in the active region
Frequency independent (static)	Intracavity gain and loss	Owing to the gain-compression factor, the gain and photon density are nonlinear, resulting in nonlinear gains and losses
	Active area spatial-hole burning	Uneven distribution of carriers and photons
	Leakage current	Defects in the internal structure of the laser cause nonlinearity between the output optical power and the injected current
Clipping effect	Clipping distortion	Undercutting distortion when the injection current is less than the laser threshold

$$\frac{dS(t)}{dt} = \Gamma g_0 \frac{[N(t) - N_0]S(t)}{1 + \varepsilon S(t)} - \frac{S(t)}{\tau_p} + \frac{\Gamma\beta}{\tau_n} N(t), \tag{3.11}$$

where $N(t)$ represents the carrier density in the active region of the semiconductor laser, and $S(t)$ represents the photon density in the semiconductor laser cavity. Because $S(t)$ is proportional to the output laser intensity, $S(t)$ can also represent the relative output power.

$I(t)$ is the current injected into the active area, q is the amount of electron charge, V_a is the volume of the active area, $g = g_0[N(t) - N_0]$ is the optical gain, g_0 is a constant, N_0 is the carrier density when $g = 0$, and ε is the gain saturation. Parameters τ_n and τ_p are the carrier and photon lifetimes, respectively, and Γ and β are the light-confinement factor and spontaneous-recombination factor, respectively.

$\frac{\Gamma\beta}{\tau_n} N(t)$ represents the rate at which spontaneous emission enters the laser mode.
$\frac{S(t)}{\tau_p}$ is the photon-loss rate induced within the resonator.
$\Gamma g_0 \frac{[N(t) - N_0]S(t)}{1 + \varepsilon S(t)}$ is the stimulated-emission photon-generation rate.
$\frac{dS(t)}{dt}$ is the photon-density increase rate.
$\frac{N(t)}{\tau_n}$ is the carrier spontaneous-recombination loss rate.
$g_0 \frac{[N(t) - N_0]S(t)}{1 + \varepsilon S(t)}$ is the carrier stimulated-recombination loss rate.
$\frac{I(t)}{qV_a}$ is the rate of the injected current.
$\frac{dN(t)}{dt}$ is the increase rate of the carrier density.

When the carrier density and photon density reach steady state values N and S, $dN(t)/dt = 0$ and $dS(t)/dt = 0$; namely,

$$\frac{I(t)}{qV_a} - g_0 \frac{[N(t) - N_0]S(t)}{1 + \varepsilon} - \frac{N(t)}{\tau_n} = 0 \tag{3.12}$$

$$\Gamma g_0 \frac{[N(t) - N_0]}{1 + \varepsilon S(t)} - \frac{S(t)}{\tau_p} + \frac{\Gamma\beta}{\tau_n} N(t) = 0. \tag{3.13}$$

When the injection current is at or below the threshold (that is, $I(t) \le I_{th}(t)$), the stimulated-emission term is negligible compared with the spontaneous-emission term, and the following can be obtained from Eq. (3.12):

$$I(t) = \frac{qV_a}{\tau_n} N(t), \quad I(t) \le I_{th}(t). \tag{3.14}$$

At the threshold (stimulated gain equals photon loss),

$$I_{th}(t) = \frac{qV_a}{\tau_n} N_{th}(t). \tag{3.15}$$

When the laser is generated, stimulated emission is dominant, β is very small ($10^{-5} \sim 10^{-2}$ orders of magnitude), the spontaneous-emission term in Eq. (3.13) can be ignored, and the gain-saturation term is $1/(1+\varepsilon S(t)) \approx 1$ (i.e., $\varepsilon S(t) \ll 1$). The following can be obtained:

$$g(N) = \frac{1}{\tau_p} = g_{th}(N_{th}). \tag{3.16}$$

Because τ_p is a constant, $g(N)$ is also a constant, and the carrier density in the cavity is locked at the saturation value of $N(t) = N_{th}$.

Above the threshold current, the gain-suppression coefficient needs to be considered, owing to the increase of the light intensity in the cavity. If the gain dispersion is ignored, the gain coefficient can be expressed as [4]

$$g_0 = \alpha(N - N_0) - BS, \tag{3.17}$$

where α is the line-width growth factor, and B is the gain-suppression coefficient.

When the number of photons is stable, the spontaneous radiation is equal to the self-excited radiation, and the number of carriers and steady-state photons at this time can be obtained by Eqs. (3.12), (3.13), (3.15), (3.16) and (3.17). For

$$N = N_{th} + \frac{B}{\Gamma \alpha q g_{th}}(I - I_{th}), \tag{3.18}$$

$$S = \frac{\Gamma \tau_p}{q V_a}(I - I_{th}). \tag{3.19}$$

It can be seen from Eqs. (3.18) to (3.19) that with the increase of the modulation frequency, the non-equilibrium carrier concentration increases, the carrier lifetime decreases, and the number of photons decreases. The generated optical power can be expressed as [5]

$$P(t) = \frac{S(t) V_\alpha \eta_D h v}{2\Gamma \tau_p}, \tag{3.20}$$

where $P(t)$ is the output optical power, η_D is the external-differential quantum efficiency, h is Planck's constant, and v is the emitted photon frequency. Let $N(t) - N_0 = n$; thus, from Eq. (3.11), we obtain

$$\left(\frac{\Gamma g_0 S(t)}{1+\varepsilon S(t)} + \frac{\Gamma \beta}{\tau_n}\right) n = \frac{dS(t)}{dt} + \frac{S(t)}{\tau_p} - \frac{\tau \beta}{\tau_n} N_0$$

$$n = \frac{\frac{dS(t)}{dt} + \frac{S(t)}{\tau_p} - \frac{\tau\beta}{\tau_n} N_0}{\frac{\Gamma g_0 S(t)}{1+\varepsilon S(t)} + \frac{\Gamma\beta}{\tau_n}}. \tag{3.21}$$

Substituting the above formula into Eq. (3.10), we can obtain

$$\frac{I(t)}{qV_a} = \frac{N_0}{\tau_n} + \frac{g_0 S(t)}{1+\varepsilon S(t)} \cdot \frac{\frac{dS(t)}{dt} + \frac{S(t)}{\tau_p} - \frac{\tau\beta}{\tau_n} N_0}{\frac{\Gamma g_0 S(t)}{1+\varepsilon S(t)} + \frac{\Gamma\beta}{\tau_n}} + \frac{d}{dt}\left[\frac{\frac{dS(t)}{dt} + \frac{S(t)}{\tau_p} - \frac{\tau\beta}{\tau_n} N_0}{\frac{\Gamma g_0 S(t)}{1+\varepsilon S(t)} + \frac{\Gamma\beta}{\tau_n}}\right]. \tag{3.22}$$

At this time, Eq. (3.22) can be expressed as the relationship between the number of photons and the modulation signal:

$$I(t) = qV_a\left\{\frac{N_0}{\tau_n} + \frac{g_0 S(t)}{1+\varepsilon S(t)} \cdot \frac{\frac{dS(t)}{dt} + \frac{S(t)}{\tau_p} - \frac{\tau\beta}{\tau_n} N_0}{\frac{\Gamma g_0 S(t)}{1+\varepsilon S(t)} + \frac{\Gamma\beta}{\tau_n}} + \frac{d}{dt}\left[\frac{\frac{dS(t)}{dt} + \frac{S(t)}{\tau_p} - \frac{\tau\beta}{\tau_n} N_0}{\frac{\Gamma g_0 S(t)}{1+\varepsilon S(t)} + \frac{\Gamma\beta}{\tau_n}}\right]\right\}. \tag{3.23}$$

The second and higher harmonic components in laser-diode (LD) intensity modulation can seriously distort the signal, thereby affecting the performance of the entire optical communication system. First, the causes of harmonic generation are analyzed and the expressions of the second and third harmonics are deduced. Let the number of carriers in the LD cavity have the following form:

$$N(t) = N + N_1 e^{j\omega_m t} + N_2 e^{j2\omega_n t} + N_3 e^{j3\omega_m t}, \tag{3.24}$$

where N is the steady-state carrier number of the LD, N_1 is the fundamental component, N_2 is the second harmonic component, and N_3 is the third harmonic component. Similarly, the number of photons has the same form:

$$S(t) = S + S_1 e^{j\omega_m t} + S_2 e^{j2\omega_m t} + S_3 e^{j3\omega_m t}. \tag{3.25}$$

In addition, the injection current is set to $I(t) = I_m e^{j\omega_m t}$, where I_m is the modulation component and ω_m is the modulation angular frequency.

We substitute Eqs. (3.24) and (3.25) into Eqs. (3.10) and (3.11), respectively, to obtain

$$\begin{aligned} &\frac{d}{dt}\left(N + N_1 e^{j\omega_m t} + N_2 e^{j2\omega_m t} + N_3 e^{j3\omega_m t}\right) \\ &= \frac{I_m e^{j\omega_m t}}{qV_a} \\ &- \frac{g_0\left(N - N_0 + N_1 e^{j\omega_m t} + N_2 e^{j2\omega_m t} + N_3 e^{j3\omega_m t}\right)\left(S + S_1 e^{j\omega_m t} + S_2 e^{j2\omega_m t} + S_3 e^{j3\omega_m t}\right)}{1 + \varepsilon\left(S + S_1 e^{j\omega_m t} + S_2 e^{j2\omega_m t} + S_3 e^{j3\omega_m t}\right)} \\ &- \frac{N + N_1 e^{j\omega_m t} + N_2 e^{j2\omega_m t} + N_3 e^{j3\omega_m t}}{\tau_n} \end{aligned} \tag{3.26}$$

$$\begin{aligned}
&\frac{d}{dt}\left(S + S_1 e^{j\omega_m t} + S_2 e^{j2\omega_m t} + S_3 e^{j3\omega_m t}\right) \\
&= \frac{\Gamma g_0\left(N - N_0 + N_1 e^{j\omega_m t} + N_2 e^{j2\omega_m t} + N_3 e^{j3\omega_m t}\right)\left(S + S_1 e^{j\omega_m t} + S_2 e^{j2\omega_m t} + S_3 e^{j3\omega_m t}\right)}{1 + \varepsilon\left(S + S_1 e^{j\omega_m t} + S_2 e^{j2\omega_m t} + S_3 e^{j3\omega_m t}\right)} \\
&- \frac{S + S_1 e^{j\omega_m t} + S_2 e^{j2\omega_m t} + S_3 e^{j3\omega_m t}}{\tau_p} \\
&+ \frac{\Gamma\beta}{\tau_n}\left(N + N_1 e^{j\omega_m t} + N_2 e^{j2\omega_m t} + N_3 e^{j3\omega_m t}\right).
\end{aligned} \tag{3.27}$$

It can be seen from Eqs. (3.26) and (3.27) that even though the carrier number $N(t)$ and the photon number $S(t)$ have only steady-state and fundamental components, their nonlinear coupling produces harmonics. By analyzing Eqs. (3.26) and (3.27), and ignoring harmonics above the third order, the following equations can be obtained:

$$\begin{cases} j\omega_m N_1 = a_{11}N_1 + a_{12}S_1 + \frac{I_m}{qV_a} \\ j\omega_m S_1 = a_{21}N_1 + a_{22}S_1 \end{cases}, \tag{3.28}$$

where the coefficients are

$$\begin{array}{ll} a_{11} = -g_0 S - \frac{1}{\tau_n} & a_{12} = -g_0(N - N_0) \\ a_{21} = \Gamma g_0 S + \frac{\Gamma\beta}{\tau_n} & a_{22} = \Gamma g_0(N - N_0) - \frac{1}{\tau_p} \end{array}.$$

After shifting Eq. (3.28), we use Cramer's rule to obtain

$$N_1 = \frac{-\frac{I_m}{qV_a}(a_{22} - j\omega_m)}{(a_{11} - j\omega_m)(a_{22} - j\omega_m) - a_{12}a_{21}} \tag{3.29}$$

$$S_1 = \frac{\frac{I_m}{qV_a}a_{21}}{(a_{11} - j\omega_m)(a_{22} - j\omega_m) - a_{12}a_{21}} \tag{3.30}$$

$$\begin{cases} j2\omega_m N_2 = a_{11}N_1 + a_{12}S_2 - g_0 N_1 S_1 \\ j2\omega_m S_2 = a_{21}N_2 + a_{22}S_2 + \Gamma g_0 N_1 S_1 \end{cases}. \tag{3.31}$$

We also obtain

$$N_2 = \frac{g_0 N_1 S_1(\Gamma a_{12} + a_{22} - j2\omega_m)}{(a_{11} - j2\omega_m)(a_{22} - j2\omega_m) - a_{12}a_{21}} \tag{3.32}$$

$$S_2 = \frac{-g_0 N_1 S_1\left[\Gamma(a_{11} - j2\omega_m) + a_{21}\right]}{(a_{11} - j2\omega_m)(a_{22} - j2\omega_m) - a_{12}a_{21}} \tag{3.33}$$

$$\begin{cases} j2\omega_m N_2 = a_{11}N_1 + a_{12}S_2 - g_0 N_1 S_1 \\ j2\omega_m S_2 = a_{21}N_2 + a_{22}S_2 + \Gamma g_0 N_1 S_1 \end{cases}. \tag{3.34}$$

Solving the above equation yields

$$N_3 = \frac{g_0(N_1S_2 + N_2S_1)(\Gamma a_{12} + a_{22} - j3\omega_m)}{(a_{11} - j3\omega_m)(a_{22} - j3\omega_m) - a_{12}a_{21}} \tag{3.35}$$

$$S_3 = \frac{-g_0(N_1S_2 + N_2S_1)\left[\Gamma(a_{11} - j3\omega_m) + a_{21}\right]}{(a_{11} - j3\omega_m)(a_{22} - j3\omega_m) - a_{12}a_{21}}. \tag{3.36}$$

Then, the intensity expressions of the second and third harmonics relative to the fundamental wave can be obtained:

$$\begin{aligned} HD_2 = \frac{S_2}{S_1} &= \frac{\frac{-g_0N_1S_1\left[\Gamma(a_{11}-j2\omega_m)+a_{21}\right]}{(a_{11}-j2\omega_m)(a_{22}-j2\omega_m)-a_{12}a_{21}}}{S_1} \\ &= \frac{-g_0I_m\left[\Gamma(a_{11} - j2\omega_m) + a_{21}\right](a_{22} - j\omega_m)}{qV_aH_1H_2} \end{aligned} \tag{3.37}$$

$$\begin{aligned} HD_3 = \frac{S_3}{S_1} &= \frac{\frac{-g_0(N_1S_2+N_2S_1)\left[\Gamma(a_{11}-j3\omega_m)+a_{21}\right]}{(a_{11}-j3\omega_m)(a_{22}-j3\omega_m)-a_{12}a_{21}}}{\frac{\frac{I_m}{qV_a}a_{21}}{(a_{11}-j\omega_m)(a_{22}-j\omega_m)-a_{12}a_{21}}} \\ &= -\left(\frac{g_0I_m}{qV_a}\right)^2(a_{22} - j\omega_m)\left[\Gamma(a_{11} - j3\omega_m) + a_{21}\right] \\ &\times \frac{\left\{a_{21}(\Gamma a_{12} + a_{22} - j2\omega_m) + (a_{22} - j\omega_m)\left[\Gamma(a_{11} - j2\omega_m) + a_{21}\right]\right\}}{H_1^2H_2H_3} \end{aligned} \tag{3.38}$$

$$\begin{aligned} H_1 &= (a_{11} - j\omega_m)(a_{22} - j\omega_m) - a_{12}a_{21} \\ H_2 &= (a_{11} - j2\omega_m)(a_{22} - j2\omega_m) - a_{12}a_{21} \\ H_3 &= (a_{11} - j3\omega_m)(a_{22} - j3\omega_m) - a_{12}a_{21} \end{aligned}$$

The starting point of the dynamic nonlinear-analysis method is the rate equation. Although the relationship between the photons, carriers, and harmonic components can be deduced, the geometric and physical parameters of the laser and the light-wave field must be known. In practice, these parameters cannot be measured directly and the values cannot be obtained accurately.

3.2.2 Static Nonlinearity

The frequency-independent static nonlinearity of a laser can be analyzed from the I–L characteristic curve. For an ideal LD model, in the region above the threshold current, the relationship between the average optical power and the driving current can be expressed as

$$P = P_{th} + (dP/dI)(I - I_{th}). \tag{3.39}$$

In the formula, I_{th} is the threshold current, and P_{th} is the average optical power when the driving current is I_{th}. If the part of the I–L characteristic curve above the threshold current is not ideally linear, the average output optical power can be expanded by a Taylor series near the threshold point:

$$\begin{aligned} P &= P_{th} + \sum_{k=1}^{\infty} \frac{1}{k!}\frac{d^k P}{dI^k}(I - I_{th})^k \\ &= P_{th} + \left(\frac{dP}{dI}\right)(I - I_{th}) + \frac{1}{2}\left(\frac{d^2P}{dI^2}\right)(I - I_{th})^2 \\ &+ \frac{1}{6}\left(\frac{d^3P}{dI^3}\right)(I - I_{th})^3 + \ldots. \end{aligned} \tag{3.40}$$

According to the laser output characteristics, we define the modulation index:

$$m = \frac{P_1}{P_0} = \frac{I_m}{P_{th}}\frac{dP}{dI}, \tag{3.41}$$

where P_0 is the average optical power when the threshold current is I_{th}, and P_1 is the output optical power caused by the total modulation current I.

Assuming that the input signal can be expressed as $I = I_{th} + I_m \cos(\omega t + \theta_n)$, I_m is the peak modulation current, and θ_n is the initial phase, then Eq. (3.40) can be transformed into

$$\begin{aligned} \frac{P}{P_{th}} &= 1 + m\cos(\omega t + \theta_n) + \frac{1}{2}m^2 P_{th}\left[\frac{d^2P}{dI^2}\Big/\left(\frac{dP}{dI}\right)^2\right][\cos(\omega t + \theta_n)]^2 \\ &+ \frac{1}{6}m^3 P_{th}^2\left[\frac{d^3P}{dI^3}\Big/\left(\frac{dP}{dI}\right)^3\right][\cos(\omega t + \theta_n)]^3 + \ldots \end{aligned} \tag{3.42}$$

Here,

$$m = \frac{P_1}{P_0} = \frac{I_m}{P_{th}}\frac{dP}{dI}.$$

It can be seen from Eq. (3.42) that, owing to the nonlinearity of the I–L curve, the optical signal is no longer a simple reproduction of the modulated signal; rather, many high-order terms appear. $\frac{d^2P}{dI^2}$ will cause a second harmonic distortion and $\frac{d^3P}{dI^3}$ will cause a third harmonic distortion. Because the second and third harmonics are directly related to the fundamental, defining the ratio of the harmonics to the fundamental evaluates the influence of the harmonics. The ratio of the second harmonic component to the fundamental component is

$$HD_2 = \frac{SH}{FH} \approx \frac{1}{4} m P_{th} \left[\frac{\frac{d^2P}{dI^2}}{\left(\frac{dP}{dI}\right)^2} \right]. \tag{3.43}$$

The ratio of the third harmonic component to the fundamental component is

$$HD_3 = \frac{TH}{FH} \approx \frac{1}{24} m^2 P_{th}^2 \left[\frac{\frac{d^3P}{dI^3}}{\left(\frac{dP}{dI}\right)^3} \right]. \tag{3.44}$$

3.2.3 *Static Model of a Semiconductor Laser*

3.2.3.1 Analysis Method

The relationship between the injected current and the output optical power in the laser-modulation process in the time domain can be characterized by an input–output mapping function. The nonlinear distortion is the product of a signal transformation; therefore, the nonlinear distortion of the laser can be analyzed from the perspective of a mapping function. Non-linear distortion can be simply summarized as follows: Under the excitation of an undistorted signal, the output-response signal produces a waveform distortion in the time domain, and harmonic components appear in the frequency domain.

If an infinite number of experimental measurements can be performed on the laser, and all allowable signals are recorded in pairs $[I(t), L(t)]$, we can construct a model that allows signal pairs to approximate the input–output relationship of the original device. Thus, the study of the device can be transformed into the study of the model [6].

Under continuous operating conditions, the second-order nonlinear distortion of the laser is mainly determined by the nonlinearity of the I–L characteristic curve [7, 8]. For lasers operating in the subcarrier frequency band of several MHz, the main source of nonlinear distortion is the frequency-independent nonlinearity of the laser. The high-power, high-efficiency, low-threshold laser itself has good linearity and can be considered as a weak nonlinear system.

3.2.3.2 Mapping Functions

The series analysis method is used to describe the nonlinear distortion of the communication system, and its response to the input signal can generally be expressed as the sum of a convolution sequence:

$$y(t) = \sum_{n=0}^{\infty} y_n(t) \tag{3.45}$$

$$y_n(t) = \int_{-\infty}^{\infty}\int_{-\infty}^{\infty} \cdots \int_{-\infty}^{\infty} h_n(\tau_1, \tau_2, \cdots, \tau_n) \cdot \prod_{j=1}^{n} x(t - \tau_i) d\tau_i. \tag{3.46}$$

Equation (3.46) is the Volterra functional expression of the input–output relationship of the nonlinear dynamic system to the input function $x(t)$, where $y_n(t)$ is the n -order output of the system, and $h_n(t)$ is the n-order Volterra kernel or n -order impulse response of the system. For an ordinary laser, the system is generally a weak nonlinear system; therefore, a response of more than three times can be ignored. Equation (3.45) can be simplified as

$$y(t) = \sum_{n=0}^{3} y_n(t). \tag{3.47}$$

The relaxation-oscillation frequency of the laser is on the order of several or dozens of GHz. Because it is an instantaneous response in the subcarrier frequency band, the Volterra kernels in Eq. (3.46) can be expressed as the following special form:

$$h_n(\tau_1, \tau_2, \ldots, \tau_n) = a_n \delta(\tau_1)\delta(\tau_2) \cdots \delta(\tau_n). \tag{3.48}$$

Equation (3.47) can be reduced to a power-series expansion:

$$y(t) = a_0 + a_1 x(t) + a_2 x^2(t) + a_3 x^3(t), \tag{3.49}$$

where $y(t)$ represents the output optical power of the laser and $x(t)$ represents the input current.

Using a power series to approximate the transmission characteristics of a laser diode is a simple way to describe it; however, it ignores the fact that the transmission characteristics vary with the frequency. From the static I–L curve of the laser, a relationship similar to Eq. (3.49) can be fitted:

$$P(t) = b_0 + b_1 I(t) + b_2 I^2(t) + b_3 I^3(t). \tag{3.50}$$

The meanings of Eqs. (3.49) and (3.50) differ: Equation (3.49) is the input–output relationship of the laser under medium- and high-frequency modulation. The influence of a modulation-frequency change is not considered. Equation (3.50) is the input–output relationship in a static operation (without modulation). The dynamic-transmission characteristic Formula (3.49) and the static characteristic Formula (3.50) of the same laser differ widely. Because the dynamic-modulation signal itself is small, the data range taken when fitting the mapping-function relation is also small. The static characteristic curve is obtained by fitting it within a larger data range.

3.2.3.3 Harmonic-Distortion Characteristics

An NC-QPSK-like sinusoidal modulation signal can be expressed as

$$S_{QPSK}(t) = I(t)\sin(\omega_c t) + Q(t)\cos(\omega_c t) \tag{3.51}$$

$$SNC\_QPSK(t) = \begin{cases} S_{QPSK}(t) & S_{QPSK}(t) > 0 \\ 0 & S_{QPSK}(t) \le 0 \end{cases}. \tag{3.52}$$

Then, the nonlinear signal can be expressed as

$$S_{NLD}(t) = a_0 + a_1 S_{NC\_QPSK}(t) + a_2\left[S_{NC\_QPSK}(t)\right]^2 + a_3\left[S_{NC\_QPSK}(t)\right]^3. \tag{3.53}$$

Substituting Eqs. (3.51) and (3.52) into Eq. (3.53), we obtain

$$S_{NLD}(t) = \begin{cases} (a_0 + a_2) + \left(a_1 + \frac{3}{2}a_3\right)\left[I(t)\sin(\omega_c t) + Q(t)\cos(\omega_c t)\right] & \\ +a_2 I(t)Q(t)\sin(2\omega_c t) + \frac{1}{2}a_3\left[I(t)\sin(3\omega_c t) - Q(t)\cos(3\omega_c t)\right] & S_{NLD}(t) > 0 \\ 0 & S_{NLD}(t) \le 0 \end{cases} \tag{3.54}$$

It can be seen from Eq. (3.54) that the output signal $S_{NLD}(t)$ not only contains a spurious DC component coefficient $a_0 + a_2$ and the fundamental-frequency component coefficient $a_1 + \frac{3}{2}a_3$, but also contains the second-harmonic component coefficient a_2 and the third-harmonic component coefficient $\frac{1}{2}a_3$.

3.2.3.4 Characteristic Curve Fitting

The data of a light–current–voltage (L–I–V) characteristic test of a laser diode (LD) are shown in Table 3.2.

(1) **Fitting method**

The degree $m(m \le n)$ polynomial can be used to fit the experimental data $(x_i, y_i)(i = 0, 1, 2, \cdots, n)$, which has a functional relationship:

Table 3.2 L–I–V characteristic test values

I (mA)	0	3	6	9	12	15	18	21
U (V)	0	0.96	0.99	1	1.02	1.03	1.04	1.05
P (mW)	0	0	44.8	173.5	297.3	426.1	549.4	675.3

$$P_m(x) = a_0 + a_1 x + a_2 x^2 + \cdots + a_m x^m = \sum_{j=0}^{m} a_j x^j. \tag{3.55}$$

The basic idea of the least-squares method used in the fitting process is to find a fitted curve $y = P_m(x)$ and minimize the sum of squares of its deviations; that is,

$$\begin{aligned}\sigma &= \sum_{i=0}^{n} R_i^2 = \sum_{i=0}^{n} \left[P_m(x_i - j_i)\right]^2 = \sum_{i=0}^{n} \left(\sum_{j=0}^{m} a_j x_i^j - y_i\right)^2 \\ &= \sum_{i=0}^{n} \left(a_0 + a_1 x_i + a_2 x_i^2 + \cdots + a_m x_i^m - y_i\right)^2\end{aligned}. \tag{3.56}$$

In the formula, x_i and y_i are known quantities, and $a_0, a_1, \cdots, a_m$ are variables to be determined. When they take different values, σ should also change accordingly. Therefore, σ is actually a multivariate function; hence, the curve-fitting problem can ultimately be attributed to the extreme-value problem of the multivariate function. To minimize the multivariate function σ, the following equation must be satisfied:

$$\frac{\partial \sigma}{\partial a_k} = 0 \quad (k = 0, 1, 2, \ldots, m). \tag{3.57}$$

Taking the partial derivative of Eq. (3.56), we obtain

$$\begin{aligned}\frac{\partial \sigma}{\partial a_k} &= \sum \sum_{i=0}^{n} 2\left[P_m(x_i - y_i)\right]\frac{\partial P_m(x_i)}{\partial a_k} = 2\sum_{i=0}^{n} \left[P_m(x_i - y_i)\right]x_i^k \\ &= 2\left[\sum_{i=0}^{m} a_j \sum_{i=0}^{m} x_i^{j+k} - \sum_{j=0}^{m} y_i x_i^k\right] = 2\left\{\sum_{i=0}^{m} a_j S_{j+k} - t_k\right\}\end{aligned}. \tag{3.58}$$

In

$$S_{j+k} = \sum_{i=0}^{n} x_i^{j+k} \quad (k = 0, 1, 2, \cdots, m; j = 0, 1, 2, \cdots, m),$$

$$t_k = \sum_{i=0}^{n} y_i x_i^k \quad (j = 0, 1, 2, \cdots, m).$$

Because $S_0 = \sum_{i=0}^{n} x_i^n = n + 1$, $\frac{\partial \sigma}{\partial a_k} = 0$ can also be expressed as

$$\sum_{j=0}^{m} a_j S_{j+k} = t_k \quad (k = 0, 1, 2, \cdots, m). \tag{3.59}$$

Therefore, Eq. (3.55) can be written as

$$\begin{cases} S_0a_0 + S_1a_1 + S_2a_2 + \cdots + S_ma_m = t_0 \\ S_1a_0 + S_2a_1 + S_3a_2 + \cdots + S_{m+1}a_m = t_1 \\ \cdots \\ S_ma_0 + S_{m+1}a_1 + S_{m+2}a_2 + \cdots + S_{2m}a_m = t_m \end{cases} . \tag{3.60}$$

Equation (3.60) is an m + 1-order linear-algebraic equation system where $a_0, a_1, \ldots, a_m$ is an unknown quantity, and the coefficient $a_j(j = 0, 1, 2, \ldots, m)$ of polynomial $P_m(x)$ can be obtained by solving the above linear-algebraic equation system.

(2) **Smoothness processing**

To reduce the random error of the experimental data, the curve needs to be smoothed. The cubic polynomial interpolation method is used, and the final fitting results are shown in Figs. 3.7 and 3.8. Among them, the seventh-order polynomial coefficient of the V–I fitting curve, the third-order polynomial coefficient of the I–L fitting curve, and the threshold current are shown in Table 3.3.

(3) **Signal loading**

As shown in Fig. 3.9, we convert the modulated subcarrier signal into a current signal and load it onto the I–L characteristic curve to simulate that the signal directly acts on the light source and modulates the light source directly.

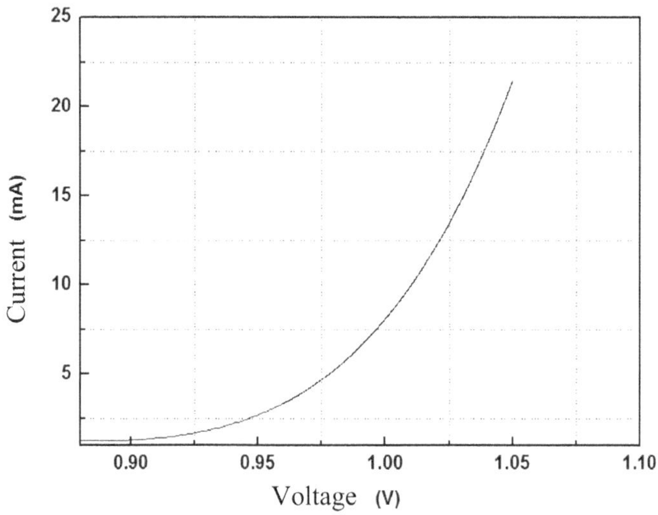

Fig. 3.7 V–I characteristic curve after fitting

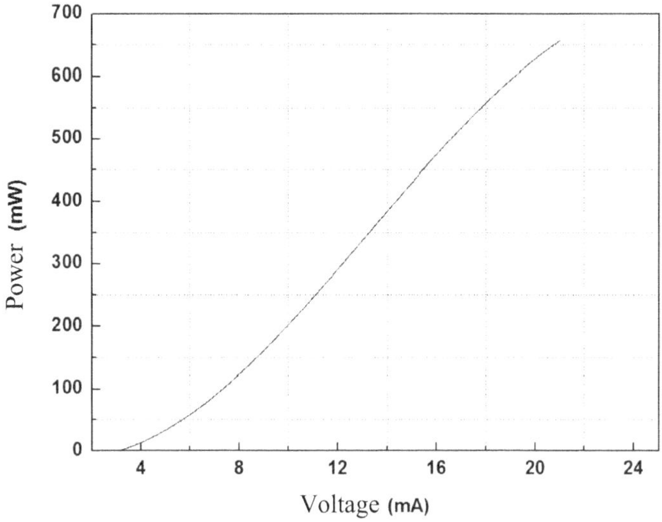

Fig. 3.8 I–L characteristic curve after fitting

Table 3.3 Fitted-curve coefficients and threshold current

V–I fitted-curve seventh-order polynomial coefficients			
$a_0 = -0.1$	$a_1 = 10.8$	$a_2 = -195.5$	$a_3 = 1534.8$
$a_4 = -5483.5$	$a_5 = -9739.3$	$a_6 = -8369.3$	$a_7 = 2771.5$
I–L fitted-curve third-order polynomial coefficients			
$a_0 = 3.8944$	$a_1 = -13.9355$	$a_2 = 4.5087$	$a_3 = -0.1126$
Threshold current $I_{th} = 4.9$ mA			

3.2.4 Linearization of Semiconductor Lasers

Common methods for suppressing laser nonlinearity include electrical (optical) negative feedback, electrical (optical) feedforward, and predistortion. The shortcomings of the negative-feedback and feedforward-compensation methods are the hysteresis compensation and the instability of the loop gain. Moreover, the modulation bandwidth is limited. Using predistortion to compensate for nonlinearity has a good effect and high stability, and is a very practical compensation technology.

3.2.4.1 Electrical (Optical) Negative-Feedback Compensation

As shown in Fig. 3.10, negative feedback feeds a part of the output signal back to the input terminal. Part of the input signal and the feedback signal produce an error

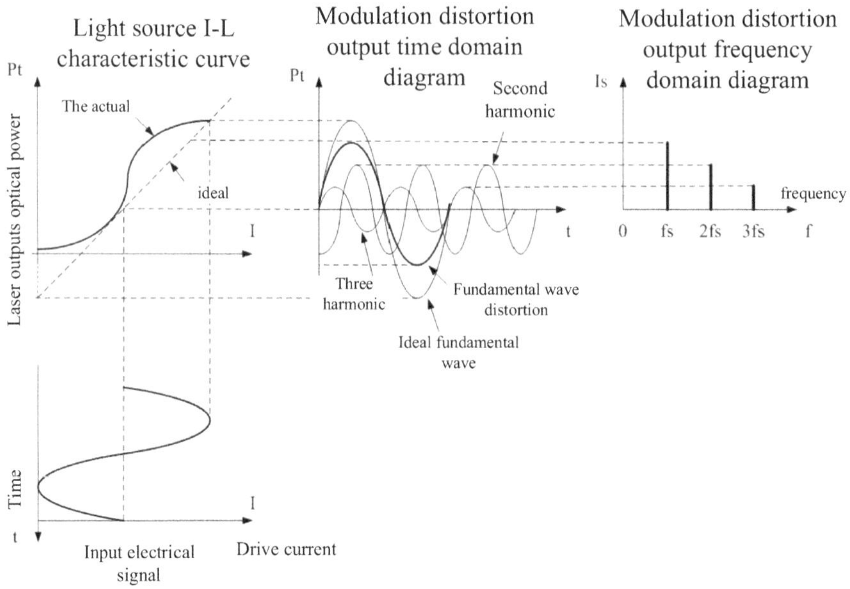

Fig. 3.9 Transmission model of the signal in the laser

signal through the comparator. The system tries to make the error between the input and output approach zero through a certain algorithm.

The error-control module generates an inverse nonlinear compensation signal and superimposes it on the driving circuit of the original laser, thereby improving the

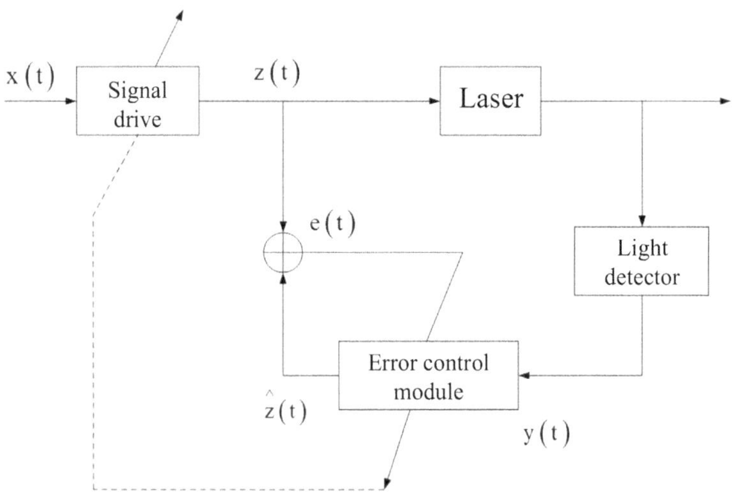

Fig. 3.10 Laser nonlinear electronegative-feedback compensation method

linearity of the signal after passing through the laser. The difference between electrical compensation and optical compensation lies in whether the nonlinear compensation signal generated by the error-control module is superimposed on the signal-driving circuit or the laser-driving circuit.

3.2.4.2 V–I Characteristic Predistortion Compensation

As shown in Fig. 3.11, the V–I characteristic curve predistortion-compensation method compensates for the predistortion by constructing a model curve opposite to the V–I characteristic of the semiconductor laser diode.

Assume the relationship between the operating voltage of the laser and the drive current is

$$i(t) = a_0 + a_1 u(t) + a_2 u^2(t) + a_3 u^3(t) + \cdots . \tag{3.61}$$

Suppose the predistorted signal generated by the predistorter is

$$i'(t) = b_0 + b_1 i(t) + b_2 i^2(t) + b_3 i^3(t) + \cdots . \tag{3.62}$$

When the coefficients of the corresponding polynomials in Eqs. (3.61) and (3.62) have the following relationship, the nonlinear distortion can be completely eliminated.

$$b_0 = a_0 \quad b_1 = \frac{1}{a_1} \quad b_2 = -\frac{a_2}{(a_1)^3} \quad b_3 = \frac{2a_2^2 - a_1 a_3}{(a_1)^5}$$

It can be seen from the V-I characteristic curve fitted in Fig. 3.12 that the amplitude range of the forward voltage is between 0.88 and 1.05 V. When the voltage is 1.05 V, it can reach the maximum allowable current in the forward direction. The maximum value of the modulation linearity is at the maximum allowable current, and the part of the input-signal amplitude below 0.88 V and above 1.05 V is clipped. After predistortion, the amplitude of the input signal can be linear in a large range.

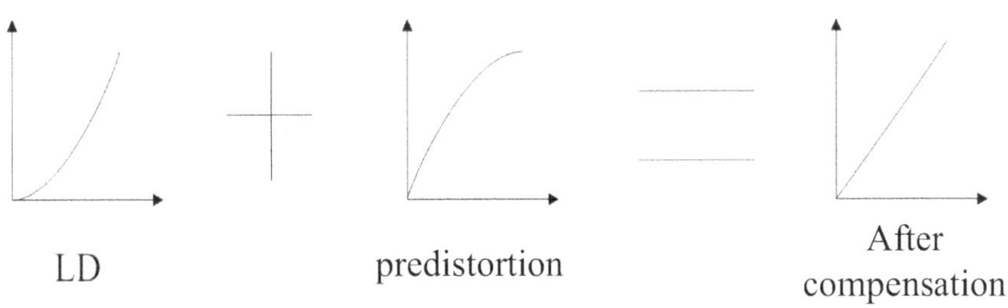

Fig. 3.11 Schematic diagram of the principle of predistortion compensation

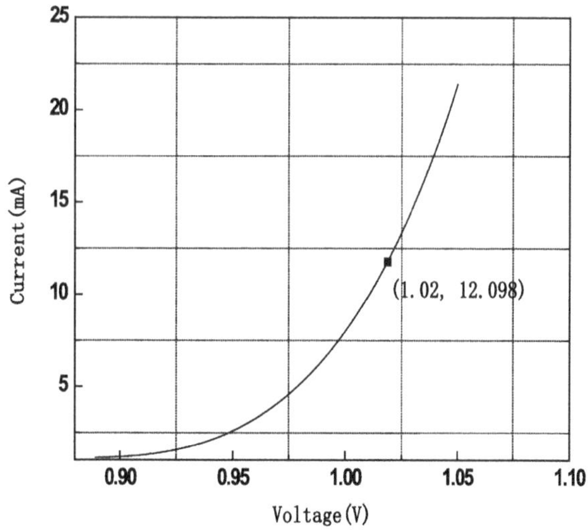

Fig. 3.12 V–I characteristic curve of an LD polynomial equation

It can be seen that the predistortion compensation of the V–I characteristic is only suitable for correcting the nonlinear part above the turn-on voltage and below the saturation voltage. The V–I characteristic predistortion method is actually a process of canceling the nonlinearity by using the predistorter and the LD as the I–V and V–I models, respectively. The corresponding characteristic curves are shown in Figs. 3.12, 3.13, 3.14, 3.15, 3.16 and 3.17.

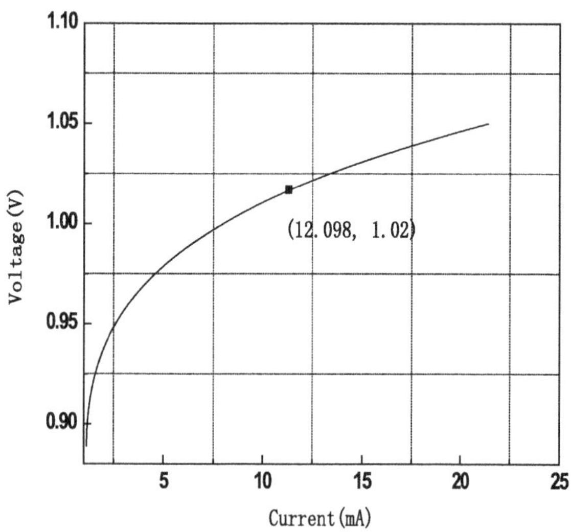

Fig. 3.13 I–V characteristic curve of an LD polynomial equation

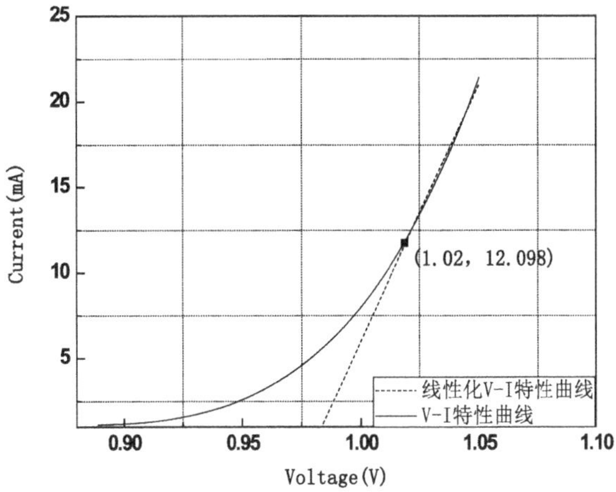

Fig. 3.14 V–I characteristic of linearized LD polynomial equation

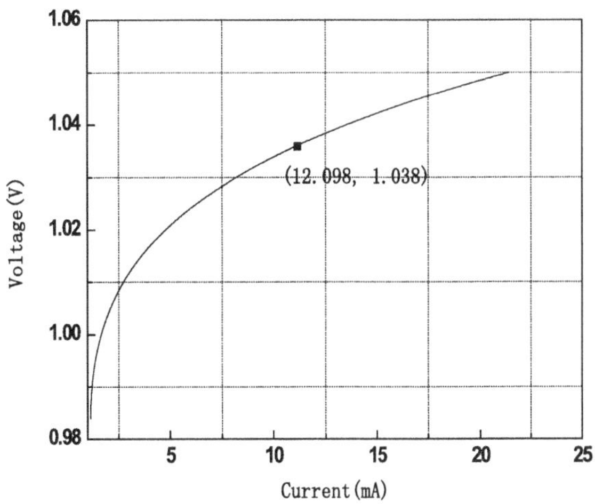

Fig. 3.15 I–V characteristic of linearized LD polynomial equation

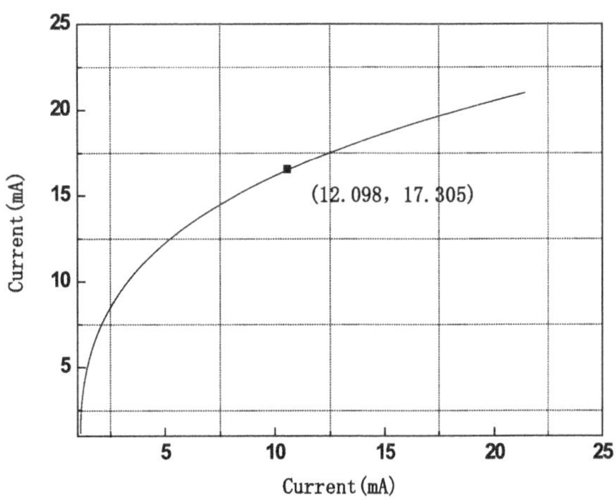

Fig. 3.16 Predistortion polynomial equation

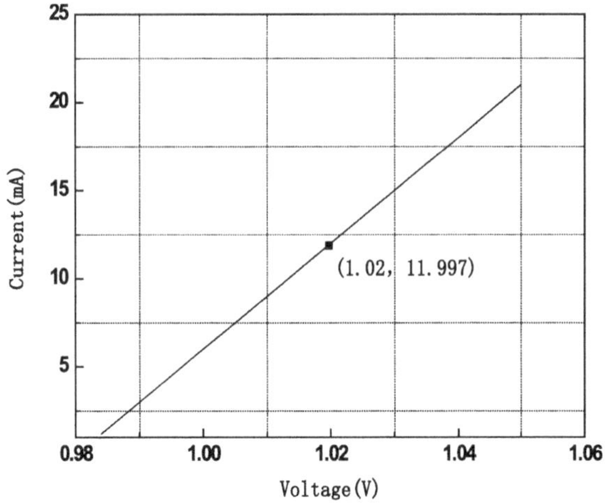

Fig. 3.17 Linear response after LD cascaded predistortion

3.2.4.3 I–L Characteristic Predistortion Compensation

Assume that the relationship between the laser output optical power and the driving current is

$$P_0 = a_0 + a_1 I + a_2 I^2 + a_3 I^3. \tag{3.63}$$

Suppose the predistorted signal generated by the predistorter is

$$I' = b_2 I^2 + b_3 I^3. \tag{3.64}$$

The basic principle of the predistorter is as follows: When the signal voltage increases, the predistortion model of the front end of the laser expedites the current through the laser, and simultaneously offsets the increase of the laser optical power when the current increases. The part with a small amplitude is a nonlinear distortion. After the combined action of the predistortion model and the laser, a linear relationship is maintained between the output optical power and the input current. It not only reduces the interference of harmonics, but also meets the needs of the signal-to-noise ratio.

As shown in Fig. 3.18, the original input signal is separated into three channels by the coupler; one is the main channel aI (processing the fundamental wave component), and the other two channels are secondary channels (processing the second and third-order distortion components). The gain of the secondary channel is the same, and the predistorter can be regarded as the signal generated by the secondary channel. The main channel directly enters the combiner through the time delay, and the secondary channel passes through the attenuator, inverter, and filter and then is output by the combiner. At this time, the output three-way combined signal passes through the laser, and the output optical power of the laser changes.

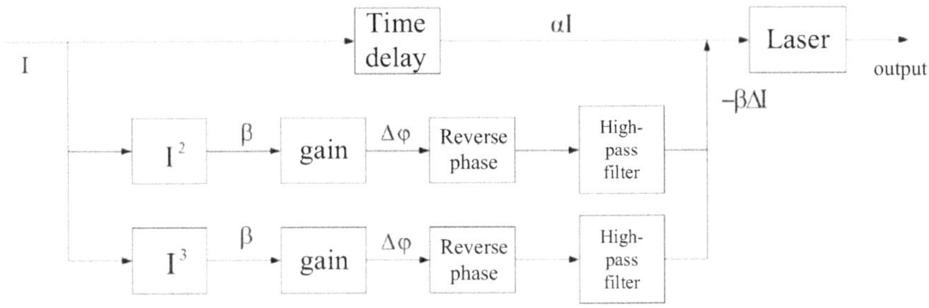

Fig. 3.18 Schematic diagram of I–L characteristic predistortion

$$P' = \sum_{m=1}^{3} a_m \left(\alpha I - \beta I'\right)^m = \sum_{m=1}^{3} a_m \left(\alpha I - \beta \sum_{n=2}^{3} b_n I^n\right)^m, \tag{3.65}$$

where α is the gain of the main channel and β is the gain of the secondary channel. Equation (3.65) can be transformed into a power-series expansion:

$$P' = \sum_{m=1}^{3} D_m I^m, \tag{3.66}$$

where

$$D_1 = \alpha a_1 \tag{3.67}$$

$$D_2 = \alpha^2 a_2 - \beta a_1 b_2 \tag{3.68}$$

$$D_3 = \alpha^3 a_3 - \beta a_1 b_3 - 2\alpha\beta a_2 b_2. \tag{3.69}$$

From Eqs. (3.67)–(3.69), the conditions for eliminating the second-order nonlinear distortion can be calculated as

$$\beta = \frac{\alpha^2 a_2}{a_1 b_2}. \tag{3.70}$$

The condition for eliminating the third-order nonlinear distortion is

$$\beta = \frac{\alpha^3 a_3}{a_1 b_3 + 2\alpha a_2 b_2}. \tag{3.71}$$

To simultaneously eliminate the second-order and third-order nonlinear distortions, it is necessary to meet

$$\alpha = \frac{a_1 a_2 b_3}{b_2 \left(a_1 a_3 - 2a_2^2\right)}. \tag{3.72}$$

3.2.4.4 Simulation and Analysis

The frequency spectrum, power spectrum, and constellation diagram of the NC-QPSK-like sine modulation with a 2.5 MHz carrier frequency are simulated and analyzed. Figure 3.19 shows the spectrum comparison diagram of the NC-QPSK-like sinusoidal-modulation signal passing through the laser directly and after the predistortion.

It can be seen from the figure that the spectrum after the LD nonlinear distortion is broadened, and there are obvious second and third harmonic components. The spectrum after predistortion is mainly concentrated at the center frequency, the DC component increases significantly, and the second and third harmonic components are improved by 23.9 dB and 18.4 dB, respectively.

Figure 3.20 shows the power spectrum of the NC-QPSK-like sinusoidal modulation signal directly passing through the laser and predistortion. The predistortion compensation effectively suppresses the in-band and out-of-band distortion, and the nonlinear compensation improves by about 3 dB and 17 dB, respectively. The suppression effect on the out-of-band distortion is obvious. The predistortion curve is below the uncompensated curve, indicating that the predistortion process comes at the cost of lower system amplification.

Figure 3.21 shows a comparison of the constellation diagram of the NC-QPSK sinusoidal-modulation signal passing directly through the laser and after predistortion (noise variance σ_n^2 is 0.2). Gain compression is the nonlinear distortion of the signal caused by excessive driving or poor performance of active devices

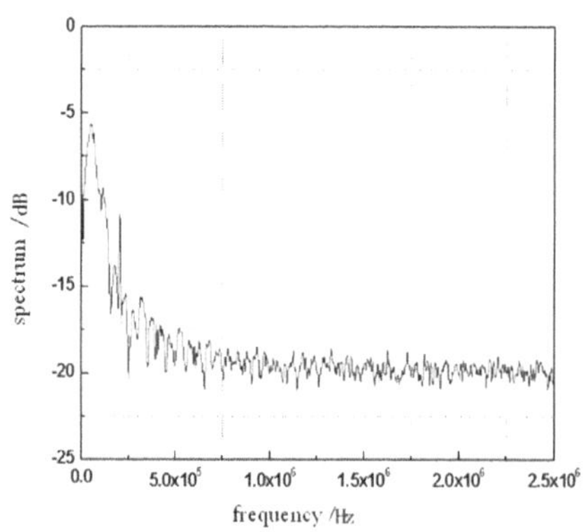

(a) Before predistortion

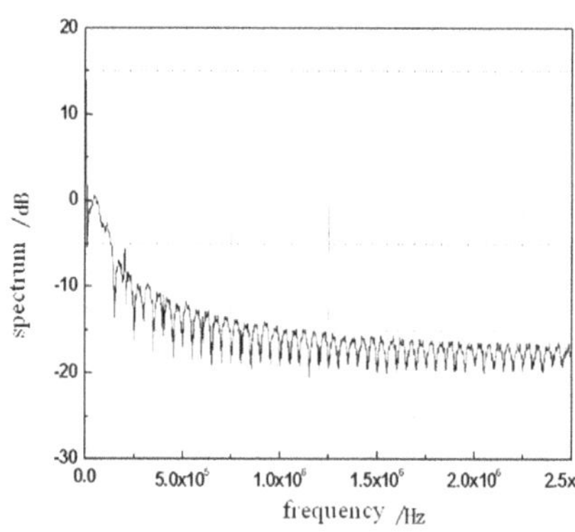

(b) After predistortion

Fig. 3.19 Spectrum of an NC-QPSK-like sinusoidal modulation signal before and after predistortion

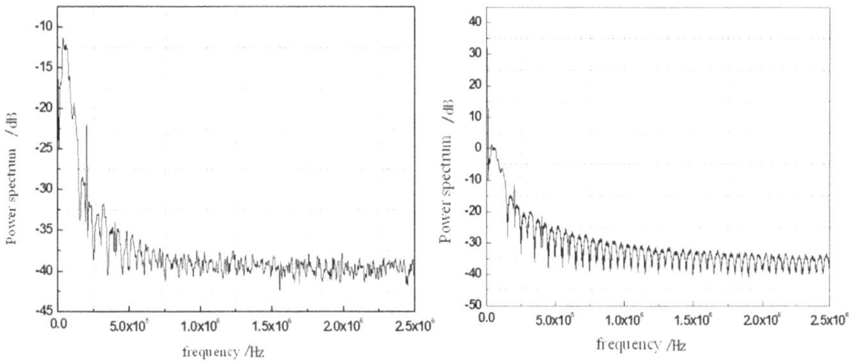

(a) Before predistortion (b) After predistortion

Fig. 3.20 Power spectrum of the NC-QPSK-like sinusoidal modulation signal before and after predistortion

(optical transmitters, optical receivers, amplifiers, etc.) in the signal-transmission path. The constellation diagram in Fig. 3.21a has obvious gain-compression distortion, showing a tendency to be drawn toward the center point on the constellation diagram. This is due to nonlinearities affecting the average power of the transmitted signal, resulting in a reduction in the minimum Euclidean distance between constellation points. The constellation diagram after predistortion correction can be restored to the signal-constellation diagram when it is not affected by nonlinearity.

It can be seen that the NC-QPSK sinusoidal modulation signal has obvious improvements in the time-domain waveform, spectrum, power spectrum, and constellation diagram after predistortion compensation, and also proves the effectiveness of the nonlinear compensation method.

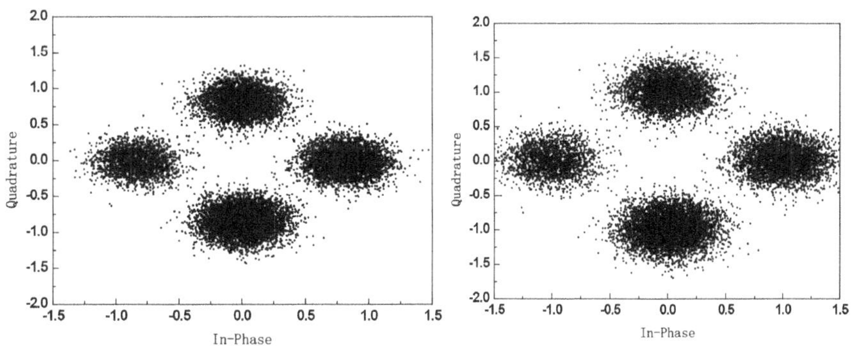

(a) Before predistortion (b) After predistortion

Fig. 3.21 Constellation diagram of an NC-QPSK-like sinusoidal modulation signal before and after predistortion

3.3 Laser Nonlinear Intermodulation Distortion in Subcarrier Modulation

The nonlinear distortion will lead to a decrease of the carrier-to-intermodulation noise ratio and an increase of the bit-error rate of the optical wireless communication (OWC) system. Based on the rate equation of the semiconductor laser, the laser nonlinearity is analyzed using the Bessel-function method. Under a gamma–gamma turbulent channel, the influence of laser nonlinear intermodulation distortion (IMD) on the carrier-intermodulation noise ratio and bit-error rate is discussed for a optical wireless multi-subcarrier binary phase-shift keying (BPSK) modulation communication system.

3.3.1 Nonlinear Intermodulation Distortion of Lasers

The nonlinear distortion of semiconductor lasers is caused by the nonlinearity of the modulation response. In an LD modulation system, the single-mode oscillation-rate equation [9–11] is used to analyze the nonlinearity of the semiconductor laser. The rate equations are as follows:

$$\frac{dS}{dt} = SnB - \frac{S}{\tau_p} \tag{3.73}$$

$$\frac{dn}{dt} = \frac{J}{qV} - SnB - \frac{n}{\tau_s}, \tag{3.74}$$

where n represents the carrier density in the active layer of the semiconductor laser and S represents the photon density in the semiconductor laser cavity. S is proportional to the output light intensity, so it can also be expressed as the relative output power. τ_p is the photon lifetime, τ_s is the carrier lifetime, and $B = \left(1/\tau_p n_s\right)$ is the Einstein coefficient. J is the injected current signal, q is an electron with a charge size equal to 1.6×10^{-19} C, and V is the volume of the electron active area. By eliminating n and dn/dt, and substituting Eq. (3.73) into Eq. (3.74), we obtain

$$\frac{d^2S}{dt^2} + \frac{dS}{dt}\left[\frac{1}{\tau_s} + SB - \frac{1}{S}\frac{dS}{dt}\right] + S\left[S\frac{S}{\tau_p} + \frac{1}{\tau_p\tau_s}\right] - \frac{JSB}{qV} = 0. \tag{3.75}$$

Assuming the photon density [9] $S = C\exp(u)$ (where C is a constant), then Eq. (3.75) becomes

$$\frac{d^2u}{dt^2} + \frac{du}{dt}\left[\frac{1}{\tau_s} + CB\exp(u)\right] + \left[\frac{1}{\tau_p\tau_s} - \frac{JB}{qV} + \frac{CB}{\tau_p}\exp(u)\right] = 0. \tag{3.76}$$

J represents the injected-signal current density:

$$J(t) = J_0 + J\sum_{n=1}^{N}\cos(\omega_n t + \varphi_n(t)), \tag{3.77}$$

where the phase $\varphi_n = \alpha_n\pi$, $\alpha_n = 0, 1$, and ω_n represents the carrier frequency. We substitute Eq. (3.77) into Eq. (3.76) to obtain

$$\frac{d^2u}{dt^2} + \frac{du}{dt}\left[\frac{1}{\tau_s} + CB\exp(u)\right] + \left[\frac{1}{\tau_p\tau_s} - \frac{BJ_0}{qV} - \frac{BJ}{qV}\sum_{n=1}^{N}\cos(\omega_n t + \varphi_n(t)) + \frac{CB}{\tau_p}\exp(u)\right] = 0. \tag{3.78}$$

Equation (3.78) represents a nonlinear equation where $u = \sum_{n=1}^{N}\alpha_n\cos(\omega_n t + \varphi_n(t))$. Using Sonine's expression [9],

$$\exp(\alpha_n\cos\omega t) = I_0(\alpha_n) + 2\sum_{q=1}^{\infty}I_q(\alpha_n)\cos q\,\omega t, \tag{3.79}$$

where $I_q(\alpha_n)$ is the modified q-order Bessel function; then, the photon density becomes

$$S = C\prod_{n=1}^{N}\left[I_0(\alpha_n) + 2\sum_{q=1}^{\infty}I_q(\alpha_n)\cos q\omega t\right]. \tag{3.80}$$

If only the input alternating current [12] is considered, the output amplitude of the equal-amplitude multi-channel subcarrier at frequency ω_r can be obtained using Eq. (3.80) as

$$S_1 = 2CI_1(\alpha)[I_0(\alpha)]^{N-1}. \tag{3.81}$$

Similarly, the amplitude of the intermodulation product factor $2\omega_x - \omega_y$ in the frequency band is

$$S_{21} = 2CI_2(\alpha)I_1(\alpha)[I_0(\alpha)]^{N-2}. \tag{3.82}$$

The amplitude of the in-band intermodulation product factor $\omega_x + \omega_y - \omega_z$ is

$$S_{111} = 2C[I_1(\alpha)]^3[I_0(\alpha)]^{N-3}. \tag{3.83}$$

Similarly, the amplitudes of other intermodulation product factors can be derived from Eq. (3.80).

Figures 3.22 and 3.23 show the second-order intermodulation-distortion product factor and the third-order intermodulation-distortion product amplitude, respectively. The input multi-channel subcarrier frequencies in Fig. 3.22 are f_1 and f_2, and the second-order intermodulation-distortion product factors are $f_2 - f_1$ and $f_1 + f_2$, respectively. The multi-channel subcarrier frequencies input in Fig. 3.23 are f_1, f_2, and f_3, and the third-order intermodulation product factors are $2f_1 + f_2$, $2f_1 - f_2$, $f_1 - 2f_2$, $f_1 + f_2 - f_3$, $f_1 - f_2 - f_3$, and $f_1 + f_2 + f_3$.

If the transmission bandwidth is one octave (the upper-limit frequency of each frequency band is twice as high as the lower-limit frequency; that is, the frequency ratio is 2), the frequency generated by the second-order intermodulation distortion is not within the frequency band, so it is not considered [9]. The intermodulation product factor in the frequency band is only the third-order intermodulation distortion. The second-order intermodulation product and the third-order intermodulation product are generated by $m^2(t)$ and $m^3(t)$, respectively. Then,

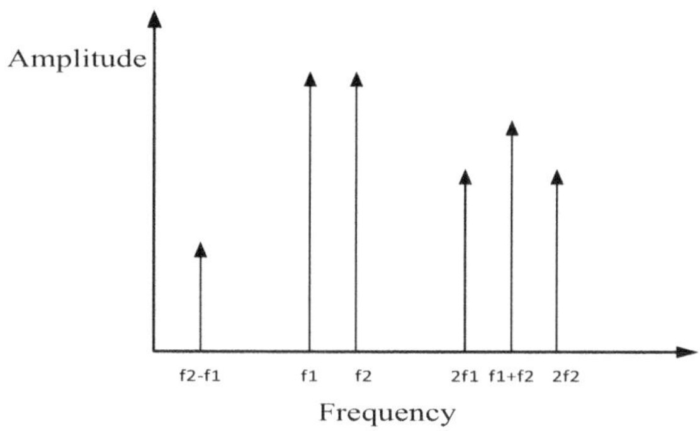

Fig. 3.22 Second-order intermodulation product amplitude

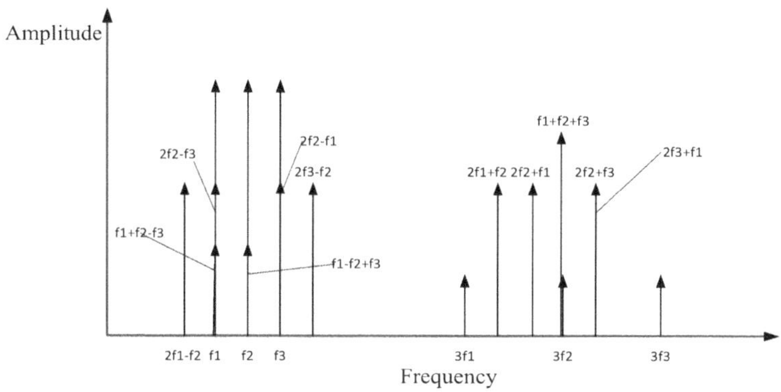

Fig. 3.23 Amplitude of third-order intermodulation products

$$m^2(t) = \frac{1}{2}\sum_{z=l}^{N}\sum_{y=l}^{N}\cos\left[(\omega_y + \omega_z)t + (\theta_y + \theta_z)\right] + \cos\left[(\omega_y - \omega_z)t + (\theta_y - \theta_z)\right] \quad (3.84)$$

$$\begin{aligned} m^3(t) = \frac{1}{4}\sum_{x=1}^{N}\sum_{y=1}^{N}\sum_{z=1}^{N} & \cos\left[(\omega_x + \omega_y + \omega_z)t + (\theta_x + \theta_y + \theta_z)\right] \\ & + \cos\left[(\omega_x + \omega_y - \omega_z)t + (\theta_x + \theta_y - \theta_z)\right] \\ & + \cos\left[(\omega_x - \omega_y + \omega_z)t + (\theta_x - \theta_y + \theta_z)\right] \\ & + \cos\left[(\omega_x - \omega_y - \omega_z)t + (\theta_x - \theta_y - \theta_z)\right]. \end{aligned} \quad (3.85)$$

At frequency ω_r, the number of third-order intermodulation product factors $2\omega_x - \omega_y$ and $\omega_x + \omega_y - \omega_z$ in the transmission band [13] are

$$N_{21} = int\left\lfloor \frac{N-k}{2} \right\rfloor + int\left\lfloor \frac{k-l}{2} \right\rfloor, l \le k \le N \quad (3.86)$$

$$\begin{aligned} N_{111} = \mathrm{int}\left\lfloor \frac{k-l}{2} \right\rfloor int\left\lfloor \frac{k-l-1}{2} \right\rfloor + (k-l)(N-k) \\ + int\left\lfloor \frac{N-k}{2} \right\rfloor int\left\lfloor \frac{N-k-1}{2} \right\rfloor, \ l \le k \le N. \end{aligned} \quad (3.87)$$

In the formula, l and N respectively represent the initial carrier and the total number of carriers; k represents the kth carrier, and $\mathrm{int}\lfloor x \rfloor$ is the rounding function.

From Eqs. (3.86)–(3.87), when $k = l = 1$ and the number of carriers is even, $N_{111} = 0.25(N-2)^2$ and $N_{21} = 0.5(N-2)$; when $k = l = 1$, the number of carriers is odd, so $N_{111} = 0.25(N-1)(N-3)$ and $N_{21} = 0.5(N-1)$. Table 3.4 shows the number of intermodulation products $2\omega_x - \omega_y$ and $\omega_x + \omega_y - \omega_z$ in the transmission band when the number of carriers $N = 16$ and k and l are different (Table 3.5).

Figure 3.24 shows the numbers, N_{111} and N_{21}, of the third-order distortion intermodulation distortion (IMD) product factors ω_{x+y-z} and ω_{2x-y} in the transmission band, where $k = l = 1$. It can be seen from the figure that as the number of subcarriers increases, the number of third-order intermodulation products N_{111} and N_{21} in the frequency band also increases gradually.

Comparing the two curves in the figure, it is found that as the number of subcarriers increases, the number of third-order intermodulation products ω_{x+y-z} in the frequency band increases significantly faster, while the changes in the number of third-order intermodulation products ω_{2x-y} in the frequency band are relatively slow. From the two curves in Fig. 3.24, it can be concluded that the number of intermodulation-product distortion factors is mainly related to the number of intermodulation product factors ω_{x+y-z} in the frequency band. The expression of the third-order intermodulation-distortion amplitude is derived from Eqs. (3.82)–(3.87)

Table 3.4 Number of intermodulation products in the bandwidth ($N = 16$)

	1	2	3	4	5	6	7	8	9	10	11	12	13	14	15	16
N_{2x-y}																
1	7	7	7	7	7	7	7	7	7	7	7	7	7	7	7	7
2		7	6	7	6	7	6	7	6	7	6	7	6	7	6	7
3			6	6	6	6	6	6	6	6	6	6	6	6	6	6
4				6	5	6	5	6	5	6	5	6	5	6	5	6
5					5	5	5	5	5	5	5	5	5	5	5	5
6						5	4	5	4	5	4	5	4	5	4	5
7							4	4	4	4	4	4	4	4	4	4
8								4	3	4	3	4	3	4	3	4
9									3	3	3	3	3	3	3	3
10										3	2	3	2	3	2	3
11											2	2	2	2	2	2
12												2	1	2	1	2
13													1	1	1	1
14														1	0	1
15															0	0
16																0

Table 3.5 Number of intermodulation products $\omega_x + \omega_y - \omega_z$ within the bandwidth ($N = 16$)

k	1	2	3	4	5	6	7	8	9	10	11	12	13	14	15	16
N_{x+y-z}																
1	49	56	62	67	71	74	76	77	77	76	74	71	67	62	56	49
2		42	49	54	59	62	65	66	67	66	65	62	59	54	49	42
3			36	42	47	51	54	56	57	57	56	54	51	47	42	36
4				30	36	40	44	46	48	48	48	46	44	40	36	30
5					25	30	34	37	39	40	40	39	37	34	30	25
6						20	25	28	31	32	33	32	31	28	25	20
7							16	20	23	25	26	26	25	23	20	16
8								12	16	18	20	20	20	18	16	12
9									9	12	14	15	15	14	12	9
10										6	9	10	11	10	9	6
11											4	6	7	7	6	4
12												1	4	4	4	1
13													1	2	2	1
14														0	1	0
15															0	0
16																0

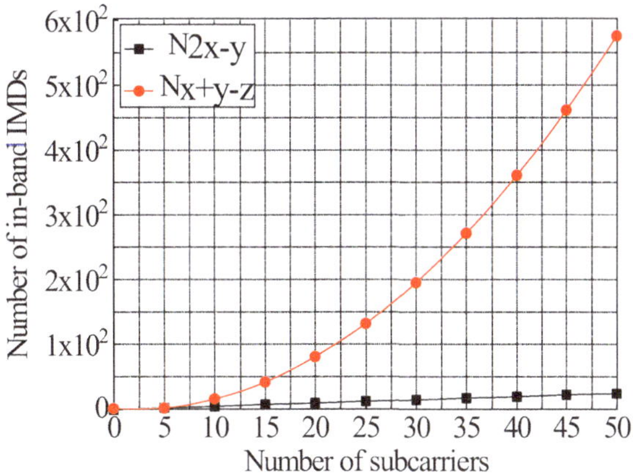

Fig. 3.24 Number of IMDs in the band

as

$$IMD3 = S_{21}N_{21} + S_{111}N_{111}. \tag{3.88}$$

The expression [14] of the variance of the third-order intermodulation distortion in the frequency band is

$$\sigma_{IMD3}^2 = \frac{1}{32}(\eta P_r)^2 m^6 (N_{21} + N_{111}), \tag{3.89}$$

where m is the laser modulation index, η is the quantum efficiency of the photodetector, and P_r is the average received optical power.

Because the power consumption is directly related to the third-order intermodulation distortion variance, it can effectively describe the strength of the intermodulation distortion power, and the expression is [15]

$$r_{imd} = 10\log\left(1 + \frac{\sigma_{IMD3}^2}{\sigma^2}\right) = 10\log\left(1 + \frac{\sigma_{IMD3}^2}{\sigma_{Sh}^2 + \sigma_{Th}^2}\right),$$

where σ_{IMD3}^2 is the third-order intermodulation distortion variance, and σ_{Sh}^2 and σ_{Th}^2 are the scattering and thermal noise, respectively.

Figure 3.25 shows the power-consumption characteristics of laser nonlinear third-order intermodulation distortion under different modulation indices. In the figure, the additive noise power is $\sigma^2 = 10^{-2}$, $\eta = 0.8$, the average received optical power is $P_r = 5.8$ dBm, $k = l = 1$, and the number of carriers N is 8, 12, and 16.

As can be seen from Fig. 3.25, with the increase of the modulation index, the power consumption of the intermodulation distortion gradually increases; this is because the size of the modulation index directly affects the variance of the third-order intermodulation distortion. The intermodulation-distortion variance increases; therefore, the intermodulation-distortion power consumption also increases.

Fig. 3.25 IMD power consumption

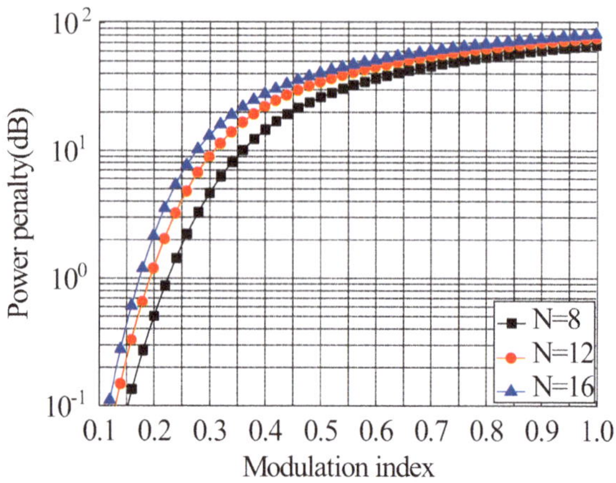

When the modulation index m reaches 0.6, the power consumption of the intermodulation distortion gradually stabilizes and reaches a maximum value of about 80 dB. As the number of carriers N increases, the power consumption of the intermodulation distortion also increases gradually. After the power consumption reaches saturation, the number of carriers has little effect on the power consumption of the IMD and tends to stabilize.

3.3.2 Nonlinear Intermodulation-Distortion Characteristics of Subcarrier Modulation

Under a gamma–gamma turbulent channel, the influence of nonlinear intermodulation distortion on the carrier intermodulation noise ratio and bit-error rate is studied for a optical wireless multi-subcarrier BPSK-modulation communication system. Its system block diagram is shown in Fig. 3.26.

The gamma–gamma probability-density distribution function is defined as

$$p(I) = \frac{2(\alpha\beta)^{\frac{\alpha+\beta}{2}}}{\Gamma(\alpha)\Gamma(\beta)} I^{\frac{\alpha+\beta}{2}-1} K_{\alpha-\beta}\left(2\sqrt{\alpha\beta I}\right), I > 0, \tag{3.90}$$

where $\Gamma()$ is the gamma function, $K_n()$ is the order-n modified Bessel function of the second kind, and α and β are the outer and inner scale parameters, respectively. SI is the scintillation index, defined as [16]

$$SI = \frac{1}{\alpha} + \frac{1}{\beta} + \frac{1}{\alpha\beta}. \tag{3.91}$$

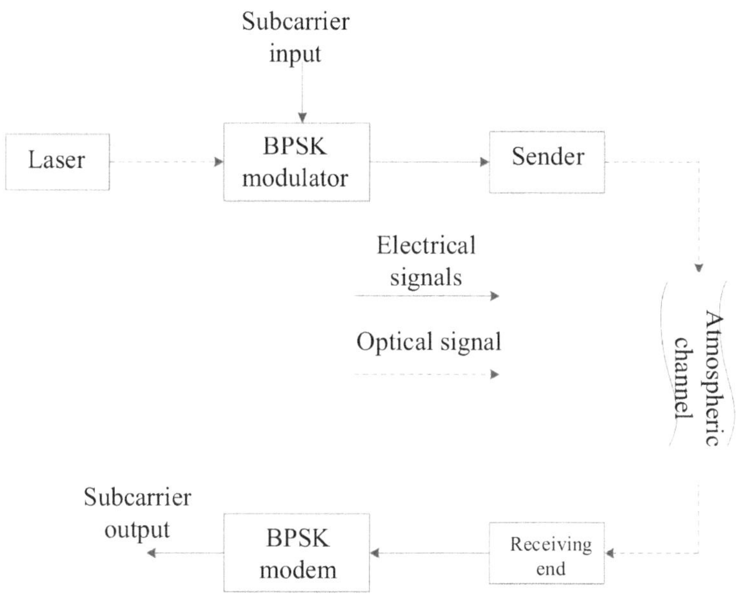

Fig. 3.26 Subcarrier BPSK modulation system

In a subcarrier-modulation wireless laser-communication system, the output power and light-intensity distribution of the laser are respectively

$$P(t) = P_T\, exp\left(m \sum_{n=1}^{N} \cos(\omega_n t + \varphi(t))\right) \tag{3.92}$$

$$I(t) = I\left[1 + mx(t) + a_2 m^2 x^2(t) + \cdots + a_i m^i x^i(t) + \cdots\right]. \tag{3.93}$$

In the formula, $\{a_n\}_{n=2}^{\infty}$ is the nonlinear coefficient of the laser, which is generally a constant. If there is no nonlinearity in the laser, $\{a_n\}_{n=2}^{\infty}$ equals zero. When using BPSK to modulate multiple subcarriers, the modulation function is $x(t) = \sum_{n=1}^{N} \cos(\omega_i t + \varphi_i)$. Considering that the transmission bandwidth is one octave, the nonlinear second-order intermodulation distortion of the laser is not within the frequency band, and only the third-order intermodulation distortion is considered for intermodulation products within the frequency band.

Because the light intensity flickers very slowly, only the alternating current signal is considered when the received light signal is photoelectrically converted into an electrical signal [11]:

$$i(t) = \eta\, Im\left(\sum_{n=1}^{N} \cos(\omega_n t + \varphi_n)\right) + n(t). \tag{3.94}$$

Substituting Eq. (3.93) into Eq. (3.94), the carrier power and additive white-noise power are obtained as

$$S_P = \frac{\eta^2 I^2}{2}\left[m + \frac{3}{4}a_3 m^3 (2N - 1)\right]^2 \quad (3.95)$$

$$\sigma^2 = \sigma_{Sh}^2 + \sigma_{Th}^2. \quad (3.96)$$

3.3.2.1 Performance Analysis of the Carrier-to-Intermodulation Noise Ratio Under Non-turbulent Conditions

The carrier-signal-to-intermodulation-distortion-and-noise ratio (CINR) expression at the receiving end of an OWC system without considering turbulence is [16]

$$CINR = \frac{S_P}{\sigma^2 + \sigma_{IMD3}^2}. \quad (3.97)$$

From Eqs. (3.88), (3.95), and (3.96), the CINR expression at the receiving end is deduced as

$$CINR = \frac{S_P}{\sigma^2 + \sigma_{IMD3}^2} = \frac{16R^2 I^2 \left[m + \frac{3}{4}a_3 m^3 (2N - 1)\right]^2}{32\sigma^2 + (\eta P_r)^2 a_3^2 I^2 m^6 (N_{21} + N_{111})}. \quad (3.98)$$

Let $d(CINR)/dm = 0$; the optimal value of the modulation index is obtained as

$$m = \left[\frac{16\sigma^2}{a_3^2 (RP_r)^2 (N_{21} + N_{111})}\right]^{1/6}. \quad (3.99)$$

Figure 3.27 shows the relationship between the modulation index and the system-carrier intermodulation noise ratio under the third-order intermodulation products of different frequency combinations. In the simulation, the number of carriers is $N = 16$, the noise power is $\sigma^2 = 10^{-2}$, the third-order nonlinear coefficient is $a_3 = 0.01$, and $k = l = 1$.

As can be seen from Fig. 3.27, if only the influence of the third-order intermodulation $2\omega_x - \omega_y$ on the communication system is considered, when the modulation index is $m = 1.0$, the CINR reaches the maximum value of 56.1 dB; when considering the third-order intermodulation $\omega_x + \omega_y - \omega_z$, when the modulation index is $m = 0.6$, the CINR reaches the maximum value of 13.6 dB.

Comparing the two curves in the figure, it is found that when the modulation index is $m = 0.6$, the CINR under the influence of the nonlinear third-order intermodulation product factor $\omega_x + \omega_y - \omega_z$ of the laser is 13.6 dB, while the CINR corresponding

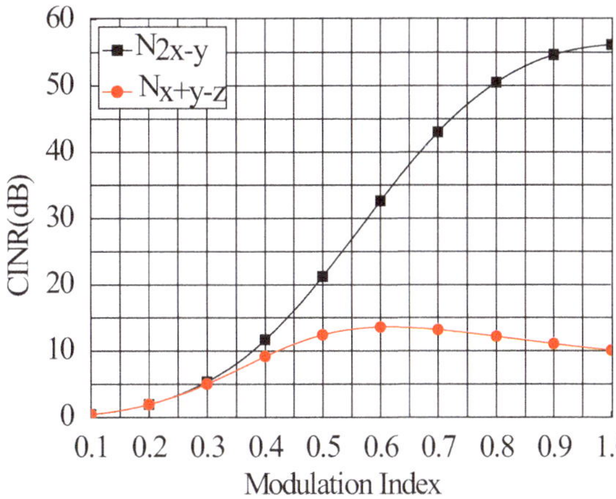

Fig. 3.27 Carrier intermodulation noise without turbulence. Relationship between the modulation-to-noise ratio and modulation index [17, 18]

to the third-order intermodulation product factor $2\omega_x - \omega_y$ is 33 dB; the former is intermodulation distortion.

The former has a significantly greater impact on the system than the latter. This is because the influence of nonlinear intermodulation distortion on the system mainly depends on the number of intermodulation product factors in the frequency band. When the number of carriers is $N = 16$, the number of third-order intermodulation products is $N_{111} \gg N_{21}$; therefore, the CINR of the system under the influence of $\omega_x + \omega_y - \omega_z$ nonlinear third-order intermodulation products is small.

Figure 3.28 shows the carrier-intermodulation noise ratio when the third-order nonlinear coefficients are different. In the simulation, a_3 is taken as 0.01, 0.1, and 1.0, the noise power is $\sigma^2 = 10^{-2}$, the number of carriers is $N = 16(N_{21} = 7, N_{111} = 49)$, and $k = l = 1$.

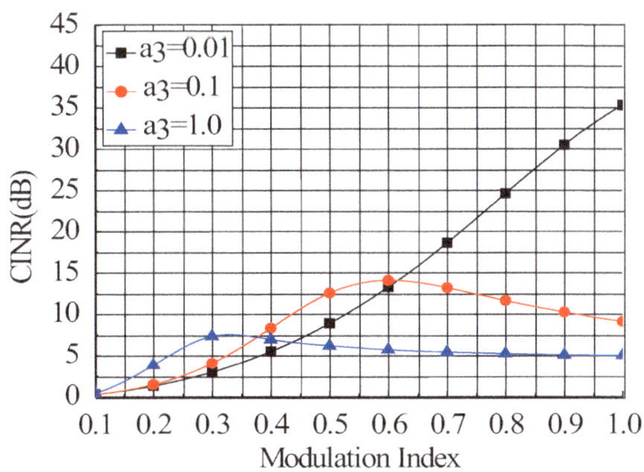

Fig. 3.28 Carrier intermodulation ratio under different third-order nonlinear coefficients $\left(N_{2x-y} = N_{21}, N_{x+y-z} = N_{111}\right)$ [17, 18]

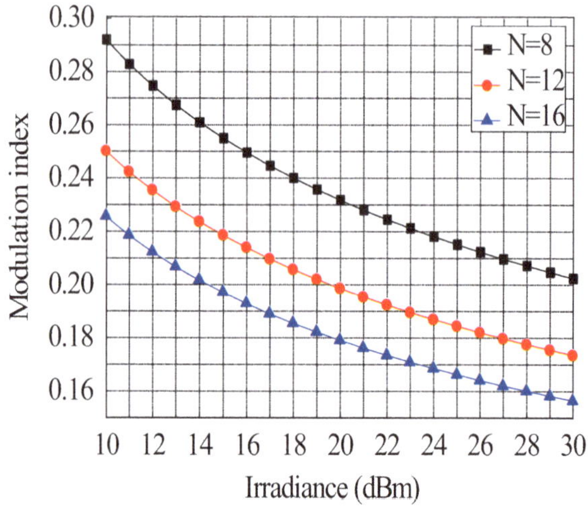

Fig. 3.29 Relationship between the modulation index and light intensity under different carrier numbers [17, 18]

It can be seen from Fig. 3.28 that with the increase of the nonlinear coefficient, the optimal modulation index corresponding to the maximum carrier-intermodulation noise-to-noise ratio decreases. When $a_3 = 0.1$, the modulation index is $m = 0.6$, and the CINR reaches the maximum value of 14.1 dB; when $a_3 = 1.0$, the modulation index is $m = 0.3$, and the CINR reaches the maximum value of 7.35 dB. When the third-order nonlinear coefficient is $a_3 = 0.01$, the modulation index is higher than 1.0, and a larger optical power output can be obtained, which increases the CINR; however, the modulation index is too large, which will cause clipping distortion.

Figure 3.29 shows the curve relationship between the modulation index and light intensity under three different carrier numbers. In the figure, the carrier numbers N are 8, 12, and 16. Comparing the three curves in the figure, it can be seen that under the same light intensity, the larger the number of carriers, the smaller the modulation index. With the same number of carriers, the light intensity gradually increases, and the modulation index gradually decreases, so that the transmission is within the dynamic range.

3.3.2.2 Performance Analysis of the Downstream Intermodulation Noise Ratio of Atmospheric-Turbulence Channels

When considering the influence of atmospheric turbulence on optical wireless communication (OWC), the system-carrier intermodulation noise ratio is recorded as $CINR_{ave}$. It can be seen from Eq. (3.27) that the expression [19, 20] is

$$CINR_{ave} = \int_0^\infty CINR \times p(I)dI. \tag{3.100}$$

The gamma–gamma model [16, 21, 22] is used for the distribution of atmospheric turbulent light-intensity fluctuations. This model is different from the lognormal distribution model and has a wider range of applications. It is not only applicable to weak turbulence, but also to moderate and strong turbulence; then, Eq. (3.100) becomes

$$CINR_{ave} = \frac{32R^2\left[m + \frac{3}{4}a_3m^3(2N-1)\right]^2(\alpha\beta)^{\frac{\alpha+\beta}{2}}}{\Gamma(\alpha)\Gamma(\beta)} \times \int_0^{\infty} \frac{I^{\frac{\alpha+\beta}{2}+1}}{32\sigma^2 + Km^6I^2a_3^2} K_{\alpha-\beta}(2\sqrt{\alpha\beta I})dI, \tag{3.101}$$

where $K = (\eta P_r)^2(N_{111} + N_{21})$.

As shown in Fig. 3.30, the carrier-to-intermodulation noise ratio under different turbulent conditions is simulated according to Eq. (3.101). We considered $\sigma^2 = 10^{-2}$, $k = l = 1$, $a_3 = 0.1$, and $N = 16(N_{21} = 7, N_{111} = 49)$. Under weak turbulence conditions, $\alpha = 11.6$, $\beta = 10.1$ ($SI = 0.2$), the modulation index $m = 0.6$, and the CINR reaches the maximum value of 12.9 dB. Under moderate turbulence conditions, $\alpha = 4.0$, $\beta = 1.9$ ($SI = 0.9$), and when the modulation index is $m = 0.55$, the CINR reaches the maximum value of 10.6 dB. Under strong turbulence conditions, $\alpha = 1.2$, $\beta = 1.1$ ($SI = 2.5$), the modulation index is $m = 0.43$, and the CINR reaches the maximum value of 8.9 dB.

It can be seen from Fig. 3.30 that under the same turbulent conditions, when the modulation index is less than the optimal value, the CINR increases with the increase of the modulation index; when the modulation index is greater than the optimal value, the CINR increases as the modulation index decreases, indicating the influence of laser nonlinear intermodulation distortion on the system.

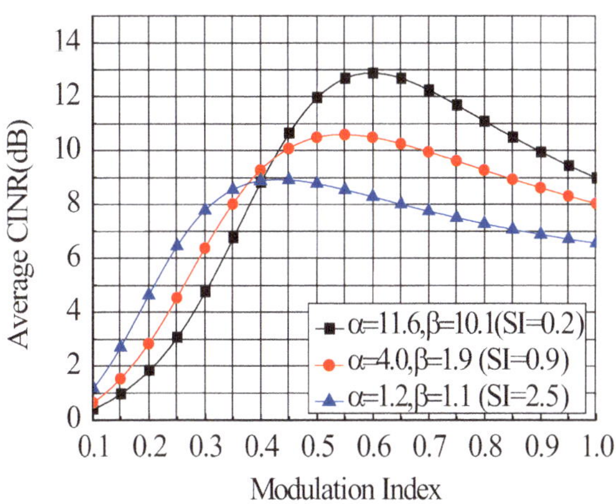

Fig. 3.30 Carrier-intermodulation noise ratio under atmospheric turbulence

The carrier-intermodulation noise ratio is closely related to the selection of the modulation index. Comparing the three curves in the figure, it can be seen that when the modulation index is higher than 0.38, the scintillation index is greater under a certain modulation index, the CINR is smaller, and the maximum carrier-to-noise ratio obtained by weak turbulence is about 4.0 dB higher than that of strong turbulence. Therefore, as the scintillation index increases, the optimal modulation-index selection corresponding to the maximum CINR of the system needs to be smaller.

Figure 3.31 analyzes the carrier-to-intermodulation noise ratio for different numbers of carriers. Let $\sigma^2 = 10^{-2}$ and k = l = 1. When considering the number of carriers as N = 8, N = 12, and N = 16, the obtained carrier-intermodulation noise ratio is shown in Fig. 3.31. When the number of carriers N = 8, the modulation index m = 0.75, and the CINR reaches the maximum value of 17.47 dB. When the number of carriers N = 12, the modulation index m = 0.65, and the best CINR value is 14.2 dB. When the number of carriers N = 16, the modulation index m = 0.6, and the best CINR value is 12.5 dB.

By analyzing the three curves in the figure, it can be seen that with the increase in the number of carriers, the carrier-intermodulation noise ratio decreases. This is because when the number of carriers increases, the intermodulation product factor in the frequency band increases, resulting in an increase in the variance of the intermodulation distortion and a decrease in the carrier-to-intermodulation noise ratio.

Figure 3.32 shows the curve relationship between the maximum CINR and the number of carriers N, where the number of carriers N varies from 8 to 16. In the figure, (SI, a_3) is set as (0.2, 1.0), (0.2, 0.3), (0.9, 1.0), and (0.9, 0.3). When the

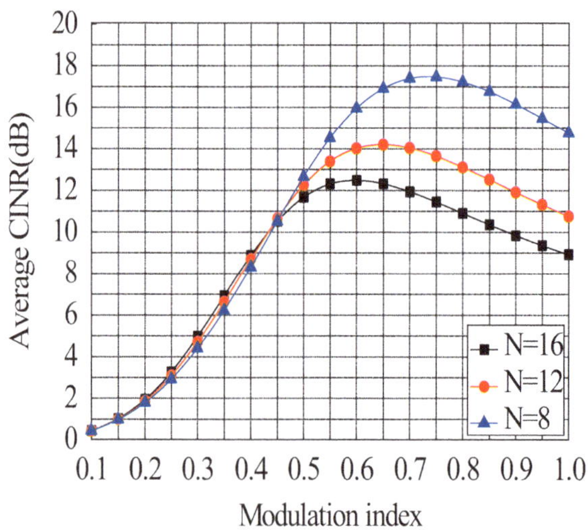

Fig. 3.31 Carrier-intermodulation noise ratio under different numbers of carriers

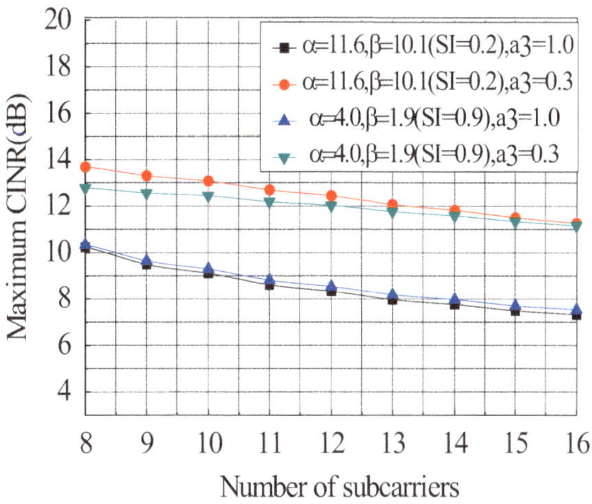

Fig. 3.32 Maximum carrier-intermodulation noise ratio under different nonlinear coefficients

turbulence intensity is the same and the third-order nonlinear system is $a_3 = 0.3$, the maximum CINR is higher than that of the corresponding maximum CINR when the linear coefficient is $a_3 = 1.0$. This is because the larger the nonlinear coefficient, the stronger the nonlinearity of the laser, resulting in a smaller CINR.

When the third-order nonlinear coefficient is the same, when the light-intensity scintillation index $SI = 0.2$, the maximum CINR of the OWC system is significantly higher than that when the scintillation index $SI = 0.9$. This is because the stronger the light-intensity scintillation of the turbulent channel, the stronger the noise of the system, and the lower the corresponding maximum CINR. For the same (SI, a_3), the larger the number of carriers N, the smaller the maximum CINR of the system.

3.3.2.3 Bit-Error Rate Performance Analysis

Under an atmospheric-turbulent channel with an OWC system using the nonlinear intermodulation distortion of the laser generated by BPSK modulation, the bit-error rate (BER) can be expressed as [23, 24]

$$\begin{aligned} P_e(m) &= \int_0^\infty Q(\sqrt{CINR})p(I)dI \\ &= \int_0^\infty Q(\sqrt{CINR})\frac{2(\alpha\beta)^{\frac{\alpha+\beta}{2}}}{\Gamma(\alpha)\Gamma(\beta)}I^{\frac{\alpha+\beta}{2}-1}K_{\alpha-\beta}(2\sqrt{\alpha\beta I})dI. \end{aligned} \tag{3.102}$$

Figure 3.33 shows the BER characteristic curve of the subcarrier BPSK modulated optical wireless communication system under different turbulence intensities. In the

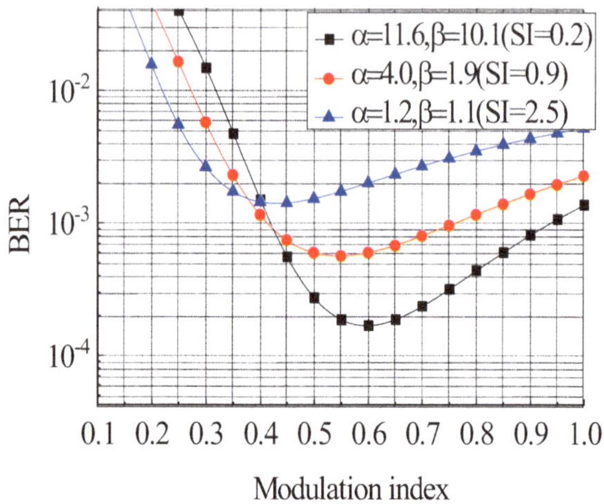

Fig. 3.33 Bit-error rate under atmospheric turbulence

simulation, $\sigma^2 = 10^{-2}$, $k = l = 1$, $a_3 = 0.1$, and $N = 16(N_{21} = 7, N_{111} = 49)$. When the modulation index is less than the optimal value, the bit-error rate decreases with the increase of the modulation index, indicating that the nonlinear distortion has little influence on the system performance; however, once the modulation index is greater than the optimal value, the nonlinear distortion of the laser has a serious impact on the system. The modulation index increases and the bit-error rate increases.

It can be seen from Fig. 3.33 that when the modulation index is equal to 0.38, the bit-error rates under the three turbulence conditions are all 1.5×10^{-3}; when the modulation index is >0.38, under the same modulation index, the larger the scintillation index, the higher the BER. A high BER seriously affects the stability and reliability of the communication system. With the further increase of the modulation index, the nonlinear distortion is saturated, and the bit-error rate also tends to be saturated.

Figure 3.34 shows the bit-error rate versus the average received optical power under different turbulence conditions: $a = 11.6$ and $\beta = 10.1$ (SI $= 0.2$) for weak turbulence; $a = 4.0$ and $\beta = 1.9$ (SI $= 0.9$) for moderate turbulence; and $a = 1.2$ and $\beta = 1.1$ (SI $= 2.5$) for strong turbulence.

It can be clearly seen from the figure that under weak turbulence, when the BER reaches 10^{-4}, the corresponding average received optical power is equal to -7 dBm; the average received optical power under moderate turbulence is -6.3 dBm, which is higher than that of weak turbulence. Under strong turbulence, the average received optical power is -5.7 dBm. The average received power efficiency is the highest under weak turbulence conditions, resulting in the smallest BER and significantly improved communication quality.

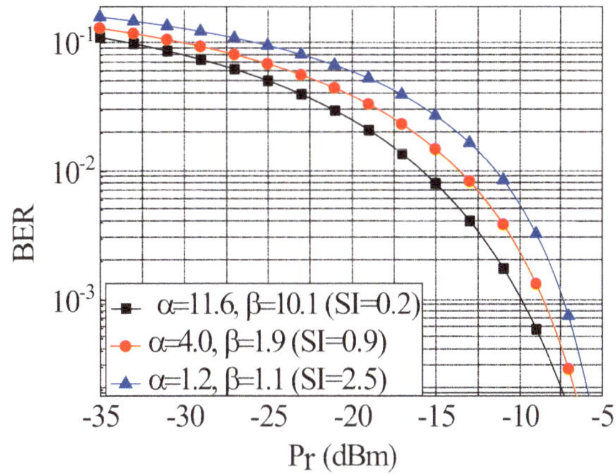

Fig. 3.34 Relationship between BER and P_r [17, 18]

3.3.2.4 Influence of Nonlinearity on the Constellation and Eye Diagrams

As shown in Fig. 3.35, taking a nonlinear distorted carrier with a modulation degree of 0.98 as an example, the correlation coefficient between the distorted carrier and the standard carrier is 0.91. The waveforms and spectra of the distorted carrier and the standard carrier are shown in the figure below. Affected by the nonlinearity of the laser, the harmonics of the distorted carrier are large.

After the QPSK signal is modulated by the distorted carrier, the demodulated constellation diagram is shown in Fig. 3.36. Owing to the large harmonic components, the useful power after the demodulated low-pass filter is reduced. Nonlinearity can cause the constellation diagram to deviate from its original position, but generally does not cause bit errors. As shown in Fig. 3.37, nonlinear distortion is reflected in the eye diagram of the demodulated signal, and the eye opening of the distorted signal is smaller than that of the normal signal. As the degree of modulation increases, the opening of the eye diagram decreases.

3.4 Semiconductor-Laser Power Control

Because the threshold current and external-differential quantum efficiency of semiconductor lasers vary with temperature and device aging, the output power of a laser operating at a given bias current will decrease accordingly. To maintain the stability of the LD output optical power, automatic power control (APC) and automatic temperature control (ATC) are typically used for protection. The APC circuit ensures that the optical signal emitted by the LD is stable within the normal range; the ATC circuit ensures that the LD has a long-term stable working environment [25].

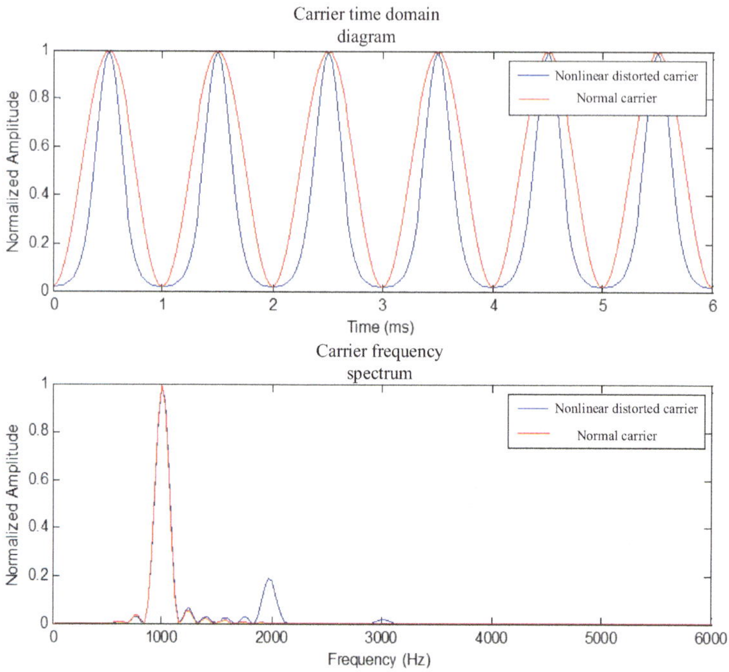

Fig. 3.35 Time domain and spectrogram of the distorted carrier

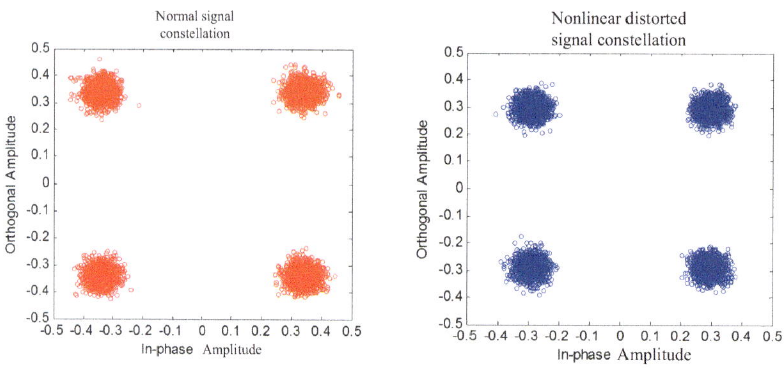

(a) Normal signal constellation (b) Nonlinear-distorted signal constellation

Fig. 3.36 Comparison of demodulation constellation diagrams [17, 18]

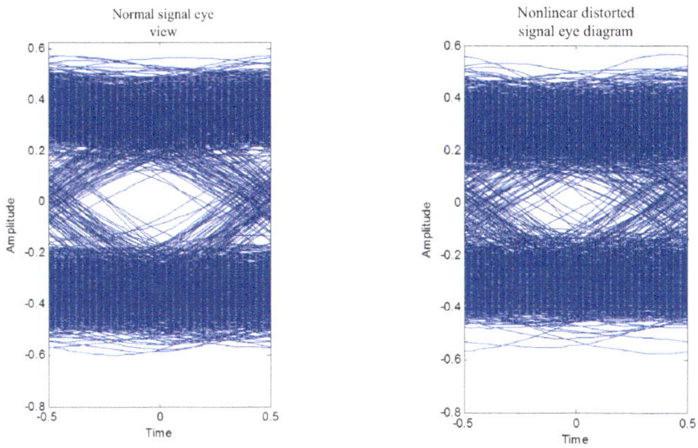

(a) Eye diagram of a normal signal (b) Eye diagram of a nonlinear distorted signal

Fig. 3.37 Eye-diagram comparison between a normal signal and nonlinear distorted signal [17, 18]

3.4.1 Effect of Temperature on Semiconductor Lasers

Owing to the junction structure of semiconductor lasers and the sensitivity of semiconductor materials to temperature, temperature has a great influence on the performance of semiconductor lasers. This influence is mainly reflected in the power stability, which basically affects their wavelength, threshold current, and output optical power, thereby affecting their working performance.

3.4.1.1 Influence of Temperature on the Laser Wavelength

The output wavelength of a semiconductor laser depends on the structural parameters of the laser itself, which are theoretically invariable. However, external factors, such as the operating current and temperature, will change the output wavelength. When a semiconductor laser is energized, its current density increases continuously; the diode will generate a thermal effect, which causes the temperature of the active area to rise.

Studies have shown that each mA change in current causes a shift of about 0.02 nm in the output wavelength [26]. In addition, because the forbidden bandwidth of the active-layer material narrows with the increase of temperature—that is, the parameters of the laser optical resonator change—the laser wavelength shifts. These parameters are material related.

In general, when the driving current is constant, the output wavelength of the semiconductor laser will be red-shifted by 0.2–0.3 nm for every 1 °C increase in temperature [27]. Therefore, to stabilize the output wavelength of the semiconductor laser, its operating temperature needs to be controlled within a suitable range.

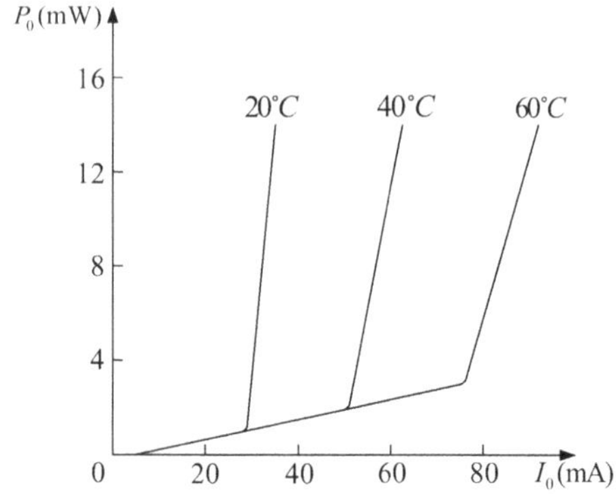

Fig. 3.38 P–I characteristics and temperature change curve of a semiconductor laser

3.4.1.2 Influence of Temperature on the Laser Threshold

As the temperature increases, the threshold current of the semiconductor laser also increases. The relationship between the temperature and threshold current can usually be expressed as

$$I_{\text{th}}(T) = I_{\text{th}}(T_0) \times \exp\left(\frac{T - T_r}{T_0}\right), \tag{3.103}$$

where T_r is the indoor temperature; T_0 is the characteristic temperature of the semiconductor laser; $I_{th}(T)$ is the threshold current when the temperature is T; $I_{th}(T_0)$ is the threshold current when the temperature is T_r. The characteristic temperature T_0 is an important parameter to characterize the temperature stability of a semiconductor laser. It is constant and depends on the laser material and device structure. The higher the characteristic temperature T_0, the lower the sensitivity and the better the laser stability.

The relationship between the temperature and threshold current can be directly observed in Fig. 3.38. It can be seen from the figure that the change in temperature has a significant impact on the threshold characteristics of the semiconductor laser. With the increase in temperature, the threshold current gradually increases. It can also be seen from the figure that the output power of a laser working under a given bias current will decrease with the increase in temperature; that is, the increase in temperature will degrade the P–I characteristics [28].

3.4.1.3 Influence of Temperature on the Laser Output Optical Power

The variation of the external-differential quantum efficiency and the threshold current with the temperature are the main elements that affect the variation of the laser output optical power. The output optical power of a laser with better linearity can be expressed as

$$P_0 = \frac{\eta_D \cdot hv}{e_0}(I - I_{\text{th}}), \tag{3.104}$$

where P_0 is the output optical power; η_D is the external micro-molecular weight efficiency; h is Planck's constant; v is the optical frequency; e_0 is a single electron; I is the injection current; and I_{th} is the threshold current. From the literature [29–38], it can be seen that the change of the external-differential quantum efficiency with temperature is not very obvious. For example, for GaAs lasers, η_D is approximately 50% at 77 K absolute temperature and approximately 30% at 300 K. It can be seen from Eq. (3.104) that the stability of the injection current and the threshold current directly affects the stability of the output optical power [28].

3.4.1.4 Failure Mechanism

Using lasers correctly under appropriate conditions can prolong their service life. However, lasers are high-efficiency energy-conversion devices, which require very high–quality working-environment and storage conditions. If used improperly, they are very easy to damage. The failure mechanisms of semiconductor lasers can be divided into internal and external modes [39].

Internal-mode failures mainly include cavity-surface damage degradation, electrode degradation in sintered parts, dark-line defects, etc. These are mainly related to the laser's fabrication process, materials, and structure. In addition, the temperature is the main reason for internal-mode failures. An increase in temperature will accelerate the oxidation of the cavity surface and the degradation of dark lines.

External-mode failures are mainly due to a sudden failure caused by the destruction of the cleavage plane of the cavity surface or the breakdown of the PN junction. The main reason for these sudden failures is an inrush (or surge) current. Generally, more than 50% of the sudden damage in semiconductor lasers is caused by inrush currents. An inrush current refers to a sudden current pulse or transient voltage with a duration of more than 8.4 ms. Inrush-current can be divided into three types: DC damage, electrostatic breakdown, and surge breakdown.

An inrush current will damage the laser instantly. To better eliminate the impact of a surge in the circuit, the following points should be observed in circuit design and experiments:

(1) Ensure that the pins of the laser or other electronic devices are in good contact, to avoid damage from time circuits;
(2) Do not plug or unplug powered electrical devices to prevent instantaneous switching of the power supply;
(3) Select qualified components to eliminate the influence of sudden changes in the electronic components' parameters and lightning induction [40].

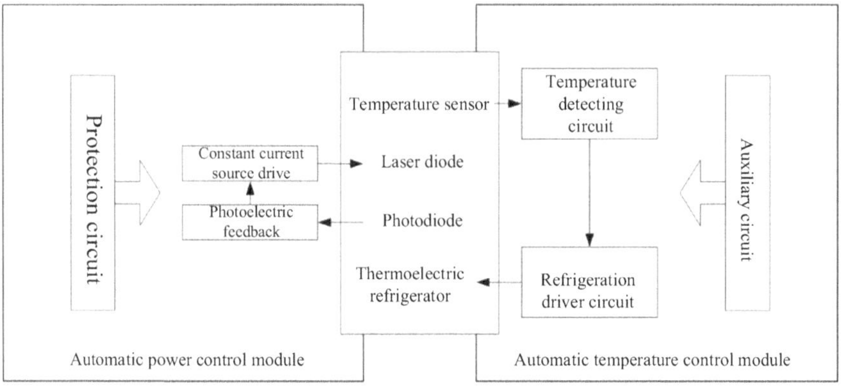

Fig. 3.39 Overall system block diagram

3.4.2 Power Control System

The semiconductor-laser power-control system can be divided into two parts: automatic power control and automatic temperature control. The overall block diagram of the system is shown in Fig. 3.39.

3.4.2.1 Principle of Photoelectric Feedback

In a circuit control system, a part or all of the system output (parameters such as voltage or current) is applied to the input loop of the system through a specific circuit form. This method is called feedback. The feedback function increases the net input as positive feedback; on the other hand, it reduces the net input as negative feedback. Negative feedback technology greatly reduces the gain of the op amp; however, its advantages are obvious. It can improve the stability, frequency bandwidth, signal-to-noise ratio, nonlinear distortion, etc., of the amplifier circuit, and is widely used [41].

A commonly used semiconductor laser packages a semiconductor laser diode (LD) with a photodiode (PD). The PD can convert the optical power signal of the LD into a current signal. Because the laser emitted by the LD includes forward and backward lasers, the wavelength and optical power of the forward and backward lasers are exactly the same; thus, the forward optical power of the LD can be obtained by measuring the backward laser [42]. Lasers can be divided into coaxial packages, butterfly packages, and dual in-line packages. The connections between the LD and PD are shown in Fig. 3.40.

The automatic optical power control has three main functions: one is to automatically track the bias current of the LD to ensure that the laser bias is in the best state; another is to control the efficiency of the P–I curve; the third is to automatically control the average power and peak-power detection method.

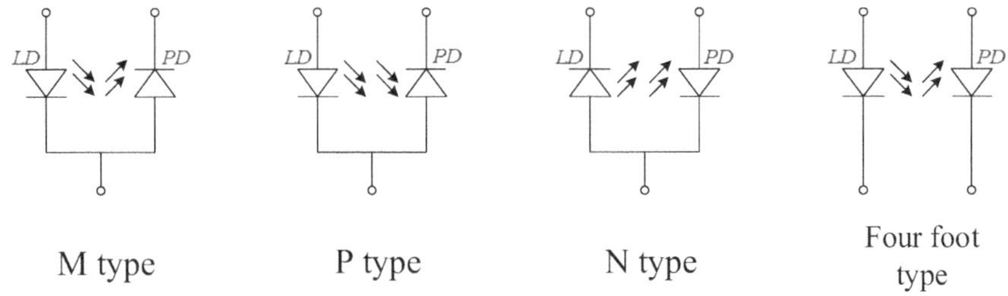

Fig. 3.40 Methods of connecting the LD and PD in a laser

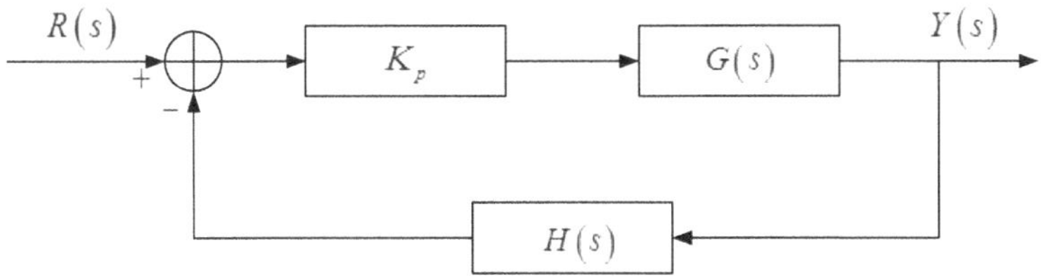

Fig. 3.41 Block diagram of the automatic power-control system

Each method has its own advantages and disadvantages for different use environments. The first method can use photoelectric feedback to monitor the LD output optical power, and adjust the injection current through feedback control to compensate for the fluctuation of the laser output power [43]. Figure 3.41 shows a block diagram of the automatic power-control system. $R(s)$ is the expected control quantity of the system; $Y(s)$ is the actual output quantity of the system; K_P is the proportional controller; $G(s)$ is the controlled object; and $H(s)$ is the feedback device.

In Fig. 3.41, the controlled object $G(s)$ is an LD, and the feedback device $H(s)$ is a photodiode packaged in a semiconductor-laser assembly. It detects a part of the optical power that linearly reflects the change of the output optical power from the back-light intensity of the laser chip. It is converted into electrical signals by photoelectric conversion. The signal is compared with the reference voltage $R(s)$, and then the DC bias current of the laser is controlled by the comparison controller K_P; therefore, the output optical power of the LD can be stabilized by changing the magnitude of the bias current.

3.4.2.2 Automatic Power-Control Circuit

The automatic power-control (APC) circuit includes a comparative-amplification module, a constant current-source module, and an optical-power feedback module. They are mainly composed of INA114 instrumentation amplifiers, OPA547 operational amplifiers, and other devices from the TI Company in the United States. The circuit has high integration, few components, and stable performance. In addition,

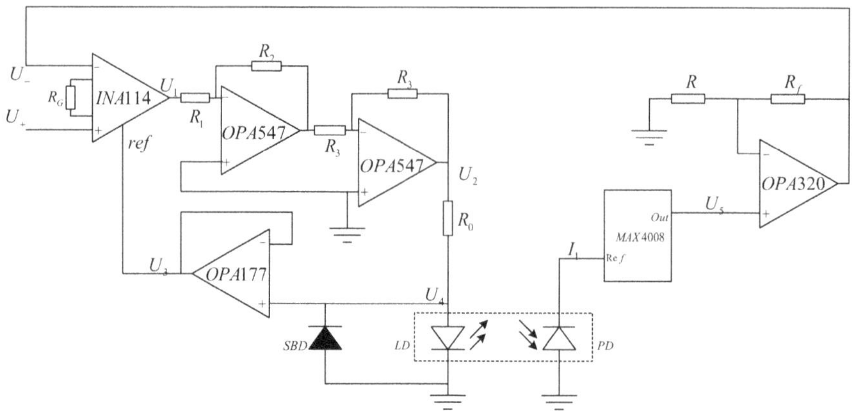

Fig. 3.42 Schematic diagram of the automatic power-control circuit design

to ensure the safe use of semiconductor lasers, a power-supply circuit and protection circuit are also designed in the APC circuit [44]. The internal structure of an M-type semiconductor laser is used, and the circuit schematic diagram is shown in Fig. 3.42.

The comparative amplification module is mainly composed of INA114, which can be observed from Fig. 3.42:

$$G = 1 + (50K\Omega/R_G) \tag{3.105}$$

$$U_1 = G(U_+ - U_-) + U_3. \tag{3.106}$$

In Eqs. (3.105) and (3.106), G is the magnification of the comparator amplifier, R_G is the external amplifier resistor, U_+ is the reference voltage, U_- is the optical-power feedback voltage, and U_1 is the output voltage of the comparator amplifier. The constant current source is formed with OPA547 and OPA177 as the core to maintain a constant current driving the semiconductor laser. The first stage of the OPA547 is a proportional link, and the second stage is an inverter; then,

$$U_2 = \frac{R_2}{R_1} U_1. \tag{3.107}$$

In the equation, R_2/R_1 is the scale factor K_p. OPA177 is a voltage follower; therefore,

$$U_3 = U_4. \tag{3.108}$$

Owing to its low bias current, the current flowing to the OPA177 is negligible. As can be seen from Fig. 3.42, the current flowing through the semiconductor laser is equal to the current flowing through the current-limiting resistor R_0. Thus, the working current of the semiconductor laser is

$$I_f = (U_2 - U_4)/R_0. \tag{3.109}$$

We substitute Eqs. (3.106), (3.107), and (3.108) into (3.109) to obtain

$$I_f = G(U_+ - U_-)/R_0. \tag{3.110}$$

It can be observed from Eq. (3.109) that the bias current flowing through the semiconductor laser is only related to the feedback voltage when the reference voltage U_+, amplification factor G, and current-limiting resistance are constant.

We use a MAX4008 and an OPA320 operational amplifier to form an optical-power feedback circuit. The MAX4008 is a high-precision current-detection chip, which is specially used to detect the current of PD or PIN photodetectors in optical-fiber applications. Its REF pin is the reference-current input pin, OUT is the detected-voltage output pin, and its voltage value is proportional to the REF pin current [45]. The relationship between the output voltage U_{out} of the MAX4008 and the reference current I_{Ref} is

$$U_{out}(\text{mV}) = I_{\text{Re}f}(\mu\text{A}). \tag{3.111}$$

In this circuit,

$$U_5(\text{mV}) = I_m(\mu\text{A}). \tag{3.112}$$

The voltage U_5 fed back by the MAX4008 is then amplified by the OPA320 operational amplifier, and the feedback voltage U_- is

$$U_- = U_5\left(1 + R_f/R\right). \tag{3.113}$$

In the equation, R_f/R is the proportional amplification factor, and R_f and R can be changed according to the parameters of specific LDs and PDs in practical applications.

When the output optical power of the laser decreases, the photocurrent flowing through the PD decreases, the feedback voltage U_- decreases, the output voltage U_1 of the INA114 increases, and the driving current I_0 of the semiconductor laser increases, thereby increasing the output power of the LD.

3.4.2.3 Stability of the Automatic Power-Control Circuit

The automatic power-control system consists of operational amplifier I, operational amplifier II, a controlled object, feedback unit, and other parts. The transfer functions of each component are listed separately.

(1) Operational amplifier I: OPA114 is a proportional amplifier; its input (reference voltage) U_+ and feedback voltage U_- deviate and are amplified; the transfer function is the amplification factor $G_1 = G = 1$ of the operational amplifier.
(2) Operational amplifier II: OPA547 constitutes a proportional link; the relationship between the output voltage and the input voltage is $U_2 = (R_2/R_1)U_1$, and its transfer function is $G_2 = R_2/R_1 = K_p$.
(3) Controlled object: According to the literature [46], the transfer function of the link from the laser output to the photodetector can be tested by sweeping the frequency of the signal source, and the relationship between the amplitude and phase of this link can be tested. Then, the transfer function is inverted. It can be seen from the test results that the laser is a first-order inertial link in the light-intensity control system, and its transfer function is

$$G_3(s) = \frac{K}{Ts + 1}. \tag{3.114}$$

Modeled by the step-response method, the transfer function of the laser unit is identified using the tangent method [45]. The step response of the process was obtained experimentally. The tangent method involves making a tangent line at the inflection point of the curve that intersects at points on the horizontal and vertical axes. The difference between the steady-state value of the system and the input value is K; the projection of the intersecting line segment on the horizontal axis is T. The laser output step-response curve is shown in Fig. 3.43.

From Fig. 3.43, $K = 10.5$ and $T = 7 - 1 = 6$. Then, the transfer function of the controlled object LD is

$$G_3(s) = \frac{10.5}{6s + 1}. \tag{3.115}$$

The forward channel of the automatic power-control loop is the series connection of two control links, and its transfer function is

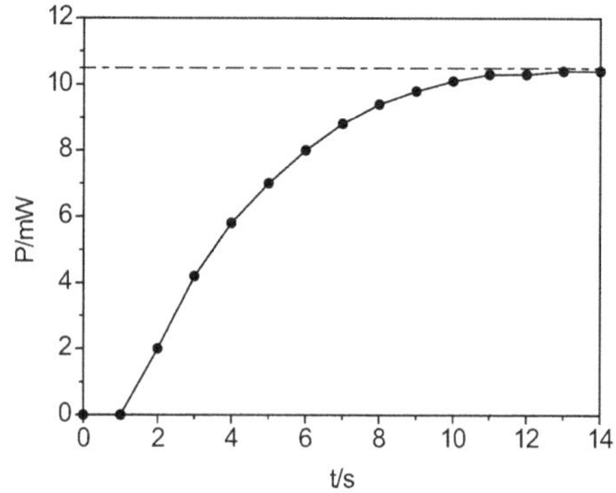

Fig. 3.43 Laser output step-response curve

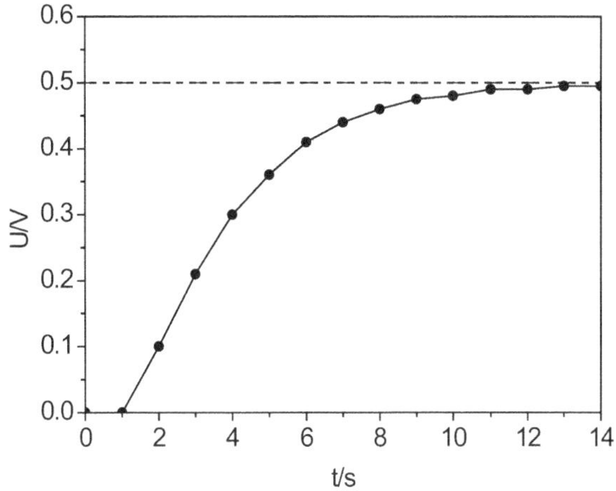

Fig. 3.44 Step-response curve of the feedback unit

$$G(s) = G_2(s)G_3(s) = \frac{1.75K_p}{s + 0.167}. \tag{3.116}$$

(4) Feedback unit: The photoelectric feedback unit is composed of a photodiode, current–voltage converter, and proportional amplifier of the same phase. The element can be approximated as a first-order inertial element, and its transfer function is

$$H(s) = \frac{K}{Ts + 1}, \tag{3.117}$$

where T is the time constant and K is the gain coefficient. Similarly, the step-response method is used for modeling, and the tangent method is used to identify the transfer function of the feedback unit. The step-response curve of the feedback unit is shown in Fig. 3.44.

From Fig. 3.44, $K = 0.5$ and $T = 5 - 1 = 4$. Then, the system identifies the transfer function of the controlled object LD as

$$H(s) = \frac{0.5}{4s + 1}. \tag{3.118}$$

The automatic power-control loop open-loop transfer function is the product of the forward-channel transfer function and the feedback-unit transfer function; namely,

$$G_0(s) = G(s)H(s) = \frac{0.2188K_p}{s^2 + 0.4167s + 0.0417}. \tag{3.119}$$

Its closed-loop transfer function is

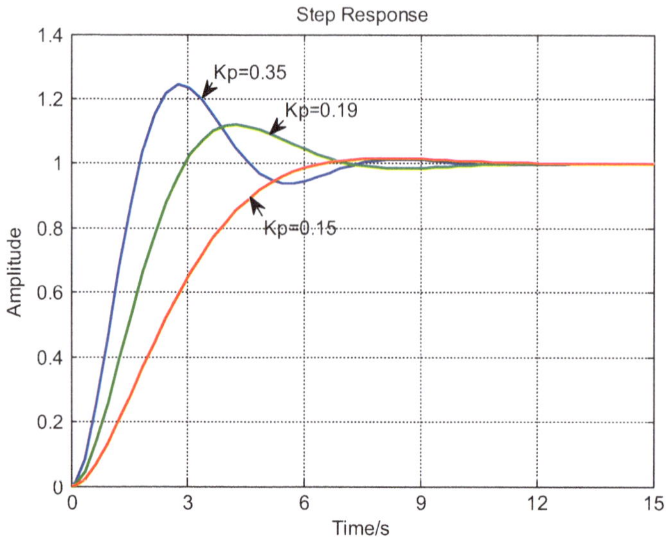

Fig. 3.45 Unit step-response curve in proportional control

$$G_0(s) = \frac{G(s)}{1 + G(s)H(s)} = \frac{(1.75s + 0.4375)K_p}{s^2 + 0.4167s + 0.0417 + 0.2188K_p}. \tag{3.120}$$

When the automatic power-control system is simulated, K_p in the model is set to 0.35, 0.19, and 0.15, and the unit step-response curve under proportional P control can be obtained, as shown in Fig. 3.45.

It can be seen from Fig. 3.45 that increasing the K_p value increases the system response and reduces the rise time; however, it also causes the system overshoot, the number of oscillations, and the adjustment time to increase. When $K_p = 0.19$, the simulation output response tends to the steady-state value $y(\infty) = 1$. At this time, the system overshoot $\sigma\% = 10\%$, rise time t_r =1.5 s, adjustment time $t_s = 6.2$ s, and steady-state error $e_{ss} = 0$.

Chapter Summary

(1) The nonlinear distortion of a laser was analyzed, including static nonlinearity, dynamic nonlinearity, and nonlinearity caused by clipping distortion. The dynamic nonlinear harmonic distortion of the laser was mainly studied.

(2) Based on the rate equation of the semiconductor laser, the amplitude expression of the nonlinear intermodulation distortion of the laser was deduced. Under a gamma–gamma turbulent channel, for optical wireless multi-channel subcarrier BPSK modulation, the influence of laser nonlinear intermodulation distortion on the performance of the carrier-intermodulation noise ratio and bit-error rate of

a wireless laser-communication system was analyzed. The optimal system was given, including the selection of the modulation index and the corresponding relationship of the nonlinear coefficient.

(3) The nonlinear intermodulation distortion of the laser was found to reduce the carrier-to-intermodulation noise ratio and increase the bit-error rate of the wireless laser-communication system. The nonlinearity existing when the laser was modulated by multiple subcarrier signals was compensated, and three compensation methods—electrical (optical) negative feedback, predistortion circuit, and positive feedback—were compared and analyzed.

(4) The photoelectric-feedback realization theory was analyzed, including a constant current-source module and the automatic power-control circuit of the photoelectric feedback module. Moreover, the effect of temperature on the performance of the semiconductor laser was analyzed.

References

1. Yan J, Cui X, Gao C (2004) Laser principle and technology, no 2. Higher Education Press, Beijing, pp 94–95
2. Jiang J (2000) Semiconductor laser. Electronic Industry Press, Beijing, pp 20–45
3. Tucker RS, Pope DJ (1983) Large signal circuit model for simulation of injection-laser modulation dynamics. IEEE J Quant Electron QE-19:1179–1183
4. Jin Y, Tian X (2004) Research on modulation characteristics of high-power semiconductor laser diode. Opt Commun Res 126(6):56–59
5. Schaer T, Rusnov R, Eagle S, Jastrebski J, Albanese S, Fernando X (2003) A dynamic simulation model for semiconductor laser diodes. CCECE Electr Comput Eng 1:293–297
6. Jiao L (1992) Theory and application of nonlinear transfer function, no 5. Xidian University Press, Xi'an, pp 107–109
7. Darcie TE, Tucker RS et al (1985) Intermodulation and harmonic distortion in InGaAsP lasers. IEEE Lett 21:665–666
8. Takemoto A, Watanabe H et al (1990) Distributed feedback laser diode and module for CATV systems. IEEE J Sel Areas Commun 8(7):1359–1365
9. Le Bihan J, Yabre G (1994) FM and IM intermodulation distortion in directly modulaed single-mode semiconductor laser. IEEE J Quant Electron 40(4):899–903
10. Yong KS, Haldar MK, Webb JF (2014) Harmonic and intermodulation distortion in direct intensity modulated quantum cascade lasers. IEEE J Lightw Technol 32(20):3735–3741
11. Mingjiang Z, Tiegen L, Jingxia L (2011) Effects of linewidth enhancement factor on the nonlinear period-one oscillation of a semiconductor laser with external optical injection. Acta Photonica Sinica 40(4):542–546.
12. Samimi H, Azmi P (2010) Subcarrier intensity modulated free-space optical communication on K-distributed turbulence channels. IEEE J Opt Commun Netw 2(8):625–632
13. Daly JC (1982) Fiber optical intermodulation distortion. IEEE Trans Commun 30(8):1954–1958
14. Huang W, Nakawaga M (1994) Nonliner effect of direct sequence CDMA in optical transmission. In: IEEE 3rd intermodulational symposium on spread spectrum techniques & applications, pp 588–592
15. Barman AD, Basu PK (2008) Incoherent in-band crosstalk induced power penalty in amplified WDM networks: a comparative study using Gaussian and chi-squared probability density functions. IET Circuits Dev Syst 2(1):139–143

16. Ghassemlooy Z, Popoola WO, Leitgib E (2007) Free-space optical communicational using subcarrier modulation Gamma–Gamma atmospheric turbulence. In: IEEE 9th international conference on transparent optical networks, pp 150–160
17. Chen D, Ke X, Zhang L (2016) Laser intermodulation distortion and characteristic under the turbulence channel. Acta Photonica Sinica 02:99–103
18. Zhang L (2014) Research on nonlinear characteristics of lasers in wireless laser communication. Xi'an University of Technology, Xi'an
19. Wakafuji K, Ohtsuki T (2005) Performance analysis of atmospheric optical subcarrier multiplexing systems and atmospheric optical subcarrier modulated code-division multiplexing system. IEEE J Lightw Technol 23(4):1676–1682
20. Xing J, Xu G, Zhang X (2005) Effect of the atmospheric turbulence on laser communication system. Acta Photonica Sinica 34(12):1850–1852
21. Uysal M, Li J (2004) Error rate performance of coded free-space optical link over Gamma-Gamma turbulence channels. In: IEEE international conference communication society, pp 3331–3335
22. Yu L, Song L, Cheng H (2013) Analysis and simulation of link power and commination performance in space-ground optical communication. Acta Photonica Sinica 42(5):543–547
23. Bekklali A, Naila CB, Kazaura K et al (2010) Transmission analysis of OFDM based wireless serivicesover turbulent ratio-on FSO links modulated by Gamma–Gamma distortion. IEEE Photon J 2(3):510–520
24. Popoola WO, Ghassemloy Z (2009) BPSK subcarrier intensity modulated free-space optical communications in atmospheric turbulence. IEEE J Lightw Technol 27(8):967–973
25. Tian G (2011) Research and analysis of AVC and ATC light transmitter circuit. Electron Des Eng 19(3):18–20
26. Li W (2013) Development of low-noise and high-stability semiconductor laser current source. Yanshan University, Qinhuangdao
27. Chen M, Chen F, Xiao Y (2011)The effect of temperature on semiconductor lasers. China Sci Technol Inf 14(3):46–47
28. O'Brien D, Mcinerney JG, White JK (2001) Temperature performance of AlGaInAs semiconductor lasers. IEEE of Lasers Electro-Opt Soc 2(3):806–807
29. Zhou Z, Wang F, Yang Z, Teng X (2013) The study of optical power control for driving circuit using the 650 nanometer low power semiconductor laser. In: IEEE international conference on measurement, information and control, vol 1, no 2, pp 222–224
30. Liang W, Qi S, Guan K (2012) Digital laser automatic power control system design based on FPGA. Electron Des Eng 20(1):151–153
31. Li T, Mi X (2012) A design of diode laser output power auto control system. Appl Laser 32(5):424–428
32. Lei H (2014) Research and design of laser drive system based on FPGA. Xi'an University of Technology, Xi'an
33. Yang G (2013) The design of laser semiconductor based on embedded system. Nanjing University, Nanjing
34. Song C (2006) Laser diode driver and modulation. Harbin Institute of Technology, Harbin
35. Menglong C, Li L, Yansong C (2010) Design of high stability digital control driving system for semiconductor laser. Opt Precis Eng 18(7):1629–1636
36. Zhou J, Yuan X, Li B (2005) Application of ADN8830 in automatic temperature control of LD[J]. Opt Optoelectron Technol 3(2):54–57
37. Cui G, Lv W, Zheng Y (2015) Design of temperature control system of high power semiconductor laser. Laser Infrared 45(5):568–570
38. Qi Z (2012) The study of temperature control and driver method about low power semiconductor laser diode. Harbin University of Science and Technology, Harbin
39. Liu C (2003) Surge damage to laser diode and its elimination. Electr Power Environ Protect 19(4):49–51
40. Yang C, Jia H, Xia G (2008) Design of the circuit to eliminate surge of the laser diode power supply. Appl Laser 28(4):310–313

41. Engell S (2007) Feedback control for optimal process operation. J Semicond J Process Control 17(3):203–219
42. Li Y, Song J, Li Z (2010) Design of power control circuit in optical transmitter based on the principle of optoelectronic feedback. J Shandong Univ 24(6):78–81
43. Fu Y, Zou W, Xiao H, Gan Y (2005) Optical power control of LD driver circuit. Infrared Laser Eng 34(5):626–629
44. Zhao J (2000) Automatic-optic-power-control circuit with three amplifiers. Foreign Electron Compon 3(10):26–27
45. Zhang Y, Zhang R, Yang Q (2014) A design of auto power control circuit for laser diode. Electron Prod World 6(1):57–62
46. Jia H, Zhang Y, Zhang Q (2014) Study on a light-intensity control system of total reflection prism laser gyros. J Opt 34(8):172–177

Chapter 4
OWC-OFDM Systems

Although orthogonal frequency-division multiplexing (OFDM) modulation technology in radio-frequency communication systems has the same principle as OFDM modulation technology in wireless laser-communication systems, there are also significant differences. For example, in an intensity modulation/direct detection (IM/DD) optical communication system, the receiving end can determine the intensity of the optical signal, but not the phase information. The channel condition of optical wireless communication (OWC)-OFDM is even worse, and the attenuation of a laser signal by the atmosphere is much greater than that of a radio-frequency signal. This chapter discusses the IM/DD-type OWC-OFDM system model.

4.1 OFDM System Principle

In a communication system, the bandwidth provided by a channel is typically much wider than that required to transmit a signal. If a channel transmits only one signal, it is extremely wasteful. To make full use of the channel bandwidth, frequency-division multiplexing is used to transmit information. Orthogonal frequency division multiplexing (OFDM) was developed from multi-carrier modulation (MCM) [1]. OFDM technology is an effective implementation method. Its modulation and demodulation are based on inverse fast Fourier transforms (IFFTs) and fast Fourier transforms (FFTs), respectively [2].

Figure 4.1 shows a schematic diagram of each OFDM subcarrier [3]. The main idea of OFDM is to divide the channel into several orthogonal subchannels, convert high-speed data signals into parallel low-speed subdata streams, and modulate them for transmission on each subchannel. The orthogonal signals can be separated by adopting a correlation technique at the receiving end, which can reduce the intersymbol interference (ISI) between the subchannels.

X. Ke, *Principles and Applications of Optical Wireless Orthogonal Frequency-Division Multiplexing*, Optical Wireless Communication Theory and Technology,
https://doi.org/10.1007/978-981-97-7973-4_4

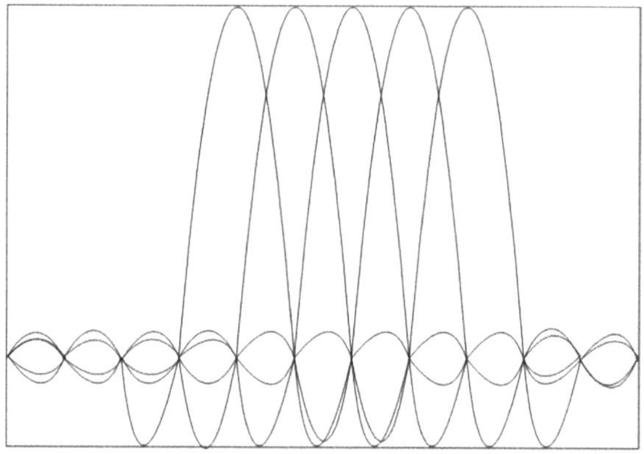

Fig. 4.1 Relationship of each subcarrier in OFDM

The signal bandwidth on each subchannel is smaller than the relevant bandwidth of the channel; thus, each subchannel can be regarded as flat fading, so that intersymbol crosstalk can be eliminated. Because the bandwidth of each subchannel is only a small part of the original channel bandwidth, the channel equalization becomes relatively easy.

4.1.1 Mathematical Model of an OFDM Signal

OFDM is a special MCM, and the general modulation method of the latter is shown in Fig. 4.2 [4].

As shown in Fig. 4.2, the structure of the complex multiplier (also called the in-phase–quadrature (IQ) modulator/demodulator) is typically used in an MCM system; therefore, the transmission signal $s(t)$ of the MCM system can be expressed as follows:

$$s(t) = \sum_{i=-\infty}^{+\infty} \sum_{k=1}^{N_{sc}} c_{ki} s_k(t - iT_s), \tag{4.1}$$

where c_{ki} is the ith information symbol on the kth subcarrier, S_k is the modulation waveform on the kth subcarrier, N_{sc} is the number of subcarriers, and T_s is the symbol period. In general, S_k can be expressed as

$$s_k(t) = \Pi(t) e^{j2\pi f_k t}, \tag{4.2}$$

where f_k is the subcarrier frequency and Π is the pulse-shaping function. For example, a commonly used rectangular window is

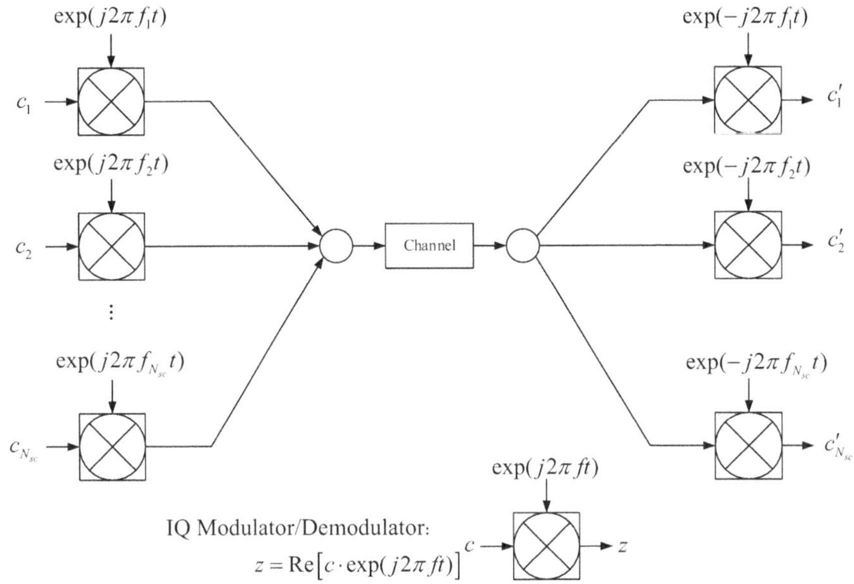

Fig. 4.2 Principle block diagram of multi-carrier modulation

$$\Pi(t) = \begin{cases} 1, & (0 < t < T_s) \\ 0, & (otherwise) \end{cases}. \tag{4.3}$$

As shown in Eq. (4.2), the best receiver for each subcarrier signal uses coherent reception to recover the signal. Therefore, the information symbol c_{ik}' detected by the correlation receiver can be expressed as

$$c'_{ik} = \frac{1}{T_s}\int_0^{T_s} r(t - rT_s)s_k^* dt - \frac{1}{T_s}\int_0^{T_s} r(t - iT_s)e^{-j2\pi f_k t} dt, \tag{4.4}$$

where $r(t)$ is the received time-domain signal. A typical MCM uses spectrally nonoverlapping signals, which requires a large number of oscillators and filters on both the modulation and receiver ends.

Therefore, the main disadvantage of this type of MCM is that it requires a large amount of redundant bandwidth. To design these filters and oscillators more economically, the channel spacing should be several times the symbol rate, thus reducing the spectrum utilization. OFDM uses sets of signals whose spectra overlap, but are orthogonal, to convey information. This orthogonality stems from the direct correlation between any two subcarriers, as follows:

$$\delta_{kl} = \frac{1}{T_s}\int_0^{T_s} s_k s_l^* dt = \frac{1}{T_s}\int_0^{T_s} \exp(j2\pi(f_k - f_l)t_s)dt. \tag{4.5}$$

From the above formula,

$$\delta_{kl} = \exp(j\pi(f_k - f_l)T_s)\frac{\sin(\pi(f_k - f_l)T_s)}{\pi(f_k - f_l)T_s}. \tag{4.6}$$

It can be seen from the above formula that the orthogonal condition is

$$f_k - f_l = m\frac{1}{T_s}. \tag{4.7}$$

At this time, the cross-correlation between any two different subcarriers is zero; that is,

$$\delta_{kl} = 0. \tag{4.8}$$

Therefore, the condition for any two subcarriers to be mutually orthogonal is that the frequency interval be an integer multiple of the reciprocal of the symbol period. These orthogonal signals whose spectra overlap each other can be completely recovered by the matched filter in Eq. (4.4) and are not affected by intercarrier interference (ICI).

4.1.2 OFDM Implementation Using DFTs

The bottleneck of a traditional OFDM system implementation is that a large number of subcarriers must be superimposed to flatten the transmission channel, which obviously leads to a large number of complex oscillators and filters at the transmitter and receiver.

Weinsten and Ebert first revealed that OFDM modulation and demodulation could be achieved using the inverse discrete Fourier transforms (IDFTs) and discrete Fourier transforms (DFTs) [5, 6]. We observe the OFDM modulation expression in Eq. (4.1); if i is omitted, N_{sc} is denoted as N, and the sampling interval is set as T_s/N, the mth OFDM symbol of the transmitted symbol sequence can be expressed as

$$s_m = \sum_{k=1}^{N} c_k \cdot e^{j2\pi f_k \frac{(m-1)T_S}{N}}. \tag{4.9}$$

Here, considering the intercarrier quadrature condition, we set f_k as

$$f_k = \frac{k-1}{T_s}. \tag{4.10}$$

Substituting into Eq. (4.9), we obtain

$$s_m = \sum_{k=1}^{N} c_k \cdot e^{j2\pi \frac{(k-1)(m-1)}{N}} \equiv IDFT\{c_k\}, \tag{4.11}$$

where IDFT is the inverse discrete Fourier transform, and the value range of m is $[1, N]$.

The above formula indicates that the OFDM transmission signal can be represented by the IDFT of the subcarrier signal. Similarly, by omitting i and denoting N_{sc} as N, the relationship between the sampled value r_m of the received signal and recovered carrier signal c'_k can be derived as

$$c'_k = \frac{1}{T_s} \int_0^{T_s} r(t) e^{-j2\pi f_k t} dt. \tag{4.12}$$

The received signal $r(t)$ is sampled at the sampling interval T_s/N to obtain $r(m)$ and the relationship between t and m is examined:

$$t = \frac{(m-1)T_s}{N} \quad \Rightarrow \quad dt = \frac{T_s}{N} dm. \tag{4.13}$$

Then, substituting the orthogonal condition between the carriers in Eq. (4.10), we can obtain

$$c'_k = \frac{1}{N} \int_1^{N} r(m) e^{-j2\pi \cdot \frac{k-1}{T_s} \cdot \frac{(m-1)T_s}{N}} dm. \tag{4.14}$$

Because $m \in [1, N]$ is an integer, the above formula can be written in the form of a summation; thus, we can obtain

$$c'_k = \frac{1}{N} \sum_{m=1}^{N} r_m \cdot e^{-j2\pi \frac{(k-1)(m-1)}{N}} \equiv DFT\{r_m\}, \tag{4.15}$$

where r_m is the sampling of the received signal and the sampling interval is T_s/N.

Equations (4.11) and (4.15) illustrate that the signal $s(t)$ transmitted by OFDM can be obtained by the N-point IDFT transform of the information symbol c_k, and the received information symbol c'_k can be obtained from the sampled value of the received signal $r(t)$; thus, the N-point DFT transform of $r(m)$ is obtained. Therefore, OFDM modulation and demodulation can be achieved using IDFT and DFT.

To realize OFDM modulation and demodulation with a discrete Fourier transform, the following two parts must be added to the system:

(1) A digital-to-analog converter (DAC), which converts the discrete transmitted signal S_m into a continuous signal $s(t)$; and
(2) An analog-to-digital converter (ADC), which discretizes the received continuous signal $r(t)$ by sampling to obtain the r_m required for demodulation.

Using the discrete Fourier transform to implement OFDM has the following two advantages [7]:

(1) The inverse and fast Fourier transforms (IFFT/FFT) can be used to implement the IDFT/DFT transforms in the process of OFDM modulation and demodulation, and the number of complex multiplications required by the IDFT and DFT transforms can be reduced from N^2 to $(N/2)\log_2 N$. In this manner, the computational complexity has an approximately linear growth relationship with the number of subcarriers, which is much simpler than using IDFT/DFT.
(2) A large number of orthogonal subcarriers can be generated by the operation, which can save many complicated radio-frequency oscillators and filters. Thus, when the number of subcarriers must be large, the structure of the entire OFDM system becomes much simpler than that built using radio-frequency devices.

Figure 4.3 [8] shows the block diagram of an OFDM system. At the transmitting end, the input serial data are first converted into parallel data and each data channel is mapped to the information symbols on the corresponding subcarriers in the OFDM symbol. The digital time-domain signal is obtained through an IFFT transformation, inserted into the protection time slot, and then the DAC is used to convert the digital signal into a real-time waveform. Adding guard time slots between symbols suppresses the intersymbol interference (ISI) caused by channel dispersion. Thus, the obtained baseband signal is upconverted to the appropriate frequency band through the IQ mixer (modulator).

At the receiving end, the OFDM signal is first down-converted to the baseband by the IQ demodulator, sampled by the ADC, and then demodulated by the FFT transform to obtain the parallel data corresponding to each subcarrier. The modulation method is demodulated and a parallel–serial conversion is performed to restore the data signal.

Observing Eq. (4.9), it can be seen that the OFDM signal S_m is only a periodic function of the variable f_k with N/T_s as the period. Therefore, any discrete set of subcarriers, as long as their frequency components are spaced apart, is equivalent for OFDM modulation. From Eqs. (4.9) and (4.10), the relationship between subcarrier frequency f_k and subcarrier label k can be summarized as follows:

$$f_k = \frac{k-1}{T_s}, \quad k \in [k_{\min}+1, k_{\min}+N], \tag{4.16}$$

where k_{min} denotes an arbitrary integer. In general, we use the following two subcarrier labels:

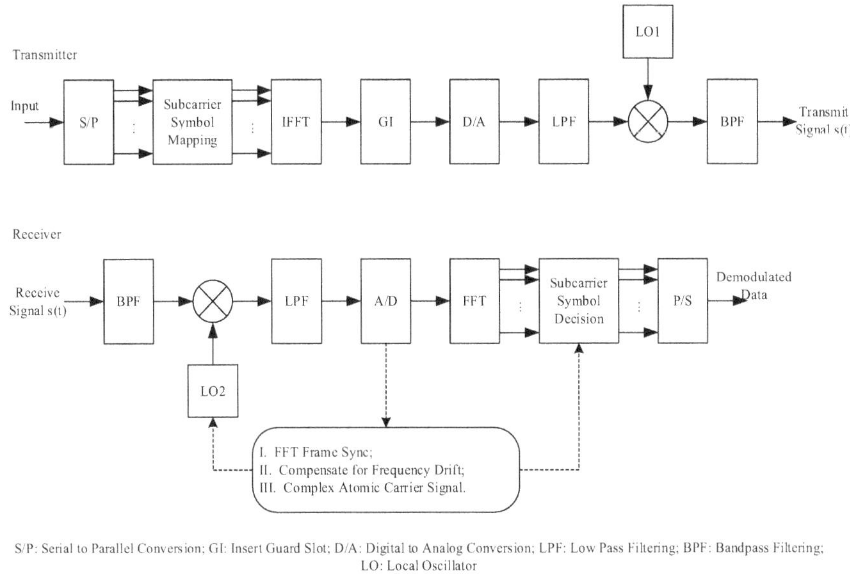

Fig. 4.3 FFT-based OFDM system block diagram [7]

$$k \in [1, N] \quad or \quad k \in \left[-N/2 + 1, N/2\right]. \tag{4.17}$$

They are mathematically equivalent.

4.2 OFDM Frequency Drift and Phase Noise

Both frequency drift and phase noise lead to OFDM intercarrier interference [9, 10] (ICI). The symbol period of an OFDM signal is much longer than that of a single-carrier signal; therefore, OFDM systems are more susceptible to frequency drift and phase noise. For an optical OFDM system, because the phase noise of the laser is larger than that of the radio-frequency device, especially for high-order modulation, the laser phase noise has a greater impact on optical OFDM systems.

The OFDM signal, including the frequency drift and phase noise, can be expressed as

$$r(t) = e^{j(2\pi \Delta f t + \phi(t))} \sum_{k=1}^{N_{sc}} c_k e^{j2\pi f_k t} + N(t), \tag{4.18}$$

where Δf is the frequency drift, $\phi(t)$ is the phase noise, and $N(t)$ is additive white Gaussian noise (AWGN). It is generally believed that $N(t)$ characterizes the background noise in a modulation system circuit. Combining with Eq. (4.4), it can be concluded that the received information symbols are

$$c'_k = \frac{1}{T_s}\int_0^{T_s} r(t)e^{-j2\pi f_k t}dt, \quad f_k = \frac{k-1}{T_s}. \tag{4.19}$$

Substituting $r(t)$ in Eq. (4.18), we obtain

$$c'_k = \frac{1}{T_s}\int_0^{T_s}\left(e^{j(2\pi\Delta f t+\phi(t))}\sum_{l=1}^{N_{sc}} c_l e^{j2\pi f_l t} + N(t)\right)e^{-j2\pi f_k t}dt. \tag{4.20}$$

Expanding the above equation, we obtain

$$\begin{aligned} c'_k &= \sum_{l=1}^{N_{sc}} c_l \cdot \frac{1}{T_s}\int_0^{T_s} e^{j2\pi(f_l-f_k)t+j(2\pi\Delta f t+\phi(t))}dt + n_k \\ &= \eta_0 c_k + \sum_{l\neq k}\eta_{k-l}c_l + n_k. \\ &\triangleq \eta_0 c_k + I_k + n_k \end{aligned} \tag{4.21}$$

The received signal, c_k, can be divided into three terms. To understand the cross-correlation between subcarriers with a label interval of m, we define the ICI coefficient η_m as

$$\eta_m = \frac{1}{T_s}\int_0^{T_s} e^{j(2\pi(f'_m+\Delta f)t+\phi(t))}dt, \quad f'_m = \frac{m}{T_s}, \quad m = -(N_{sc}-1), ..., 0, 1, ..., N_{sc}-1. \tag{4.22}$$

η_0 and η_{k-l} in Eq. (4.21) can be understood as the autocorrelation and cross-correlation between each subcarrier, and $\eta_0 c_k$ can be understood as a useful part of the received signal. The remaining two items I_k and n_k are

$$I_k = \sum_{l\neq k}\eta_{k-l}c_l, \quad l = 1, 2, ..., N_{sc}. \tag{4.23}$$

$$n_k = \frac{1}{T_s}\int_0^{T_s} N(t)e^{-j2\pi f_k t}dt. \tag{4.24}$$

Correspondingly, I_k can be understood as the ICI noise in the received signal, which also includes the influence of frequency drift Δf and phase noise $\phi(t)$; n_k can be understood as the influence of the circuit background noise (which can be considered AWGN) at the transmitting end in the received signal. We investigate the effects of frequency drift and phase noise separately.

4.2.1 Effect of the Frequency Drift on OFDM

When only considering the influence of frequency drift, we assume $\phi(t) = 0$ and incorporate it into Eq. (4.22) to obtain:

$$\eta_m = \frac{1}{T_s} \int_0^{T_s} e^{j2\pi(f_m' + \Delta f)t} dt. \tag{4.25}$$

Integrating the above equation, we can obtain

$$\sigma_{ICI}^2 = \sigma_c^2 \sum_{m=1}^{N_{sc}} \eta_m^2 \triangleq \chi \cdot \sigma_c^2, \tag{4.26}$$

where $\delta \triangleq \Delta f \cdot T_s$ can be regarded as the normalized frequency-offset coefficient. From Eqs. (4.21) and (4.23), we can calculate the variance σ_{ICI}^2 of the signal crosstalk caused by the frequency drift as

$$\sigma_{ICI}^2 = \sigma_c^2 \sum_{m=1}^{N_{sc}} \eta_m^2 \triangleq \chi \cdot \sigma_c^2, \tag{4.27}$$

where σ_c^2 is the variance of the information symbols on each subcarrier. We define χ as the sum of all ICI noise terms, which can be expressed as

$$\chi = \sum_{m=1}^{N_{sc}} \eta_m^2. \tag{4.28}$$

Assuming that the interchannel interference is AWGN, the signal-to-noise ratio (SNR) γ' can be calculated using the following formula:

$$\gamma' = \frac{\eta_0^2 \delta_c^2}{\sigma_{ICI}^2 + \sigma_w^2} = \frac{\eta_0^2}{x + \gamma^{-1}}, \tag{4.29}$$

where $\gamma = \sigma_c^2 / \sigma_w^2$ is the received signal-to-noise ratio without the influence of frequency drift. For example, for an OFDM system in which the subcarriers are modulated by QPSK, the bit-error rate (BER) is

$$P_e = \frac{1}{2} erfc\left(\sqrt{\frac{\gamma'}{2}}\right). \tag{4.30}$$

The system parameters can be set through various bit-error rate formulas similar to the above and the signal-to-noise ratio/bit-error rate curve of the OFDM system can be calculated under the influence of the frequency drift under various subcarrier modulation modes.

4.2.2 Phase Noise in OFDM Systems

In contrast to the previous section, here, we only consider the effect of phase noise. Assuming the frequency drift $\Delta f = 0$ in Eq. (4.22), we obtain [10]

$$\eta_m = \frac{1}{T_s}\int_0^{T_s} e^{j(2\pi f'_m t + \phi(t))} dt, \quad f'_m = \frac{m}{T_s}, \quad m = -(N_{sc} - 1), \ldots, 0, 1, \ldots, N_{sc} - 1. \tag{4.31}$$

Note that $\eta_0 = |\eta_0| e^{j\phi_0}$. ϕ_0 is also called the common phase error (CPE). Substituting Eq. (4.31) into Eq. (4.21), we obtain

$$\tilde{c}_k = c'_k e^{-j\phi_0} = |\eta_0| c_k + e^{-j\phi_0} \sum_{l \neq k} \eta_{k-l} c_l + e^{-j\phi_0} n_k, \tag{4.32}$$

where $\tilde{c}_k$ is the received information symbol after the CPE is removed. Therefore, it can be considered that the phase noise has two main effects on the OFDM system:

(1) ϕ_0 rotates the constellation diagram of the received signal. A skewed constellation can be corrected by rotating it, as shown in Eq. (4.32).
(2) The effect of the phase noise on the ICI term can be observed from the second term in Eq. (4.32). To estimate the size of the ICI caused by the phase noise, it can be further assumed that the phase noise $\phi(t)$ is a Wiener process, which satisfies

$$E\left[(\phi(\tau + t) - \phi(t))^2\right] = 2\pi\beta\tau, \tag{4.33}$$

where $E[\cdot]$ is the statistical average and β is the 3-dB linewidth of the laser (or strictly speaking, the integrated linewidth of the laser transmitter and receiver). The subcarrier autocorrelation $|\eta_0|$ can be expressed as

$$|\eta_0| = a + \zeta, \tag{4.34}$$

where $a = \langle|\eta_0|\rangle$ and ζ is the residual amplitude noise of each OFDM symbol. Therefore, the signal-to-noise ratio derived from $\tilde{c}_k$ can be re-expressed using Eq. (4.34):

$$\gamma' = \frac{a^2\sigma_k^2}{n^2 + n_p^2}, \tag{4.35}$$

where σ_k^2 and n_k^2 are the variances of information and AWGN on the kth subcarrier, respectively, and n_p^2 is the noise variance including ICI and $|\eta_0|$ (which can be regarded as signal-amplitude disturbance). To simplify the calculation, it can be considered that $\sigma_k^2 = 1$. Using Eq. (4.32), n_p^2 can be expressed as

$$n_p^2 = E[\zeta^2] + E\left[\sum_{l \neq k} |\eta_{k-l}|^2\right]. \tag{4.36}$$

We define $a_1 = E[|\eta_0|^2]$; therefore, we obtain

$$E[\zeta^2] = E[|\eta_0|^2] - E^2[|\eta_0|] = a_1 - a^2. \tag{4.37}$$

In addition,

$$E[\sum_{l \neq k} |\eta_{k-l}|^2] = 1 - E[|\eta_0|^2] = 1 - a_1. \tag{4.38}$$

Substituting Eqs. (4.37) and (4.38) into Eq. (4.36) and substituting the result into Eq. (4.35), we obtain the SNR expression, including the effect of the phase noise:

$$\gamma' = \frac{a^2}{\gamma^{-1} + 1 - a^2} = \gamma \frac{a^2}{1 + (1 - a^2)\gamma}, \tag{4.39}$$

where $\gamma = \delta_k^2 / n_k^2$ is the received signal-to-noise ratio without the influence of phase noise. It is further possible to calculate the ratio of the SNR to a phase-noise-free system (expressed in decibels) as

$$\Delta\gamma(dB) = 10\log_{10}\left(\frac{a^2}{1 + (1 - a^2)\gamma}\right). \tag{4.40}$$

$\Delta\gamma$ can also be understood as the additional signal-to-noise ratio (SNR) required to achieve the same signal-to-noise ratio or bit-error rate. Substituting into the laser linewidth β, $a = \langle|\eta_0|\rangle$ can be approximated as

$$a \approx 1 - \frac{11}{60} \cdot 4\pi\beta T_s\gamma. \tag{4.41}$$

In the formula, keeping the first-order term constant, $\Delta\gamma$ is proportional to βT_s; therefore,

$$\Delta\gamma(dB) = \frac{10}{\ln 10} \cdot \frac{11}{60} \cdot 4\pi\beta T_s\gamma = 10\beta T_s\gamma. \tag{4.42}$$

The following can be seen from Eq. (4.42):

(1) The loss value of the logarithmic signal-to-noise ratio is proportional to the symbol period T_s. Therefore, the effect of phase noise on OFDM signals is greater than that on single-carrier signals.
(2) The loss of the logarithmic signal-to-noise ratio is proportional to γ; thus, it is also related to γ'. This implies that higher-order modulation may be more sensitive to phase noise. In other words, to achieve the same effective signal-to-noise ratio, higher-order modulation systems require a higher received signal-to-noise ratio.
(3) The loss of the logarithmic signal-to-noise ratio caused by phase noise is also proportional to the laser linewidth. To limit the influence of phase noise on the system within the allowable range, the linewidth of the laser must also be limited within a certain range.

4.3 OWC-OFDM System Structure

The OWC-OFDM system divides a high-speed data string into several low-speed data substrings and transmit them in parallel via many narrowband subcarriers to obtain a higher transmission rate. The transmission of low-speed data streams on narrowband subcarriers produces far less distortion than high-speed data streams, and no equalization is required.

Signal-processing circuits in the RF domain are also much more mature than those in optical devices. For example, the frequency selectivity of the filter in the radio-frequency domain and the frequency stability of the oscillator are better than those of the corresponding optical devices. The phase-noise parameter of the radio-frequency oscillator is much smaller than that of the commonly used distributed feedback (DFB) laser. This means that it is much easier to correlate signals in the radio-frequency domain than in the optical domain. Therefore, many existing mature correlation-detection devices in wireless communication can be used, if the signal can be processed in the radio-frequency domain [11].

Based on the above considerations, we adopted the OWC-OFDM system structure shown in Fig. 4.4. Figure 4.4a–c show the transmitter, receiver, and OWC-link structures of the system [4], respectively.

In the OWC-OFDM transmitter shown in Fig. 4.4a, one channel of high-speed serial data is first divided into N channels of parallel low-speed data, which are respectively mapped by a constellation map (using M-ary phase-shift keying (M-PSK) or M-ary quadrature amplitude modulation (M-QAM)).

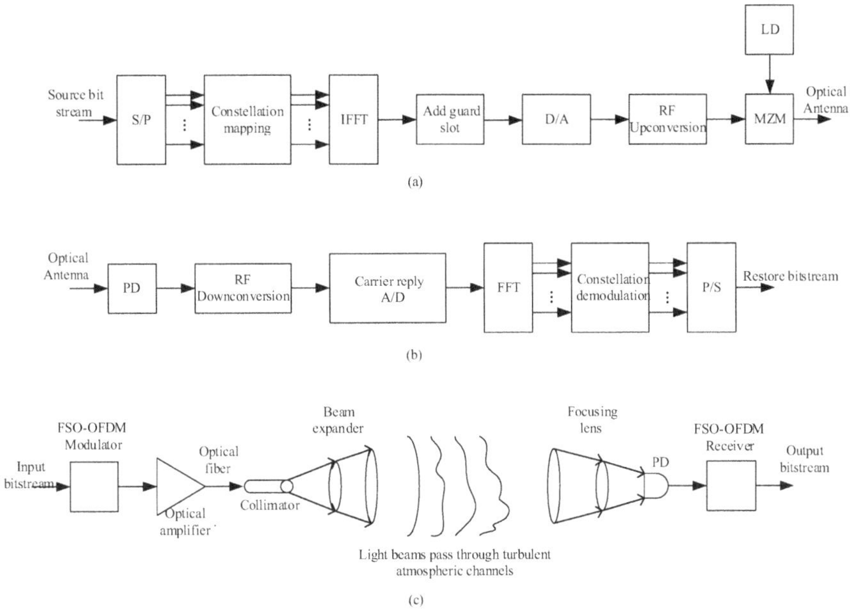

Fig. 4.4 OWC-OFDM system-structure block diagram

After the complex signal, an N-point IFFT operator is input in parallel to generate a multi-carrier digital OFDM signal. After the serial input of the IFFT-converted signal is converted into an analog signal by the DAC, it is up-converted by an RF converter to become a radio-frequency signal, and a DC offset is added. The negative part of the signal is set and trimmed, and then the non-negative analog signal obtained in this manner is used to drive a dual-drive Mach-Zehnder modulator (MZM) together with a laser local oscillator. Finally, the signal is transmitted using an optical antenna.

If the symbol period of the corresponding single-carrier system is T, the signal symbol period of the OFDM system, as shown in Fig. 4.4, is $T_s = NT$. If the number N of subcarriers in OFDM is sufficiently large, its symbol period will be much larger than the width of the channel-pulse broadening, and the ISI that the entire system is subjected to will be greatly reduced.

As shown in Fig. 4.4a, the OWC-OFDM transmission signal modulated by the MZM can be expressed as follows:

$$s(t) = s_{OFDM}(t) + D, \tag{4.43}$$

where S_{OFDM} is the OFDM symbol and D is the DC offset. The duration of the OFDM signal satisfies

$$t \in [kT - T_G/2 - T_{win},\ kT + T_{IFFT} + T_G/2 + T_{win}]. \tag{4.44}$$

In this formula, S_{OFDM} can be written as follows [12]:

$$S_{OFDM}(t) = \mathrm{Re}\left\{ \sum_{k=-\infty}^{\infty} (t - kTs) \sum_{i=1}^{N_{IFFT}} X_{i,k} e^{j\frac{2\pi i(t-kTs)}{T_{IFFT}}} e^{j2\pi f_{RF} t} \right\}, \tag{4.45}$$

where $X_{i,k}$ is the kth OFDM symbol on the ith subcarrier, $w(t)$ is the window function, f_{RF} is the radio-frequency carrier frequency, T_s is the OFDM symbol period, T_{IFFT} is the IFFT sequence period, T_G is the protection-slot (cycle-extension) duration, T_{win} is the window-function period. The details of the cyclic extension and window-function conditions of the OFDM signal symbols are shown in Fig. 4.5.

The steps for OFDM symbol generation are as follows: After the N_{QAM}-bit QAM signal symbol is filled with zeros, the input sequence of the IFFT transform of N_{IFFT} (2^m, $m > 1$) bits is obtained, and the N_G bit-sampling point is inserted as the guard

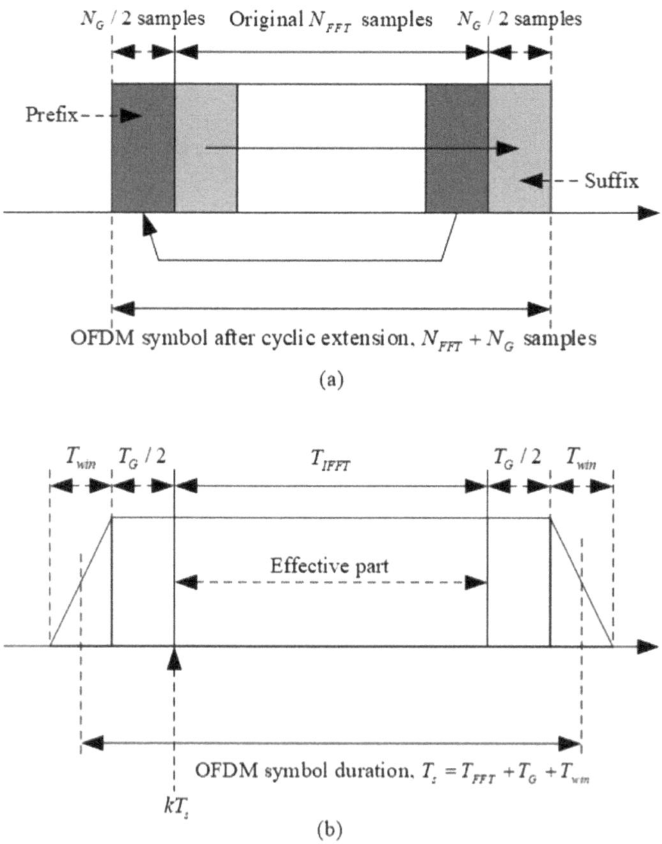

Fig. 4.5 Schematic diagram of a cyclic extension and window-function shaping of an OWC-OFDM signal sequence

interval T_G. Finally, the obtained OFDM code element is multiplied by the window function to obtain the final output. Some commonly used window functions, such as the raised-cosine, Kayser, and Blackman windows, can be used for OFDM signal conditioning.

Cyclic extension is introduced to maintain orthogonality between subcarriers when adjacent OFDM symbols partially overlap. As shown in Fig. 4.5, the so-called cyclic extension is used to copy the last $N_G/2$ points of the original IFFT sequence (total N_{IFFT} points, the period is T_{IFFT}) as the frame prefix of the sequence; $N_G/2$ points are copied as the frame suffix of the sequence. The OFDM symbol sequence after a cyclic extension is the $N_{IFFT} + N_G$ point sequence.

A window function is added to suppress the spectral leakage of the transmitted signal. When the number of carriers in the OFDM system is small, it is more efficient to add a window function, because the smaller the number of orthogonal subcarriers, the higher the power of the out-of-band spectrum.

After the generated OFDM symbol sequence undergoes D/A conversion and RF up-conversion, the radio-frequency signal must be converted into an optical signal by a laser modulator. According to the speed of the signal rate, two different modulation methods are often used:

(1) For an OFDM signal whose signal rate is <10 Gbaud/s, the DFB laser can be directly driven (internal modulation) to obtain the OWC-OFDM signal;
(2) For ultra-high-speed OFDM signals, whose rates exceed 10 Gbaud/s, a double-end-driven MZM (external modulation) is used to obtain OWC-OFDM signals.

D in Eq. (4.43) is the DC offset in the OWC-OFDM signal. Inserting the DC offset into the OFDM signal directly detects and recovers the carrier signal using a non-correlated method, such as M-PSK and M-QAM signals. Because the optical signal is non-negative, according to the size of the inserted DC component, the OWC-OFDM signal generation can be divided into three different types: DC-biased, clipped, and non-clipped.

4.3.1 DC-Biased OFDM System

The light-intensity signal is non-negative, so the negative part of the OFDM signal cannot be modulated into it. To solve this problem, we add a DC component, D, to S_{OFDM} before the OFDM signal drives the laser modulator, satisfying

$$D \geq \max\{S_{OFDM}(t)\}. \tag{4.46}$$

Thus, the bipolar OFDM signal can be converted into a nonnegative unipolar signal. In general, D is an estimated relatively large value, and the OFDM signal can be moved to the horizontal axis in the time domain. The advantage of this is that it is convenient for signal processing: a fixed DC-offset value is added to the transmitting part, which is well implemented in the RF conditioning circuit.

At the receiving end, the biased OFDM (B-OFDM) signal can also be directly demodulated, thereby simplifying the receiver structure, which is beneficial for the realization of the system. The disadvantage of B-OFDM is also evident: adding a constant DC component to the transmitted signal requires additional energy, which reduces the power efficiency of the entire system.

4.3.2 Clipped OFDM System

Compared with the B-OFDM scheme, the clipped OFDM system (C-OFDM) scheme improves the power efficiency of the system. In the C-OFDM scheme, the DC component D added to the S_{OFDM} signal satisfies

$$\frac{1}{2}\max\{S_{OFDM}(t)\} < D < \max\{S_{OFDM}(t)\}. \tag{4.47}$$

The transmitted OFDM signal is the positive part of the OFDM signal with an added DC offset; such an OFDM signal can also be demodulated. The added DC offset is smaller than that required in the B-OFDM scheme; therefore, the power efficiency of the C-OFDM system is better than that of the former.

The problem to be solved by the C-OFDM scheme is to determine a suitable offset D to obtain the optimal laser-transmission power. A known optimal solution for D is to take 50% of the power allocated to the carrier before the signal is clipped to negative values so that the original signal can be recovered without severely distorting the signal.

4.3.3 Unclipped OFDM Systems

Unclipped OFDM, or U-OFDM, is another approach for addressing the power efficiency of OWC-OFDM signals. If an external-modulation method is used, the effect of distortion caused by clipping can be eliminated. Therefore, the core of the U-OFDM scheme is to use an MZM containing $LiNbO_3$ crystals.

Unlike B-OFDM and C-OFDM, which directly modulate the light intensity, U-OFDM modulates the signal to the phase of the optical carrier. In this manner, the negative part of the OFDM signal can be transmitted to the photodetector (PD) at the receiving end. The signal distortion caused by the square-law characteristic of the PD device can be eliminated by an appropriate filter so that the recovered signal distortion is very small.

The power efficiency of the U-OFDM scheme is not as good as that of C-OFDM because it does not transmit the energy of the negative part of the signal. Moreover, because the nonlinearity of the PD in the U-OFDM scheme is compensated by the filter at the front end of the detector, this filter also reduces some of the useful signal energy.

Nevertheless, the power efficiency of U-OFDM is higher than that of B-OFDM. We note that the essence of adding a DC offset D to S_{OFDM} is to increase the average voltage of the signal. Another intuitive understanding of the low power efficiency of B-OFDM is that this scheme requires a much larger voltage offset than that required for C-OFDM. All S_{OFDM} are raised above zero.

4.3.4 *Demodulation of an OWC-OFDM Signal*

As shown in Fig. 4.4c, the OWC-OFDM communication system consists of a transmitter, propagation medium (atmospheric channel), and receiver. At the transmitting end, the modulated beam is sent out through an optical antenna; at the receiving end, another optical antenna collects the beam and focuses it on a photodetector. Generally, the output current of a PD is proportional to its irradiated optical power.

In the OWC-OFDM system, the photodetector is generally composed of a PIN (positive-intrinsic-negative, intrinsic semiconductor doping) photodiode plus a preamplifier or an avalanche diode. After the emission beam propagates through the atmosphere, the atmospheric turbulence, absorption, and rocking of buildings change the amplitude and phase of the optical signal. The general formula for the PD output current can be expressed as

$$\begin{aligned} i(t) &= R_{PD}|a(t)(s_{OFDM}(t) + D)|^2 \\ &= R_{PD}\left[|a(t)s_{OFDM}(t)|^2 + |a(t)D|^2 + 2\mathrm{Re}\left\{|a(t)|^2 s_{OFDM}(t)D\right\}\right], \end{aligned} \tag{4.48}$$

where R_{PD} is the photosensitivity of the photodiode and $|a(t)|^2$ is the light-intensity fluctuation function caused by atmospheric turbulence.

The output current of the detector is down-converted by RF, and the signal after proper filtering can be expressed as

$$r(t) = [i(t)k_{RF}\cos(\omega_{RF}t) * h_e(\tau) + n(t)], \tag{4.49}$$

where k_{RF} is the RF down-conversion coefficient, ω_{RF} is the mixing frequency, "$*$" is the convolution symbol, $h_e(t)$ is the impulse response of the low-pass filter (LPF), the corresponding transfer function is $H_e(jw)$, and $n(t)$ is the electrical noise of the receiver, generally AWGN.

The demodulation process of the OFDM signal is as follows: The signal undergoes an A/D conversion, removes the cyclic extension, and obtains the modulated signal of N channels of subcarriers through an N-point FFT operation. It then obtains the parallel signal through the corresponding M-PSK or M-QAM demodulation, obtains the data sequence, and finally combines the N-way parallel data into the final high-speed serial data sequence.

4.4 OFDM Signal Structure

4.4.1 Guard Interval and Cyclic Prefix

One of the main reasons for introducing OFDM modulation technology to OWC is that it can effectively suppress the delay spread caused by atmospheric scattering and multipath phenomena. By distributing the input data stream in parallel to N parallel subchannels, the symbol period of each OFDM can be expanded to N times the original data-symbol period; thus, the ratio of the delay spread to the symbol period is reduced by N times.

In an OFDM system, to eliminate the intersymbol interference to the greatest extent, a guard interval should be inserted between each OFDM symbol. However, it should not interfere with the next symbol. In the OWC-OFDM system, the length T_g of the guard interval is greater than the transmission delay caused by dispersion between the subchannels. Usually, to completely eliminate the intersymbol interference (ISI) in optical communication, the length of the guard interval is defined as [13]

$$\frac{c}{f^2}|D_t| \cdot N \cdot \Delta f \leq T_g, \tag{4.50}$$

where f is the optical carrier, c is the speed of light, D_t is the cumulative dispersion, the unit is *ps/pm*, and N is the total number of subcarriers.

No signal can be inserted into the guard interval; that is, it is an idle transmission period. However, in this case, owing to the influence of multipath propagation, interference between subchannels will occur; that is, the orthogonality between subcarriers will be destroyed, and interference will occur between the different subcarriers.

As shown in Fig. 4.6, to eliminate the intercarrier interference (ICI) caused by multipath propagation, the OFDM symbol with the original width T is periodically extended, and the guard interval is filled with the extended signal. The signal inside is called the cyclic prefix. The signal in the cyclic prefix is the same as that of the OFDM symbol tail width T_g. The signal after adding the cyclic prefix is

$$s_g(t) = \frac{1}{N}\sum_{k=1}^{N} x(k)\exp(j2\pi kt/T)\mathrm{Rect}_g(t). \tag{4.51}$$

At this time, the corresponding rectangular-window function becomes

$$\mathrm{Rect}_g(t) = \begin{cases} 1 & (-T_g < t < T) \\ 0 & otherwise. \end{cases} \tag{4.52}$$

When $\tau_m \leq T_g$, as shown in Fig. 4.7, the received signal is

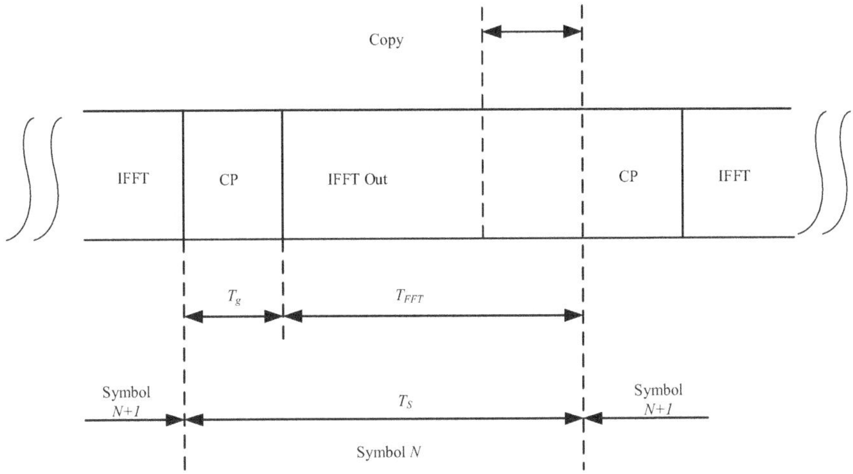

Fig. 4.6 Cyclic prefix

$$y(t) = \begin{cases} \tilde{s}_l(t) \otimes h(t) & lT_{\text{OFDM}} \le t \le (l+1)T_{OFDM} - T_g \\ T_{l-1,u}(t) \otimes h(t) + T_{l,u}(t) \otimes h(t) & lT_{\text{OFDM}} - T_g \le t \le lT_{OFDM} \end{cases}. \tag{4.53}$$

It can be seen that all ISIs fall within the guard interval, and the influence of ISI can be eliminated as long as the receiving end removes the cyclic prefix.

When $\tau_m \ge T_g$, as shown in Fig. 4.8, the received signal is

$$y(t) = \begin{cases} \tilde{s}_l(t) \otimes h(t) & lT_{OFDM} - T_g + \tau_m \le t < (l+1)T_{OFDM} - T_g \\ T_{l-1,u}(t) \otimes h(t) + T_{l,u}(t) \otimes h(t) & lT_{OFDM} - T_g \le t < lT_{OFDM} - T_g + \tau_m \end{cases}. \tag{4.54}$$

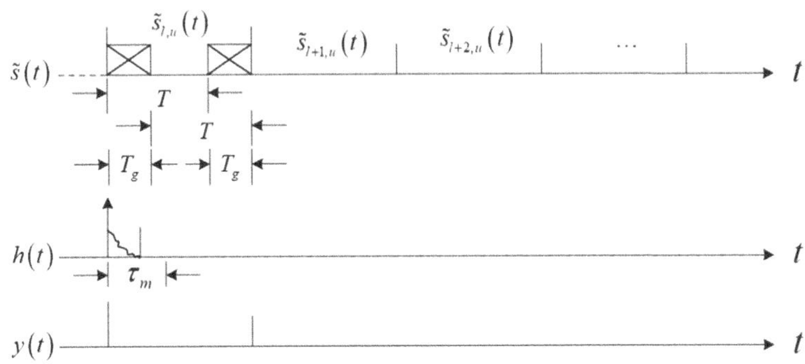

Fig. 4.7 Time relationship between the transmitted signal, channel response, and received signal, for $\tau_m \le T_g$

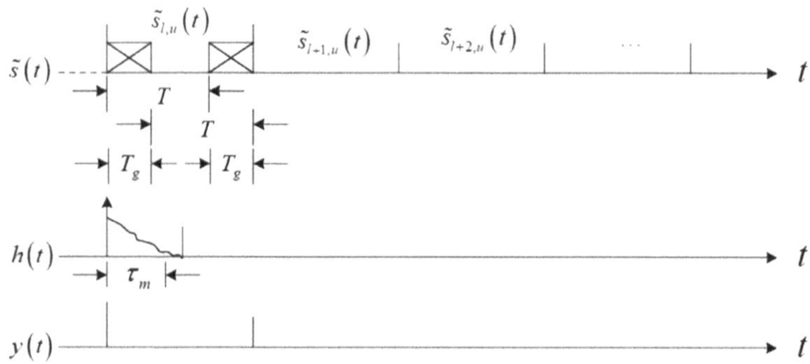

Fig. 4.8 Time relationship between the transmitted signal, channel response, and received signal, for $\tau_m \geq T_g$

It can be seen that after the receiving end removes the cyclic prefix, some ISIs still fall within the useful signal range, so that the subchannels of the OFDM signal are no longer orthogonal.

Adding a cyclic prefix to OFDM has two main functions: one is to act as a guard interval to eliminate or at least greatly reduce the ISI, and the other is to maintain the orthogonality between subchannels and reduce the ICI. Figure 4.9 shows the effects of adding a cyclic prefix and not adding it on the system performance in an OFDM system.

Although the introduction of a cyclic prefix can reduce the ISI and ICI and maintain the orthogonality between subchannels, the cyclic prefix is a type of redundant information relative to the original information data; thus, it requires a certain amount

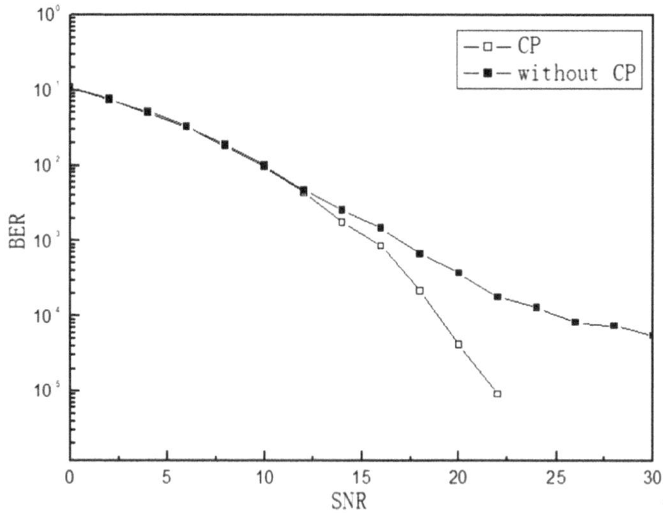

Fig. 4.9 Impact of the cyclic prefix on system performance

of additional spectrum and power resources. Therefore, the introduction of a cyclic prefix results in a loss of signal-to-noise ratio, power, and information rate. The loss of the signal-to-noise ratio is defined as

$$SNR_{\text{loss}} = -10\lg\left(\frac{T_{OFDM} - T_g}{T_{OFDM}}\right) = -10\lg\left(1 - \frac{T_g}{T_{OFDM}}\right). \tag{4.55}$$

That is, the longer the cyclic prefix, the greater the loss of the signal-to-noise ratio. The power loss is defined as

$$\eta_p = 10\log_{10}\left(\frac{T'}{T_{OFDM}}\right) = 10\log_{10}\left(\frac{N+L}{N}\right). \tag{4.56}$$

The loss of information rate is defined as

$$\eta_R = \frac{T_g}{T'} = \frac{L}{N+L}. \tag{4.57}$$

It can be seen from Eqs. (4.55), (4.56), and (4.57) that when the cyclic prefix accounts for 20% of the OFDM symbol period, the power loss is <1 dB; however, the information-rate loss is 20%. Fortunately, inserting a cyclic prefix can eliminate the effect of the ICI caused by chromatic dispersion.

4.4.2 Oversampling

In an actual system, to avoid aliasing effects during digital signal processing, it is generally necessary to oversample OFDM symbols. Implementing oversampling helps to use discrete data to more accurately reflect changes in OFDM symbols. Therefore, it is necessary to perform oversampling on the OFDM symbols.

In general, L is used to represent the oversampling rate, and $[n]$ is used to represent the Nyquist sampling rate in the discrete time domain, such that the oversampled IFFT output is

$$s(n/L) = \frac{1}{N}\sum_{k=1}^{N} x(k)\exp(j2\pi kn/NL)\text{Rect}(n/L). \tag{4.58}$$

After the above formula is disassembled and integrated, the following equation can be derived:

$$s(n/L) = \text{IFFT}\left(\sqrt{L}[x(1)\cdots x(N/2)\underbrace{0\cdots\cdots 0}_{N(L-1)}x(N/2+1)\cdots x(N)]^T\right)$$

$$= \text{IFFT}\Big(\sqrt{L}x_L(k)\Big). \tag{4.59}$$

At this point, the rectangular window function is:

$$\text{Re}ct(n/L) = \begin{cases} 1 \; (0 < n < NL - 1) \\ 0 \qquad others \end{cases}, \tag{4.60}$$

where $x_L(k)$ refers to the L-fold oversampling of the original symbol (M-QAM or M-PSK), that is, inserting $N(L-1)$ zero-valued data into the original sampled data $x(k)$.

4.4.3 Windowed

With the increase in the number of subcarriers, the shape of the power spectrum approaches a rectangle; thus, it approaches the characteristics of an ideal low-pass filter. For the OFDM signal defined by Eq. (4.7), the out-of-band attenuation of the power spectrum is quite slow. Even in the case of $N = 256$ subcarriers, the -40 dB bandwidth is still four times the -3 dB bandwidth. Therefore, to increase the falling speed of the out-of-band part of the power spectrum of the OFDM signal, it is necessary to perform windowing processing on each OFDM symbol so that the amplitude value at the edge of the symbol period is rapidly attenuated to zero.

The commonly used type of window function is the raised cosine function, which is defined as

$$\varsigma(t) = \begin{cases} 0.5 + 0.5\cos(\pi + t\pi/(\beta T_s)) & 0 \le t \le \beta Ts \\ 1.0 & \beta Ts \le t \le T_s \\ 0.5 + 0.5\cos((t - T_s)\pi/(\beta T_s)) & T_s \le t \le (1+\beta)T_s \end{cases}, \tag{4.61}$$

where T_s represents the length of the symbol before windowing, and the length of the symbol after windowing is $(1+\beta)T_s$; thus, there is a mutually covering area between adjacent symbols. The OFDM symbol after loading the cyclic prefix and raised cosine windowing is shown in Fig. 4.10 [14].

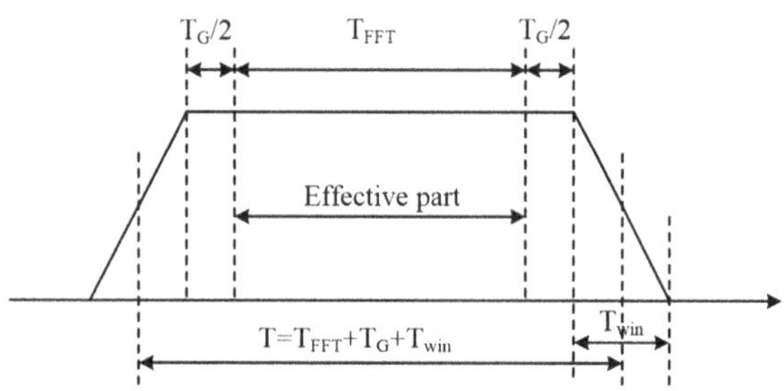

Fig. 4.10 Windowed OFDM symbols

4.4.4 *Noise Characteristics of OWC-OFDM Signals*

The influence of noise on OWC-OFDM signals is discussed in the following two cases:

(1) The influence of multiplicative noise in the gamma–gamma channel on the signal constellation, and
(2) The influence of mixed noise in the gamma–gamma channel on the signal constellation.

4.4.4.1 Multiplicative Noise

The characteristics of OFDM modulation and single-carrier QPSK modulation on the demodulated signal constellation at the receiver end in a gamma–gamma-distribution atmospheric-turbulence channel model were analyzed under multiplicative noise [15].

When $\sigma_R = 0.2$, the constellation diagrams of the demodulated signals at the receiving end under OFDM and single-carrier QPSK modulation under weak turbulence are shown in Fig. 4.11a, b, respectively; in the case of medium turbulence, $\sigma_R = 1$, the constellation diagrams of the demodulated signals at the receiving end are shown in Fig. 4.11c, d, respectively. In the case of strong turbulence, $\sigma_R = 2$, the constellation diagrams of the demodulated signals at the receiving end are shown in Fig. 4.11e, f, respectively.

As shown in Fig. 4.11b, after the modulated signal is affected by multiplicative noise, the constellation diagram of the single-carrier QPSK demodulated signal appears to expand in the radial direction. If it is a multilayer constellation, the communication performance deteriorates. Figure 4.11a shows a constellation diagram of the demodulated OFDM signal. The constellation diagram shows a Gaussian distribution.

As shown in Fig. 4.11, the stronger the turbulence intensity, the more significant the radial expansion of the constellation diagram of the single-carrier QPSK demodulation signal. The convergence performance of the constellation diagram using the OFDM demodulation signal is clearly improved, and the distribution of the constellation points follows a two-dimensional Gaussian distribution. It can be seen that the constellation diagram of the OFDM signal has the best convergence state under weak turbulence, and the constellation diagrams of the OFDM signal with moderate turbulence and strong turbulence are more scattered than that of weak turbulence.

4.4.4.2 Mixed Noise

We consider a mixed-noise channel model with gamma–gamma distributed atmospheric turbulence and Gaussian white noise and analyze the influence of OFDM modulation and single-carrier QPSK modulation on the demodulated signal constellation at the receiver [17].

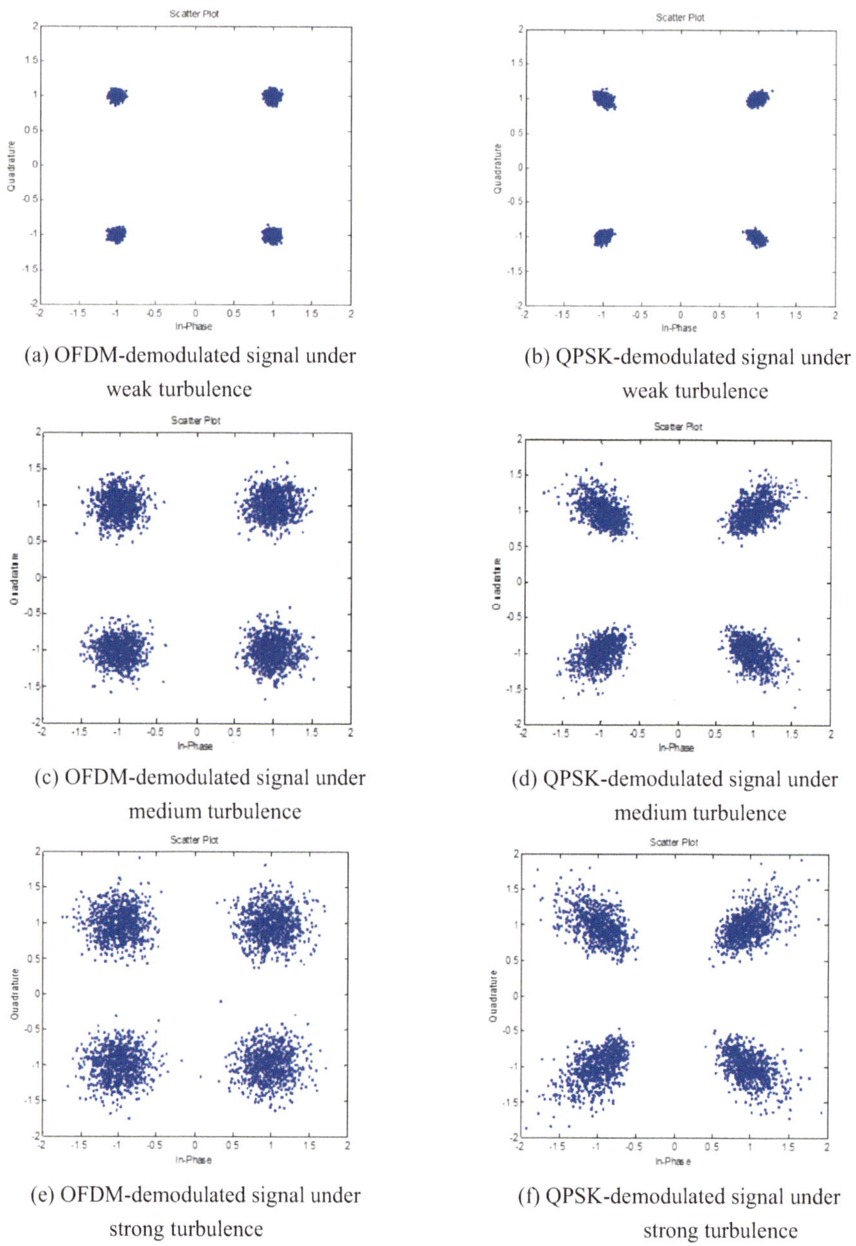

(a) OFDM-demodulated signal under weak turbulence

(b) QPSK-demodulated signal under weak turbulence

(c) OFDM-demodulated signal under medium turbulence

(d) QPSK-demodulated signal under medium turbulence

(e) OFDM-demodulated signal under strong turbulence

(f) QPSK-demodulated signal under strong turbulence

Fig. 4.11 Constellation diagrams of OFDM and single-carrier QPSK demodulated signals under the influence of multiplicative noise [16, 17]

Assuming the additive noise SNR = 0, when $\sigma_R = 0.2$, the constellation diagrams of the demodulated signals at the receiving end under OFDM and single-carrier QPSK modulation are shown in Fig. 4.12a, b, respectively; when $\sigma_R = 1$, the constellation diagrams of demodulated signals at the receiving end are shown in Fig. 4.12c, d, respectively; when $\sigma_R = 2$, the constellation diagrams of the demodulated signals at the lower receiver are shown in Fig. 4.12e, f, respectively.

It can be seen from Fig. 4.12a, b that in the case of weak turbulence, when the signal-to-noise ratio is small, the constellation diagram of the demodulated signal is more represented by the joint influence of additive and multiplicative noise. As shown in Fig. 4.12, as the turbulence intensity increases, the constellation diagram of the single-carrier QPSK demodulation signal expands more significantly in the radial direction, and the influence of multiplicative noise becomes increasingly evident.

Under any turbulence intensity, the constellation diagram of the OFDM-demodulated signal has a clearer phase than that of the QPSK-demodulated signal, which significantly improves the convergence. The distribution of the constellation points follows a Gaussian distribution.

Assuming the additive noise SNR = 10, when $\sigma_R = 0.2$, the constellation diagrams of the demodulated signals at the receiving end under OFDM modulation and single-carrier QPSK modulation are shown in Fig. 4.13a, b, respectively; when $\sigma_R = 1$, the constellation diagrams of demodulated signals at the receiving end are shown in Fig. 4.13c, d, respectively; when $\sigma_R = 2$, the constellation diagrams of the demodulated signals at the receiver are shown in Fig. 4.13e, f, respectively.

It can be seen from Fig. 4.13a, b that in the case of weak turbulence, when the signal-to-noise ratio is large, the demodulated signal constellation diagram shows more multiplicative-noise effects. From Fig. 4.13, it can be seen that with the increase in turbulence intensity, the constellation diagram of the single-carrier QPSK-demodulation signal expands more significantly in the radial direction. However, the demodulated signal-constellation diagram of the OWC-OFDM system does not show radial expansion; its convergence is good, and the constellation point distribution follows the Gaussian distribution.

By comparing the demodulated signal constellations of the OWC-OFDM and OWC-QPSK systems under different turbulence conditions when SNR = 0 and SNR = 10, as shown in Figs. 4.12 and 4.13, we find that when the signal-to-noise ratio is small, the influence of the gamma–gamma channel on the OWC-QPSK system is not very obvious; however, when the signal-to-noise ratio is large, the OWC-QPSK system shows a greater influence of the gamma–gamma channel. In general, the OWC-OFDM system has a suppressing effect on the atmospheric turbulence of the gamma–gamma channel, and the distribution of the constellation diagram is a two-dimensional Gaussian distribution.

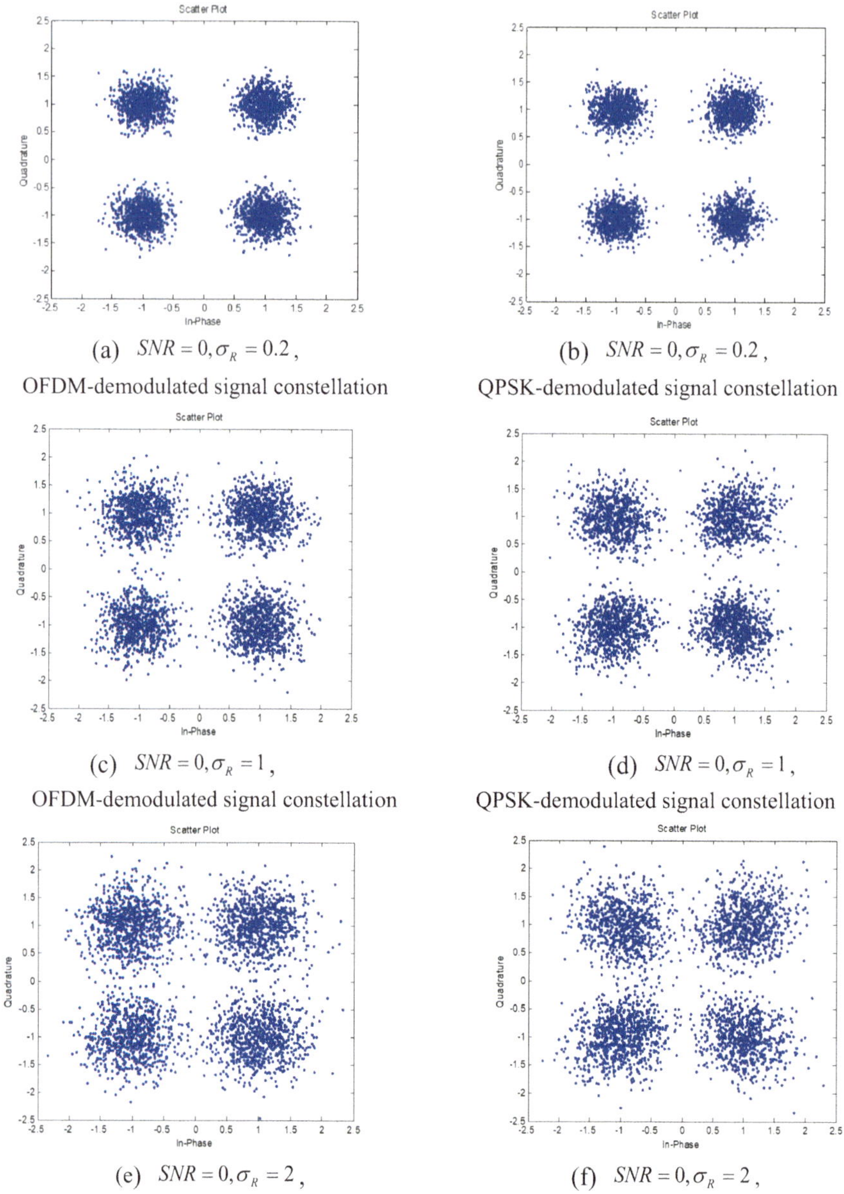

(a) $SNR = 0, \sigma_R = 0.2$,
OFDM-demodulated signal constellation

(b) $SNR = 0, \sigma_R = 0.2$,
QPSK-demodulated signal constellation

(c) $SNR = 0, \sigma_R = 1$,
OFDM-demodulated signal constellation

(d) $SNR = 0, \sigma_R = 1$,
QPSK-demodulated signal constellation

(e) $SNR = 0, \sigma_R = 2$,
OFDM-demodulated signal constellation

(f) $SNR = 0, \sigma_R = 2$,
QPSK-demodulated signal constellation

Fig. 4.12 Constellation diagrams of OFDM and single-carrier QPSK demodulated signals under different turbulence intensities [16, 17]

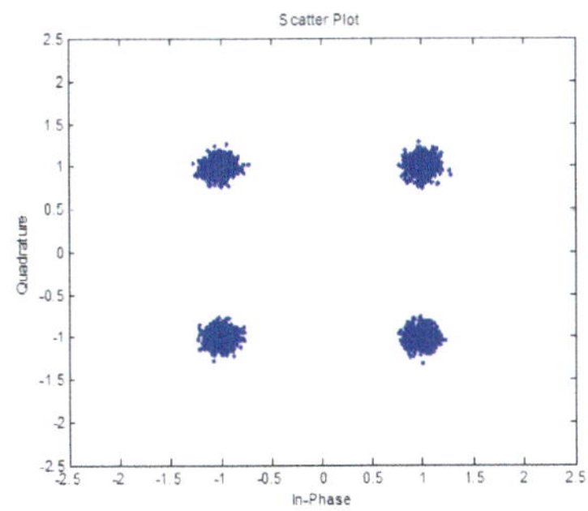

(a) $SNR = 10, \sigma_R = 0.2$,
OFDM-demodulated signal constellation

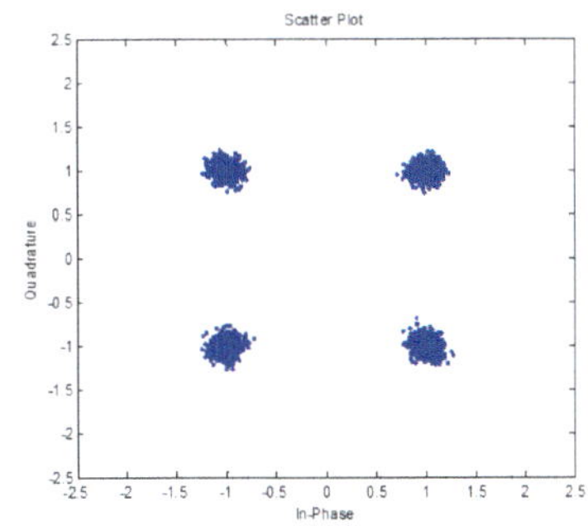

(b) $SNR = 10, \sigma_R = 0.2$,
QPSK-demodulated signal constellation

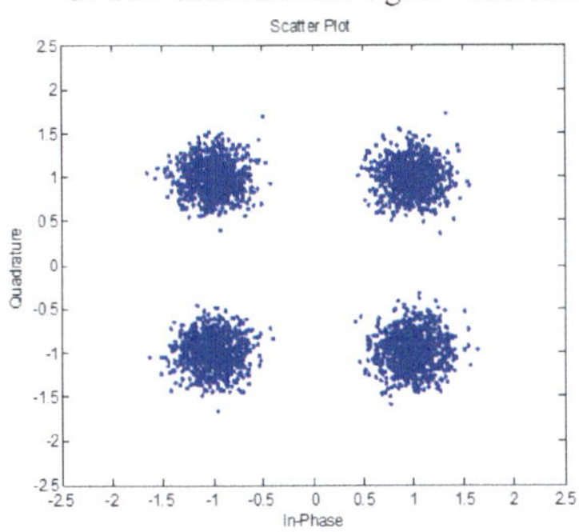

(c) $SNR = 10, \sigma_R = 1$,
OFDM-demodulated signal constellation

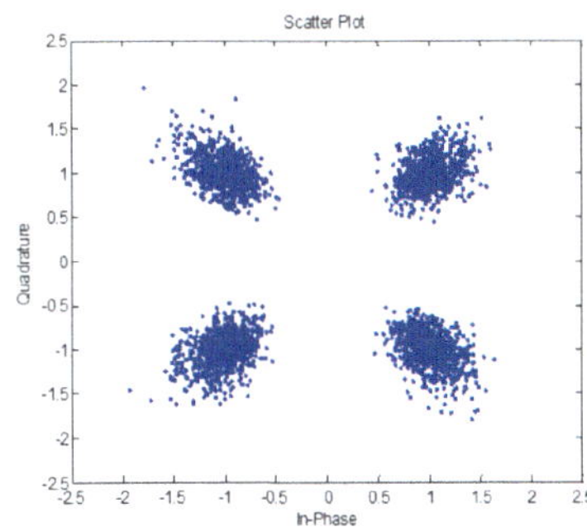

(d) $SNR = 10, \sigma_R = 1$,
QPSK-demodulated signal constellation

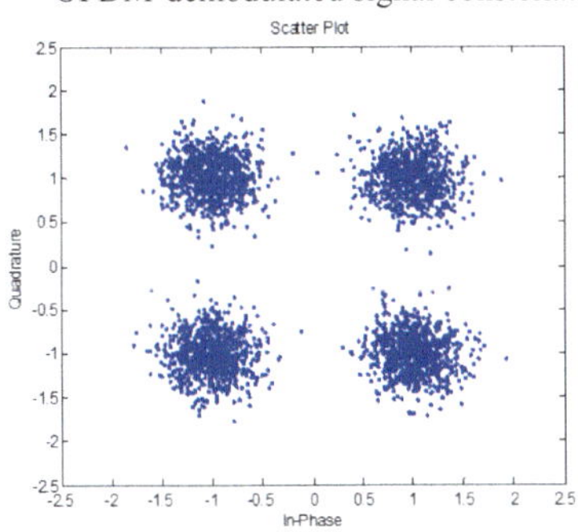

(e) $SNR = 10, \sigma_R = 2$,
OFDM-demodulated signal constellation

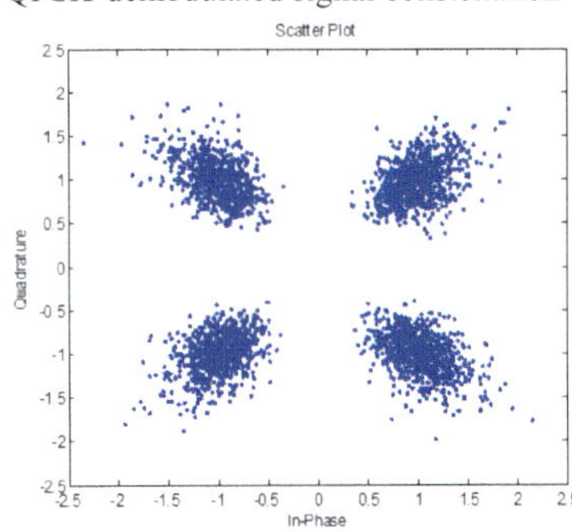

(f) $SNR = 10, \sigma_R = 2$,
QPSK-demodulated signal constellation

Fig. 4.13 Constellation diagrams of OFDM and single-carrier QPSK demodulated signals under different turbulence intensities [16, 17]

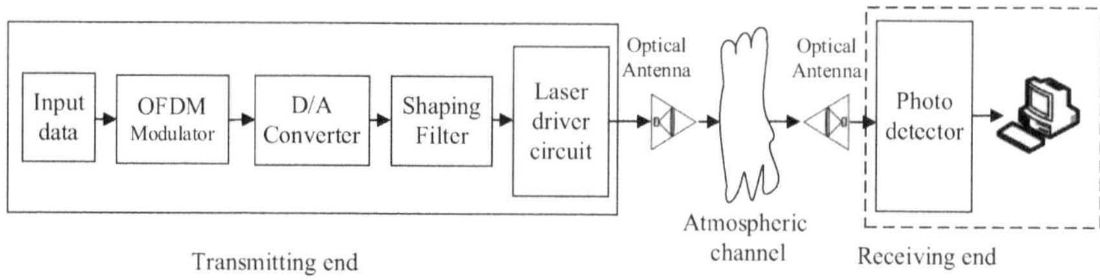

Fig. 4.14 Schematic diagram of the OWC-OFDM experimental-measurement setup [16, 17]

4.4.5 *Experimental Research on OWC-OFDM*

Figure 4.14 shows a schematic diagram of the OWC-OFDM experiment [18]. In the experiment, the OFDM data frame has 32 subcarriers, which are modulated by QPSK. First, the OFDM modulation module performs algorithm processing, such as symbol conversion, serial-to-parallel conversion, and constellation mapping. Then, the digital signal is converted to analog by the D/A converter, and the modulation system is completed using a high-speed digital-to-analog conversion circuit board (AD9788 EVB).

After the digital-to-analog conversion, an electro-optical conversion is performed on the OFDM signal through the signal-processing module, and an appropriate amount of DC bias is added to it to for transmitting wireless-laser channels. Finally, the modulated OFDM signal is loaded onto a semiconductor laser (LQA-850E), and the optical OFDM signal is transmitted through the transmitting antenna.

The signal is sent to the receiving end through the atmospheric channel. The receiving end couples the optical signal into the optical fiber through the receiving antenna, uses the photodetector to convert the optical signal into an electrical signal, and collects the received signal on the computer. Finally, the data are demodulated and analyzed using MATLAB on a computer.

Figure 4.15 show the constellations of the demodulated signal of the OWC-OFDM system obtained from the experiment, and the demodulated signal of the single-carrier OWC-QPSK system

Observing Fig. 4.15a, the constellation points have no aliasing, and the distribution is relatively concentrated and clear; observing Fig. 4.15b, it can be found that the constellation points are more distributed than those of Fig. 4.15a. To more clearly observe the constellation diagrams of the demodulated signals of the OWC-OFDM and OWC-QPSK systems, we performed a statistical analysis of the probability density of the constellation points on the constellation diagrams of the two systems. We obtained the probability-density curves shown in Fig. 4.16.

As shown in Fig. 4.16, the constellation-point distribution of the demodulation constellation diagram of the OWC-OFDM system is more concentrated than that of the OWC-QPSK system. From the above analysis of the demodulation constellation diagram of the signal at the receiving end, we can conclude that the performance of the OWC-OFDM system is better than that of the OWC-QPSK system. This is

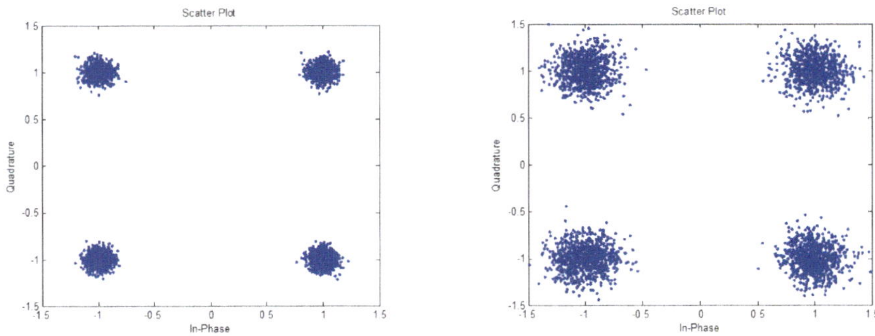

(a) OFDM-demodulated signal constellation (b) QPSK-demodulated signal constellation.

Fig. 4.15 Constellation diagrams of OWC-OFDM and single-carrier OWC-QPSK demodulated signals [16, 17]

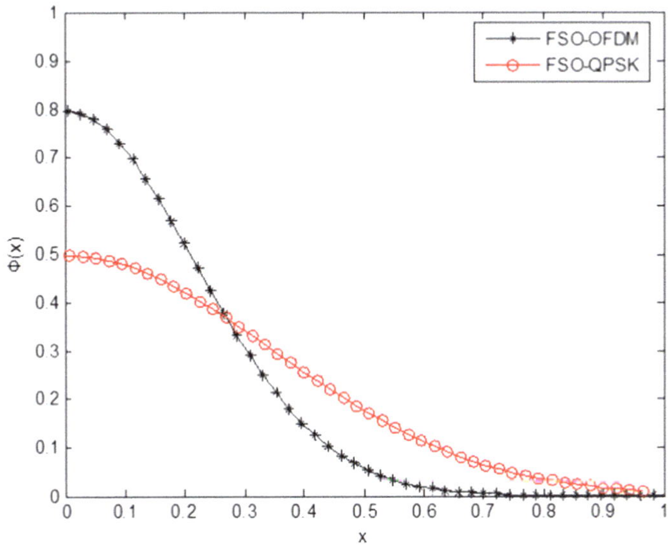

Fig. 4.16 Constellation-point distribution comparison of the OWC-OFDM and OWC-QPSK demodulated signal constellations [16, 17]

because OFDM has a high spectral efficiency and good resistance to intersymbol interference.

Experiments showed that the OWC-OFDM system has a better anti-intersymbol interference performance, and its bit-error rate is lower than that of a single-carrier modulation system. The OWC-OFDM system not only breaks through the bottleneck restricting the development of optical communication but also improves the information-transmission rate and reliability of optical communication, which has high practical value.

Chapter Summary

The basic principle of OFDM was introduced, and the IDFT/DFT implementation of OFDM and the effect of a cyclic prefix on OFDM were discussed.

The effect of OWC-OFDM modulation on the signal constellation and eye diagram was experimentally studied, the feasibility of OWC-OFDM to realize point-to-point communication was verified, and the influence of the wireless laser channel on an optical OFDM signal was analyzed.

The results showed the following:

(1) The bit-error rate of the OWC-OFDM system was lower than that of the single-carrier QPSK system; the impact of intersymbol crosstalk on the OWC-OFDM modulation system was lower than that of the single-carrier QPSK modulation system.
(2) The wireless laser channel had a certain influence on the optical OFDM signal transmission. Owing to the effects of atmospheric attenuation and turbulence, this influence became increasingly obvious as the SNR increased.

References

1. Tan Z, Nie X (2005) Key technologies and applications of OFDM [M]. Southwest Jiaotong University Press, Chengdu
2. Hirosaki B (1981) An orthogonally multiplexed QAM system using the discrete fourier transform. IEEE Trans Commun 29(7):982–989
3. Zhao L, Ke X, Liu J (2009) Modeling and LS channel estimation research of band FSO-OFDM system. Laser J 30(3):38–39
4. Shieh W, Djordjevic I (2010) OFDM for optical communications
5. Weinstein SB, Ebert PM (1971) Data transmission by frequency division multiplexing using the discrete fourier transform. IEEE Trans Commun Technol 19(5):628–634
6. Zhao L, Ke X, Liu J (2009) An improved baseband model for FSO-OFDM system. Semicond Optoelectron 30(02):277–279+283
7. Duhamel P, Hollmann H (1984) Split radix' FFT algorithm. Electron Lett 20(1):14–16
8. Zhao L, Ke X, Wang H (2009) Free-space optical communication orthogonal frequency division multiplexing system model based on time-frequency coding. Chin J Lasers 36(10):2757–2762
9. Pollet T, Van Bladel M, Moeneclaey M (1995) BER sensitivity of OFDM systems to carrier frequency offset and Wiener phase noise. IEEE Trans Commun 43(234):191–193
10. Armada AG (2001) Understanding the effects of phase noise in orthogonal frequency division multiplexing (OFDM). IEEE Trans Broadcast 47(2):153–159
11. Zhang X (1989) Progress of coherent optical fiber communication technology. Study Opt Commun 33(1):1–11
12. Nee RV, Prasad R (2000) OFDM for wireless multimedia communications. Artech House, Boston
13. Zhao L (2009) Research on key technologies of FSO-OFDM system. Xi'an University of Technology
14. Xie M (2012) Research on the influence of atmospheric dispersion on FSO-OFDM system. Xi'an University of Technology

15. Zhu X, Kahn JM (2002) Free-space optical communication through atmospheric turbulence channels. IEEE Trans Commun 50(8):1293–1300
16. Ke X, Lei S, Li B (2015) Experimental research on optical wireless orthogonal frequency division multiplexing system. Chin J Lasers (12):144–152
17. Li BL (2015) Experimental research on FSO-OFDM systems. Xi'an University of Technology, Xi'an
18. Xie M, Ke X (2013) Research on the impact of atmospheric turbulence on signal-to-noise ratio in optical wireless communication systems. Laser Optoelectron Progr 50(11):61–67

Chapter 5
Atmospheric Channels

When a laser signal passes through an atmospheric channel, atmospheric molecules and aerosols absorb and scatter it. Atmospheric absorption reduces the received optical power, and the scattering of the laser signal by the atmosphere causes a multipath laser-transmission effect, which causes the laser-pulse signal to broaden over space and time. In the receiver, this appears as intersymbol crosstalk.

The random fluctuation of the optical refractive index caused by atmospheric turbulence causes the laser signal to exhibit light-intensity fluctuation, arrival-angle fluctuation, beam drift, beam spread, and other phenomena in the transmission process. It also increases the communication bit-error rate and may even cause short-term communication interruptions. This seriously affects the stability and reliability of atmospheric optical communications. In this chapter, the transmission and scattering of laser signals in an atmospheric channel and the influence of the atmospheric environment on optical-signal transmissions are analyzed.

5.1 Atmospheric Scattering and Attenuation in Laser Propagation

5.1.1 Atmospheric Scattering

Atmospheric scattering is a phenomenon in which the incident energy of light is redistributed in all directions according to a certain law. In essence, particles, such as atmospheric molecules or aerosols, generate electric dipole or multipole oscillation under the action of incident electromagnetic waves, and radiate wavelets with the same frequency as the incident wave; that is, they scatter waves. The energy distribution of the scattered wave is related to the wavelength and intensity of the incident wave, as well as the size, shape, and refractive index of the particles.

X. Ke, *Principles and Applications of Optical Wireless Orthogonal Frequency-Division Multiplexing*, Optical Wireless Communication Theory and Technology,
https://doi.org/10.1007/978-981-97-7973-4_5

The atmospheric-scattering effect specifically refers to the influence of atmospheric particles on the laser transmission. The diameter distribution of atmospheric particles is wide, ranging from 10^{-4} to 10 μm in orders of magnitude [1]. When the gas-molecular spacing is more than ten times the molecular diameter, the spacing of aerosol particles or suspended particles in the atmosphere is much larger than the particle diameter; thus, it meets the conditions for single-particle scattering. In other words, when the particle spacing is more than three times the particle diameter, the scattering of each particle is approximately independent of the others. Therefore, although many types of particles exist in the atmosphere, the single-particle scattering theory is always applicable to light propagation.

The attenuation coefficient of the laser energy generated by single-particle scattering is β [2]:

$$\beta = \beta_m + \beta_p, \tag{5.1}$$

where β_m is the molecular-scattering coefficient, $\beta_m = \sigma_m \cdot n$; β_p is the particle-scattering coefficient, σ_m is the molecular-scattering cross section, which is inversely proportional to the wavelength λ, and n is the molecular density of the atmosphere.

5.1.1.1 Atmospheric Molecular Scattering

Molecular scattering is also known as Rayleigh scattering. Rayleigh scattering occurs when the laser wavelength is larger than the scattering particle size. Rayleigh scattering mainly occurs in the ultraviolet wave band or at high altitudes with few suspended particles. This scattering is inversely proportional to the fourth power of the wavelength of scattered light. The empirical publicity of the Rayleigh scattering coefficient is [1]

$$\sigma_m = 0.827 \times N \times A^3/\lambda^4 \tag{5.2}$$

where A is the cross-sectional area of the scattering element (cm^2), N is the number of particles per unit volume (cm^{-3}), λ is the density of the scattering element and the laser wavelength.

It can be seen from Eq. (5.2) that as the radius of the scattering molecules increases, the scattering increases; as the wavelength increases, the scattering decreases. Visible light is stronger than infrared light and blue light is stronger than infrared light. Molecular scattering plays a major role in clear-sky conditions with few other large particles. Because blue-light scattering is the strongest, a clear sky is blue.

The volume-scattering coefficient of Rayleigh scattering is [1]

$$\alpha_m(\lambda) = \frac{8\pi^2}{3} \cdot \frac{\left(n^2 - 1\right)^2}{N^2\lambda^4} \tag{5.3}$$

where n denotes the refractive index of the medium. In dry and fresh air, the Rayleigh scattering coefficient [1] is

$$\alpha_m(\lambda) = 1.09 \times 10^{-3}\lambda^{-4.05}\ \text{km}^{-1}. \tag{5.4}$$

Generally, for particles with radius $r < 0.03$ μm, if the light wavelength is around 1 μm, the calculation error of the Rayleigh scattering coefficient is $\tilde{n} < 1\%$[3]; therefore, the above formula can accurately describe the scattering intensity of molecules in the atmosphere.

5.1.1.2 Atmospheric Particle Scattering

Mie scattering occurs when the light wavelength is comparable to the size of the particles. Mie scattering depends mainly on the size, density distribution, and refractive-index characteristics of the scattered particles. The dependence on wavelength is far weaker than Rayleigh scattering, and its coefficient is σ_n [1]:

$$\sigma_n = N(r)\pi r^2 Q_s(X_r, m), \tag{5.5}$$

where $N(r)$ is the number of particles per unit volume (cm^{-3}), r is the radius of the particle (cm), Q_s is the scattering efficiency, which is defined as the ratio of the particle's scattered energy to the energy of the geometric section of the incident particle, $X_r = 2\pi r/\lambda$ is a function of the relative scale, $m = n - iK_a$ is the complex refractive index of the particle, n and K_a are the real and imaginary parts of the complex refractive index of the particle, respectively, and λ is the laser wavelength.

As the size distribution of aerosol particles is empirical, the following approximate formula σ is generally used to calculate the total attenuation coefficient of aerosol particles [4]:

$$\sigma = \sigma_m \sigma_n = \frac{3.912}{V}\left[\frac{0.55}{\lambda}\right]^q \tag{5.6}$$

where V is the visibility (km). q is the coefficient related to the size and density distribution of particles in the atmosphere; when there is high visibility ($V > 50$ km), q is 1.6; when visibility is general (6 km $< V <$ 50 km), q is 1.3; when visibility is low ($V < 6$ km), q is $0.85V^{1/3}$.

5.1.2 Atmospheric Attenuation

Atmospheric attenuation is the influence of the atmosphere on the absorption and scattering of light waves in a laser transmission. The atmospheric transmittance can be used to measure the degree of atmospheric attenuation. The atmospheric transmittance of a monochromatic wave can be expressed as follows [2].

Horizontal uniform path transmission:

$$T_{atm}(\lambda) = \exp[-k_e(\lambda)L]. \tag{5.7}$$

Oblique path transmission:

$$T_{atm}(\lambda) = \exp\left(-\sec\varphi\int_0^z k_e(\lambda, r)dr\right), \tag{5.8}$$

where $n = 1 + \left(2876.04 + \frac{3\times16.288}{\lambda^2} + \frac{5\times0.136}{\lambda^4}\right) \times 10^{-7}$ is the atmospheric transmittance at wavelength λ, L is the horizontal transmission distance, Z is the vertical height of the oblique path, φ is the zenith angle of the oblique path, and $k_e(\lambda) = k_s + k_a$ is the atmospheric-extinction coefficient.

For oblique-path transmissions, the atmospheric-extinction coefficient varies with altitude; thus, path integration is required to calculate the transmittance. If the initial light intensity is I_0, the light intensity $I(L)$ after the beam travels a certain distance L is [5]

$$I(L) = I_0 T_{atm}(\lambda) \tag{5.9}$$

The optical thickness n_s of the atmospheric optical transmission channel can be used to characterize the attenuation characteristics of the channel [6]:

$$\tau = \tau_a + \tau_s \tag{5.10}$$

where $n_s = 1 + \left(287.6155 + \frac{1.62887}{\lambda^2} + \frac{0.01360}{\lambda^4}\right) \times 10^{-6}$ is the scattering optical thickness and τ_s is the absorption optical thickness.

Because of the uncertainty of atmospheric motion, the atmospheric-extinction coefficient varies widely under different weather conditions. In the near-surface atmosphere, the influence of molecular scattering is very small, and the attenuation of light energy is mainly caused by the Mie scattering of large-scale particles.

When designing an atmospheric optical wireless communication link, the wavelength in the atmospheric window is usually selected such that the contribution of atmospheric absorption to the laser-energy attenuation is relatively small, and the atmospheric attenuation mainly comes from atmospheric scattering. For near-ground laser atmospheric transmissions, atmospheric attenuation can be calculated using visibility. The empirical formula between visibility and the atmospheric extinction coefficient is [5]

$$k_e(\lambda, R_v) = \frac{3.912}{R_v}\left(\frac{550}{\lambda}\right)^q \tag{5.11}$$

Table 5.1 International visibility levels [7]

Grade	Weather status	Visibility (km)	Scattering coefficient	Grade	Weather status	Visibility (km)	Scattering coefficient
0	Extremely dense fog	<0.05	>78.2	5	Haze	2–4	1.960–0.954
1	Thick fog	0.05–0.2	78.2–19.6	6	Light haze	4–10	0.954–0.391
2	Medium fog	0.2–0.5	19.6–7.82	7	Sunny	10–20	0.391–0.196
3	Mist	0.5–1	7.82–3.91	8	Very sunny	20–50	0.196–0.078
4	Light mist	1–2	3.91–1.96	9	Extremely sunny	>50	0.0141

$$q = \begin{cases} 1.6 & R_v > 50\ \text{km} \\ 13 & 6\ \text{km} < R_v < 50\ \text{km} \\ 0.585R_v^{1/3} & R_v < 6\ \text{km} \end{cases} \tag{5.12}$$

where R_v is the atmospheric visibility (km) and λ is the laser wavelength (nm). $k_e(\lambda, R_v)$ is the extinction coefficient in km^{-1}.

In the aviation industry, visibility is defined as the maximum distance at which a person with normal vision can clearly see the contour of a target under prevailing weather conditions. Visibility is closely related to the weather conditions at that time. When rainfall, fog, haze, sandstorms, and other weather processes occur, the atmospheric transparency is low and visibility is poor.

Generally, a visual-inspection method can be used to measure atmospheric visibility, and atmospheric-transmission instruments, laser-visibility automatic measuring instruments, and other measuring instruments can be used for testing. Meteorology divides visibility into ten levels, as shown in Table 5.1.

5.2 Laser Propagation in Atmospheric Turbulence

A random change in atmospheric temperature produces a random change in atmospheric density, which leads to random fluctuations in the atmospheric refractive index. The cumulative effect of random fluctuations leads to an uneven distribution of the atmospheric refractive index. The nature of this fluctuation, caused by atmospheric-turbulence motion, is that the optical wave parameters (amplitude and phase) produce random fluctuations, resulting in the flicker, bending, splitting, expansion, and spatial coherence reduction of a beam. These effects are important factors that limit the full performance of optical wireless communication systems.

5.2.1 Statistical Characteristics of Atmospheric Turbulence

According to the Kolmogorov theory [8], a change in the turbulence average velocity causes the turbulence to obtain energy. The random fluctuation $n(\mathrm{r})$ of the atmospheric refractive index is mainly caused by the random microstructure of the temperature spatial distribution. This change in microstructure is due to the large-scale temperature inhomogeneity caused by different degrees of solar heating in different regions of the Earth's surface. This large-scale temperature inhomogeneity, in turn, leads to large-scale refractive index inhomogeneity. These atmospheric refractive-index inhomogeneities are usually called turbulent "vortices."

Turbulence can be characterized using two scales. In the atmospheric boundary layer, the largest scale vortex that can be observed and analyzed is L_0; also referred to as the outer turbulence scale. L_0 is usually within the range of tens to hundreds of meters, while the smallest scale is l_0, also referred to as the inner turbulence scale; it is only a few millimeters [9], as shown in Fig. 5.1 [10].

When the beam propagates through these vortices, large-scale turbulent vortices mainly produce refraction effects on the beam and small-scale vortices mainly produce diffraction effects on the beam. Factors related to the formation of atmospheric turbulence, such as the temperature, atmospheric refractive index, and distribution of aerosol plasmid, have a turbulence-doping effect, while the refractive index of light waves propagating in the turbulent atmosphere is only related to the position of two points in space; it can be expressed as [11]

$$n(r, t) = n_0 + n_1(r, t) \tag{5.13}$$

where $n(r, t)$ is the refractive index dependent on the time and position parameters, n_0 represents the refractive index in optical wireless (without turbulence), and $n_1(r, t)$ represents the random fluctuation around the average value n_0, which is caused by turbulence in the atmosphere.

We must determine the statistical characteristics of the random fluctuation of the refractive index in a turbulent flow field. The atmospheric refractive index is a

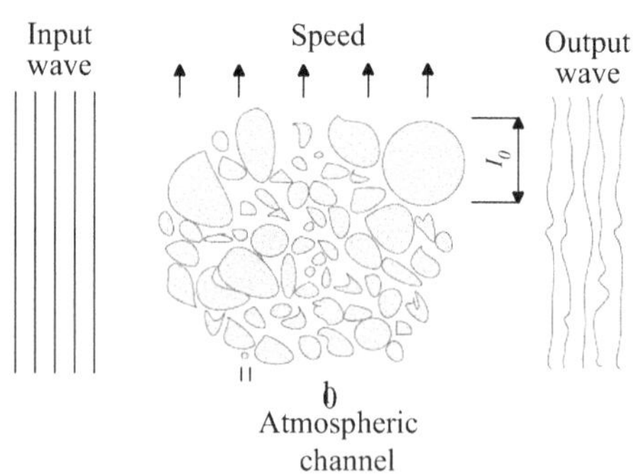

Fig. 5.1 Turbulent vortex in an atmospheric channel [10]

function of temperature and its structural function is closely related to temperature fluctuations.

5.2.1.1 Refractive-Index Fluctuation Power Spectrum

The refractive-index fluctuation power spectrum of atmospheric turbulence Φ_n is mainly based on the Kolmogorov turbulence theory. Typical models include the Kolmogorov spectrum, von Karman spectrum (also known as the Tatarskii spectrum) [12], Hill spectrum [13], and modified Hill spectrum [14] proposed by Andrews. A non-Kolmogorov spectrum has also been reported [15].

Atmospheric turbulence leads to random fluctuations in the atmospheric optical refractive index. The atmospheric refractive index is controlled by changes in the air temperature and pressure; therefore, it is a function of pressure and temperature. Its expression is [11]

$$n = 1 + 77.6\left(1 + 7.52 \times 10^{-3}\lambda^{-2}\right)\frac{P}{T_e} \times 10^{-6} \tag{5.14}$$

where λ is the wavelength (μm), P is the pressure ($100P_a$), and T is the thermodynamic temperature (K). Kolmogorov theory holds that [16] for locally uniform and isotropic turbulence, the change in refractive index $D_n(r)$ can be characterized by the refractive-index structure function, which is proportional to the power of the scalar distance $r^{2/3}$:

$$D_n(r) = \left\langle [n(r + r_1) - n(r)]^2 \right\rangle = C_n^2 r^{2/3},\ l_0 < r < L_0 \tag{5.15}$$

where $D_n(r)$ is the ensemble average of the refractive-index increment between the two observation points, $n(r)$ represents the refractive index, and C_n^2 is a constant coefficient of the refractive-index structure function $D_n(r)$. C_n^2 is called the atmospheric refractive-index structure constant, which is a physical quantity used to measure the intensity of optical turbulence. Based on the "2/3 power law" of the atmospheric refractive-index structure function proposed by Kolmogorov, the power spectrum describing the fluctuation of the refractive index caused by atmospheric turbulence can be obtained.

The Kolmogorov spectrum can be written as [17]

$$\Phi_n(k) = 0.033 C_n^2 k^{-11/3} \tag{5.16}$$

where $k = 2\pi/l$ is the spatial wavenumber and l is the scale of the turbulent vortex. The Kolmogorov spectrum only holds in the inertial subinterval, that is, $2\pi/L_0 < k < 2\pi/l_0$. When the Kolmogorov spectrum is used to calculate the laser atmospheric-turbulent propagation, it is often assumed that the outer turbulence scale is infinite, and the inner turbulence scale can be ignored.

For the dissipative zone, Tatarskii describes $\Phi_n(k)$, the rapid decline phenomenon, with the following model [17]:

$$\Phi_n(k) = 0.033C_n^2 k^{-11/3} \exp\left(-k^2/k_m^2\right) \tag{5.17}$$

where, if $k_m = 5.92/l_0$, the above formula is true for $k > k_0$.

In the Kolmogorov and Tatarskii spectra, when $k \to 0$ and $\Phi_n \to \infty$, the rapid-decline phenomenon was impossible in the limited Earth atmosphere at that time. To overcome this defect, the von Karman spectrum is often used to approximate the value [17]:

$$\Phi_n(k) = 0.033C_n^2 \left(k^2 + k_0^2\right)^{-11/6} \exp\left(-k^2/k_m^2\right) \tag{5.18}$$

where $k_0 = 2\pi/L_0$; the above formula is valid within the range $0 \le k < \infty$.

If Eq. (5.18) is used to describe the turbulence spectrum in the input region, it can only be an approximate value, because the turbulence in this region is usually anisotropic. The von Karman spectrum does not consider bump phenomena in the high wavenumber region. Hill proposed an accurate numerical model [14]:

$$\Phi_n(k) = 0.033C_n^2 \left(k^2 + k_0^2\right)^{-11/6} \exp\left(-k^2/k_1^2\right)\left[1 + a_1(k/k_1) - a_2(k/k_1)^{7/6}\right] \tag{5.19}$$

where $a_1 = 1.802$, $a_2 = 0.254$ and $k_l = 3.3/l_0$. From the above Eq. (5.19), it can be seen that after $a_1 = a_2 = 0$ and the substitution of the combination $k_1 = k_m$, the modified Hill spectrum is simplified to the von Karman spectrum. When $k_0 = l_0 = 0$, it degenerates into the Kolmogorov spectrum. The distribution curves of the different spectral models shown in Fig. 5.2 can be obtained by ignoring the external scale effect (value $k_0 = 0$) and normalizing other spectral models with the Kolmogorov spectral model.

It can be seen from Fig. 5.2 that the von Karman spectrum is basically the same as the Kolmogorov spectrum for $kl_0 < 0.5$ (inertial sub-region). With an increase in the wavenumber, the von Karman spectrum decreases sharply and monotonically, which well explains the rapid decline of the dissipation region $k < k_m = 5.92/l_0$ observed in the experiment.

The Hill spectrum and modified Hill spectrum are significantly different from the Kolmogorov spectrum in the region $0.1 < kl_0 < 4$ (high-wavenumber region of the inertial sub-region), which reflects the influence of a sudden change in the high-wavenumber region on an optical transmission. In the dissipative region, the decline rate of the Hill spectrum and modified Hill spectrum is faster than that of the von Karman spectrum, and all three spectral models tend to zero at that time. The Hill spectrum and modified Hill spectrum reach their maximum values near $kl_0 = 1.1$.

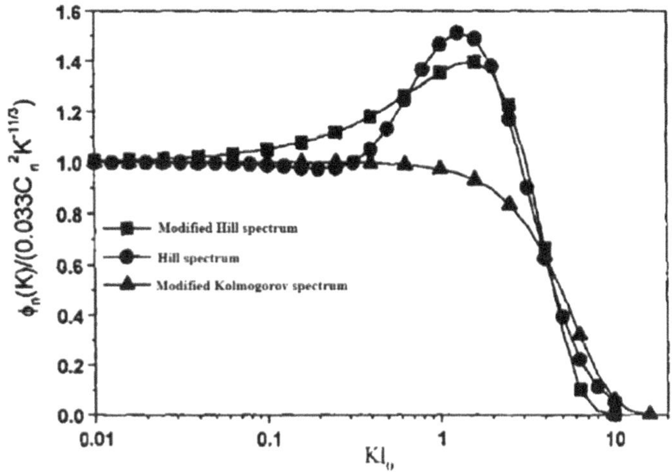

Fig. 5.2 Distribution curves of different refractive-index fluctuation power-spectrum models [18]

5.2.1.2 Atmospheric Refractive-Index Structure Constant

A change in atmospheric temperature will cause a change in the atmospheric refractive index. With a 1 °C change in air temperature, the refractive index will change about 1×10^{-6}. The temperature change can be characterized by the structure function of the temperature field [18]:

$$D_T(r) = \left\langle [T(r + r_1) - T(r)]^2 \right\rangle = C_T^2 r^{2/3} \tag{5.20}$$

where C_T^2 denotes the atmospheric temperature structure constant. The relationship between the atmospheric refractive-index structure constant and the atmospheric temperature structure constant is [18]

$$C_n^2 = \left[\frac{10^{-6}}{T} \left(\frac{77.6p}{T} \right) + \frac{0.584p}{T\lambda^2} \right]^2 C_T^2 \tag{5.21}$$

where T is the atmospheric temperature, P is the atmospheric pressure, and λ is the optical wavelength.

For heights ranging from the ground up to 100 m, C_n^2 is usually $10^{-15}\text{m}^{-2/3} \sim 10^{-13}\text{m}^{-2/3}$. The refractive-index structure constant near the ground is affected by the topography, weather, geographical location, and other factors. Because the atmospheric condition itself changes with time, the refractive-index structure constant C_n^2 also varies with time and changes significantly with the altitude. The common atmospheric-refractive index structure constant models include the following [19–21].

(1) **Hufnagel model**

$$C_n^2(h) = 2.72 \times 10^{-16}\left[3\langle v\rangle^2\left(\frac{h}{10}\right)^2 \exp(-h) + \exp(-h/1.5)\right] \quad (5.22)$$

where $\langle v\rangle$ is the average wind speed (m/s) and h is the height from the ground (km). The model can be used to predict the atmospheric refractive-index structure constant in a region with an average altitude of more than 3 km at night and in the infrared band. Hughes provided the variation relationship between wind speed and altitude on the basis of experimental observations [15]:

$$v(h) = 3 + 17\exp\left[-\frac{(h-12.5)^2}{16}\right] \quad (5.23)$$

where $v(h)$ is the wind speed (m/s), h is the height from the ground (km), and the model is applicable to areas below 20 km.

(2) **Hufnagel–Valley model**

$$\begin{aligned} C_n^2 &= 0.00594(v/27)^2\left(10^{-5}h\right)^{10}\exp(-h/1000) \\ &+ 2.7 \times 10^{-16}\exp(-h/1500) + \hat{A}\exp(-h/100) \end{aligned} \quad (5.24)$$

where v is the vertical-path wind speed (m/s), $h(m)$ is the system signal-transmission height, and the value of $\hat{A}$ depends on the ground value $C_n^2(0)$. The Hufnagel–Valley model is suitable for predicting the atmospheric refractive-index structure constant during the daytime in inland areas. The values range from strong turbulence ($10^{-12}m^{-2/3}$) to weak turbulence ($10^{-17}m^{-2/3}$), with a typical average value of $10^{-15}m^{-2/3}$.

HV-21 model

$$\begin{aligned} C_n^2 &= 5.94 \times 10^{-53}\left(\frac{21}{27}\right)^2 h^{10}\exp(-h/1000) \\ &+ 2.7 \times 10^{-16}\exp(-h/1500) + 1.74 \times 10^{-14}\exp(-h/100) \end{aligned} \quad (5.25)$$

where h is the height from the ground (m). The model is based on the Hufnagel–Valley model when $v = 21m/s$, $\hat{A} = 1.7 \times 10^{-14}m^{-2/3}$.

(3) **Modified HV model**

$$\begin{aligned} C_n^2 &= 8.16 \times 10^{-54}h^{10}\exp(-h/1000) \\ &+ 3.02 \times 10^{-17}\exp(-h/1500) + 1.9 \times 10^{-15}\exp(-h/100) \end{aligned} \quad (5.26)$$

where h is the height from the ground (m). The model is applicable for predicting the atmospheric refractive-index structure constant at night.

5.2.2 *Influence of Atmospheric Turbulence on Laser Propagation*

The influence of atmospheric turbulence on beam propagation is closely related to the ratio of the beam diameter d_B to the turbulent vortex scale l. When $d_B/l << 1$, the beam diameter is much smaller than the scale of the turbulent vortex, and the role of the turbulent vortex is to randomly deflect the beam as a whole.

On the distant receiving plane, the projection point (i.e., the spot position) of the beam center takes the statistical average position as the center, resulting in a rapid random jump. This phenomenon is called beam drift. When $d_B/l \approx 1$, the beam diameter is equivalent to the scale of the turbulent vortex. The turbulent vortex causes the beam wavefront to randomly deflect and the arrival angle on the receiving plane to fluctuate, resulting in the image point jittering on the receiving focal plane.

When $d_B/l \gg 1$, that is, the beam diameter is much larger than the scale of the turbulent vortex, the beam section contains multiple turbulent vortices, and each vortex independently scatters and diffracts the part of the beam passing through it. This results in random fluctuations of the beam intensity in time and space, and light-intensity flickering [22–24].

(1) **Intensity fluctuations (atmospheric scintillation)**

When the turbulence is weak and the transmission distance is short, the logarithmic intensity variance of scintillation is expressed as [27]

$$\sigma_x^2 = C_0 k^{7/6} L^{11/6} C_n^2 \tag{5.27}$$

where L is the laser transmission distance, C_0 is a constant, and $k = 2\pi/\lambda$ is the wavenumber. C_n^2 is the atmospheric refractive-index structure constant. When the laser beam is horizontally transmitted near the ground, when σ_x^2 reaches 1–2, it does not increase with the increase in turbulence intensity or transmission distance, but may decrease. This phenomenon is known as the scintillation-saturation effect. When the visible-band laser passes upward or downward through the atmosphere, σ_x^2 is approximately 0.02. This intensity fluctuation does not have a significant impact on laser applications.

(2) **Beam drift**

Generally, the diameter of the laser beam is smaller than the external turbulence scale, and a vortex larger than the diameter of the laser beam causes the mass center of the beam cross-section to drift randomly. This may cause the laser beam to leave the receiver's field of view in serious cases. Therefore, beam drift is an important factor that affects the performance of atmospheric optical communication.

Clifford provides an expression for calculating the mean-square value of the beam drift when a plane wave propagates [25]:

$$\langle p_l^2 \rangle = 2.2 C_n^2 l_0^{-1/3} L^3 \tag{5.28}$$

where l_0 is the internal turbulent scale. Ishimaru derived an expression for the mean-square value of the beam drift when a Gaussian collimated beam propagates. The time scale of the beam drift is generally on the same order of magnitude as the ratio of the beam diameter to the radial wind speed. When the turbulence effect is very strong, the laser beam breaks up because of the loss of coherence of the laser wavefront. At this time, the concept of beam drift is insignificant.

(3) **Beam spread**

Vortices smaller than the beam diameter cause the beam to expand, causing the beam to spread faster in a turbulent atmosphere than in a vacuum. Therefore, for the same communication distance, to ensure the received signal-to-noise ratio, the required laser-emission power in a turbulent atmosphere must be greater than that in a vacuum. For the atmospheric-turbulent propagation of a Gaussian beam in a weak turbulent region, the beam diameter on the receiving plane is [26]

$$D_T = D_0\left(1 + 1.33\sigma_1^2\Lambda^{5/6}\right)^{1/2} \tag{5.29}$$

In a strong turbulent region, the beam diameter on the receiving plane is [26]

$$D_T = D_0\left(1 + 1.63\sigma_l^2\Lambda^{5/6}\right)^{1/2} \tag{5.30}$$

where D_0 is the beam diameter on the receiving plane when propagating in a vacuum, σ_l^2 is the Rytov variance, and Λ is the Gaussian beam parameter [19].

(4) **Phase fluctuations**

The random characteristics of atmospheric turbulence not only cause light-intensity flicker, but also cause the laser phase to fluctuate in time and space. The optical-wave phase fluctuation reduces the focusing performance of the optical-receiving telescope on the signal light, and increases the spot area on the focal plane of the receiver. Relative to the diffraction-limit spot, the laser phase fluctuation causes the spot area on the focal plane to increase by $(D/r_0)^2$ times, where D is the aperture of the receiving telescope; it is the Fried parameter.

The output signal-to-noise ratio of photodetectors is typically approximately proportional to the data rate and detection area. To eliminate the influence of phase fluctuations, the detection area must be increased, which reduces the performance of the optical communication link multiple times.

5.3 Atmospheric-Turbulence Model

The geometric size of gas molecules, water mist, snow, aerosols, and other particles in the atmosphere is close to or even smaller than the laser wavelength, which causes light absorption and scattering. Especially in the case of strong turbulence, the light signal is significantly disturbed.

The atmospheric-turbulence effect caused by uneven atmospheric temperature and pressure leads to random fluctuations in the light intensity on the receiving surface with time and space, which is the so-called "intensity scintillation effect." The random fluctuation of the light intensity at the receiving end is an important manifestation of the atmospheric-turbulence effect, and is also a major factor affecting the performance of optical-communication systems based on direct detection.

Taylor proposed the following [7]: Under certain conditions, when turbulence flows through the sensor, it can be considered to be frozen. The coherent time t_0 of the atmospheric turbulence is on the order of milliseconds, which is very different from a typical data-symbol time. Therefore, the atmospheric-turbulence channel can be used as a "slow-fading channel" to statically describe the duration of data symbols [27].

5.3.1 Lognormal Turbulence Model

A random distribution in the turbulent atmosphere causes a small disturbance to the light wave that has been propagating stably. Therefore, after the light wave travels a certain distance in a turbulent atmosphere, its amplitude and phase will randomly fluctuate. The Maxwell equation for light propagation in a medium is

$$\nabla^2 \vec{E} + k^2 n^2 \vec{E} = 0 \tag{5.31}$$

where $k = 2\pi/\lambda$ is the spatial wavenumber and n is the refractive index at spatial position r. This equation differs from the conventional wave equation only in that $n(r)$ in the second-term coefficient is a function of position r; therefore, it is difficult to accurately solve this basic atmospheric light-transmission equation. For atmospheric-turbulent media, the fluctuation of the refractive index function $n(r)$ is a random process in time; therefore, it needs to be described by statistical theory [28].

Because the three components of the electric-field vector follow the same wave equation, the scalar equation can be used to replace the vector equation shown in Eq. (5.31), and the following is obtained:

$$\nabla^2 \tilde{u} + k^2 n(r) \tilde{u} = 0 \tag{5.32}$$

where $\tilde{u}$ represents any field component, E_x, E_y, or E_z. Applying the Rytov transformation to the above equation, $\tilde{u}$ [29],

$$\psi = \text{In}[\tilde{u}] \tag{5.33}$$

Then, Eq. (5.32) is transformed into the Riccati equation [29]:

$$\nabla^2 \Psi(r) + [\nabla \Psi(r)]^2 + k^2 n^2(r) = 0 \tag{5.34}$$

For the Earth's atmosphere, $n(r) = 1 + n_1(r)$; hence, the above formula can be written as

$$\nabla^2\Psi + [\nabla\Psi]^2 + k^2[1 + n_1(r)]^2 = 0 \tag{5.35}$$

We set $\Psi = \Psi_0 + \Psi_1 + \Psi_2 + \cdots$; Ψ_0 must meet the following:

$$\nabla^2\Psi_0 + [\nabla\Psi_0]^2 + k^2 = 0 \tag{5.36}$$

Let a, ψ_1, and ignore all terms higher than $\psi = \psi_0 + \psi_1$; consider Eq. (5.36):

$$\nabla^2\psi_1 + \nabla\psi_1(2\nabla\psi_0 + \nabla\psi_1)^2 + 2k^2n_1(r) + k^2n_1^2(r) = 0 \tag{5.37}$$

For atmospheric turbulence $n_1(r) \ll 1$, if $|\nabla\psi_1| \ll |\nabla\psi_0|$, the second-order small-sum $(\nabla\psi_1)^2$ and $k^2n_1^2(r)$ in the above formula can be ignored. Finally, the following equation [29] is obtained:

$$\nabla^2\psi_1 + 2\nabla\psi_1\nabla\psi_0 + 2k^2n_1(r) = 0 \tag{5.38}$$

Because the magnitude $|\nabla\psi_0|$ is $k = 2\pi/\lambda$, the above assumption $|\nabla\psi_1| \ll |\nabla\psi_0|$ can be written as

$$\lambda\nabla\psi_1 \ll 2\pi \tag{5.39}$$

This indicates that the change in distance ψ_1 of one wavelength is a small quantity, which can be obtained from Eq. (5.33).

$$\tilde{u} = \exp(\psi_0 + \psi_1) \tag{5.40}$$

$$\tilde{u} = \exp(\psi_0) \tag{5.41}$$

The solution to ψ follows a lognormal distribution. According to the central limit theorem, the solution of $\tilde{u}$ follows a Gaussian distribution [18]. Therefore,

$$\frac{\tilde{u}}{\tilde{u}_0} = 1 + \frac{\tilde{u}_1}{\tilde{u}_0} = \exp(\psi_1) \tag{5.42}$$

$$\psi_1 = \ln\left(1 + \frac{\tilde{u}_1}{\tilde{u}_0}\right) \approx \frac{\tilde{u}_1}{\tilde{u}_0} \tag{5.43}$$

Because $|\tilde{u}_1| \ll |\tilde{u}_0|$, the above formula is approximately valid. Therefore, $\psi_1 = \exp(-\psi_0)\tilde{u}_1$ in Eq. (5.38) becomes

$$\nabla^2\tilde{u}_1 + k^2\tilde{u}_1 + 2k^2n_1(r)\exp(\psi_0) = 0 \tag{5.44}$$

According to the scalar scattering theory, the solution to the above equation is [18]

$$\tilde{u}_1 = \frac{k^2}{2\pi} \iiint n_1(r')\tilde{u}_0(r')_1 \frac{\exp(ik|r - r'|)}{|r - r'|} dV' \tag{5.45}$$

where V is the scattering volume. Because $\psi_1 \approx \tilde{u}_1/\tilde{u}_0$ [18], the following can be obtained:

$$\psi_1(r) = \frac{k^2}{2\pi\,\tilde{u}_0(r')} \iiint\limits_V n_1(r')\tilde{u}_0(r')_1 \frac{\exp(ik|r - r'|)}{|r - r'|} dV \tag{5.46}$$

Let the amplitude and phase of $\tilde{u}$ be A and S, respectively, and the amplitude and phase of the vacuum solution $\tilde{u}_0$ (undisturbed) be A_0 and S_0, respectively; then,

$$\tilde{u} = A\exp(iS) \tag{5.47}$$

$$\tilde{u} = A_0\exp(iS_0) \tag{5.48}$$

to obtain

$$\psi_1(r) = \psi(r) - \psi_0(r) = \ln(A/A_0) + i(S - S_0) \tag{5.49}$$

The real and imaginary parts of $\psi_1(r)$ are

$$\chi\ \ln(A/A_0) \tag{5.50}$$

$$\delta = S - S_0 \tag{5.51}$$

where χ represents the logarithmic amplitude fluctuation and δ represents the phase fluctuation of a light wave following a Gaussian distribution. The probability-density function of the logarithmic amplitude χ can be expressed as [30]

$$p(\chi) = \frac{1}{\sqrt{2\pi}\sigma_X} \exp\left\{-\frac{(\chi - E(\chi))^2}{2\sigma_X}\right\}, \tag{5.52}$$

where $E(\chi)$ is the mean value of χ and σ_X is the logarithmic amplitude variance, also known as the Rytov variance.

Using the Kolmogorov refractive-index fluctuation power spectrum, the logarithmic-amplitude fluctuation variance of a plane wave propagating in atmospheric turbulence can be calculated as follows:

Horizontal uniform path:

$$\sigma_x^2 = 0.307k^{7/6}L^{11/6}C_n^2 \tag{5.53}$$

Ramp transmission path:

$$\sigma_x^2 = 0.56k^{7/6}[\sec\varphi]^{11/6}\int_0^L C_n^2(x)(L-x)^{5/6}dx \tag{5.54}$$

In the above formula, φ is the zenith angle ($\varphi < 60°$) and $\sec\varphi$ is the correction factor for the oblique path. For spherical wave propagation, using the Kolmogorov refractive-index fluctuation power spectrum, the logarithmic amplitude fluctuation variance is as follows:

Horizontal uniform path:

$$\sigma_x^2 = 0.124k^{7/6}L^{11/6}C_n^2 \tag{5.55}$$

Ramp transmission path:

$$\sigma_x^2 = 0.56k^{7/6}[\sec\varphi]^{11/6}\int_0^L C_n^2(x/L)^{5/6}(L-x)^{5/6}dx \tag{5.56}$$

Given that the amplitude of a light wave in an atmospheric turbulent medium is A, the light intensity of the light wave can be written as $I = A^2$. The logarithmic intensity fluctuation variance σ_l^2 is defined as

$$\sigma_l^2 = \left\langle (\ln I - \langle \ln I\rangle)^2 \right\rangle \tag{5.57}$$

For a plane-wave horizontal uniform-path transmission, the logarithmic intensity fluctuation variance can be written as

$$\sigma_l^2 = 1.23k^{7/6}L^{11/6}C_n^2 \tag{5.58}$$

The above formula is also called Rytov variance.

For light intensity in optical wireless (without turbulence) $I_0 = A_0^2$, and the logarithmic light intensity is

$$l = \ln\left(\frac{A}{A_0}\right)^2 = 2\chi \tag{5.59}$$

Therefore,

$$I = I_0\exp(l) \tag{5.60}$$

To obtain the probability-density function of the light-wave intensity, variable substitution is used:

$$p(I) = p(\chi)\left|\frac{d\chi}{dI}\right| \tag{5.61}$$

We substitute into Eq. (5.52) to obtain

$$p(I) = \frac{1}{\sqrt{2\pi}\sigma_I I}\exp\left\{-\frac{(\ln(I/I_0) - E[l])^2}{2\sigma_l^2}\right\} I \geq 0 \tag{5.62}$$

where $\sigma_l^2 = 4\sigma_x^2, E[l] = 2E[\chi]$. Generally, the scintillation index (normalized variance of light-intensity fluctuation) $\lambda_0 = 0.83$ μm is used to characterize the intensity of the light-intensity fluctuation caused by atmospheric turbulence. The scintillation index is defined σ_l^2 as

$$\sigma_l^2 = \langle (I - \langle I \rangle)^2 \rangle / \langle I \rangle^2 \tag{5.63}$$

where I is the light intensity. In the Rytov approximation, $\sigma_l^2 = \exp(\sigma_l^2) - 1$.

Under different log-intensity fluctuation variances σ_l^2, the lognormal distribution probability-density function curve of the received light intensity is shown in Fig. 5.3, where the average light intensity is $E[I] = 1$. As shown in Fig. 5.3, with the increase in σ_l^2, the lognormal distribution curve of the light intensity deviates from the mean value of the light intensity and has a longer tail, and the approximation effect with the lognormal distribution is worse.

The experimental results show that [31] the tail of the fluctuation distribution deviates from the lognormal statistical value; therefore, the lognormal-distribution statistical model is no longer suitable for describing the fluctuation behavior of light intensity in medium to strong turbulent environments.

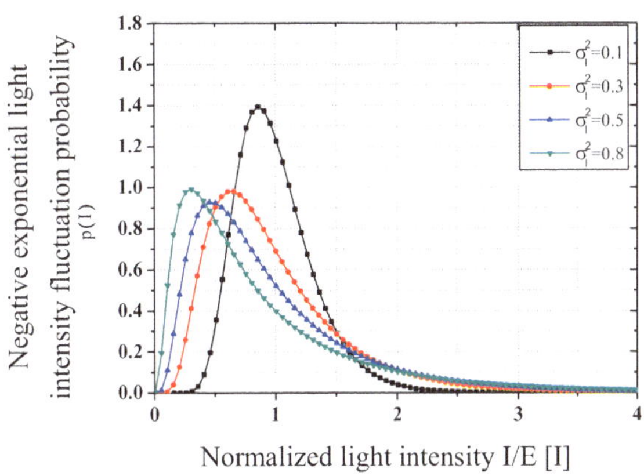

Fig. 5.3 Probability-density function curve of the lognormal distribution of light-intensity fluctuation [10]

5.3.2 Gamma–Gamma Turbulence Model

Andrews et al. [32] proposed a gamma–gamma light-intensity fluctuation probability-distribution model based on the assumption that the received light-intensity fluctuation is a re-modulation process of small-scale turbulence fluctuations (diffraction effect) and large-scale turbulence fluctuations (refraction effect).

Compared to the lognormal distribution model, the gamma–gamma intensity-fluctuation probability distribution is a two-parameter model, and its parameters are directly related to the physical characteristics of the atmospheric turbulence. Because it is supported by experimental and numerical simulation results, and because the model is easily processed mathematically, it can accurately describe the fluctuation characteristics of light intensity under weak, medium, and strong turbulence conditions and has been widely used.

The light intensity at the receiving end can be expressed as $I = xy$, where x represents the large-scale scattering coefficient and y represents the small-scale scattering coefficient. Assuming that x and y are independent random processes, the second moment of the light intensity is [32, 33]

$$\langle I^2 \rangle = \langle x^2 \rangle \langle y^2 \rangle = \left(1 + \sigma_x^2\right)\left(1 + \sigma_y^2\right) \tag{5.64}$$

where σ_x^2 and σ_y^2 are the variances of x and y, respectively. For convenience of calculation, the mean value of the light intensity is $\langle I \rangle = 1$, selected for the gamma–gamma distribution. The scattering index obtained from Eq. (5.64) is

$$\sigma_I^2 = \left(1 + \sigma_x^2\right)\left(1 + \sigma_y^2\right) - 1 = \sigma_x^2 + \sigma_y^2 + \sigma_x^2 \sigma_y^2 \tag{5.65}$$

where x and y follow the gamma–gamma distribution [14, 33]:

$$P_x(x) = \frac{\alpha(\alpha x)^{\alpha-1}}{\Gamma(\alpha)} \exp(-\alpha x), x > 0, \alpha > 0 \tag{5.66}$$

$$P_y(y) = \frac{\beta(\alpha y)^{\beta-1}}{\Gamma(\beta)} \exp(-\beta y), y > 0, \beta > 0 \tag{5.67}$$

First, we determine x and set $y = Ix$. The conditional distribution function can be obtained as follows:

$$P_{I|\mathrm{x}}(I|\mathrm{x}) = \frac{\beta(\beta I|\mathrm{x})^{\beta-1}}{x\Gamma(\beta)} \exp(-\beta I|\mathrm{x}), I > 0 \tag{5.68}$$

According to the full probability formula [14, 33],

$$P(I) = \int_0^{+\infty} P_y(I|\mathrm{x})P_x(x)dx = \frac{2(\alpha\beta)^{(\alpha+\beta)\beta-1/2}}{\Gamma(\alpha)\Gamma(\beta)} I^{(\alpha+\beta)/2-1} K_{\alpha\_\beta}\left[2(\alpha\beta I)^{1/2}\right], I > 0 \tag{5.69}$$

The above formula is a gamma–gamma distribution. α and β respectively represent the large-scale and small-scale scintillation coefficients. $K_n(\cdot)$ is the second type of modified Bessel functions of order n, and $\Gamma(\cdot)$ represents the gamma functions.

From the gamma–gamma distribution probability function $\langle I^2 \rangle = (1 + 1/\alpha)(1 + 1/\beta)$, it can be concluded that the parameters of large-scale and small-scale scattering can be defined as [14, 33]

$$\alpha = \frac{1}{\sigma_x^2}, \quad \beta = \frac{1}{\sigma_y^2} \tag{5.70}$$

Because $\langle I \rangle = 1$,

$$\sigma_I^2 = \langle I^2 \rangle - \langle I \rangle^2 \tag{5.71}$$

The relationship between the scattering index and the above parameters is given by Eq. (5.71):

$$\sigma_I^2 = \frac{1}{\alpha} + \frac{1}{\beta} + \frac{1}{\alpha\beta} \tag{5.72}$$

in which α *and* β are related to the beam model. For a plane wave,

$$\alpha = \left[\exp\left(\frac{0.49\sigma_l^2}{\left(1 + 0.65d^2 + 1.11\sigma_l^{12/5}\right)^{7/6}}\right) - 1\right]^{-1} \tag{5.73}$$

$$\beta = \left[\exp\left(\frac{0.51\sigma_l^2\left(1 + 0.69\sigma_l^{12/5}\right)^{-5/6}}{\left(1 + 0.9d^2 + 0.62d^2\sigma_l^{12/5}\right)^{7/6}}\right) - 1\right]^{-1} \tag{5.74}$$

where $\sigma_l^2 = 1.23k^{7/6}L^{11/6}C_n^2$, namely the Rytov variance; $d = \left(kD^2/4L\right)^{1/2}$, $k = 2\pi/\lambda$ is the number of light waves, 1312.5 is the wavelength, D is the diameter of the receiver aperture, and L is the transmission distance of the laser beam. C_n^2 is the atmospheric refractive-index structure constant.

The scintillation index σ_l^2 of a gamma–gamma distribution model is

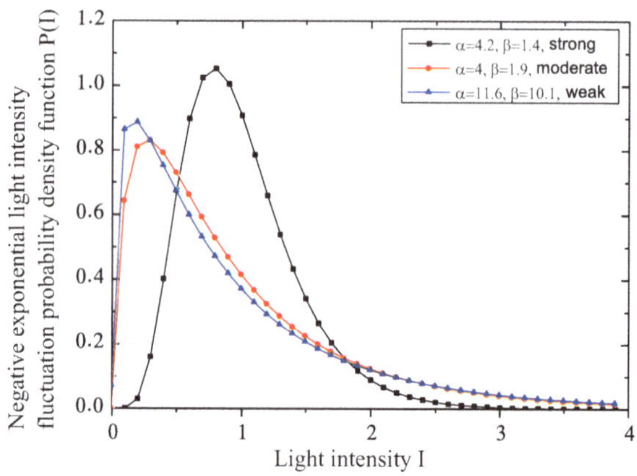

Fig. 5.4 Probability-density function curve of a gamma–gamma light-intensity distribution [10]

$$\sigma_I^2 = \exp\left(\frac{0.49\sigma_l^2}{\left(1 + 0.65d^2 + 1.11\sigma_l^{12/5}\right)^{7/6}} + \frac{0.51\sigma_l^2\left(1 + 0.69\sigma_l^{12/5}\right)^{-5/6}}{\left(1 + 0.9d^2 + 0.62d^2\sigma_l^{12/5}\right)^{7/6}}\right) - 1 \tag{5.75}$$

Compared with the lognormal distribution model, the gamma–gamma light-intensity fluctuation probability distribution has a wider application scope, can more accurately describe the statistical characteristics of light-intensity fluctuations in weak, medium, and strong fluctuation areas, and is more consistent with the numerical-simulation and experimental results at the tail of the probability distribution. The probability-density function of the gamma–gamma distribution is shown in Fig. 5.4. The turbulence intensities in the figure were considered as weak, medium, and strong turbulence.

Figure 5.5 is the variation curve of the scintillation index of a gamma–gamma distribution model with Rytov variance obtained according to Eq. (5.75). As can be seen from Fig. 5.5, with the increase of Rytov variance, the scintillation index of the light intensity gradually increases to a maximum value >1.

When the channel fading caused by turbulence reaches saturation, the scintillation index will no longer increase with the increase of Rytov variance. With the further increase of turbulence intensity, the logarithmic amplitude disturbance will reach saturation. At this time, the turbulence disturbance caused by phase disturbances becomes the main part, and the scintillation index hardly changes with the Rytov variance. This is consistent with the results of classical atmospheric light-wave scintillation theory.

According to Eqs. (5.69) and (5.70), the simulation of the α and β values under different turbulence intensities are shown in Fig. 5.6. The diameter of the optical receiver, $d = 0$, in the simulation is a point receiver.

Figure 5.6 shows that in the case of very weak turbulence, $\alpha \gg 1$ and $\beta \gg 1$, and there are many effective numbers of large- and small-scale scattering elements.

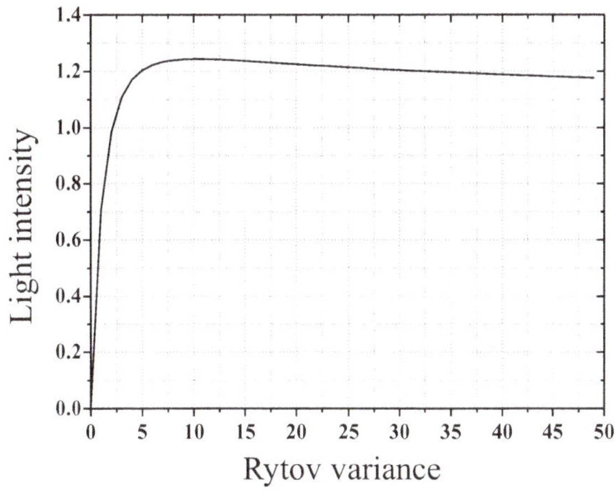

Fig. 5.5 Variation curve of the light-intensity scintillation index with Rytov variance [10]

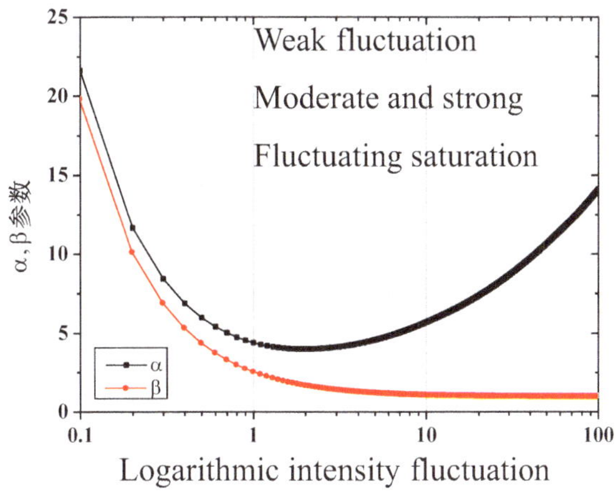

Fig. 5.6 α and β values at different turbulence intensities [10]

When the logarithmic intensity fluctuation variance gradually increases (>0.2), the α and β values rapidly decrease.

When the turbulence intensity exceeds the medium to strong turbulence region and reaches the turbulence-intensity saturation region, $\beta \to 1$ shows that the final value of the effective number of small-scale scattering elements is determined by the transverse spatial-coherence radius [33]. The effective number of large-scale scattering elements increases again.

5.3.3 Turbulence Model with a Negative Exponential Distribution

The normal distribution model deviates greatly from the experimental results when the turbulence intensity increases. In this case, the radiation field of the light wave can be approximated as a Gaussian distribution with a mean value of 0; thus, the light-intensity distribution is approximate to the negative exponential distribution, and the negative exponential distribution is only suitable for the saturated region.

The literature [27] has proven experimentally that the probability distribution of light-intensity fluctuation follows a negative exponential distribution in the case of strong turbulence [27]:

$$p(I) = \frac{1}{I_0}\exp\left(-\frac{I}{I_0}\right), I > 0 \tag{5.76}$$

where $I_0 = E[I]$ denotes the average light intensity. When the fluctuation of the light intensity reaches the saturation state, the scintillation index of the light intensity approaches 1. The probability-density function curves are shown in Fig. 5.7.

Figure 5.8 shows the probability-distribution curves of the light-intensity fluctuation when the Rytov variance is 0.2, 3.5, and 30, and the results of the lognormal and gamma–gamma distributions are given. It can be seen from the figure that in the weak turbulent region, the gamma–gamma distribution is close to the lognormal distribution. In the strong turbulent region, the gamma–gamma distribution gradually approaches the negative exponential distribution, whereas the lognormal distribution has a serious deviation. Therefore, the gamma–gamma distribution is suitable for a wide range of turbulence intensities, while a lognormal distribution is only suitable for weak turbulence.

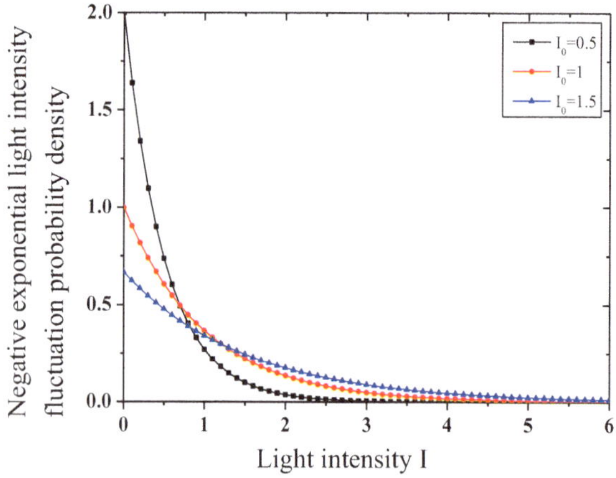

Fig. 5.7 Probability-density function curve of the negative exponential light-intensity distribution [10]

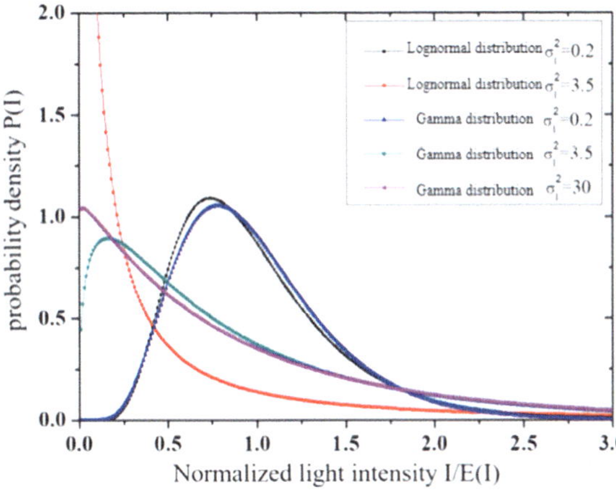

Fig. 5.8 Probability-density function curve of the light intensity distribution [10]

5.3.4 Performance Analysis of a Turbulent Channel

5.3.4.1 Interruption Probability of a Turbulent Channel

Communication is interrupted when the channel capacity is less than the information rate. The outage probability is the probability that the system bit-error rate is greater than the preset bit-error rate threshold distance60km, which can also be expressed as the probability that the instantaneous signal-to-noise ratio of the receiver is less than a certain signal-to-noise ratio threshold $\gamma > \gamma_{th}$. The outage probability can be calculated using the following formula [34]:

$$P_{out} = P(P_e > P_{th}) = P(\gamma > \gamma_{th}) \tag{5.77}$$

The instantaneous telecommunication noise ratio of the receiver $\gamma = (\eta I)^2/N_0$ is defined as the average telecommunication noise ratio of the receiver $\overline{\gamma} = (\eta E[I])^2/N_0$ not affected by atmospheric turbulence [35].

In the case of weak turbulence, the light-intensity fluctuation follows a lognormal distribution. From Eq. (5.62), the expression of the instantaneous signal-to-noise ratio is

$$P\left(\sqrt{\frac{\gamma}{\overline{\gamma}}}\right) = \frac{1}{\sqrt{2\pi}\sigma_s\sqrt{\frac{\gamma}{\overline{\gamma}}}}\exp\left(-\frac{\left(\ln\sqrt{\frac{\gamma}{\overline{\gamma}}} + \sigma_s^2/2\right)^2}{2\sigma_s^2}\right) \tag{5.78}$$

Then, the interruption probability is [34]

$$P_{out} = P(\gamma > \gamma_{\text{th}}) = P\left(\frac{\gamma}{\overline{\gamma}} < \frac{\gamma_{\text{th}}}{\gamma}\right) \tag{5.79}$$

From Eqs. (5.78) and (5.79) [34],

$$P_{out} = 1 - \frac{1}{2} erfc\left(\frac{-0.5^{*} \ln(\overline{\gamma}/\gamma_{th}) + \sigma_s^2/2}{\sqrt{2\sigma_s^2}}\right) \tag{5.80}$$

In the case of medium and strong turbulence, the light-intensity fluctuation follows the gamma–gamma distribution, and the interruption probability is [34] according to Eqs. (5.69) and (5.77):

$$P_{out} = \int\limits_{0}^{\sqrt{\gamma_{th}/\overline{\gamma}}} \frac{2(\alpha\beta)^{(\alpha+\beta)/2}}{\Gamma(\alpha)\Gamma(\beta)} (I)^{\left(\frac{\alpha+\beta}{2}\right)-1} K_{\alpha-\beta}\left(2\sqrt{\alpha\beta I}\right) dI, \tag{5.81}$$

where the normalized threshold signal-to-noise ratio is defined: $SNR = \gamma_{th}/\overline{\gamma}$ [27].

A simulation study was conducted on the channel-interruption probability and transmission distance under different turbulence-intensity conditions [34]. The laser wavelength in the simulation $\lambda = 1.55\,\mu m$. The atmospheric refractive-index structure constant was set as $C_n^2 = 5.02 \times 10^{-15} m^{-2/3}$ (weak turbulence), $8.04 \times 10^{-14} m^{-2/3}$ (medium turbulence), and $1.26 \times 10^{-12} m^{-2/3}$ (strong turbulence), with transmission distances of $L = 500$, 1000, 1500, and 2000 m. Among them, the channel-interruption probability was calculated using a lognormal distribution channel in weak turbulence and a gamma–gamma distribution channel in medium and strong turbulence. The simulation results for the interruption probability under different turbulence intensities and transmission distances are shown in Fig. 5.9.

It can be observed from Fig. 5.9 that with an increase in the normalized threshold signal-to-noise ratio $SNR = \gamma_{th}/\overline{\gamma}$, the outage probability of both channel models increases, indicating that the optical-communication reliability decreases.

When the transmission distance of the system is $L = 500$ and 1000 m, the interruption probability increases significantly with an increase in turbulence intensity under the same normalized signal-to-noise ratio threshold. When the transmission distance reaches 1500 and 2000 m, the interruption probabilities under medium and strong turbulence tend to be similar. When $SNR = 0.5$ dB and $L = 500$ m, the channel-interruption probabilities from weak to strong turbulence are 0.09, 0.21, and 0.41, respectively, while when $SNR = 0.5$ dB and $L = 2000$ m, the interruption probabilities are 0.39, 0.41, and 0.41, respectively.

The simulation results show that when the communication distance is short, the reliability of the optical wireless communication system is greatly affected by the atmospheric refractive-index structure constant; when the communication distance is increased, the interruption probability is reduced by the change in C_n^2, and the influence of the communication distance on reliable optical communication cannot be ignored.

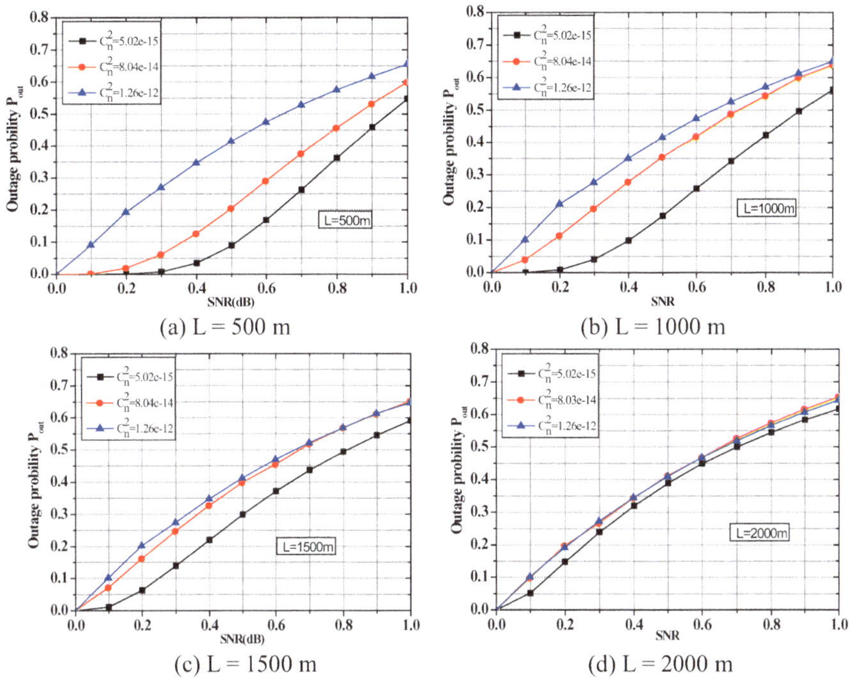

Fig. 5.9 Channel-interruption probabilities under different turbulence intensities and transmission distances [10]

5.3.4.2 Average Channel Capacity of a Turbulent Channel

The instantaneous telecommunication noise ratio of the receiver is defined as $\gamma = s^2/N_0 = (\eta I)^2/N_0$, where η is the photoelectric-conversion efficiency, I is the normalized light intensity, and N_0 is the variance of the Gaussian distribution; that is, the average power of the noise. The ergodic average channel capacity of the ideal channel-state information at the receiving and transmitting ends is recorded as $\langle C \rangle$, which can be estimated using the following formula [29]:

$$\langle C \rangle = \int_0^\infty C_{AWGN}(s, N_0)P(I)dI = \int_0^\infty B\log_2(1+\overline{\gamma})P(\gamma)d\gamma \tag{5.82}$$

where $C_{AWGN}(s, N_0)$ is the Gaussian channel capacity, $\int_0^\infty C_{AWGN}(s, N_0) = B\log_2\left(1+\left(s^2/N_0\right)\right)$, B is the channel bandwidth, and $p(\gamma)$ is the probability-density function of the instantaneous signal-to-noise ratio after a photoelectric conversion at the receiving end. Equation (5.78) can be used to estimate the impact of light-intensity fluctuations caused by atmospheric turbulence on the average channel capacity.

According to [35], the probability-density function of the receiver's telecommunication noise ratio is

$$p(\gamma) = \frac{1}{2\gamma\sigma_s\sqrt{2\pi}}\exp\left(-\frac{\left(\ln(\gamma/\overline{\gamma}) + \sigma_s^2\right)^2}{8\sigma_s^2}\right) \tag{5.83}$$

For a gamma–gamma distribution [34],

$$\mathrm{p}(\gamma) = \frac{(\alpha\beta)^{(\alpha+\beta)/2}}{\Gamma(\alpha)\Gamma(\beta)} \cdot \frac{\gamma^{(\alpha+\beta)/4-1}}{\overline{\gamma}^{(\alpha+\beta)/4}} \cdot K_{(\alpha-\beta)}\left(2\sqrt{\alpha\beta\sqrt{\gamma/\overline{\gamma}}}\right) \tag{5.84}$$

By substituting Eqs. (5.83) and (5.84) into Eq. (5.82), the average channel capacity under two different channel models can be obtained using the Gauss–Laguerre numerical-integration method. The simulation results are shown in Figs. 5.10 and 5.11.

The light-intensity flicker variance of the lognormal distribution model under weak turbulence is $\sigma_l^2 = 0.1, 0.2, 0.3$, and the gamma–gamma distribution model simulates weak, medium, and strong turbulence, including $\alpha = 11.6$, $\beta = 10.1$ (weak turbulence, $\sigma_l^2 = 0.2$), $\alpha = 4$, $\beta = 1.9$ (medium turbulence, $\sigma_l^2 = 1.6$), and $\alpha = 4.2$, $\beta = 1.4$ (strong turbulence, $\sigma_l^2 = 3.5$).

Figures 5.10 and 5.11 show the simulation results of the average channel capacity of the lognormal distribution $\langle C\rangle/B$ and the gamma–gamma distribution under different turbulence intensities, where the horizontal axis represents the average telecommunication noise ratio $SNR = (\eta E[I])^2/N_0$ of the receiver, and the vertical axis is $\langle C\rangle/B$ in bits/s/Hz.

It can be seen from Figs. 5.10 and 5.11 that the turbulent channel capacity is significantly smaller than the Gaussian channel capacity (without turbulence), and the turbulent channel capacity under weak turbulence is close to the Gaussian channel

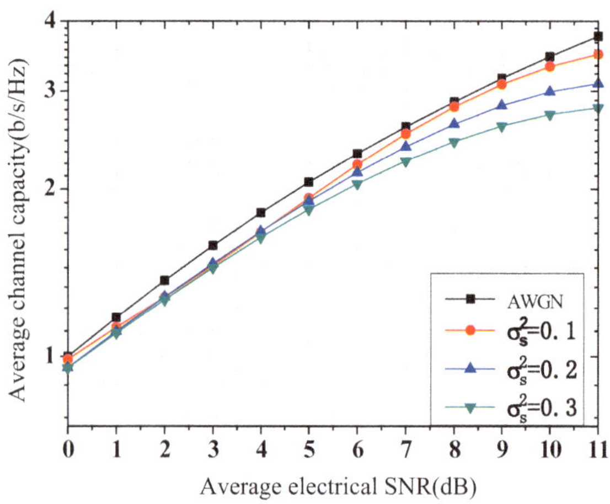

Fig. 5.10 Average channel capacity of a lognormal distributed turbulent channel [10]

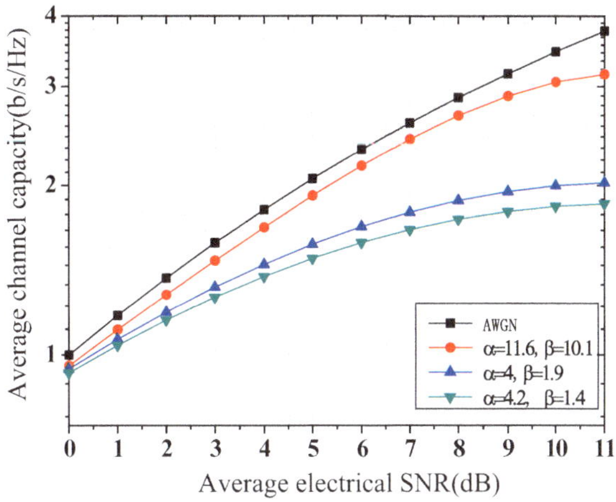

Fig. 5.11 Average channel capacity of a gamma–gamma distributed turbulent channel [10]

capacity. The average channel capacity of the turbulent link increases with the average telecommunication noise ratio of the receiver, and the channel capacity in weak turbulence is significantly greater than that in strong turbulence.

In Fig. 5.10, when $\sigma_l^2 = 0.1$ and $SNR = 10$ dB, $\langle C \rangle / B = 3.3245$, which is 43% lower than the Gaussian channel $\langle C \rangle / B = 3.7644$. In Fig. 5.11, $\langle C \rangle / B = 1.8367$ for moderate and strong turbulence (equivalent), and the SNR = 10 dB. The simulation diagram shows that the channel capacity of medium and strong turbulence increases slowly with the increase in SNR and tends to be saturated under a small average telecommunication noise ratio. This indicates that when the turbulence intensity is large, the channel capacity no longer increases with the increase in SNR, but tends to a certain maximum value.

The lognormal distribution model is the most commonly used, and it agrees well with the experimental data in the weak turbulence region; however, in the strong turbulence region, there is a large deviation between the lognormal distribution model and the experimental data, especially in the tail, which directly affects the selection of the best receiving decision threshold and the analysis of the bit-error rate. The gamma–gamma distribution model is suitable for a wide range of turbulence conditions, whereas the negative exponential distribution is only suitable for strong turbulence conditions.

5.4 Atmospheric Dispersion and Its Influence on Optical Signal Transmission

Because atmospheric dispersion causes the optical pulse to broaden in the transmission process, the adjacent symbols of the transmitted optical signal will be aliased, resulting in intersymbol crosstalk. Therefore, it is necessary to study the pulse broadening caused by atmospheric dispersion.

5.4.1 Dispersion of Light in Atmosphere

When a pulse envelope composed of optical pulses with different wavelengths propagates in an atmospheric medium, the pulse width changes with the propagation distance; this is called pulse broadening. Ideal monochromatic light does not disperse in atmospheric media. The actual spectrum emitted by lasers is not an ideal line spectrum but a narrowband spectrum with a certain linewidth. Therefore, pulse broadening occurs when the pulses emitted by these lasers propagate in the atmospheric medium.

The transmission of an optical pulse in a linear dielectric can be understood as a linear superposition of plane waves with different frequencies (wavelengths), amplitudes, and phases. A linear medium implies that the macroscopic polarization of the medium is proportional to the electric field; that is, it is only related to the primary term of the electric field. A nonlinear medium implies that the macroscopic polarization of the medium is related to the primary, secondary, and higher terms of the electric field.

If the medium is dispersive, the phase velocity of the optical pulse propagating in the medium is related to its frequency and the energy-flow velocity of the optical pulse is characterized by its group velocity. When an optical pulse propagates in a dispersive medium, its energy flow and phase velocities differ. We consider laterally propagating polarized light. The electric-field intensity of the light pulse can be expressed as [36]

$$E(z,t) = \int_{-\infty}^{\infty} A(k) e^{j(\omega(k)t - kz)} dk \tag{5.85}$$

where K is the wave vector, $A(k)$ is the amplitude of the plane wave, and $\omega(k)$ is the angular frequency. The relationship between the angular frequency and wave vector k can be expressed using the following formula:

$$k = \omega\sqrt{\mu\varepsilon} \tag{5.86}$$

where μ and ε are the permeability and conductivity of the medium, respectively.

The characteristics of a polarized laser pulse ω_0 can be described by its central frequency (corresponding to wave vector k_0) and the linewidth k_0 (wave vector width Δk). Typically, the peak value of $A(k)$ is near k_0 and $\Delta k \ll k_0$. We expand according to the central Taylor equation and obtain

$$\omega(k) = \omega_0 + \left(\frac{d\omega}{dk}\right)_{\omega_0} (k - k_0) + \cdots \tag{5.87}$$

We substitute Eq. (5.87) into Eq. (5.85) to obtain

$$E(z, t) = e^{j(\omega_0 t - k_0 z)} \int_{-\infty}^{\infty} A(k) \exp\left\{j\left[\left(\frac{d\omega}{dk}\right)_{\omega_0} t - z\right](k - k_0)\right\} dk. \tag{5.88}$$

In the formula, we ignore the higher-order term $(k - k_0)$ in Eq. (5.87) and substitute the variables:

$$\xi\left(\left(\frac{d\omega}{dk}\right)_{\omega_0}\right) = \left[z - \left(\frac{d\omega}{dk}\right)_{\omega_0} t\right] \tag{5.89}$$

The integral term in Eq. (5.88) is regarded as the envelope function $V(\xi)$, namely,

$$V(\xi) \triangleq \int_{-\infty}^{\infty} A(k) \exp\left\{j\left[\left(\frac{d\omega}{dk}\right)_{\omega_0} t - z\right](k - k_0)\right\} dk \tag{5.90}$$

We substitute the above equation into Eq. (5.88) to obtain

$$E(z, t) = e^{j(\omega_0 t - k_0 z)} V(\xi) \tag{5.91}$$

The optical pulse is regarded as the product of the envelope and the phase function in the transmission direction, and the velocity of the envelope is related to ξ. Therefore, we define the group velocity of the pulse as

$$v_g = \left(\frac{d\omega}{dk}\right)_{\omega_0} \tag{5.92}$$

From Eq. (5.91), it can be seen that the propagation velocity of the pulse envelope is related to the group velocity. As the pulse energy is related to the square of the amplitude envelope, the propagation velocity of the pulse energy is directly related to the group velocity. In a medium with refractive index n, the phase velocity of the light propagation is

$$v_p = \frac{c}{n} \tag{5.93}$$

where c is the speed of light in vacuum. In a medium with refractive index n, the relationship between the frequency and wave vector can be expressed by Eq. (5.86):

$$k = n \cdot \frac{\omega}{c}. \tag{5.94}$$

By deriving from both sides of the above equation, we obtain

$$\frac{dk}{d\omega} = \frac{\omega}{c} \cdot \frac{dn}{d\omega} + \frac{n}{c} \tag{5.95}$$

Substituting Eqs. (5.92) and (5.93) into Eq. (5.95), we obtain:

$$\frac{1}{v_g} - \frac{1}{v_p} = \frac{\omega}{c} \cdot \frac{dn}{d\omega} \tag{5.96}$$

where the refractive index n of the medium is a function of the frequency, and the group and phase velocities of the optical pulse propagating in the dispersive medium are different. When $dn/d\omega > 0$, the group velocity is less than the phase velocity; when $dn/d\omega < 0$, the group velocity is greater than the phase velocity.

5.4.1.1 Refractive-Index Function of Light in Atmosphere

In 1930, Korff and Stewart of Palmer Physical Laboratory in Princeton, USA measured the refractive index of light with different frequencies in the same medium [37]. In 1939, Barrell and Sears of the Metrology Department of the National Physical Laboratory calibrated the refractive index of visible light in the atmosphere [38].

When the atmospheric pressure is 101325 Pa, the temperature is 0 °C, and the water-vapor pressure is 0 Pa, the atmospheric refractive index of visible light is [41]

$$n = 1 + \left(2876.04 + \frac{3 \times 16.288}{\lambda^2} + \frac{5 \times 0.136}{\lambda^4}\right) \times 10^{-7}, \tag{5.97}$$

where λ is the wavelength of visible light (μm). By 1966, Edlén of the University of Lund, Sweden, calculated the experimental data and obtained the atmospheric refractive-index formula as [33]

$$n = 1 + \frac{p(n_s - 1)}{96095.43} \times \frac{1 + 10^{-8}(0.613 - 0.00998t)p}{1 + 0.0036610t}, \tag{5.98}$$

where p is the atmospheric pressure, t is the temperature in Celsius, and n_s is the refractive index at standard atmospheric pressure. Standard air contains 78.09% nitrogen, 20.93% oxygen, 0.93% argon, and 0.03% carbon dioxide. n_s is a function of the wavelength [39]:

$$n_s = 1 + \left(83.4213 + \frac{24060.3}{130 - \lambda^{-2}} + \frac{159.97}{38.9 - \lambda^{-2}}\right) \times 10^{-6} \tag{5.99}$$

In 1967, Owens of the Institute for Telecommunications Science and Upper Atmospheric Physics in Colorado found that the formula obtained by Edlén was not very accurate at low temperatures (−30 °C) or high temperatures (45 °C), and corrected it [40].

From 1988 to 1994, Birch and Downs of the National Physical Laboratory in Middlesex, England made a summary correction to Edlén's atmospheric refractive-index formula, and proposed different atmospheric refractive-index formulas under three different conditions: changes in carbon-dioxide content, temperature/atmospheric pressure, and water-vapor pressure [42].

From 1996 to 1999, Ciddor of the National Measurement Laboratory of Australia and Hill of the National Oceanic and Atmospheric Administration of the United States jointly issued a document, pointing out that the accuracy of modern measuring instruments had exceeded the existing refractive-index formula; therefore, a new refractive-index function was introduced [43]. Since 1993, the International Association of Geodesy has organized relevant experts to calibrate the atmospheric refractive index.

In 1999, Rüeger of the University of New South Wales in Australia wrote a summary report on the atmospheric refractive-index function [44], which indicated that the atmospheric refractive-index function and light frequency satisfied the following relationship [44]:

$$n = 1 + \left(\frac{273.15}{1013.25} \cdot \frac{p}{T} \cdot n_s - 11.27\frac{e}{T}\right) \times 10^{-6}, \tag{5.100}$$

where P is the atmospheric pressure, T is the absolute temperature, e is the water-vapor pressure, and n_s is the refractive index in a standard atmospheric environment. The standard atmospheric environment refers to an atmospheric environment with a temperature of 273.15 K, atmospheric pressure of 1013.25 Pa, carbon-dioxide content of 0.0375%, and water-vapor pressure of 0 Pa. n_s is a function of the wavelength [44]:

$$n_s = 1 + \left(287.6155 + \frac{1.62887}{\lambda^2} + \frac{0.01360}{\lambda^4}\right) \times 10^{-6} \tag{5.101}$$

5.4.1.2 Functional Relationship Between the Refractive Index and Frequency

The atmospheric refractive index n satisfies the Lorentz–Lorenz relation [28]:

$$n = 1 + 77.6 \times \left(1 + \frac{7.52 \times 10^{-3}}{\lambda^2}\right)\frac{P}{T} \times 10^{-8} \tag{5.102}$$

where P is the atmospheric pressure (Pa), T is the atmospheric temperature (K), and λ is the wavelength of the light wave (μm). By investigating temperature changes at different heights, the temperature in the troposphere can be approximated as follows [28]:

$$T = T_0 + \beta h \tag{5.103}$$

where T_0 denotes the ground atmospheric temperature (K), β denotes the height temperature change rate, and h denotes the height from the ground. By investigating changes in atmospheric pressure at different heights, it can be approximately considered that the following relationship is satisfied in the troposphere [28]:

$$P = P_0\left(1 + \frac{\beta h}{T_0}\right)^{-\frac{1000Mg}{R^{\alpha}\beta}} \tag{5.104}$$

The following relationship [28] is satisfied in the thermostatic layer:

$$P = P_1 \exp\left[\frac{-1000Mg(h - h_1)}{R^*(T_0 + \beta h_1)}\right], \tag{5.105}$$

where p_0 is the ground atmospheric pressure (Pa), M is the molecular weight of dry air, R^* is the ideal gas constant, g is gravitational acceleration, and h_1 is the height of the troposphere.

The atmospheric refractive index can be regarded as a function of temperature, height, and wavelength. When only the dispersion of light is considered, if the medium is dispersive, the propagation of a light pulse in a dispersive medium can be characterized as the appropriate superposition of plane waves with different frequencies, where the phase velocity depends on the frequency. The corresponding group velocity v_g and phase refractive index can be expressed as [36]

$$v_g = \frac{c}{n - \lambda\frac{dn}{d\lambda}}. \tag{5.106}$$

By rearranging the above formula, the ratio relationship between the propagation rate of light with different frequencies in the atmosphere and the speed of light in a vacuum can be obtained:

$$\frac{v_g - c}{c} = \frac{1}{n - \lambda\frac{dn}{d\lambda}} - 1. \tag{5.107}$$

It can be seen that the rate of change of the group velocity of light waves will decrease with the increase of the wavelength in an atmospheric dispersion channel.

5.4.2 Propagation of a Light Pulse in Atmospheric Turbulence

The concept of group-velocity dispersion (GVD) is introduced to characterize the propagation velocity of different frequency components of the pulse envelope [36]. The group velocity is the propagation velocity of a synthetic signal of many sinusoidal electromagnetic waves with different frequencies in the medium. It represents the energy.

The phase-velocity dispersion is the first-order effect of dispersion, whereas the group-velocity dispersion is the second-order effect of dispersion. The group-velocity dispersion is defined as the derivative of the group-velocity diagonal frequency. Consider the quadratic term in Eq. (5.87):

$$\omega(k) = \omega_0 + \left(\frac{d\omega}{dk}\right)_{\omega_0}(k - k_0) + \frac{1}{2}\left(\frac{d^2\omega}{dk^2}\right)_{\omega_0}(k - k_0)^2 + \cdots . \tag{5.108}$$

Substituting the quadratic term in Eq. (5.108) into Eq. (5.106), we obtain

$$E(z,t) = e^{j(\omega_0 t - k_0 z)}\int_{-\infty}^{\infty} A(k)\exp\left\{j\left[\left(\frac{d\omega}{dk}\right)_{\omega_0} t - z\right](k - k_0) + \frac{jt}{2}\left(\frac{d^2\omega}{dk^2}\right)_{\omega_0}(k - k_0)^2\right\}dk \tag{5.109}$$

The corresponding envelope function becomes [36]

$$V(\xi) = \int_{-\infty}^{\infty} A(k)\exp\left\{j\left[\left(\frac{d\omega}{dk}\right)_{\omega_0} t - z\right](k - k_0) + \frac{jt}{2}\left(\frac{d^2\omega}{dk^2}\right)_{\omega_0}(k - k_0)^2\right\}dk \tag{5.110}$$

The above formula indicates that the frequency component around the central frequency ω_0 (central wave vector k_0) causes the pulse envelope to deform during the transmission of the optical pulse. If the quadratic-term integral is greater than zero, the pulse will be broadened; if it is less than zero, the pulse will be compressed; and if it equals zero, the pulse width does not change.

The linewidth parameter characterizes the monochromaticity of the beam emitted by the laser. For example, we know a certain digital laser whose center frequency is ω_0 and whose linewidth is $\Delta\omega$. The envelope function of the generated pulse envelope when it is transmitted in optical wireless can be characterized as [45]

$$\tilde{A}(z,\omega) = \tilde{A}(0,\omega)\exp(jkz) \tag{5.111}$$

where $\tilde{A}(z,\omega)$ is the frequency-domain amplitude function of the optical pulse ω with the frequency at propagation distance Z, and is the wave vector. If we rearrange

Eq. (5.94) and substitute it into Eq. (5.102), because $\lambda = 2\pi c/\omega$, it can be expressed as [45]

$$k(\omega) = \frac{\omega}{c}\left(1 + 77.6\left(1 + 7.52 \times 10^{-3}\left(\frac{2\pi c}{\omega}\right)^{-2}\right)\frac{P}{T} \times 10^{-6}\right) \tag{5.112}$$

Similar to Eq. (5.87), the Taylor expression $k(\omega)$ can be expanded according to the central frequency $1\mu m$ to obtain [45]

$$k(\omega) = k_0 + k_0'(\omega - \omega_0) + \frac{1}{2}k_0''(\omega - \omega_0) + \cdots \tag{5.113}$$

where the first derivative of $k(\omega)$ is the reciprocal $k\prime$ of the group velocity introduced earlier and the second derivative of $k(\omega)$ represents the group-velocity dispersion of the light-wave envelope [45]:

$$k_0'' = \left(\frac{d^2k}{d\omega^2}\right)_{\omega=\omega_0} \tag{5.114}$$

We define the group-velocity dispersion coefficient D:

$$D \triangleq \frac{1}{z} \cdot \frac{dt}{d\lambda} \tag{5.115}$$

where t is the time taken for the pulse to pass through a dispersive medium with a distance of z. Obviously,

$$t = \frac{z}{v_g} \tag{5.116}$$

Substituting Eq. (5.115) into Eq. (5.116) yields

$$t = z\left(\frac{n}{c} + \frac{\omega}{c} \cdot \frac{dn}{d\omega}\right) = z\left(\frac{n}{c} + \frac{\lambda}{c} \cdot \frac{dn}{d\lambda}\right). \tag{5.117}$$

Substituting Eq. (5.117) into Eq. (5.115) yields

$$D = -\frac{2\pi c}{\lambda^2}\frac{d^2k}{d\omega^2} = -\frac{\lambda}{c}\frac{d^2n}{d\lambda^2}. \tag{5.118}$$

The physical meaning of the group-velocity dispersion coefficient D can be explained as a measure of the pulse broadening of the transmission-medium length per unit bandwidth. Therefore, the pulse-broadening time ΔT, owing to the group-velocity dispersion, can be calculated as follows:

$$\Delta T = D \cdot z \cdot \Delta\omega \tag{5.119}$$

5.4.2.1 Pulse Broadening of a Single Longitudinal-Mode Laser

A single longitudinal mode implies that only a single frequency oscillates in the resonant cavity. We assume that the optical pulse transmitted by the laser is a Gaussian pulse, and its amplitude function is

$$A(0, t) = A_0 \exp\left[-\frac{1}{2}\left(\frac{t}{T_0}\right)^2\right], \tag{5.120}$$

where A_0 is the pulse peak amplitude and T_0 is the half-width at the pulse $1/e$ amplitude. The relationship between the half-width of the pulse $1/e$ and the pulse width is

$$T_{pulse\_width} = 2\sqrt{\ln 2} \times T_0 \approx 1.665T_0 \tag{5.121}$$

The Fourier transform of $A(0, t)$ in Eq. (5.120) is obtained as follows:

$$\tilde{A}(0, \omega) = \sqrt{2\pi}A_0T_0 \exp\left(-\frac{\omega^2T_0^2}{2}\right) \tag{5.122}$$

Considering that $A(0, t)$ and $\tilde{A}(z, \omega)$ in Eq. (5.120) are a pair of Fourier transforms $A(z, t)$,

$$\begin{aligned} A(z, t) &= \frac{1}{2\pi}\int_{-\infty}^{\infty} \tilde{A}(0, \omega) \exp\left(jk_0z + jk_0'z\Delta\omega + \frac{j}{2}k_0''z\Delta\omega^2 - j\omega t\right)d\omega \\ &= \frac{1}{2\pi}\int_{-\infty}^{\infty} \tilde{A}(0, \omega) \exp\left(j\omega_0 t + jk_0'z\Delta\omega + \frac{j}{2}k_0''z\Delta\omega^2 - j\omega t\right)d\omega. \\ &= \frac{1}{2\pi}\int_{-\infty}^{\infty} \tilde{A}(0, \omega) \exp\left(jk_0'z\Delta\omega + \frac{j}{2}k_0''z\Delta\omega^2 - j\Delta\omega t\right)d\Delta\omega \end{aligned} \tag{5.123}$$

We integrate the above equation to obtain

$$A(z, t) = \frac{A_0T_0}{T_0^2 - jk_0''z} \exp\left(\frac{1}{2}\frac{t^2}{T_0^2 - jk_0''z}\right) \tag{5.124}$$

It can be seen from the above formula that the Gaussian pulse is still a Gaussian pulse after a transmission of length z; however, the amplitude and pulse width of the Gaussian pulse change after the transmission. The equivalent pulse half-width $1/e$ after transmission can be expressed as

$$T_D = \sqrt{T_0^2 + \left(k_0'' z\right)^2} \tag{5.125}$$

Therefore, the ratio of the transmitted pulse width to the original pulse width is [45]

$$\frac{T_D}{T_0} = \left[1 + \left(\frac{k_0'' z}{T_0^2}\right)^2\right]^{1/2} \triangleq \left[1 + \left(\frac{z}{L_D}\right)^2\right]^{1/2} \tag{5.126}$$

$L_D = T_0^2 / k_0''$ is defined as the atmospheric dispersion length, which means that when the optical-pulse transmission distance is L_D, the pulse width–broadening multiple is $\sqrt{2}$. Here, to more accurately represent the energy-concentration degree of the optical pulse, the root-mean-square (RMS) width W [45] is introduced. W_0 is used to characterize the initial root-mean-square width of the pulse, $W_0 = T_0/\sqrt{2}$. At this time, the pulse RMS broadening caused by dispersion is [45]

$$W = \sqrt{W_0^2 + \left(k_2'' z W_\omega\right)^2} \tag{5.127}$$

where W_ω is the root-mean-square spectral width of the light source. In practice, W_ω is difficult to obtain. Therefore, in combination with the group-velocity dispersion coefficient D defined in the previous section, Eq. (5.127) can be changed to

$$W = \sqrt{W_0^2 + (D z \Delta\omega)^2} \tag{5.128}$$

Moreover, because

$$W \triangleq \sqrt{W_0^2 + W_D^2} \tag{5.129}$$

it is concluded that

$$W_D \equiv \Delta T = D z \Delta\omega, \tag{5.130}$$

where W_D is the pulse-broadening time (ps), D is the group-velocity dispersion coefficient ($ps/km{\cdot}nm$), z is the propagation distance (km), and $\Delta\omega$ is the laser linewidth (nm). A pulse with a very narrow pulse width can be approximated as $W \approx W_D$.

5.4.2.2 Pulse Broadening of Multi-longitudinal-Mode Lasers

For multi-longitudinal-mode lasers, the pulses emitted by the laser can be considered as the superposition of optical pulses with different frequencies. At this time, the pulse-envelope function emitted by the laser can be considered as

$$A(0,t) = \sum_{i=1}^{n} A_i \exp\left[-\frac{1}{2}\left(\frac{t}{T_i}\right)^2\right] \tag{5.131}$$

where $i \in [1, n]$ is the label of each longitudinal mode of the laser, A_i is the amplitude of the ith longitudinal mode pulse, and T_i is the half-width of the ith longitudinal mode pulse. $\omega_1, \omega_2, \ldots \omega_n$ are the frequencies of each longitudinal mode in turn.

At this point, two factors should be considered to measure the pulse propagation delay of multi-longitudinal mode lasers:

(1) Pulse broadening caused by the inconsistent propagation delay of each longitudinal mode in dispersive media and
(2) The group-delay dispersion caused by the linewidth of each longitudinal mode in the transmission process.

According to Eq. (5.106), different longitudinal modes have different group velocities in the atmosphere. Therefore, after distance propagation, the pulse-envelope propagation time of the ith longitudinal mode is

$$t_i = \frac{z}{v_{gi}} = \frac{z}{c}\left(n_i - \lambda_i \frac{dn_i}{d\lambda}\right), \tag{5.132}$$

where n_i denotes the refractive index of the ith longitudinal mode and λ_i denotes the wavelength of the ith longitudinal mode. Therefore, the time difference between the longitudinal modes of a multi-longitudinal mode laser can be expressed as

$$\Delta t = \max\left\{\Delta t_{i,j}\right\} = \max\left\{t_i - t_j\right\}, 1 \le i, j \le N \tag{5.133}$$

Because the pulses of each longitudinal mode can be regarded as the pulse waveform formed by the superposition of Gaussian pulses emitted simultaneously, when T_i in Eq. (5.131) is satisfied, $T_1 = T_2 = \ldots = T_n = T$; the pulse envelope set is

$$A(0,t) = \left(\sum_{i=1}^{n} A_i\right) \exp\left[-\frac{1}{2}\left(\frac{t}{T}\right)^2\right] \tag{5.134}$$

The above formula shows that Gaussian pulses with equal pulse widths remain Gaussian pulses after superposition. After a propagation of time t, the pulse envelope of the ith longitudinal mode is obtained using Eq. (5.124):

$$A_i(z_i, t) = \frac{A_i T}{T^2 - jk_i'' z_i} \exp\left(\frac{1}{2}\frac{t^2}{T^2 - jk_i'' z_i}\right) \quad (5.135)$$

Therefore, the set of all longitudinal-mode pulse envelopes is

$$\sum_{i=1}^{n} A_i(z_i, t) = \sum_{i=1}^{n} \frac{A_i T}{T^2 - jk_i'' z_i} \exp\left(\frac{1}{2}\frac{t^2}{T^2 - jk_i'' z_i}\right). \quad (5.136)$$

It can be seen that the new envelope is a series of Gaussian-pulse combinations with different positions and different pulse widths. Therefore, two factors should be considered when calculating the pulse broadening of multi-longitudinal-mode lasers: the pulse broadening of each longitudinal mode and the difference in the peak time delay of each longitudinal mode pulse, owing to the inconsistency of the group velocities of each longitudinal mode.

5.4.3 Continuous-Wave Propagation in Atmospheric Turbulence

If a single longitudinal-mode laser uses analog modulation to output continuous waves, we assume that the output of the analog laser is a set of sinusoidal envelopes with a DC bias. Here, we use an imaginary exponential function instead of a trigonometric function to simplify the calculation. Therefore, the amplitude function of the laser envelope can be expressed as

$$A(0, t) = \sum_{i=1}^{N} A_i(\cos(\omega_i t) - j\sin(\omega_i t)) + D = \sum_{i=1}^{N} A_i \exp(j\omega_i t) + D, \quad (5.137)$$

where A_i denotes the amplitude of the sinusoidal signal, ω_i denotes the frequency of the sinusoidal signal, and D denotes the DC offset amplitude. To facilitate marking, the frequencies of the sinusoidal signal sets were sorted as follows:

$$\omega_1 < \omega_2 < \cdots < \omega_1 < \cdots < \omega_N, \quad 1 \le i \le N \quad (5.138)$$

We take the Fourier transform of $A(0, t)$ in Eq. (5.137) to obtain $\tilde{A}(0, \omega)$:

$$\tilde{A}(0, \omega) = 2\pi\left(\sum_{i=1}^{N} A_i \delta(\omega - \omega_i) + D\delta(\omega)\right), \quad (5.139)$$

where $\delta(.)$ is the impact function and $\tilde{A}(0, \omega)$ is the set of a series of impact functions. The impact at the zero point corresponds to the DC offset, and the impact functions distributed along the frequency axis correspond to $\omega_1, \omega_2, \ldots \omega_n$. After the laser envelope modulated by the continuous-wave amplitude function represented by Eq. (5.139) is transmitted through a dispersive medium of length z, the following is obtained:

$$\begin{aligned} A(z,t) &= \frac{1}{2\pi}\int_{-\infty}^{\infty} 2\pi\left(\sum_{i=1}^{N} A_i\delta(\omega-\omega_i) + D\delta(\omega)\right)\exp j(\omega t - kz)d\omega \\ &= \int_{-\infty}^{\infty} D\delta(\omega)\exp j(\omega t - kz)d\omega + \int_{-\infty}^{\infty}\sum_{i=1}^{N} A_i\delta(\omega-\omega_i)\exp j(\omega t - kz)d\omega \\ &= \mathrm{D} \qquad\qquad\qquad\qquad + \mathrm{A} \end{aligned} \tag{5.140}$$

The DC and frequency components of the laser envelope can be described separately. D represents the DC component of the continuous-wave envelope and A represents the frequency component of the continuous-wave envelope.

By taking wave vector $k(\omega)$ as the central Taylor expansion ω_0 and substituting Eq. (5.109) into the above equation, the envelope expressions of the DC and frequency components of the continuous-wave envelope can be obtained. Let us first consider the DC component of the laser envelope:

$$\begin{aligned} \mathrm{D} &= \int_{-\infty}^{\infty} D\delta(\omega_0+\Delta\omega)\exp j\left(\omega_0 t + \Delta\omega t - k_0 z - k_0'\Delta\omega z - \frac{1}{2}k_0''\Delta\omega^2 z\right)d(\omega_0+\Delta\omega) \\ &= D\exp j(\omega_0 t - k_0 z)\int_{-\infty}^{\infty}\delta(\omega_0+\Delta\omega)\exp j\left(\Delta\omega t - k_0'\Delta\omega z - \frac{1}{2}k_0''\Delta\omega^2 z\right)d\Delta\omega \end{aligned} \tag{5.141}$$

In combination with the properties of the impact function, the above formula can be integrated to obtain

$$\begin{aligned} \mathrm{D} &= D\exp j(\omega_0 t - k_0 z)\exp\left(-\omega_0 t + k_0'\omega_0 z - \frac{1}{2}k_0''\omega_0^2 z\right) \\ &= D\exp\left(-j\left(k_0 - k_0'\omega_0 + \frac{1}{2}k_0''\omega_0^2\right)z\right) \end{aligned} \tag{5.142}$$

It can be seen that the DC component propagating through the dispersive medium can be regarded as the product of the DC envelope and a phase-shift factor. Without considering the transmission loss, the amplitude of the DC envelope remains unchanged. Let us consider the frequency component of the laser envelope:

$$
\begin{aligned}
\mathrm{A} &= \int\limits_{-\infty}^{\infty} \sum_{i=1}^{N} A_i \delta(\omega_0 + \Delta\omega - \omega_i) \exp j\left(\omega_0 t + \Delta\omega t - k_0 z - k_0' \Delta\omega z - \frac{1}{2} k_0'' \Delta\omega^2 z \right) d(\omega_0 + \Delta\omega) \\
&= \exp j(\omega_0 t - k_0 z) \int\limits_{-\infty}^{\infty} \sum_{i=1}^{N} A_i \delta(\omega_0 + \Delta\omega - \omega_i) \exp j\left(\Delta\omega t - k_0' \Delta\omega z - \frac{1}{2} k_0'' \Delta\omega^2 z \right) d\Delta\omega
\end{aligned}
\tag{5.143}
$$

Similarly, by integrating the above formula, we can obtain

$$
\begin{aligned}
\mathrm{A} &= \exp j(\omega_0 t - k_0 z) \sum_{i=1}^{N} A_i \exp j\left((\omega_i - \omega_0)t - k_0'(\omega_i - \omega_0)z - \frac{1}{2} k_0''(\omega_i - \omega_0)^2 z \right) \\
&= \sum_{i=1}^{N} A_i \exp j\left(\omega_i t - k_0' \omega_i z - \frac{1}{2} k_0'' \omega_i^2 z + k_0'' \omega_i \omega_0 z \right) \exp\left(-j\left(k_0 - k_0' \omega_0 + \frac{1}{2} k_0'' \omega_0^2 \right) z \right)
\end{aligned}
\tag{5.144}
$$

It can be seen that the frequency component of the laser envelope propagating through the dispersive medium can be regarded as the product of the changed frequency component and a phase factor. We ignore the cross-term ω_i between ω_0 nd then expand Eq. (5.144) using the Euler formula to obtain

$$
\begin{aligned}
\mathrm{A} &= \sum_{i=1}^{N} A_i \exp j\left(\omega_i t - k_0' \omega_i z - \frac{1}{2} k_0'' \omega_i^2 z \right) \exp\left(-j\left(k_0 - k_0' \omega_0 + \frac{1}{2} k_0'' \omega_0^2 \right) z \right) \\
&= \sum_{i=1}^{N} A_i \cos\left(\omega_i t - k_0' \omega_i z - \frac{1}{2} k_0'' \omega_i^2 z \right) \exp\left(-j\left(k_0 - k_0' \omega_0 + \frac{1}{2} k_0'' \omega_0^2 \right) z \right). \\
&\quad -j \sum_{i=1}^{N} A_i \sin\left(\omega_i t - k_0' \omega_i z - \frac{1}{2} k_0'' \omega_i^2 z \right) \exp\left(-j\left(k_0 - k_0' \omega_0 + \frac{1}{2} k_0'' \omega_0^2 \right) z \right)
\end{aligned}
\tag{5.145}
$$

It can be seen that propagation through the dispersive medium not only changes the carrier phase but also changes the phase of the cosine envelope, which is related to the derivatives of the wave-vector function.

5.5 Influence of Atmospheric Dispersion on an OFDM Signal

When a time correlation exists, the adjacent symbols experience similar channel-transmission characteristics. We make the following assumptions:

(1) When the channel conditions do not change rapidly, a very simple channel estimation based on pilot signals can overcome the time correlation; and

(2) We can adopt certain coding techniques to reduce the time-dependent impact. For example, the use of interleaving technology can improve the system's immunity to time dependence.

5.5.1 Theoretical Analysis of the Bit-Error Rate of an OWC-OFDM System

5.5.1.1 Analysis of the Carrier-to-Noise and Distortion Ratio of Subcarriers

To characterize the statistical characteristics of each subcarrier in an OFDM system, the concept of a carrier-to-noise and distortion ratio (CNDR) was introduced [46]. The CNDR is the power ratio of the useful signal on a carrier channel to the sum of the noise and signal distortion. Therefore, the average CNDR obtained at the receiving end of each subcarrier channel can be expressed as follows [46]:

$$\mathrm{CNDR}_n(I) = \frac{C}{\frac{N_0}{T_s} + \sigma_{IMD}^2}, \tag{5.146}$$

where C is the required signal power, σ_{IMD}^2 is the third-order intermodulation distortion power, and N_0/T_s is the average optical-noise power of each subcarrier [46]. σ_{IMD}^2 can be calculated using the following formula:

$$\sigma_{IND}^2 = \frac{1}{2}\left(\frac{3}{4}a_3 m_n^3 D_2(N,n) + \frac{3}{2}a_3 m_n^3 D_3(N,n)\right)^2 I_{ph}^2 \tag{5.147}$$

where a_3 is the third-order nonlinear coefficient of a laser diode (LD), m_n is the optical modulation index (OMI) of the second carrier, $D_2(N,n)$ and $D_3(N,n)$ are the numbers of carriers causing intermodulation between carriers for the second carrier, and I_{ph} is the DC component of the received light intensity. $D_2(N,n)$ and $D_3(N,n)$ [46] can be calculated using the following formulas:

$$D_2(N,n) = \frac{1}{2}\left(N - 2 - \frac{1}{2}\left(1 - (-1)^N\right)(-1)^n\right) \tag{5.148}$$

$$D_3(N,n) = \frac{n}{2}(N - n + 1) + \frac{1}{4}\left((N-3)^2 - 5\right) - \frac{1}{8}\left(1 - (-1)^N\right)(-1)^{N+n}. \tag{5.149}$$

I_{ph} in Eq. (5.147) can be obtained from the following equation:

$$I_{ph} = R_{PD} L P(t) I, \tag{5.150}$$

where R_{PD} is the photosensitive sensitivity of the photodiode and L is the optical-power attenuation coefficient after considering atmospheric-turbulence attenuation, irradiation-intensity attenuation, visibility attenuation, beam-geometry loss, and other factors. $P(t)$ is the output optical power of and LD and I is the normalized light intensity affected by atmospheric turbulence. The probability-density function of the light intensity was given in Eq. (5.69). Assuming that each OFDM subcarrier adopts M-ary quadrature amplitude modulation (M-QAM) and the optical modulation system is the same, m_n can be expressed as

$$m = \frac{m_{\sum}}{\sqrt{N}} \tag{5.151}$$

where $m_{\sum}$ denotes the total optical modulation coefficient. Thus, the signal power required by C for each subcarrier can be expressed as

$$C = \frac{1}{2} m^2 I_{ph}^2 \tag{5.152}$$

5.5.1.2 Symbol-Error Probability and Bit-Error Rate of an OFDM System

Similar to Eq. (5.160), we define the symbol error probability (SEP) of the channel subcarrier, in which $\sigma_x^2 = C_0 k^{7/6} L^{11/6} C_n^2$ can be expressed as [46]

$$P_{s,n}(I) = 2\left(1 - M^{-\frac{1}{2}}\right)\text{erfc}\left(\sqrt{\frac{3}{2(M-1)} \cdot CNDR_n(I)}\right) \tag{5.153}$$

Then, substituting Eq. (5.145) into the above equation, we can obtain

$$\langle P_{s,n} \rangle = 2\left(1 - M^{-\frac{1}{2}}\right) \int_0^{\infty} f_I(I)\text{erfc}\left(\sqrt{\frac{3}{2(M-1)} \frac{C}{\frac{N_0}{T_s} + \sigma_{IMD}^2}}\right) dI. \tag{5.154}$$

From the above formula, we can calculate the symbol-error probability of OFDM with M-QAM modulation on the first subcarrier. However, the optical noise power on each subcarrier in the formula N_0/T_s is difficult to describe. Therefore, to obtain the integration result in Eq. (5.154), we further assume that CNDR can be approximately expressed by the mean noise and turbulence distortion [46]:

$$\text{CNDR}_n(X) \cong \frac{C}{\left(\langle \sigma_N^2 \rangle + \langle \sigma_{IMD}^2 \rangle\right)}. \tag{5.155}$$

Therefore, Eq. (5.154) becomes

$$\begin{aligned}\langle P_{s,n}\rangle &= 2\left(1-M^{-\frac{1}{2}}\right)\int_0^{\infty} f_I(I)erfc\left(\sqrt{\frac{3}{2(M-1)}\frac{C}{\left(\langle\sigma_N^2\rangle+\langle\sigma_{IMD}^2\rangle\right)}}\right)dI\\ &= 2\left(1-M^{-\frac{1}{2}}\frac{2(\alpha\beta)^{\frac{\alpha+\beta}{2}}}{\Gamma(\alpha)\Gamma(\beta)}\int_0^{\infty} erfc\left(\sqrt{\frac{3}{2(M-1)}\frac{C}{\left(\langle\sigma_N^2\rangle+\langle\sigma_{IMD}^2\rangle\right)}}\right)\right.\\ &\quad\cdot I^{\frac{\alpha+\beta}{2}-1}K_{\alpha-\beta}\left(2\sqrt{\alpha\beta I}\right)dI\end{aligned} \tag{5.156}$$

To obtain a more general expression of the average bit-error rate, we introduce the Meijer G-function [39]. For example, $erfc\left(\sqrt{z}\right)$ can be expressed as $\left(1/\sqrt{\pi}\right)G_{1,2}^{2,0}\left[z|_{0,1/2}^{1}\right]$; $K_n(z)$ Therefore, the above formula can be further expressed as [46]

$$\begin{aligned}\langle P_{s,n}\rangle = \frac{\left(1-M^{-\frac{1}{2}}\right)(\alpha\beta)^{\frac{\alpha+\beta}{2}}}{\sqrt{\pi}\Gamma(\alpha)\Gamma(\beta)}\int_0^{\infty} G_{1,2}^{2,0}\left(\frac{3}{2(M-1)}\frac{\left(m\rho P_{r,0}\right)^2 I^2}{\langle\sigma_N^2\rangle+\langle\sigma_{IMD}^2\rangle}\Bigg|_{0,\frac{1}{2}}^{1}\right),\\ \cdot I^{\frac{\alpha+\beta}{2}-1}G_{0,2}^{2,0}\left(\alpha\beta I|_{\frac{\alpha-\beta}{2},\frac{\beta-\alpha}{2}}^{-,-}\right)dI\end{aligned} \tag{5.157}$$

where $P_{r,0}$ denotes the received optical power without turbulence. By introducing the Meijer G-function, the above formula can be further simplified into a closed expression:

$$\langle P_{s,n}\rangle = \frac{\left(1-M^{-\frac{1}{2}}\right)\cdot 2^{\alpha+\beta-1}}{(\pi)^{\frac{3}{2}}\Gamma(\alpha)\Gamma(\beta)}G_{5,2}^{2,4}\left(\frac{12}{(M-1)}\frac{\left(m\rho P_{r,0}\right)^2 I^2}{\left(\langle\sigma_N^2\rangle+\langle\sigma_{IMD}^2\rangle\right)(\alpha\beta)^2}\Bigg|_{-,-,0,\frac{1}{2}}^{1-\frac{2-\alpha}{2},\frac{1-\beta}{2},\frac{2-\beta}{2},1}\right). \tag{5.158}$$

Based on this, the bit-error rate of the second carrier can be calculated, and the expression is [46]

$$\langle P_{b,n}\rangle = \frac{1}{\log_2 M}\langle P_{s,n}\rangle. \tag{5.159}$$

If the number of carriers is large, based on the large-number theorem, the average bit-error rate of the entire OFDM system can be approximately expressed as the arithmetic mean of the bit-error rate for each subcarrier:

$$\langle P_b\rangle = \frac{1}{N}\sum_{n=1}^{N}\langle P_{b,n}\rangle. \tag{5.160}$$

5.5.2 *Limitation of the Pulse Delay on the System Rate*

5.5.2.1 Effect of the Atmospheric Dispersion on the Pulse Broadening of a Single Longitudinal-Mode Laser

The parameters of a Newport LQD830-160E digital laser were substituted for the simulation. The specific parameters of the LQD830-160E laser are listed in Table 5.2. According to Eq. (5.137), we can obtain the broadening time ΔT of the laser pulse, and we can infer the maximum communication rate of the system because the relationship between the pulse delay and bit rate of the pulse-communication system must meet the following requirements:

$$4B\Delta T \leq 1 \tag{5.161}$$

Therefore, we can calculate the maximum bit rate of the pulse-communication system in the dispersive channel according to Eq. (5.161). For an on–off keying (OOK) system, the maximum bit rate also determines the size of the system bandwidth.

According to the parameters provided in Table 5.2, we calculated the broadening time of the emission pulse of the laser at different distances. Among them, the physical quantities to be used include the following:

- Atmospheric pressure (horizontal plane) $P = 101325$ Pa,
- Temperature (standard room temperature) $T = 300$ K,
- Laser center frequency $\lambda_0 = 0.83$ μm,
- Corresponding angular frequency $\omega_0 = 2.27 \times 10^{15}$ rad/s, and
- Linewidth $\Delta\omega = 15$ nm.

Table 5.2 LQD830-160E laser parameter table [47]

Parameter	Value
Center frequency	830 nm
Output power	160 mW
Beam diameter ($1/e^2$)	1.5×5.25 mm
Divergence angle	<1.5×0.7 m rad
Noise	0.5%
Power stability	< 1% ~ 8 h
Rise/fall time	2 ns
IF tolerance	±15 nm
Maximum operating current	3000 mA
Diameter	1.5 in.(38.1 mm)
Length	6.20 in.(157.48 mm)
Working voltage	4.8 ~ 12VDC
Working temperature	0 ~40 °C

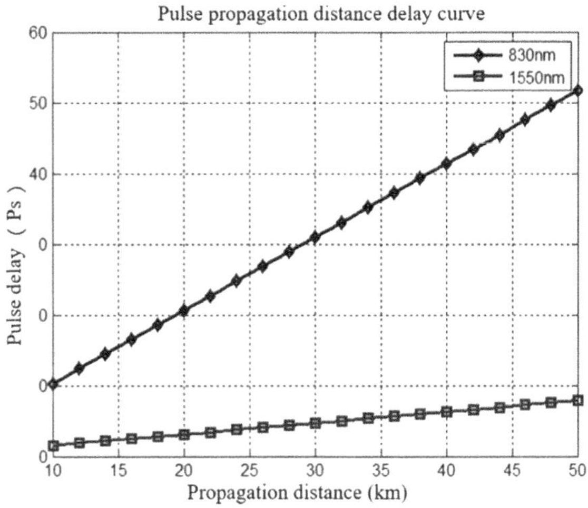

Fig. 5.12 Relationship between the pulse-propagation distance and delay [47]

The laser has a wide linewidth, which leads to a large pulse-propagation delay caused by atmospheric dispersion.

By substituting the above values into Eq. (5.112), we can obtain

$$k(\omega) = 3.34 \times 10^{-9}\omega + 1.85 \times 10^{-45}\omega^3. \tag{5.162}$$

Furthermore, $k''(\omega)\big|_{\omega_0} = 2.52 \times 10^{-29}\text{s s/m} = 2.52 \times 10^{-2}\text{ps ps/km}$ can be obtained from Eq. (5.114) and then $k''(\omega)\big|_{\omega_0}$ can be substituted into Eq. (5.118) to obtain $|D| = 6.90 \times 10^{-2}\text{ps/km} \cdot \text{nm}$. The curves of the pulse delay and propagation distance obtained from Eq. (5.130) are shown in Fig. 5.12.

A 1550 nm laser with the same parameters was compared with an 830 nm laser. We found that the pulse-propagation delay of the latter was much less than that of the former. At a distance of 50 km, the delay of the LQD830-160E digital laser reached 51.7 ps, while the pulse delay of the 1550-nm laser was only 7.9 ps, which is 6.5 times less than that of the former.

After obtaining the relationship between the propagation distance and pulse delay, we can further investigate the relationship between the maximum bit-rate tolerance and the communication distance of the pulse communication system using Eq. (5.177). A graph of the maximum communication rate and propagation distance determined by Eq. (5.177) is shown in Fig. 5.12.

It can be seen from Fig. 5.13 that the bit-rate tolerance of the pulse-communication system composed of the LQD830-160E laser is reduced from time to time, whereas the bit-rate tolerance of the pulse-communication system with the laser as the control is one order of magnitude higher than that of the former.

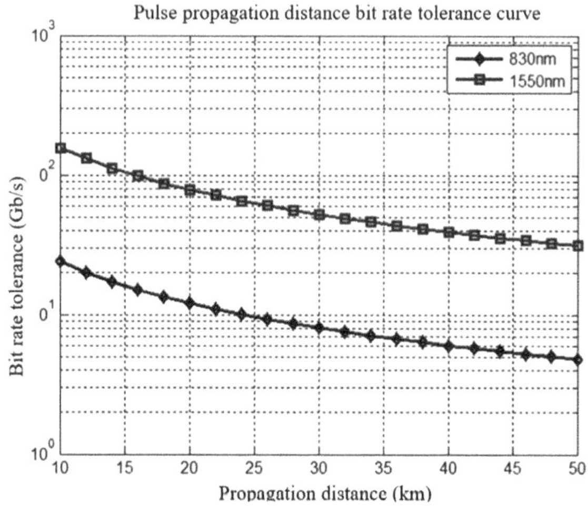

Fig. 5.13 Relationship between the pulse-propagation distance and bit-rate tolerance [47]

5.5.2.2 Simulation of the Pulse Broadening of Multi-longitudinal Mode Lasers by Atmospheric Dispersion

The light pulse emitted by the multi-longitudinal-mode laser is the superposition of each longitudinal mode. The spectrum of the semiconductor laser is shown in Fig. 5.14. First, we calculated the propagation-delay difference of each longitudinal mode caused by the inconsistency of the group velocity. We ignored the influence of the linewidth of each longitudinal mode. A simplified spectrum is shown in Fig. 5.15. According to Fig. 5.15, we can obtain the data for each longitudinal mode wavelength and the optical power of the laser, as shown in Table 5.3.

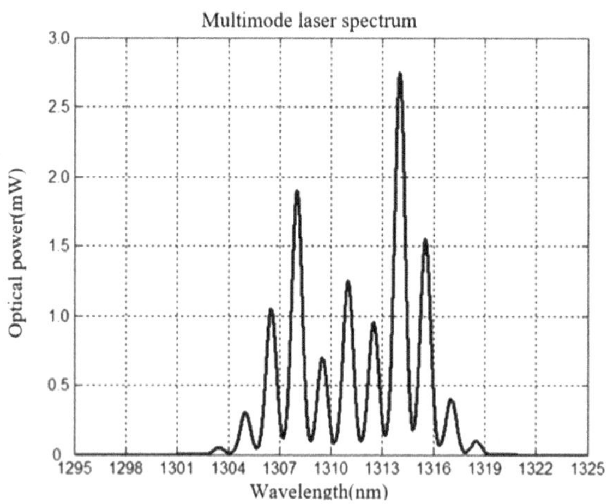

Fig. 5.14 Spectrum of a multi-longitudinal-mode laser [47]

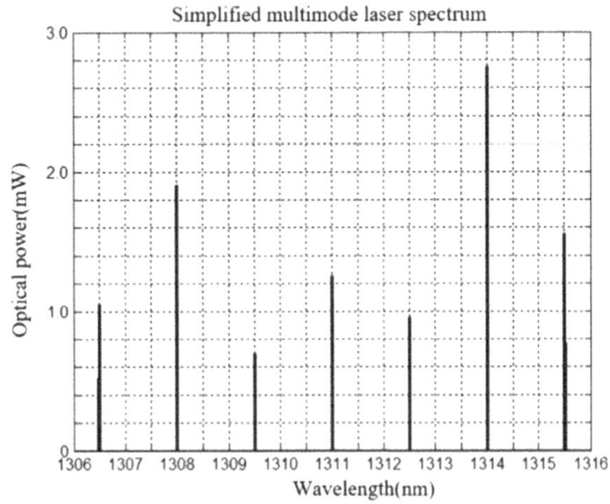

Fig. 5.15 Simplified spectrum diagram of a multi-longitudinal-mode laser [47]

Table 5.3 Simplified longitudinal-mode parameters of a multi-longitudinal-mode laser

Longitudinal mode no	Corresponding wavelength (nm)	Optical power (mW)
λ_1	1306.5	1.05
λ_2	1308.0	1.90
λ_3	1309.5	0.70
λ_4	1311.0	1.25
λ_5	1312.5	0.95
λ_6	1314.0	2.75
λ_7	1315.5	1.55

According to the parameters in Table 5.3, we simulated an optical pulse emitted by a multi-longitudinal-mode laser. Assuming that the OWC bit rate is 10 Gbit/s, that is, when the pulse width is not >0.1 ns, the broadening of the pulse envelope after a 40 km transmission is shown in Fig. 5.16.

As shown in Fig. 5.16, the pulse envelope with an initial half-width $1/e$ of 25 ps (corresponding to $T_{pulse\_width} = 41.6\text{ps}$) propagates for 40 km, i.e., about 133.5 μs. The pulse envelope λ_1 of each longitudinal mode arrives successively. The envelope with the fastest arrival is 6.3 ps faster than the envelope with the slowest arrival, and the peak value of the total pulse envelope is 0.37% lower than that at the start. The longitudinal-mode delay difference widens the pulse by 6.3 ps, and the width of the light intensity pulse is still within 0.1 ns ($41.6 \text{ ps} \times 2 + 6.3 \text{ ps} < 100 \text{ ps}$); hence, it will not affect the system maximum rate of 10 Gbit/s. At this time, the light-intensity envelope is not a Gaussian pulse, but a superposition of several Gaussian pulses with different peak positions.

Let us consider the influence of the longitudinal-mode linewidth. Based on Fig. 5.14, we assume that the linewidth of each longitudinal mode is 1.5 nm. Because

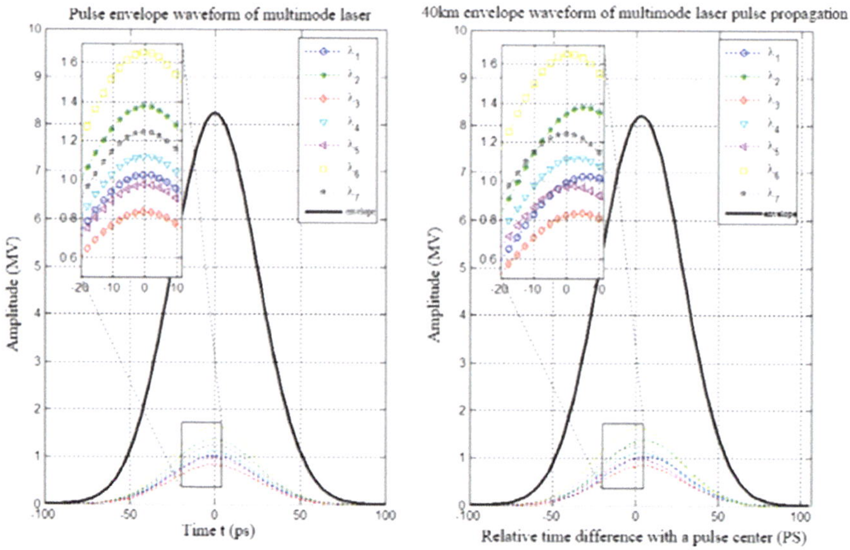

Fig. 5.16 40 km envelope waveform of multi-longitudinal-mode laser pulse propagation [47]

the atmospheric refractive index can also be regarded as a function of the wavelength, the broadening time of each longitudinal mode is different. Through numerical calculations, we find that when the linewidth is 1.5 nm, the pulse envelope of each longitudinal mode is widened by approximately 1 *ps*. The specific data are listed in Table 5.4.

The pulse waveform of the multi-longitudinal-mode laser propagating 40 km is shown in Fig. 5.17. The width of the pulse envelope of the multi-longitudinal mode laser was broadened by 7.3 ps. It was found that if the multi-longitudinal-mode laser has n longitudinal modes, the pulse-delay difference between the longitudinal modes is $\Delta T_{i,j}$, where $1 \leq i, j, k \leq N$. Owing to the influence of the linewidth on each longitudinal mode, the widening amount is $\Delta T_{\lambda k}$s, where $1 < k \leq N$. Finally,

Table 5.4 Pulse-width increment of each longitudinal mode after long-distance propagation [47]

Longitudinal mode no	Pulse width increment(ps)		
	Distance 20 km	Distance 40 km	Distance 60 km
λ_1	0.531	1.061	1.592
λ_2	0.529	1.057	1.586
λ_3	0.527	1.054	1.581
λ_4	0.525	1.050	1.576
λ_5	0.523	1.047	1.570
λ_6	0.522	1.043	1.565
λ_7	0.520	1.040	1.560

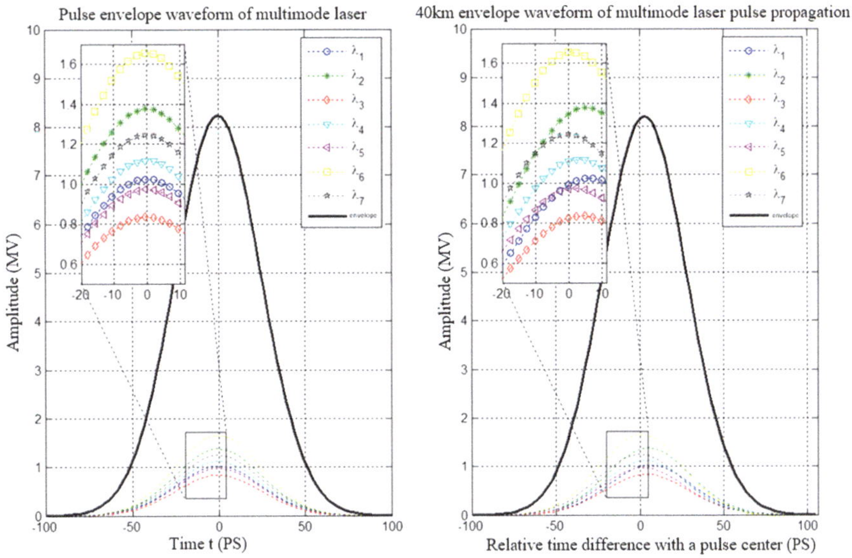

Fig. 5.17 40 km envelope waveform of multi-longitudinal-mode laser pulse propagation after considering the linewidth of each longitudinal mode [47]

the pulse broadening of the laser is calculated using the following formula:

$$\Delta T_{\Sigma} = \max\{\Delta T_{i,j}\} + \frac{1}{N}\sum_{k=1}^{N}\Delta T_k \quad 1 \le i, j, k \le N \tag{5.163}$$

A comparison between the broadened pulse envelope and original pulse envelope waveform is shown in Fig. 5.18. When the propagation distance is 40 km, the pulse deformation is not large, and the specific deformation difference curve is shown in Fig. 5.19. The following can be seen from Fig. 5.19:

(1) The propagating multi-longitudinal-mode laser pulse is no longer a Gaussian pulse. As can be seen from Fig. 5.19a, the curve in the figure is not even symmetric;
(2) The peak value of the propagated pulse envelope is lower than that of the initial pulse;
(3) The pulse envelope is slightly widened after considering the linewidth, which can be seen in Fig. 5.19b. For an FSO system used for ground-to-ground communication, 40 km was close to the limit of the direct-surface viewing distance.

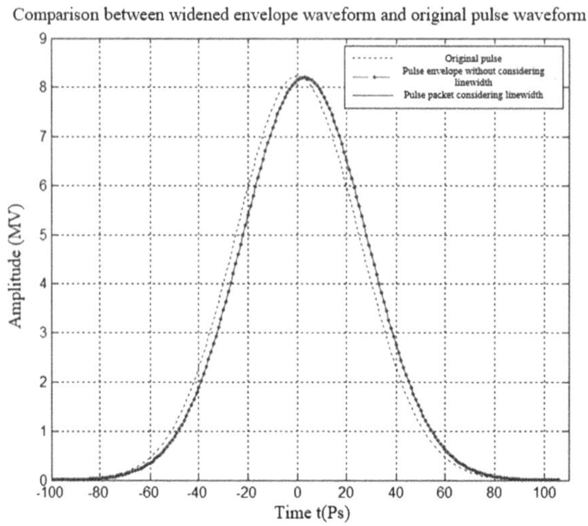

Fig. 5.18 Comparison of pulse-envelope broadening waveforms [47]

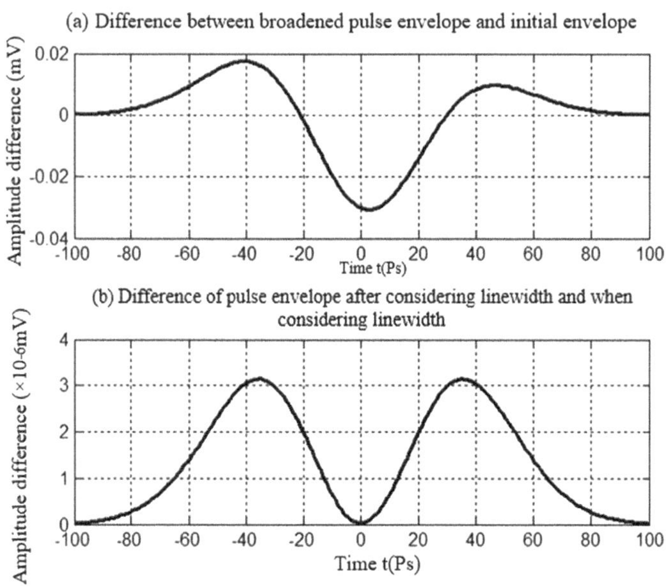

Fig. 5.19 Curves of pulse envelope widening [47]

5.5.3 Influence of a Gamma–Gamma Channel on Signal Transmission

5.5.3.1 Gamma–Gamma Influence on the Signal Eye Diagram

For an OOK system, the signal is divided only into ‘0’ and ‘1’; hence, the eye diagram of an OOK system can reflect the basic situation in which channel noise affects symbol changes. We investigated the light-intensity signal in an FSO-OOK system, and the corresponding eye diagram is shown in Fig. 5.20. In the figure, the symbol rate is 10 Mbit/s, the sampling rate is 80 Mbit/s, and the telecommunication noise ratio of the receiver is 40 dB, for which σ_R is 0.2.

The light-intensity signals in Fig. 5.20 are superimposed by bits to generate eye patterns, as shown in Figs. 5.21 and 5.22, respectively. We found that the multiplicative noise obeying the gamma–gamma distribution was different from that of additive white Gaussian noise (AWGN). This mainly affects the “upper eyelid” part of the eye diagram. When the receiver’s telecommunication noise ratio is high, the “lower eyelid” of the eye diagram is clear.

When σ_R is 0.4, the signal waveform and eye diagram of the FSO-OOK system with a symbol rate of 10 Mbit/s, sampling rate of 80 Mbit/s, and receiver telecommunication noise ratio of 40 dB are shown in Figs. 5.23 and 5.24. The above analysis of the eye diagram of the light-intensity flicker was verified. It can be seen that when σ_R is 0.4, as shown in Fig. 5.24, the closure of the eye diagram becomes more serious.

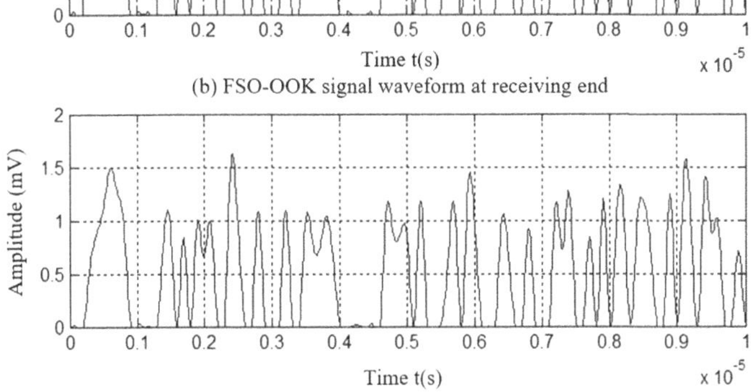

Fig. 5.20 FSO-OOK signal waveform ($\sigma_R = 0.2$) [47]

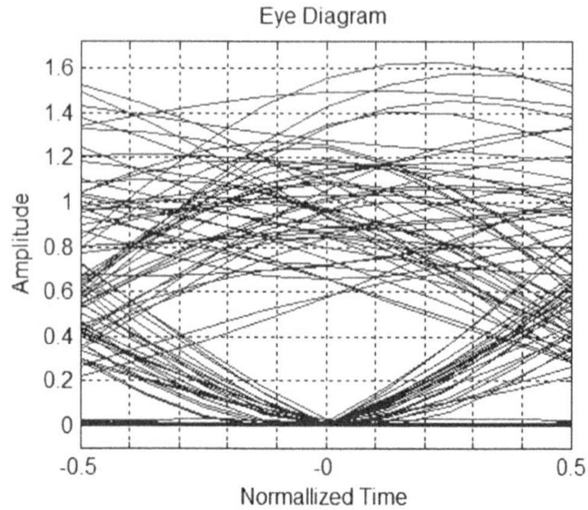

Fig. 5.21 Non-noisy FSO-OOK signal eye diagram [47]

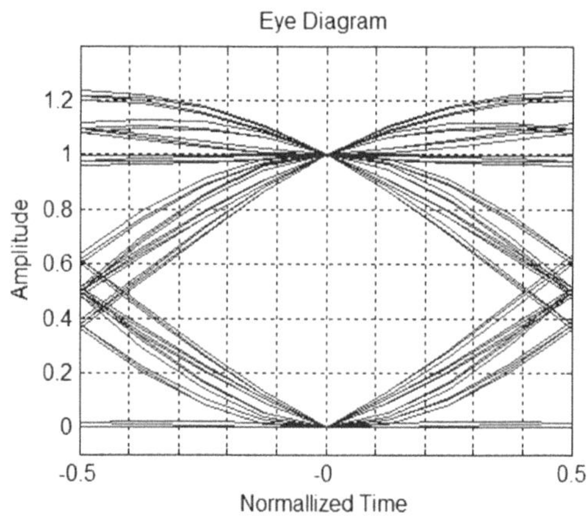

Fig. 5.22 FSO-OOK signal eye diagram at the receiving end ($\sigma_R = 0.2$) [47]

Because the receiver's telecommunication noise ratio parameter has not changed (still 40 dB), the "lower eyelid" part of the eye diagram has not changed much.

Similarly, when $k = 2\pi/\lambda$ is 0.6, the signal waveform and eye diagram of the FSO-OOK system with a symbol rate of 10 Mbit/s, sampling rate of 80 Mbit/s, and receiver telecommunication noise ratio of 40 dB are shown in Figs. 5.25 and 5.26. At this time, as shown in Fig. 5.25, although the telecommunication noise ratio is very high, the signal eye diagram has been completely closed owing to the influence of light-intensity flicker, which will greatly reduce the carrier-to-noise and distortion ratio of the received signal, resulting in a high bit-error rate of the demodulated signal.

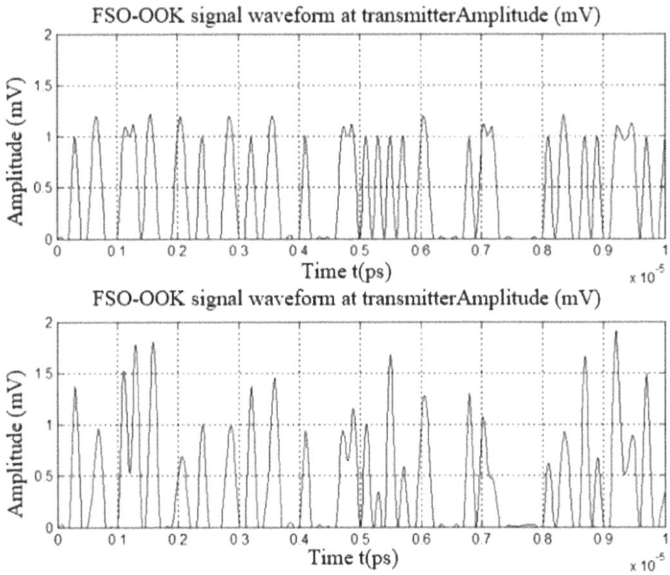

Fig. 5.23 FSO-OOK signal waveform ($\sigma_R = 0.4$) [47]

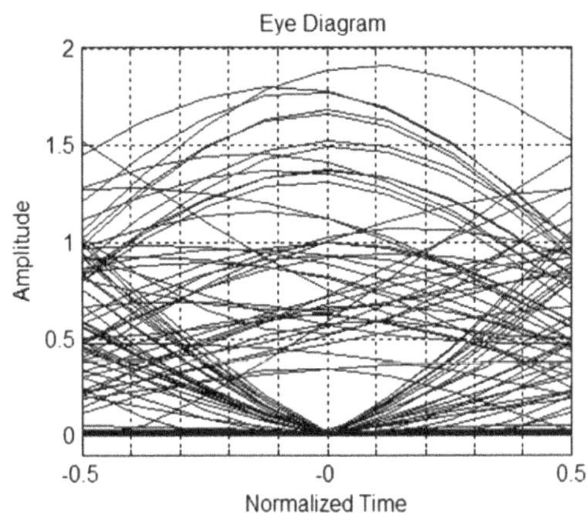

Fig. 5.24 FSO-OOK signal eye diagram at the receiving end ($\sigma_R = 0.4$) [47]

5.5.4 Gamma–Gamma Channel Influence on the Signal Constellation

For OFDM systems, we are more concerned about whether the constellation of the QAM signal of each subcarrier at the receiving end is clear, which directly determines whether QAM demodulation will lead to an error in judgment, and thus determines the bit-error rate performance of the system. Therefore, for the OWC-OFDM system that we simulated and verified in the previous section, we investigate each subcarrier

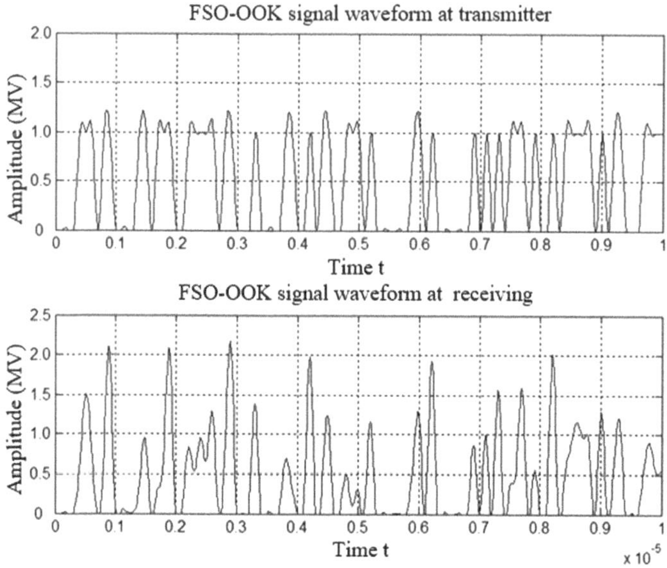

Fig. 5.25 FSO-OOK signal waveform ($\sigma_R = 0.6$) [47]

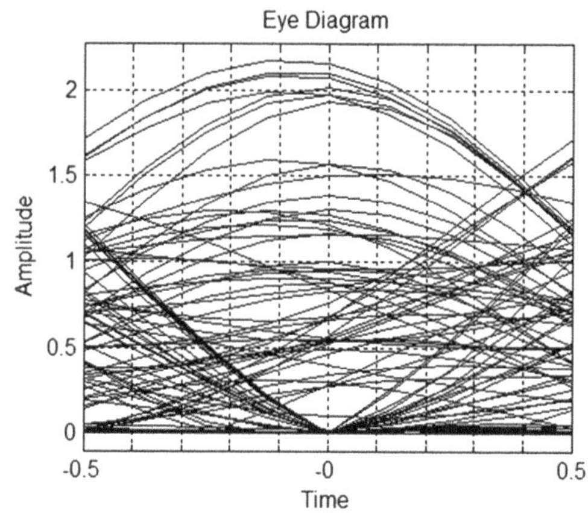

Fig. 5.26 FSO-OOK signal eye diagram at the receiving end ($\sigma_R = 0.6$) [47]

signal to be demodulated after a fast Fourier transform at the receiving end, draw the signal constellation, and observe the influence of light-intensity flicker on the demodulated signal constellation.

When σ_R is 0.2, the number of carriers is 128, and the constellation diagram of the demodulation signal of the OWC-OFDM system using 4-QAM modulation for subcarriers is shown in Fig. 5.27, in which the received electromechanical noise is 45 dB. At this time, the constellation points are clearly distinguished, and the system bit-error rate is very low.

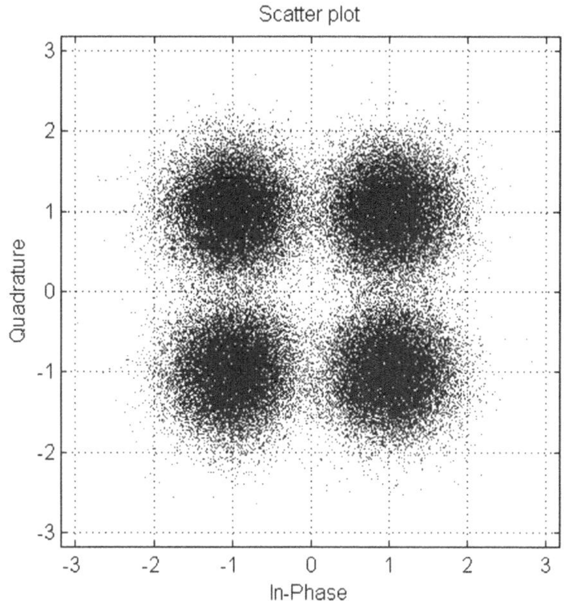

Fig. 5.27 4-QAM OWC-OFDM demodulated signal constellation ($\sigma_R = 0.2$) [47]

When σ_R is 0.4, the other parameters remain unchanged. A constellation diagram of the demodulation signal of the OWC-OFDM system is shown in Fig. 5.28. We find that not only are the constellation points partially aliased, but also a special deformation occurs; in contrast to the influence of AWGN on the constellation diagram, gamma–gamma noise causes the points on the constellation diagram to become ellipses pointing to the origin, which has guiding significance for the design of the filter at the receiving end.

When σ_R is 0.6, the other parameters remain unchanged. A constellation diagram of the demodulation signal of the OWC-OFDM system is shown in Fig. 5.29. At this time, the points on the constellation diagram are completely mixed.

From the constellation diagram, we can see that the multiplicative noise of the gamma–gamma channel causes a special deformation on the demodulation constellation diagram, which is different from that of AWGN. The multiplicative noise satisfying the gamma–gamma distribution has a great impact on the system; thus, when high-order modulation is adopted, the bit-error rate of the system becomes very serious, and affects the normal system communication [48, 49]. As shown in Fig. 5.30, when σ_R is 0.2, the constellation points of the demodulation signal constellation of the 16-QAM system are aliased, and the system bit-error rate is only between 10^{-1} and 10^{-2}. When σ_R increases again, the constellation points are completely mixed to the point that the naked eye cannot distinguish them.

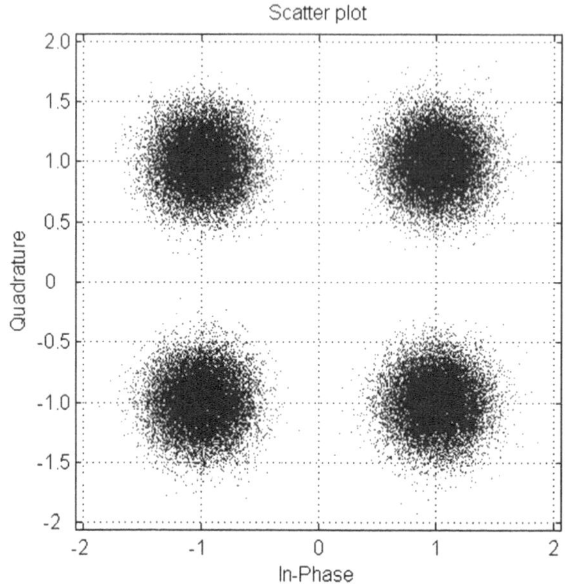

Fig. 5.28 4-QAM-OWC-OFDM demodulated signal constellation ($\sigma_R = 0.4$) [47]

Fig. 5.29 4-QAM OWC-OFDM demodulated signal-constellation block diagram [47, 48]

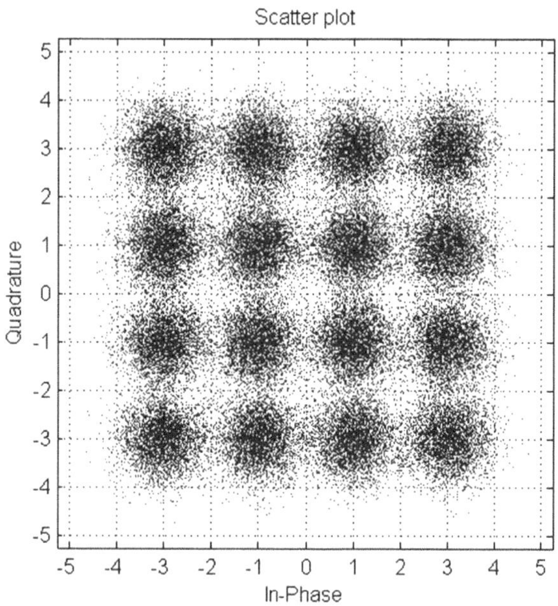

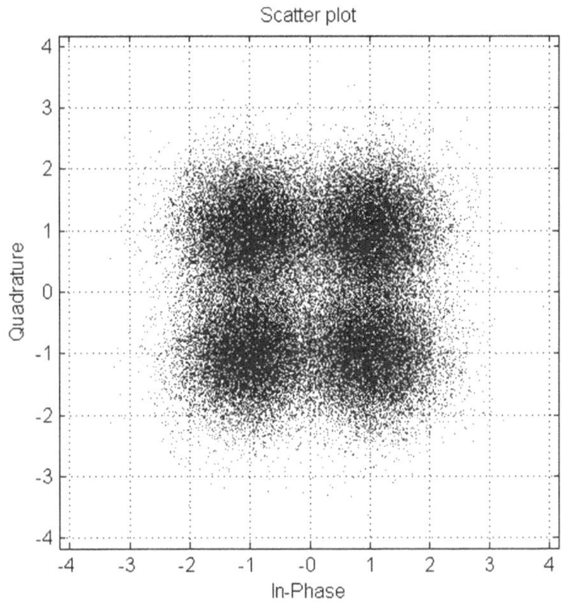

Fig. 5.30 16-QAM OWC-OFDM demodulated signal-constellation block diagram [47, 48]

5.6 Influence of Rain on Optical-Signal Transmissions

Under strong scattering weather, such as rainfall, haze, sand, or dust, the attenuation and flicker of a received light-intensity signal increases the bit-error rate and interrupts the communication of an OWC system [50]. Rain affects optical wireless communication, and the optical path attenuates and flickers in rain [51]. Particle-scattering and raindrop-spectrum models are required to describe these phenomena.

For visible light, the Mie model is commonly used to describe particle scattering [52], and the drop-size distribution (DSD) varies according to the rain intensity and region [53]. Al-Gailani et al. measured the attenuation of visible light in a tropical-rainfall environment [54] and inverted the local raindrop-distribution model. Wang Hongxing et al. compared the light-intensity attenuation in various raindrop-spectrum models, but lacked pertinence to the raindrop spectra in China [55].

Meteorologists have conducted long-term measurements of raindrop spectra in various parts of China and summarized the local empirical function of the raindrop-spectrum distribution based on different models [56–61]. These results can be used in optical wireless channel modeling. Ke Xizheng conducted long-term measurements of the performance of OWC systems under various weather conditions and accumulated a large amount of measured data [62, 63]. However, previous studies have mainly focused on the attenuation of the received light intensity.

A simple study of attenuation cannot explain the phenomenon in which the light transmittance is large and the OWC bit-error rate is also increased in heavy rain. For OWC, the received light-intensity variance also affects the system performance [64]. The relationship between the received light intensity and variance was calculated using the Mie model and raindrop spectra in China, and the model was verified using measured data.

5.6.1 Average Received Light Intensity

When light propagates in the line of sight between the transmitter and the receiver, the mean value of the signal intensity detected by the receiver can be divided into the sum of the coherent and incoherent fields. The intensity attenuation is caused by the absorption of the medium [65]:

$$\frac{I_t}{I_0} = \frac{I_c + I_i}{I_0} = \exp(-\rho\sigma_a L) \tag{5.164}$$

where ρ is the particle number density, σ_a is the absorption cross section of the medium; its value is the total cross section of particles σ_t and the scattering cross section difference of σ_s; L is the propagation distance. According to the Mie scattering theory, the total cross section and scattering cross section of a single spherical particle can be expressed as follows:

$$\sigma_t = 2\pi \sum_{n=1}^{\infty} (2n+1)\mathrm{Re}(a_n + b_n) \tag{5.165}$$

$$\sigma_s = 2\pi \sum_{n=1}^{\infty} (2n+1)\left(|a_n|^2 + |b_n|^2\right) \tag{5.166}$$

$$\sigma_a = \sigma_t - \sigma_s, \tag{5.167}$$

where $\gamma = (\eta I)^2/N_0$ and $\gamma = (\eta I)^2/N_0$ are the nth terms of two infinite sequences. The complex refractive index of the water medium corresponding to each wavelength required in the calculation can be found in the literature [66].

The ratio of the cross-sectional area S of a particle is recorded as its scattering coefficient. Therefore, the scattering cross-sectional coefficient of the particle can be calculated for the scattering cross section σ_s, extinction cross section σ_t, and absorption cross section σ_a. The relationship between the coefficients and the particle size x is shown in Figs. 5.31 and 5.32.

The discontinuity σ_a in Figs. 5.31 and 5.32 is due to the numerical error caused by calculating the sum of σ_s and σ_t. Linear fitting is performed to obtain

$$\alpha_{a\_\mathrm{fit}}^{550} = 6.655 \times 10^{-9}x + 1.161 \times 10^{-7} \tag{5.168}$$

$$\alpha_{a\_\mathrm{fit}}^{650} = 5.564 \times 10^{-8}x + 3.577 \times 10^{-7}. \tag{5.169}$$

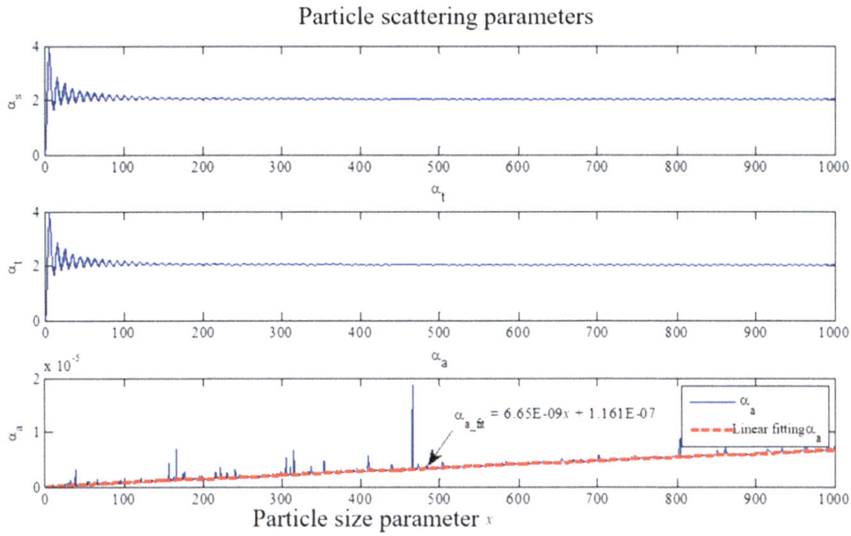

Fig. 5.31 Particle-scattering parameters of 550 nm light

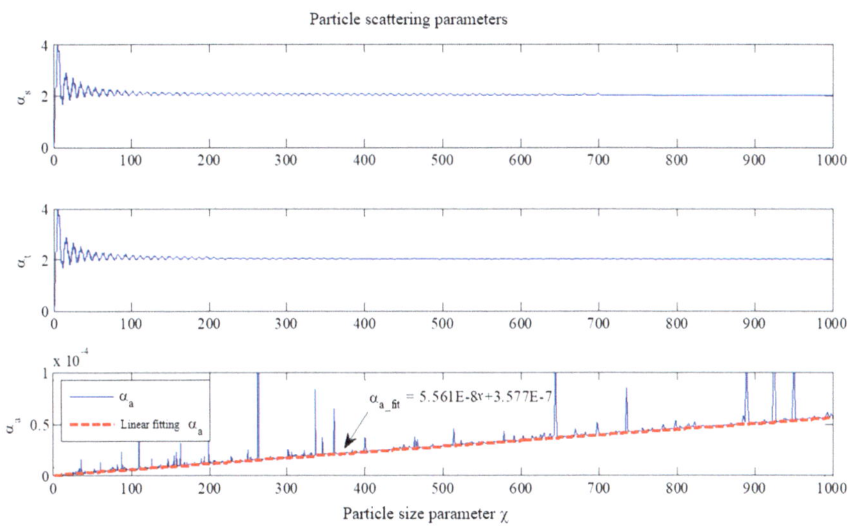

Fig. 5.32 Particle-scattering parameters of 650 nm light

The calculation can be simplified by substituting α_a from Eqs. (5.167) and (5.169) into the calculation. If the ratio of I_t and I_0 is converted to the light-intensity attenuation coefficient per unit length in dB/km [67],

$$\frac{I_t}{I_0}(\mathrm{dB}) = 4.343 \times 10^3 \times \frac{\pi}{4} \int_{D_{\min}}^{D_{\max}} \alpha_a(x, m) D^2 N(D) \mathrm{d}D, \tag{5.170}$$

where x is the particle size, M is the complex refractive index of the medium, n is the particle-size distribution function, and D is the particle diameter.

Many fitting formulas conform to the actual local situation for a raindrop particle-size distribution function under various rainfall conditions. The common raindrop particle-size distribution is an exponential distribution, including the Marshall–Palmer (M-P) and gamma distributions.

Chinese meteorologists have measured rain spectra for a long time, accumulated a large amount of data, fitted them, and obtained different empirical functions of the raindrop spectra. At present, the raindrop-scale distribution in the literature includes the cities of Guangzhou, Qingdao, Shenyang, Changchun, Xi'an, and some parts of Ningxia [56–61]. See Table 5.5 for the raindrop spectral-distribution functions in various regions.

By substituting various raindrop spectra into Eq. (5.169), the total light-intensity attenuation under different rainfall rates was calculated, as shown in Figs. 5.33 and 5.34.

It can be seen from Figs. 5.33 and 5.34 that the light-intensity attenuation caused by raindrop absorption increases with the increase in rainfall rate. Compared with the M-P distribution, the light-intensity attenuation of rainfall in different regions can be divided into two groups: The light-intensity attenuation per unit distance in Guangzhou and Qingdao is greater than the M-P distribution, and the light-intensity attenuation per unit distance in the other regions is less than the M-P distribution. This indicates that raindrops with the same rainfall rate in Guangzhou and Qingdao have a stronger light-absorption capacity.

Comparing Figs. 5.33 and 5.34, because the imaginary part of the complex refractive index of water increases with an increase in wavelength in the visible light band, 650 nm red light is more absorbed by the water medium than 550 nm green light, and the attenuation per unit distance is greater.

Table 5.5 Raindrop spectral-distribution functions

Distribution area	Function (individual/(m^3 mm))
Guangzhou City	$N(D) = 230600\, p^{0.364}\, D^{-0.274} \exp(-7.411\, p^{-0.0527}\, D^{0.452})$
Changchun City	$N(D) = 2020 p^{-0.2} \exp(-3.18\, p^{-0.28}\, D)$
Shenyang City	$N(D) = 13{,}829.7 p^{0.1} \exp(-4.56 p^{-0.16}\, D)$
Qingdao City	$N(D) = 1387 p^{0.4052} \exp(-2.38 p^{-0.0877}\, D)$
Parts of Ningxia	$N(D) = 5688.2 p^{0.3996} \exp(-4.05 p^{-0.1213}\, D)$
Xian City	$N(D) = 257 p^{0.315}\, (8315 p^{-0.202})^2 D/2\mathrm{K}_0(8315 p^{-0.202} D/2)$
Marshall–Palmer distribution	$N(D) = 8000 \exp(-4.1 p^{-0.21}\, D)$

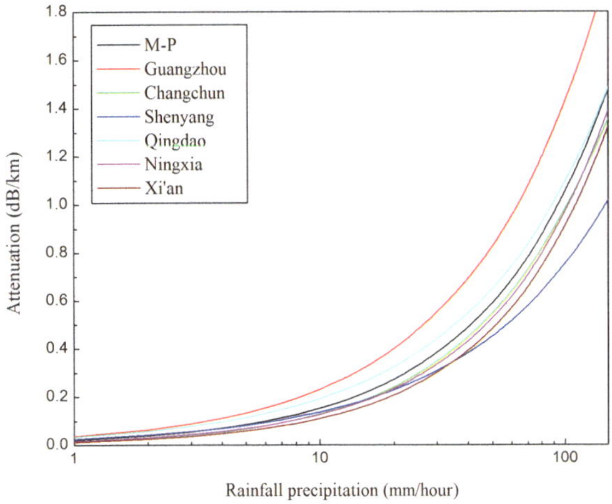

Fig. 5.33 Variation of the total attenuation of a 550 nm-wavelength optical transmission with local rainfall rate

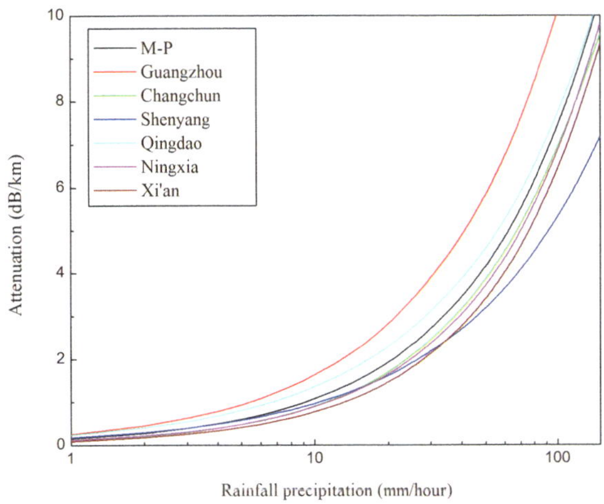

Fig. 5.34 Variation of the total attenuation of a 650 nm-wavelength optical transmission with local rainfall rate

5.6.2 *Statistical Properties of Coherent and Incoherent Fields*

The attenuation of a coherent field can be considered as an equal proportion of light attenuation generated by the optical thickness of the medium on the propagation path. According to the Beer–Lambert theorem,

$$\frac{I_c}{I_0} = \exp(-\rho\sigma_t L). \tag{5.171}$$

Similar to Eq. (5.169), we convert the ratio of I_c and I_0 to the light-intensity attenuation coefficient in dB/km, and the following is obtained:

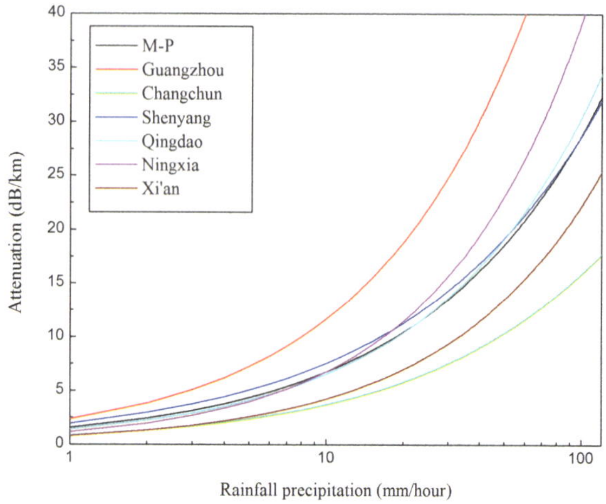

Fig. 5.35 Variation of the coherent field attenuation of a 650 nm optical transmission with the local rainfall rate

$$\frac{I_c}{I_0}(\mathrm{dB}) = 4.343 \times 10^3 \times \frac{\pi}{4} \int_{D_{\min}}^{D_{\max}} \alpha_t(x, m) D^2 N(D) \mathrm{d}D \tag{5.172}$$

According to the Mie scattering theory, the light-transmission attenuation under different rainfall rates σ_t is shown in Fig. 5.35, by substituting the raindrop spectra of various places.

It can be seen from Fig. 5.35 that the light-intensity attenuation increases with an increase in the rainfall rate. For visible light, the sizes of the raindrop particles are generally large, and the extinction cross-section coefficient of most particles is approximately 2; therefore, the wavelength has little effect on the coherent field attenuation. The attenuation of the coherent field of the 550 nm optical transmission was almost identical to that of the 650 nm transmission in Fig. 5.35.

We compared the raindrop spectra and M-P distributions in different regions. The extinction rate of the rain in Qingdao and Shenyang was very close to that of the M-P distribution; the extinction rate of the rain in Guangzhou was larger than that of the M-P distribution, and when the rainfall rate was >11 mm/h, the extinction rate of the rain in some areas of Ningxia was also greater than that of the M-P distribution. When the rainfall rate was <11 mm/h, the extinction rate was slightly less than that of the M-P distribution, and the rain extinction in Xi'an and Changchun was obviously smaller than that of the M-P distribution.

Figure 5.35 also shows that under the same rainfall rate, to achieve the same received signal-to-noise ratio, greater transmission power is required in Guangzhou than in Changchun. As the light-intensity signal is non-negative, the light-intensity fluctuation caused by scattering also has an average value, and the average value of the light-intensity fluctuation is [67]

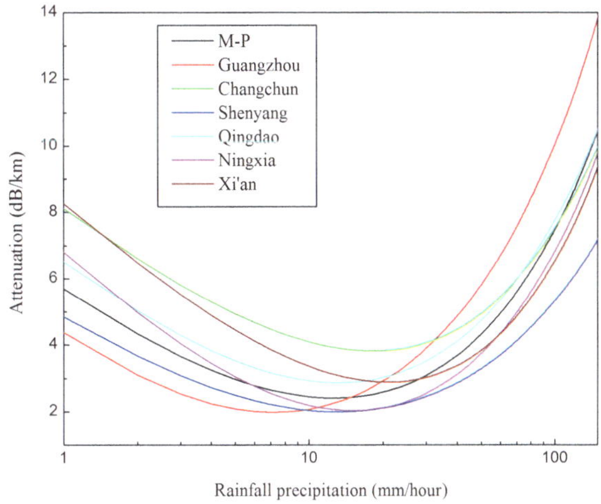

Fig. 5.36 Variation of the incoherent field attenuation of a 650 nm-wavelength optical transmission with the local rainfall rate

$$\frac{I_{\mathrm{i}}}{I_0} = \exp(-\rho\sigma_{\mathrm{a}}L) - \exp(-\rho\sigma_{\mathrm{t}}L). \tag{5.173}$$

The above equation is the difference between Eqs. (5.170) and (5.172). The incoherent field attenuation of 650 nm red light in rain was calculated, as shown in Fig. 5.36. The mean value of the scattered light intensity first increased and then decreased with increasing rainfall rate. This may be because when the rain changes from light rain to moderate rain, the radius of the raindrop particles in the raindrop distribution spectrum increases; however, the number density does not increase significantly.

At this time, the forward-scattering intensity increases with the increase in particles. When the rainfall rate continues to increase, the radius of the raindrop particles increases indistinctly, and the number density increases sharply. The increase in the total optical thickness of the raindrop particles reduces the mean value of the scattered light intensity.

The mean value of the scattered light intensity first increases and then decreases; thus, there is an extreme value, which is related to the wavelength and raindrop spectrum. The complex refractive indices of raindrops with different wavelengths are different. The influence of different raindrop spectra can be seen in Fig. 5.37.

For 650 nm red light, the maximum scattered light intensity in rain of the M-P distribution appears at 12 mm/h, and the rainfall rates at extreme points of the raindrop-spectrum distribution in other regions are 7 mm/h in Guangzhou, 13 mm/h in Shenyang and Qingdao, 15 mm/h in Ningxia, 18 mm/h in Changchun, and 21 mm/h in Xi'an. Similarly, the incoherent field attenuation of 550 nm green light is calculated, as shown in Fig. 5.37.

Because of the different complex refractive indices, the absorption capacities of the media for different wavelengths of light are different, and the corresponding extreme points also change. Comparing Figs. 5.36 and 5.37, it can be seen that the

Fig. 5.37 Variation of the incoherent field attenuation of a 550 nm-wavelength optical transmission with the local rainfall rate

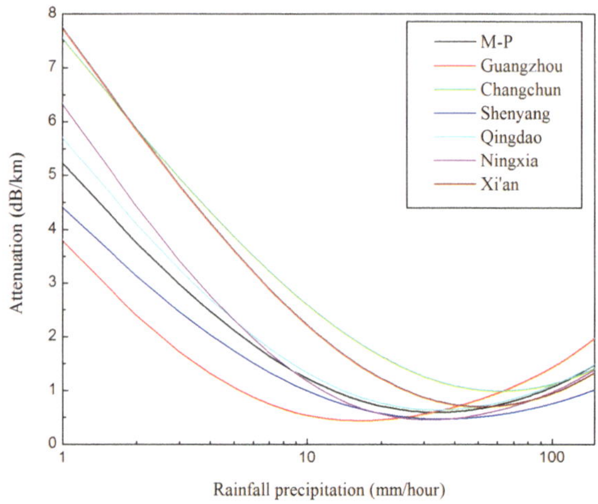

attenuation of the scattered light intensity of green light in rain is less than that of red light. The results also show that the received scattered light intensity will first increase and then decrease with an increase in the rainfall rate, and that the specific extreme point is related to the transmission wavelength and different raindrop spectral distributions.

Although the incoherent field caused by scattering transmits energy from the transmitter to the receiver, the random motion of the scattering medium causes the incoherent field to change randomly. The intermittent extinction and enhancement of the light-intensity signal in the scattering medium measured in an actual environment can be attributed to the scattering of the moving medium. The scattered light intensity in an actual environment varies, and the incoherent field intensity changes randomly with time.

5.6.3 Spectral Characteristics and Variances of Incoherent Fields

The moving medium produces a time-varying scattering field, and the differential cross section at rest is $\sigma_l^2 = 0.2$. When a single particle D moves at velocity V, its time-dependent differential scattering cross section can be written as [65]

$$\sigma_{\mathrm{d}}(\tau) = \sigma_{\mathrm{d}}\langle \exp(i\boldsymbol{k}_{\mathrm{s}} \cdot \boldsymbol{V}\tau)\rangle, \tag{5.174}$$

where k_s is the scattered wave vector and τ is the time difference. Velocity V is decomposed into the sum of the average velocity u and velocity fluctuation V_f:

$$\sigma_{\mathrm{d}}(\tau) = \sigma_{\mathrm{d}}\langle \exp(i\boldsymbol{k}_{\mathrm{s}} \cdot \boldsymbol{V}_{\mathrm{f}}\tau)\rangle \exp(i\boldsymbol{k}_{\mathrm{s}} \cdot \boldsymbol{U}\tau). \tag{5.175}$$

The relationship between the differential scattering cross section and total scattering cross section is

$$\sigma_{\mathrm{s}} - \int_{4\pi} \sigma_{\mathrm{d}} \mathrm{d}\omega \tag{5.176}$$

where ω is the solid angle of space. The time-correlation function of the scattering field strength of the moving particles B_u is

$$\begin{aligned} B_{\mathrm{u}}(\tau) &= \langle u_{\mathrm{f}}(\boldsymbol{r}, t_1) u_{\mathrm{f}}^*(\boldsymbol{r}, t_2) \rangle \\ &= \int_V \sigma_{\mathrm{d}} \langle \exp(i\boldsymbol{k}_{\mathrm{s}} \cdot \boldsymbol{V}_{\mathrm{f}}\tau) \rangle \exp(i\boldsymbol{k}_{\mathrm{s}} \cdot \boldsymbol{U}\tau) \exp(-\gamma_0 - \gamma') \rho \mathrm{d}V. \end{aligned} \tag{5.177}$$

A Fourier transform of $B_u(\tau)$ is performed to obtain the power spectrum of the scattered field strength:

$$W_{\mathrm{u}}(\omega) = 2 \int_{-\infty}^{\infty} B_{\mathrm{u}}(\tau) \mathrm{e}^{\mathrm{j}\omega\tau} \mathrm{d}\tau = (4\rho\sigma_{\mathrm{s}} L) \left(\frac{\pi\alpha}{k^2 U^2} \right)^{\frac{1}{2}} \exp\left(-\gamma - \frac{\alpha_{\mathrm{p}} \omega^2}{k^2 U^2} \right). \tag{5.178}$$

Substituting the raindrop scale distribution into Eq. (5.177), we obtain

$$W_{\mathrm{u}}(\omega) = \int_{D_{\min}}^{D_{\max}} N(D) \mathrm{d}D 4\sigma_{\mathrm{s}} L \left(\frac{\pi\alpha_{\mathrm{p}}(D)}{k^2 U^2(D)} \right)^{\frac{1}{2}} \exp\left(-\int_{D_{\min}}^{D_{\max}} N(D) \sigma_{\mathrm{t}} L \mathrm{d}D - \frac{\alpha_{\mathrm{p}}(D) \omega^2}{k^2 U^2(D)} \right), \tag{5.179}$$

where $\alpha_p(D) = 2.77/(1.02\lambda/D)2$. There is an approximate relationship between the raindrop velocity and diameter, as follows [64]:

$$U(D) = 200.8 \left(\frac{D}{2} \right)^{\frac{1}{2}} \tag{5.180}$$

For example, assuming a rainfall rate of $p = 12.7$ mm/h, and substituting various raindrop spectra into Eq. (5.179), the scattered field-intensity spectra under different rainfall conditions are shown in Fig. 5.38. The spectrum bandwidth under an M-P distribution is 12.43 kHz. The models with smaller bandwidths than the M-P distribution are Changchun with 10.1 kHz, Qingdao with 11.16 kHz, and Xi'an with 11.8 kHz. The models with larger bandwidths than the M-P distribution include Guangzhou with 13.38 kHz, Ningxia with 13.9 kHz, and Shenyang with 14 kHz.

Different models estimate the number of raindrop particles, resulting in different bandwidth-calculation results. The main factor affecting spectral broadening in rain is the velocity of heavy particles in the raindrop spectrum.

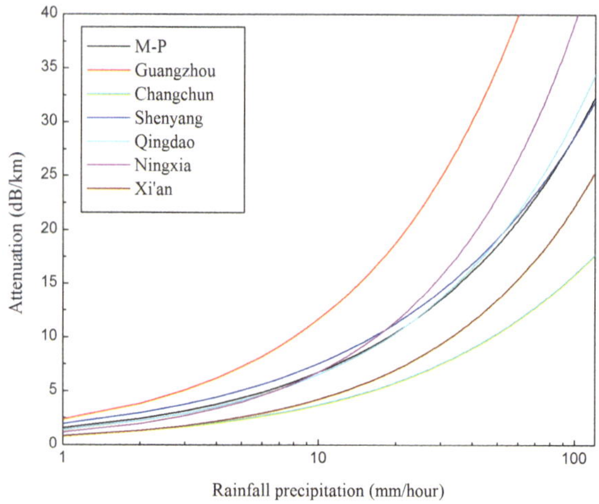

Fig. 5.38 Scattered light spectra of 650 nm-wavelength light in various rainfall environments [68]

The field-strength variance of the scattered field can be obtained from the spectrum of the scattered field when ω = 0 according to Eq. (5.15):

$$W_{\mathrm{u}}(0) = 2\int_{-\infty}^{\infty} B_{\mathrm{u}}(\tau)\mathrm{d}\tau = 2\int_{-\infty}^{\infty} \langle u_{\mathrm{f}}(\boldsymbol{r}, t)u_{\mathrm{f}}^{*}(\boldsymbol{r}, t)\rangle \mathrm{d}\tau. \tag{5.181}$$

For the discrete light-intensity signal collected by the receiver, Eq. (5.18) can be expressed as

$$W_{u}(0) = 2\sum u_{\mathrm{f}}^{2}. \tag{5.182}$$

The field-strength variance of the scattered field can be expressed as

$$D(u_{\mathrm{f}}) = E\left(u_{\mathrm{f}}^{2} - E^{2}(u_{\mathrm{f}})\right) \tag{5.183}$$

In a weak turbulent environment, the mean field strength is 0, and the field-strength variance is directly proportional to $W_{\mathrm{u}}(0)$:

$$D(u_{\mathrm{f}}) \propto W_{\mathrm{u}}(0) \tag{5.184}$$

When order ω = 0, the scattering-field intensity variance of 650 nm red light propagating under different rainfall rates was calculated using Eq. (5.177), as shown in Fig. 5.39. It can be seen from Fig. 5.39 that the variance of the scattering field first increases and then decreases with an increase in the rainfall rate.

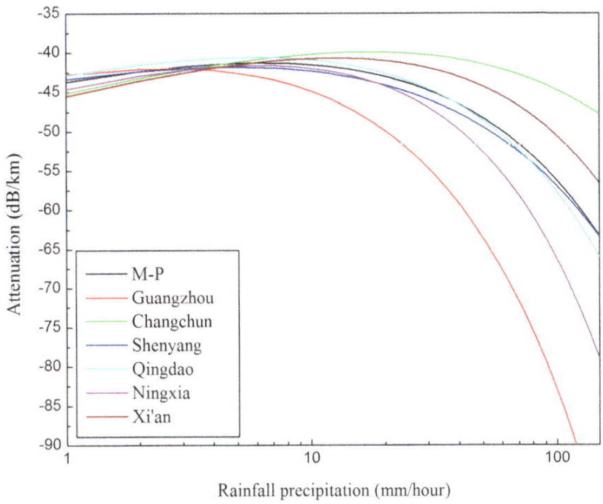

Fig. 5.39 Reception variance of 650 nm-wavelength light propagating under various rainfall conditions [68]

From the raindrop-spectrum model, with the increase in rainfall rate, the scattering is first enhanced, owing to the increase in small raindrops with significant weight, and then the raindrops no longer increase when they increase to a certain extent. At this time, the number is increased to increase the rainfall rate. An increase in the number of particles results in an increase in the optical thickness, resulting in a decrease in the scattering-field variance.

Comparing Figs. 5.34 and 5.39, we can see the relationship between the received light-intensity mean value, field-intensity variance, and rainfall rate. For an OWC, the received light intensity without a recorded field rainfall rate, received light intensity, and field-intensity variance can be linked.

The 650-nm red OWC measured data [62, 63] under rainfall weather in Xi'an from July to September 2007 were used to verify the relationship between the mean value of light intensity and variance, as shown in Fig. 5.40. Because most of the variance values in the figure are very small, they are converted into logarithms by referring to the definition in Eq. (5.181).

The limited experimental data show that the light-intensity variance decreases with an increase in the mean light intensity, and this trend is consistent with the M-P and local distributions. These experimental data can be used to verify the relationship between the mean light intensity and the variance in each model. However, owing to the existence of circuit noise, alignment errors, and other factors in the experimental data, the results are scattered to a certain extent. Therefore, distinguishing the differences between the models requires further study.

In addition, the variance value of the measured data in the figure changes significantly when the mean value of the light intensity remains unchanged because the received light will fade or enhance for a short time in the rainfall environment. Owing to the short duration of occurrence, it has little impact on the statistical mean value;

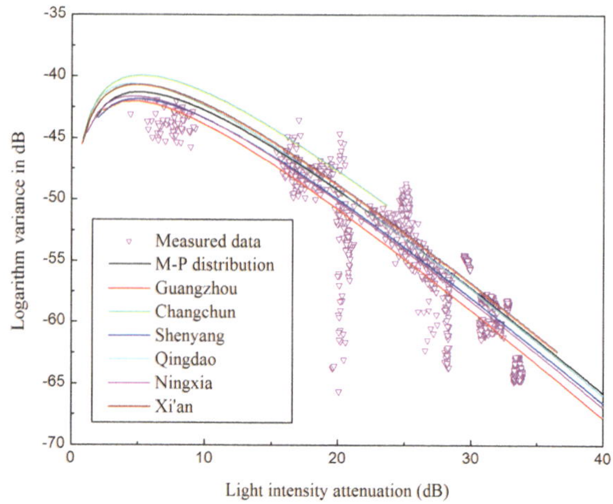

Fig. 5.40 Relationship between the measured light-intensity attenuation and variance at 650 nm [68]

however, it will cause the variance to deviate from the median value. This reflects the time-varying non-stationary characteristics of light propagation in turbulent media.

The results show that the mean value of the scattered light intensity first decreases and then increases with an increase in the rainfall rate, and the extreme point is related to the wavelength and different raindrop distributions. The spectrum of the received light-intensity signal was calculated using the correlation function of the scattering field, and the spectrum broadening of the received light intensity was obtained by substituting the velocity of the raindrop particles. Zero-frequency attenuation under different rainfall rates was calculated, and the variance of the scattered-field intensity was qualitatively analyzed.

The results show that the variance of the scattered field intensity also depends on different raindrop spectrum distributions, and the receiving variance first increases and then decreases with the increase in rainfall rate. Finally, the model was verified by experimental data, which verified the relationship between the mean and variance of the light intensity in the above model.

Chapter Summary

Atmospheric influence is divided into three main forms: attenuation, scattering, and the atmospheric-turbulence effect. The absorption of light waves by atmospheric molecules depends on the wavelength. Atmospheric scattering can be divided into the scattering of atmospheric molecules and the scattering of atmospheric particles, according to the scale number of scattering particles.

The atmospheric absorption and scattering effects reduce the laser power density reaching the optical receiver; that is, atmospheric attenuation. The atmospheric extinction coefficient is equal to the sum of the atmospheric absorption and scattering coefficients. In engineering calculations, it is difficult to obtain actual atmospheric absorption and scattering coefficients. Thus, the atmospheric visibility is typically used to approximate the atmospheric extinction coefficients.

In addition to reducing the laser power density, repeated atmospheric scattering causes photons of the same laser pulse to reach the receiving end through different paths, resulting in a long-time tailing of the laser pulse at the receiving end; that is, the time broadening of the laser pulse. The extent of the laser-pulse time broadening represents the transmission bandwidth of the atmospheric channel. When the bandwidth of the data signal is larger than the channel-transmission bandwidth, the data signal is distorted and forms intersymbol crosstalk.

The characteristics and application scope of the lognormal distribution, gamma–gamma distribution, and negative exponential distribution models were analyzed. Among these, the lognormal distribution model is the most commonly used. In a weak turbulence region, the model agreed well with the experimental data. However, in a strong turbulence region, the lognormal distribution model had a large deviation from the experimental data, particularly in the tail, which directly affects the selection of the best reception decision threshold and the analysis of the bit-error rate. The gamma–gamma distribution model is suitable for a wide range of turbulence conditions, whereas the negative exponential distribution is only suitable for strong turbulence conditions.

The transmission performance of the OFDM signal in the atmosphere was analyzed, and the wide applicability of the gamma–gamma model to the description of atmospheric turbulence was explained:

(1) The BER performance of an OFDM system with low-order modulation was better than that of the OOK system;
(2) The multiplicative noise introduced by the gamma–gamma channel model is different from ordinary Gaussian noise;
(3) The influence of rain attenuation on wireless laser communication was established.

References

1. Gregory JP (1991) Trellis codes for the optical direct-detection channel. IEEE Trans Commun 39(8):1182–1183
2. Wu J, Le SX (1998) Theory of light propagation in random media. Chengdu Institute of Telecommunications Engineering Press
3. Yang CH (1990) Laser and infrared technology manual. National Defense Industry Press, Beijing
4. Deng DZ (2004) Influence of atmospheric random channel on wireless laser communication. South West Jiao tong university, Chengdu
5. Kurse PW (1962) Elements of infrared technology: generation, transmission and detection. Wiley, New York

6. Lou Y, Giang HL, Chen CY (2010) Numerical simulation and experimental analysis of atmospheric channel for space optical communication. In: Proceedings of the 2010 Academic conference on system simulation technology and its applications, pp 150–153
7. Sheng PX, Mao JT, Li JG et al (2003) Atmospheric physics. Peking University Press, Beijing
8. Frederic GS (1993) Atmospheric propagation of radiation. In: The infrared & electro-optical systems handbook. SPIE Press, Bellingham, Washington, vol 2
9. Moffatt HK (2002) GK batchelor and the homogenization of turbulence. Annu Rev Fluid Mech 34:19–35
10. Chen D (2011) Research on optical wireless subcarrier modulation and atmospheric impact suppression technology. Xi'an University of technology, Xi'an
11. Karp S, Gagliardi RM, Moran SE, Stotts LB (1998) Optical channels: fibers, clouds, water and the atmosphere. Plenum Press, NewYork
12. Tatarskii VI (1961) Wave propagation in a random medium. McGraw-Hill Book Company, New York
13. Hill RJ et al (1978) Modified spectrum of atmospheric temperature fluctuations and its application to optical propagation. J Opt Soc Am 68:892
14. Andrews LC (1992) Analytical model for the refractive index power spectrum and its application to optical scintillations in the atmosphere. J Mod Opt 39:1849–1850
15. Stribling BE (1994) Laser beam propagation in non-Kolmogotov atmospheric turbulence. Master's thesis of Air Force Institute of Tech Wright-PATTERSON AFB OH School of Engineering
16. Kolmogorov AN (1962) A refinementof previous hypotheses concerning local structure of turbulence in viscous incompressible fluid at high Reynolds number. J Fluid Mech 13(1):82–85
17. Williams RM, Paulson CA (1977) Microscale temperature and velocity spectrain the atmospheric boundary layer. J Fluid Mech 83(3):547–567
18. Chen CY (2009) Research on atmospheric impact mechanism and suppression technology in optical wireless communication. Changchun University of technology
19. Wu J, Yang CP, Liu JB (2005) optical transmission theory in the atmosphere. Beijing University of Posts and Telecommunications Press, Beijing
20. Chen CY, Yang HM, Zhong SF et al (2008) Modeling and power analysis of aircraft to giant star laser communication link. J Commun 29(11):125–131
21. Zhang XF, Yu X, Yan GX et al (2005) Influence of turbulence on image resolution of optical system. Opt Technol 31(2):263–265
22. Zhu X, Kahn J (2003) Performance bound for coded free-space optical communication through turbulence channel. IEEE Trans Commun 51(8):1233–1239
23. Rao RZ, Wang SP, Liu XC et al (2000) Experimental study on laser beam drift in turbulent atmosphere. China Laser 27(11):1011–1015
24. Nuber RM, Shanmugan KS (1989) Modeling the atmosphere as an unguided optical communications channel. In: IEEE international conference on communications, vol 3, 1480–1484
25. Clifford SF (1978) The classical theory of wave propagation in a turbulent medium. In: Laser beam propagation in the atmosphere. Springer, Berlin, pp 9–41
26. Andrews LC, Phillips RL (2005) Laser beam propagation through random media, 2nd edn. SPIE Press
27. Osche GR (2002) Optical detection theory for laser applications. Wiley, New Jersey
28. Zhang YX, Chi ZY (1997) Transmission and imaging of light waves in the atmosphere. National Defense Industry Press, Beijing
29. Tatarski VI (1961) Wave propagation in a turbulent medium. McGraw-Hill, New York
30. Flatte SM, Bracher C, Yu Wang G (2003A) Probability density functions of irradiance for waves in atmospheric turbulence calculated by numerical simulations. J Opt Soc Am 11:2080–2092
31. Wu HL, Li XY, Yan HX (2008) Analysis of bit error characteristics of atmospheric optical communication system in gamma gamma turbulent channel. Acta Optica Sinica 12(28):99–104
32. Andrews LC, Phillips RL, Hopen CY (2001) Laser beam scintillation with applications. SPIE, Bellingham

33. Al-Habash MA, Andrews LC, Phillips RL (2001) Mathematical model for the irradiance probability density function of a laser beam propagating through turbulent media. Opt Eng 40:1554–1562
34. Chen D, Ke X, Sun Q (2011) Outage probability and average capacity research on optical wireless communication over turbulence channel. In: IEEE 2011 10th international conference on electronic measurement and instruments, ICEMI 2011, pp 19–23
35. Zhu X, Kahn JM (2002) Free-space optical communication through atmospheric turbulence channels. IEEE Trans Commun 50(8):1293–1300
36. Amnon Y, Pochi Y (2009) In: Chen HM, Shi WH, Wang GL et al (trans) Photonics—Modern communication optoelectronics. Electronic Industry Press, Beijing
37. Korff SA, Stewart JQ (1930) A sensitive method for determining refractive indices. Rev Sci Instrum 1(6):341–346
38. Barrell H, Sears JE (1939) The refraction and dispersion of air for the visible spectrum. Philos Trans R Soc London Ser A Math Phys Sci 238(786):1–64
39. Edlén B (1966) The refractive index of air. Metrologia 2(2):71–80
40. James C (1967) Owens optical refractive index of air: dependence on pressure, temperature and composition. Appl Opt 6(1):51–59
41. Peck ER, Reeder K (1972) Dispersion of air. JOSA 62(8):958–962
42. Birch KP, Downs MJ (1993) An updated Edlén equation for the refractive index of air. Metrologia 30(3):155–162
43. Ciddor PE, Hill RJ (1999) Refractive index of air: 2. Group Index Appl Opt 38(9):1663–1667
44. Rüeger JM (1999) Refractive indices of light, infrared and radio waves. In: The atmosphere, report of the ad-hoc working party of the IAG special Commission SC3 on Fundamental Constants, 1993–1999, 22nd General assembly of IUGG Birmingham, UK, 25 p
45. Wolfram Function Site (2011)
46. Abudelmoula B et al (2010) Transmission analysis of OFDM-based wireless services over turbulent radio-on-FSO links modeled by gamma-gamma distribution. Photon J IEEE 2(3):510
47. Xie MQ (2012) research on the influence of atmospheric dispersion on FSO-OFDM system. Xi'an University of technology, Xi'an
48. Zhao L, Ke XZ, Ren AH (2010) Research and simulation of scattering effect on FSO-OFDM system. Laser Technol 02:185–188
49. Zhao L, Lei ZY, Ke XZ (2011) Performance analysis of FSO-OFDM system based on QAM. Infrared Laser Eng 40(07):1323–1327
50. Ghassemlooy Z, Arnon S, Uysal M et al (2015) Emerging optical wireless communications-advances and challenges. IEEE J Sel Areas Commun 33(9):1738–1749
51. Wu XJ, Wang HX, Li BF et al (2015) Analysis of the influence of atmospheric turbulence on the fading characteristics of optical wireless communication under different transmission environments. China Laser 42(5):22–29
52. Sun XM, Han YP (2006) Absorption and scattering characteristics of visible light by ice water mixed clouds. Chin J Phys 55(2):682–687
53. Grabner M, Kvicera V (2014) Multiple scattering in rain and fog on free-space optical links. Lightwave Technol J 32(3):513–520
54. Al-Gailani SA, Mohammad AB et al (2014) Determination of rain attenuation parameters for free space optical link in tropical rain. Optik 125:1575–1578
55. Song B, Wang HX, Liu M (2012) Applicability analysis of raindrop spectrum model for rain attenuation calculation. Laser Infrared 42(3):310–313
56. Zhao ZW (1995) Raindrop size distribution model and Rain Attenuation Prediction in Guangzhou area. J Radio Wave Sci 10(4):33–37
57. Chen DL, Gu SF (1989) Study on the average spectrum of raindrops in heavy rain. J Meteorol 47(1):124–127
58. Zheng JH, Chen BJ (2007) Selection of raindrop spectral distribution function: a comparative study of M-P and gamma distributions. Meteorol Sci 27(1):17–25
59. Huang J, Hu DZ (1991) Size distribution model of Qingdao area. J Radio Wave Sci (Supplement 1):177–180

60. Niu SJ, Sang JR (2004) Observational study on the characteristics of summer rainfall spectral distribution parameters in Ningxia under different weather systems. Plateau Meteorol 21(1):37–44
61. Meng SW, Wang YP (1995) Application of K-distribution raindrop spectrum. J Radio Wave Sci 10(3):15–19
62. Ke XZ, Yang LH, Ma DD (2008) Transmission attenuation of laser signal in rain. Infrared Laser Eng 37(6):1021–1024
63. Zhu YL, An R, Ke XZ (2012) Influence of rainfall on wireless laser communication. Acta Optica Sinica 32(12):1206003
64. Xie M, Ke X (2013) Study on the influence of atmospheric turbulence on signal-to-noise ratio of optical wireless communication system. Progr Laser Optoelectron 50(11):110102
65. Zhang Y (1986) Propagation and scattering of waves in a random medium. Science Press, , Beijing, pp 132–141
66. McCartney EJ (1998) In: Pannaixian et al (trans) Atmospheric optical molecular and particle scattering. Science Press, Beijing, pp 237–342
67. Dong QS, Zhao ZW, Cong HJ (1996) Millimeter wave attenuation caused by dust. J Radio Wave Sci 11(2):29–32
68. Xie MQ, Ke XZ (2016) Study on characteristics of light intensity attenuation and variance in rainfall in different regions of China. J Xi'an Univ Technol 01:1–7

Chapter 6
OFDM Synchronization Technology

Owing to the multipath effect in a channel, the transmission of digital signals will cause a delay spread, resulting in the overlapping of symbols before and after the received signal. This results in intersymbol interference (ISI) and incorrect decisions, and seriously affects the transmission quality. When the symbol rate is high, the delay spread spans multiple symbols, resulting in significant intersymbol interference.

When the symbol rate is high, the signal bandwidth is wide. When the signal bandwidth approaches or exceeds the coherent bandwidth of the channel, the time dispersion of the channel causes frequency-selective fading of the received signal. Reliable data transmission is impossible without accurate synchronization. This is the premise for reliable information transmission. This chapter discusses time synchronization in orthogonal frequency-division multiplexing (OFDM) systems.

6.1 Time Synchronization in OFDM

Figure 6.1 shows the synchronization concept of a optical wireless communication (OWC)-OFDM system.

Figure 6.1 shows three types of synchronization and their positions in the OWC-OFDM system. Symbol synchronization distinguishes the boundaries of each symbol so that the start and end times for the inverse fast Fourier transform (IFFT) and fast Fourier transform (FFT) are consistent.

Compared with a single-carrier system, an OFDM system has a strong ability to suppress frequency-selective fading. However, owing to the orthogonality between subcarriers, the carrier-frequency offset will lead to interference between subchannels, resulting in a very serious floor effect on the system performance. Therefore, the system performance degradation caused by unsynchronized carriers between the transmitter and the receiver is much greater than that of a single carrier.

X. Ke, *Principles and Applications of Optical Wireless Orthogonal Frequency-Division Multiplexing*, Optical Wireless Communication Theory and Technology,
https://doi.org/10.1007/978-981-97-7973-4_6

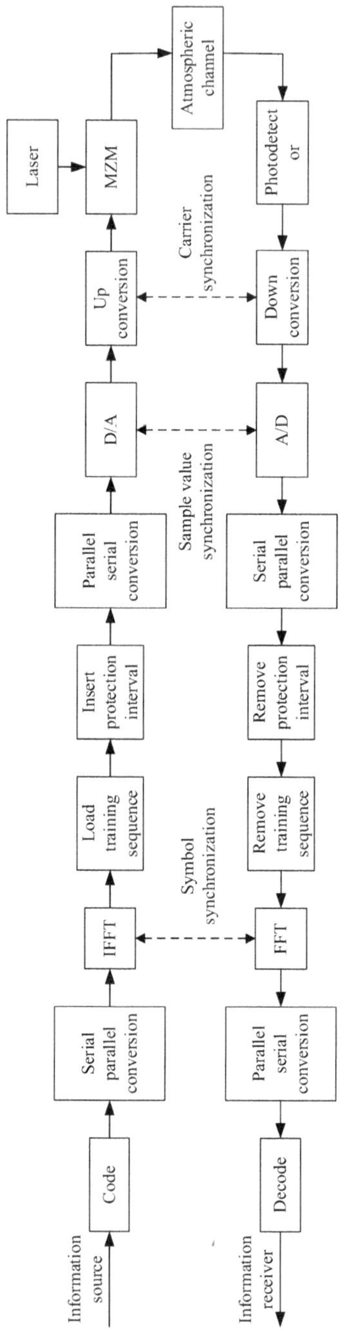

Fig. 6.1 OWC-OFDM system-synchronization concept diagram

Carrier synchronization is necessary for signal demodulation, and the signal frequency at the receiving end must be in the same frequency and phase as the carrier at the transmitting end. Sample synchronization makes the sampling time at the receiving end completely consistent with that at the sending end.

6.2 Impact of Synchronization Deviation on OFDM System Performance

6.2.1 Impact of Symbol-Timing Deviation on the System Performance

After a signal is transmitted through the channel, owing to the influence of the channel characteristics, inconsistencies will occur between the FFT observation window and the transmission window at the receiving end. These lead to intercarrier interference (ICI) and intersymbol interference (ISI).

To correctly demodulate the data, it is necessary to find the starting position of the OFDM symbol, align the starting position of the FFT window with the starting point of the OFDM symbol so that the FFT window contains n samples of the current OFDM symbol, and demodulate correctly. If the timing estimation is inaccurate and the starting position of the FFT window is not on the first sample of the current OFDM symbol, the FFT window contains samples from two adjacent OFDM symbols, resulting in intersymbol interference [1].

Figure 6.2 indicates that the ideal starting position of the FFT window is the first sample value after the cyclic prefix (CP) of the OFDM symbol. If the start position of the FFT window falls within the cyclic prefix (see "case 1" in Fig. 6.2), the intercepted signal does not contain information from other known data symbols, so intersymbol interference will not occur. In addition, because the information of the wrong cyclic prefix corresponds to useful data information intercepted by the error, the FFT window still includes n samples of the current OFDM symbol and it simply makes a cyclic shift relative to the correct signal.

Because of the nature of the Fourier transform, the output result of the FFT is only equivalent to making each subcarrier produce a phase deflection without destroying the orthogonality between the subcarriers. This phase rotation can be expressed as

$$\varphi_k = 2\pi f_k \tau = 2\pi \frac{k}{N} \tau, \tag{6.1}$$

where f_k is the carrier frequency of the kth subcarrier, τ is the timing deviation, and N is the number of subcarriers. After FFT demodulation, the received frequency-domain signal becomes

$$Y(k) = y(k) e^{j2\pi \frac{k}{N} \tau}, \tag{6.2}$$

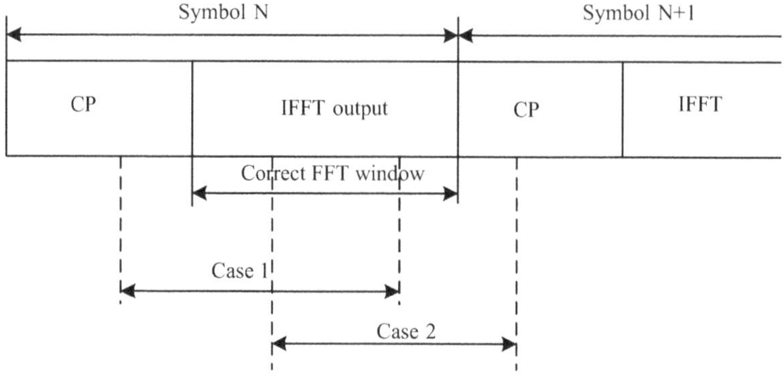

Fig. 6.2 Relationship between the timing synchronization and cyclic prefix

where $Y(k)$ represents the signal received by T, and $y(k)$ represents the original transmission signal. This phase deflection can be eliminated using differential modulation or phase estimation.

If the starting position of the FFT window falls in the data area (see "case 2" in Fig. 6.2) and this offset is greater than the sampling interval, the FFT window contains the contents of two adjacent symbols, causing intersymbol interference (ISI). The influence of symbol-timing deviation on the demodulation performance in this case is analyzed in detail below [1].

Assuming that the timing deviation of the OFDM system is τ, the FFT window contains samples from the ith OFDM symbol (τ to $N - 1$) and samples from the $i + 1$st OFDM symbol (0 to $\tau - 1$) for the FFT transformation, including

$$\begin{aligned} Y_{i,k} &= \frac{1}{N}\left(\sum_{n=0}^{\tau-1} r_{i+1,N-L-l+n} \cdot e^{-j2\pi nk/N} + \sum_{n=\tau}^{N-l} r_{i,n} \cdot e^{-j2\pi nk/N}\right) \\ &= \frac{1}{N}\left(\sum_{n=0}^{N-1} r_{i,n} \cdot e^{-j2\pi nk/N} + \sum_{n=0}^{\tau-l} \left(r_{i+1,N-L-1+n} - r_{i,n}\right) \cdot e^{-j2\pi nk/N}\right), \\ &= y_{i,k} + I_k + W_k \end{aligned} \tag{6.3}$$

where $Y_{i,k}$ represents the FFT output value of the kth subcarrier of the ith OFDM symbol, $y_{i,k}$ is the original transmission signal, and W_k is a white-noise term. I_k is the ISI caused by the timing deviation τ:

$$I_k = \sum_{n=0}^{\tau-1} \left(\hat{r}_{i+1,N-L-1+n} - \hat{r}_{i,n}\right) \cdot e^{-j2\pi nk/N}, \tag{6.4}$$

where $\hat{r}_{i,n}$ refers to the signal after subtracting the white noise from $r_{i,n}$. Timing deviations will cause the useful signal phase to rotate by $2\pi\tau k/N$ and generate an ISI interference term, which reduces the signal-to-noise ratio (SNR) at the demodulation

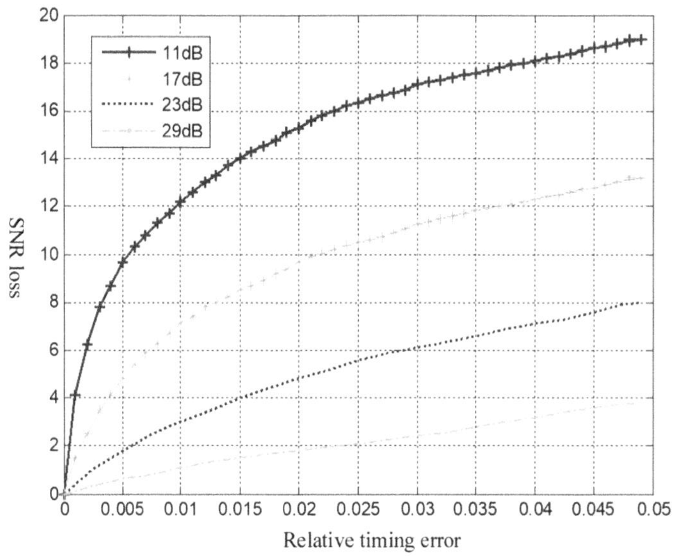

Fig. 6.3 Signal-to-noise ratio loss in the case of timing errors

end. If the correlation between $\hat{r}_{i,n}$ and $\hat{r}_{i+1,N-L-1+n}$ is not considered, the SNR with a symbol timing deviation can be approximated as follows:

$$SNR = \frac{E_c}{N_0 + \frac{2\tau}{N}E_c} = \frac{\frac{E_c}{N_0}}{1 + \frac{2\tau}{N}\frac{E_c}{N_0}} \tag{6.5}$$

$$SNR = 10\log 10\big(1 + (2\tau/N)\cdot(E_0/N)\big). \tag{6.6}$$

According to Eq. (6.6), Fig. 6.3 shows the signal-to-noise ratio of the OFDM system when there is a timing deviation.

It can be seen from Fig. 6.3 that the existence of timing deviations causes serious ISI and degrades the demodulation performance. Therefore, the timing deviation must be accurately estimated. Cyclic prefixes are generally added between OFDM symbols. Cyclic prefixes not only enable OFDM symbols to avoid the ISI caused by multipath interference but also reduce the requirements for timing estimation [2].

When the timing-estimation position deviates, as long as it is in the cyclic-prefix interval, the FFT window still contains n samples of the current OFDM symbol, and the timing deviation only causes the phase of the useful signal to rotate. For a phase rotation, the influence of the phase deflection can be eliminated by differential modulation or phase estimation.

In a multipath fading channel, some cyclic-prefix intervals are damaged by ISI. Therefore, the timing-estimation position must fall in the part of the cyclic prefix area that has not been damaged by ISI to be unaffected. When the starting position of the FFT window falls in other positions, the window will contain samples of two adjacent symbols, resulting in intersymbol interference.

6.2.2 Influence of Carrier-Frequency Deviation on the System Performance

OFDM systems require each subcarrier to maintain a strict orthogonality with the others. When a carrier at the receiving and transmitting ends has a frequency offset, the orthogonality between OFDM subcarriers is damaged, causing serious intercarrier interference (ICI). The data on each subcarrier will be interfered with by the data on the other subcarriers, which will significantly deteriorate the demodulation performance.

The frequency offset is generally expressed by the relative frequency offset ε (the ratio of the actual frequency offset to the subcarrier frequency interval). It can be divided into two parts: integer and decimal. If the relative frequency offset is an integer k, although the orthogonality between subcarriers can still be maintained, the frequency-sampling value shifts the position of k subcarriers, resulting in a data-symbol error rate of 0.5 mapped in the OFDM spectrum. If the relative frequency offset is not an integer, energy will "leak" between subcarriers, resulting in damage to the orthogonality between subcarriers. This causes interference between subcarriers and deteriorates the bit-error performance of the system [3].

Figure 6.4 shows a spectrum diagram of the OFDM signal. Figure 6.4a shows the case with no frequency deviation and Fig. 6.4b shows the case with a frequency deviation. It can be seen from the figure that when there is no frequency deviation, there is no interference between the subcarriers, and when there is a frequency deviation, there is mutual interference between the subcarriers.

Because the integer-frequency offset causes a periodic shift in the demodulation result, the influence of the decimal-frequency offset on demodulation performance is analyzed in detail below.

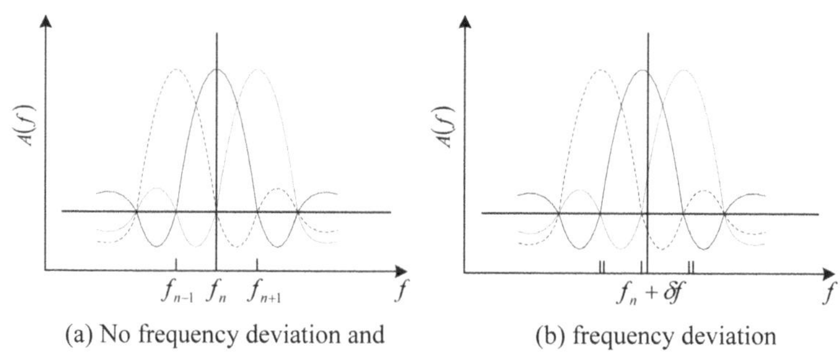

Fig. 6.4 Graphs of carrier synchronization and carrier non-synchronization

When a frequency deviation occurs, the result of the FFT at the receiving end can be expressed as [4]

$$Y(k) = s(k)\frac{\sin(\pi\varepsilon)}{N\sin\left(\frac{\pi\varepsilon}{N}\right)} \cdot e^{j\pi\varepsilon\left(1-\frac{1}{N}\right)} + I_k + W_k k = 0, 1, \ldots, N-1. \tag{6.7}$$

The first item is the useful signal; I_k is the ICI caused by ε:

$$I_k = \frac{1}{N}\sum_{i=0, j\neq k}^{N-1} s(i)\frac{\sin(\pi(i+\varepsilon-k))}{N\sin\left(\frac{\pi(i+\varepsilon-k)}{N}\right)} \cdot e^{j\pi(i+\varepsilon-k)\left(1-\frac{1}{N}\right)}. \tag{6.8}$$

W_k is the result of Gaussian white-noise demodulation:

$$W_k = \frac{1}{N}\sum_{m=0}^{N-1} W(m) \cdot e^{-j\frac{2\pi kn}{N}}. \tag{6.9}$$

It can be seen that when there is a decimal-frequency offset, the useful signal amplitude is attenuated by $\frac{\sin(\pi\varepsilon)}{N\sin(\pi\varepsilon/N)}$, with a phase offset $\pi\varepsilon\left(1-1/N\right)$. In addition, the orthogonality of the subcarriers is destroyed, which causes ICI and reduces the signal-to-noise ratio. Reference [5] gives a formula for estimating the signal-to-noise ratio:

$$SNR \leq \left(\frac{E_c}{N_0}\right) \cdot \left\{\frac{\sin(\pi\varepsilon)}{\pi\varepsilon}\right\}^2 / \left\{1 + 0.5947\frac{E_c}{N_0}(\sin(\pi\varepsilon))^2\right\} |\varepsilon| \leq 0.5. \tag{6.10}$$

In Eq. (6.10), E_c/N_0 represents the original signal-to-noise ratio. When ε is small, the result of Eq. (6.10) is close to the actual SNR. Therefore, when frequency offset ε exists, the SNR lost at the demodulation end can be obtained as follows:

$$SNR = 20\log 10\left(\sin(\pi\varepsilon)/(\pi\varepsilon)\right) - 10\log 10\left(1 + 0.5947\left(E_0/N_0\right)(\sin(\pi\varepsilon))^2\right). \tag{6.11}$$

According to Eq. (6.10), Fig. 6.5 shows the SNR loss of the demodulation terminal under different ε when E_c/N_0 is 11 dB, 17 dB, 23 dB, and 29 dB.

It can be seen from Fig. 6.5 that when the same frequency offset exists, the higher the SNR, the greater the SNR loss. Under the same SNR, the greater the frequency offset, the greater the SNR loss. When the frequency offset is >4%, the SNR loss increases. Therefore, it is generally necessary to reduce it to within 4% to obtain better performance. It needs to be further reduced and controlled within 1%.

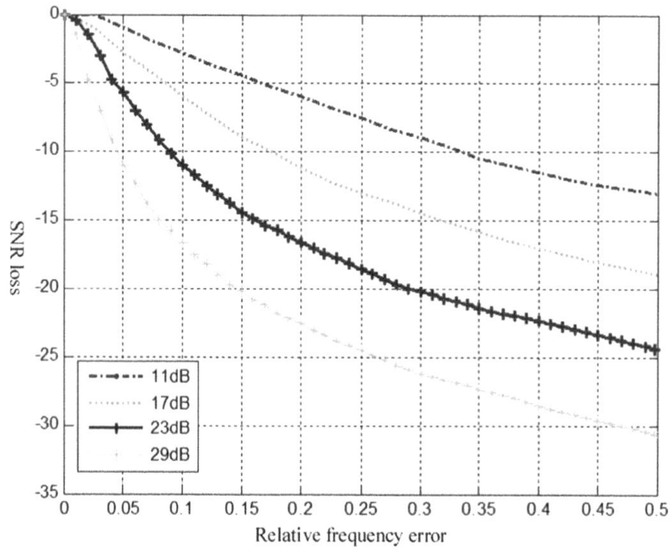

Fig. 6.5 SNR loss with frequency offset

6.2.3 Impact of Sampling-Clock Deviation on the System Performance

After the symbols and carriers are synchronized, a sampling rate error still exists in the system. This causes ICI and leads to symbol-timing drift and the deterioration of symbol synchronization. The sampling period at the receiving end must be consistent with the actual OFDM sample interval to correctly receive n samples and recover the transmitted data.

In practice, owing to the unstable frequency of the oscillator, the frequency of the sampling clock deviates, and the frequency deviation also causes phase deviation. Therefore, the sampling clock must be synchronized. The purpose of sampling-clock synchronization is to make the sampling-clock frequency at the receiving end consistent with that at the transmitting end.

Usually, on the basis of symbol synchronization, the estimated value of the sampling-rate error is obtained using the data after an FFT, and the output of the voltage-controlled crystal oscillator (VCXO) is controlled by a phase-locked loop to adjust the sampling frequency at the receiving end. This method is usually referred to as the direct method.

In practical applications, the indirect-compensation method can also be used to synchronize the sampling rate [6]. The influence of the sampling-rate error on the signal includes timing-phase and frequency deviations. The influence of phase deviation is similar to that of symbol deviation; therefore, it can be merged into symbol synchronization. The influence of frequency deviation $\beta = \Delta t / T$ (T is the sampling interval at the transmitter) on the signal can be expressed by the following formula:

$$Y_{n,k} = e^{j2\pi nk\beta(I_S/T_u)} X_{n,k} \cdot H_{n,k}. \tag{6.12}$$

It can be observed that the sampling-rate frequency deviation introduces a carrier-frequency offset of $\Delta f = k\beta$ into the kth subcarrier. Therefore, the frequency offset of each subcarrier of the received signal in the frequency domain is estimated and compensated, which reduces the impact of the out-of-step sampling rate.

6.2.4 OFDM Synchronization-Algorithm Principle

The cross-correlation function of signals $x(t)$ and $y(t)$ is defined as

$$R_{x,y}(\tau) = \lim_{T\to\infty} \frac{1}{T} \int_0^T x(t)y(t+\tau)dt, \tag{6.13}$$

where T denotes the average time. The cross-correlation function is equivalent to the average product of the time of two signals, one of which has moved τ seconds in time; τ is called the time difference. Generally, Eq. (6.13) can be replaced by the following equation:

$$R_{x,y}(\tau) = \frac{1}{T} \int_0^T x(t)y(t+\tau)dt. \tag{6.14}$$

The cross-correlation function of two signals can be used to understand the degree of correlation between two unknown signals or the time relationship between two known signals. If the two are related, the maximum value of $R_{x,y}(\tau)$ can be obtained by adjusting the time difference between them to obtain the time delay between them. If the signal is correlated with itself, the autocorrelation function is

$$R_{x,x}(\tau) = \frac{1}{T} \int_0^T x(t)x(t+\tau)dt. \tag{6.15}$$

For a signal with an autocorrelated structure, the starting position of the correlation component can be determined by calculating the peak value of its autocorrelation function. The synchronization in an OFDM system is based on this correlation principle. In fact, the effective data to be transmitted in the OFDM system do not have this correlation characteristic; therefore, some data must be introduced to ensure that the transmitted OFDM signal is partially autocorrelated.

6.2.4.1 Design of the Training Sequence

The correlation characteristics of the training sequence play a decisive role in the performance of a synchronization scheme based on a correlation search. Therefore, starting from the training sequence is also a breakthrough in improving the performance of the synchronization scheme. It is hoped that this synchronization sequence can have a sharp autocorrelation, as small as possible cross-correlation, and be easy to implement in engineering.

6.2.4.2 m-Sequence

The maximum-length linear shift-register sequence (*m*-sequence) is an important class of pseudo-random sequences, and is currently the most deeply studied [7]. *m*-sequences have ideal pseudo-random characteristics; the autocorrelation function is sharp, and it is a binary function.

$$R_x(l) = \begin{cases} N & l = 0(\bmod / N) \\ -1 & l \neq 0(\bmod / N) \end{cases} \tag{6.16}$$

When the *m*-sequence was used in the symbol-synchronization algorithm, it was found that the sequence had a poor ability to suppress the frequency offset. When the normalized frequency offset increased, the sequence correlation worsened.

For the synchronization algorithm, no circular dependency should exist between synchronization links. For example, the frequency-synchronization algorithm requires the symbol-timing synchronization to be completed first; the symbol-synchronization algorithm takes frequency synchronization as its premise, which leads to a circular dependency. Therefore, we hope to find a sequence with high suppression of the frequency offset.

6.2.4.3 FH Sequence

Frank and Heimiller [8] proposed a method to construct sequences with length $L = N2$ (N is the number of phases), which are called FH sequences. First, we take a primitive unit root $\omega = e^{j\frac{2\pi M}{N}}$, where M and N are coprimes, as the array:

$$F_N = \left[\omega^{ik(\bmod N)}\right] = \begin{bmatrix} \omega^0 & \omega^0 & \omega^0 & \cdots & \omega^0 \\ \omega^0 & \omega^1 & \omega^2 & \cdots & \omega^{N-1} \\ \omega^0 & \omega^2 & \omega^4 & \cdots & \omega^{2(N-1)} \\ \vdots & \vdots & \vdots & \ddots & \vdots \\ \omega^0 & \omega^{N-1} & \omega^{2(N-1)} & \cdots & \omega^{(N-1)^2} \end{bmatrix} \cdot i, k = 0, 1, 2, \cdots, N-1, \tag{6.17}$$

where $\omega^N = \omega^0 = 1$. The above array is serially arranged by row or column to obtain a sequence with length $L = N^2$; that is, the FH sequence.

A polyphase sequence is represented by a finite-field sequence. If there is a polyphase sequence u, there is a corresponding finite-field sequence a, and $u_i = \omega^{a_i}$, $i = 0, 1, 2, ..., L-1$. The time delay of the FH-sequence autocorrelation function at zero can be expressed as

$$R_a(0) = \sum_{i=0}^{L-1} \left|\omega^{a_i}\right|^2 = N^2 = L. \tag{6.18}$$

The side lobe of the autocorrelation function is

$$\begin{aligned} R_a(l) &= \sum_{i=0}^{L-1} \omega^{a_i - a_i + 1} = \sum_{k=0}^{N-1} \sum_{i=kN}^{(k+1)N-1} \omega^{ik-(i+l)k} \\ &= \sum_{k=0}^{N-1} \sum_{i=kN}^{(k+1)N-1} \omega^{-lk} = N \sum_{k=0}^{N-1} \omega^{-lk} = N \frac{\left(1-\omega^{-Nl}\right)}{\left(1-\omega^{-l}\right)} = 0. \end{aligned} \tag{6.19}$$

This indicates that the autocorrelation sidelobes of the FH sequence are equal to 0. Figure 6.6 shows the autocorrelation functions of several common training sequences.

As can be seen from Fig. 6.6, when the signal-to-noise ratio (SNR) is 0 dB, for general pseudo-noise (PN) sequences and m-sequences, when the frequency offset exceeds 0.5, the correlation peak is almost unrecognizable. The peak value of the FH sequence is still obvious even when the frequency offset is large.

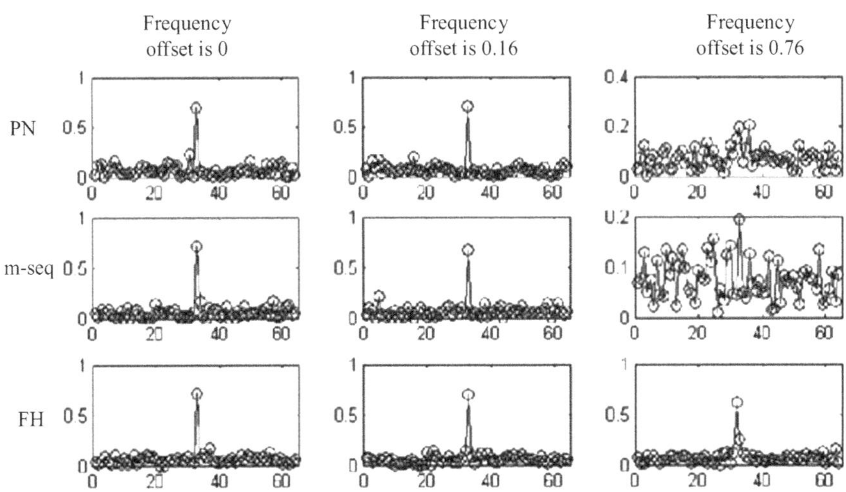

Fig. 6.6 Autocorrelation function graphs of common training sequences when the SNR = 0 dB

6.3 OWC-OFDM Symbol Synchronization

Synchronization algorithms for estimating the frequency and timing offsets can be divided into two categories: one is based on a training sequence and the other is blind; that is, it is not based on a training sequence, but depends completely on the OFDM signal itself or its spectral characteristics.

OFDM-synchronization algorithms based on training sequences can be roughly divided into two types. The first uses a cyclic prefix (CP) [9, 10], which is mainly used in continuous-transmission systems, such as digital-audio broadcasting (DAB). The second uses the algorithm of the training sequence [11–15]. It is more suitable for burst-transmission systems, such as wireless local area networks (WLANs).

In 1994, Moose [5] first used a repeated training sequence to estimate the maximum likelihood of the carrier. The transmitter sends the repeated sequence and calculates the frequency offset in the frequency domain after the FFT. When the length of a part is $N/2$ (where N is the length of the OFDM signal), the frequency offset within a ± 1 subcarrier interval can be estimated. By shortening the length of the part, the frequency-offset estimation range can be expanded; however, the accuracy of the estimation will be reduced because the average number of sampling points is small.

Classen [12] proposed a synchronization algorithm for joint time and frequency offsets; however, it requires extensive computation because it needs to search in the entire frequency-acquisition range until the correct frequency offset is found. Schmidl [13] simplified Classen's method. In the Schmidl method, the repeated signal in an OFDM symbol is obtained by transmitting data on an even carrier and no data on an odd carrier.

Two training sequences are used to obtain time and frequency synchronizations in two steps, and the two training sequences are placed at the beginning of the frame. The time is synchronized by searching for the correlation between the first and second halves of the sequence, while the frequency offset is partially corrected (1/2 subcarrier interval). Then, the former sequence is correlated with the latter sequence in the frequency domain through an FFT to obtain the frequency offset of an integer times the subcarrier interval.

Minn [14] improved Schmidl's algorithm by using the negative sign of the second part to reduce the variance of the timing algorithm. Park [15] obtained a timing spike similar to the impulse function by designing a time-domain symmetric training sequence; however, there were two relatively high secondary peaks, which affected the timing performance.

The synchronization methods using a cyclic prefix includes the joint maximum-likelihood time and frequency-offset estimation method proposed by Van de Beek [16], which can be used without pilots. A simulation with additive Gaussian white noise (AGWN) shows that frequency-offset estimation can be used for the tracking mode (the frequency offset is less than one-half of a carrier interval), and timing estimation can be used for the acquisition mode. This estimation assumes the channel state in advance. Because the guard interval is affected by ISI, the estimation performance is poor in a dispersive channel.

Blind-estimation methods include the Tureli's Esprit method [17] and Liu's MUSIC method [18], both of which are based on the signal subspace. The multiple signal classification (MUSIC) method uses the orthogonality of the subcarriers to equate the frequency-offset estimation problem to the problem of finding the roots of polynomials. Both these methods have super-resolution performance; however, their extensive computation affects their engineering practicality.

6.3.1 Traditional Symbol-Synchronization Algorithms

The following introduces two classical symbol synchronization algorithms: the Schmidl and Cox [11] and the Minn [12].

6.3.1.1 Schmidl and Cox Algorithm

A special characteristic is used in the Schmidl and Cox algorithm. Half of the sample values are copies of the other half. This algorithm can be used in burst data-transmission mode to determine the starting position of the burst transmission. The key to the algorithm is finding a character with the same value in the time domain. After this character passes through the channel, the front and back halves remain the same, except for the phase difference caused by the frequency offset.

A pseudo-random sequence can be transmitted at odd frequency points, a zero value can be uploaded at even frequency points, and a training character corresponding to the same two half-sample values in the time domain can be obtained through an IFFT transformation. Another method can be used to generate these training characters. An IFFT with half the length of the training characters is used to transform the PN sequence, and the transformed characters are repeated to obtain the required training characters.

Considering that the first and second halves of the training characters in the time domain are the same, their received characters in the time domain are also the same, except for the phase difference caused by the frequency offset. If any sample value is taken in the first half of the received character and its conjugate is multiplied by the corresponding sample value in the second half, the influence of the channel can be eliminated, and the phase of the multiplied result is approximately $\phi = \pi T \Delta f$.

At the beginning of the frame, each pair of corresponding samples is multiplied, and the phases of the results are approximately equal; thus, the amplitude of the sum will be large. The equivalent time-domain structure of the training sequence is shown in Fig. 6.7.

Assuming that half the length of the training sequence (except the cyclic prefix) is L, the sum of the products of each pair of corresponding samples is

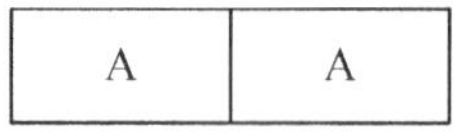

Fig. 6.7 Time-domain structure of the training sequence

$$p(d) = \sum_{m=0}^{L-1} \left(r_{d+m}^{*} r_{d+m+L} \right). \tag{6.20}$$

Equation (6.20) can be calculated iteratively with the following equation:

$$p(d+1) = p(d) + \left(r_{d+L}^{*} r_{d+2L} \right) - \left(r_{d}^{*} r_{d+L} \right), \tag{6.21}$$

where d represents the time subscript corresponding to the first sample in a window of L samples. When the receiver searches for the first training character, the window slides over time. The receiving energy of the second segment of characters is defined as

$$R(d) = \sum_{m=0}^{L-1} \left| r_{d+m+L} \right|^2. \tag{6.22}$$

Equation (6.22) can also be calculated using iterations. $R(d)$ can be used as part of an automatic gain-control loop (AGC). We define the following time-scaling formula:

$$M(d) = \frac{|p(d)|^2}{(R(d))^2}. \tag{6.23}$$

When $M(d)$ reaches the maximum value, the corresponding d is $d_{\max}$; then, $d_{\max}$ is the timing-synchronization position to be searched.

Figure 6.8 shows the time-scaling curve of the algorithm when the number of subcarriers in the AWGN channel is 1024, the time delay is 529, and the carrier frequency offset is 0.25. In the Schmidl and Cox algorithm, the timing is synchronized by determining the maximum value of the timing measurement.

The rationality of this method can be explained using $P(d)$. When noise is received, because the correlation value of the noise signal is 0, the output $P(d)$ will be a random variable with a zero mean value, which will ensure a low $M(d)$ level before the data frame arrives. When the data frame begins to be received, $P(d)$ is the correlation value of the two halves before and after the first symbol, which causes $M(d)$ to quickly jump to a large value.

However, it can be seen from the figure that the time-scaling curve has a flat section with no ISI effect. The length of the flat section is equal to the length of the cyclic prefix minus the length of the channel impulse response. Therefore, the existence of this gentle section causes the timing synchronization to deviate. In other words, the longer the channel-response length, the shorter the flat section and the more accurate

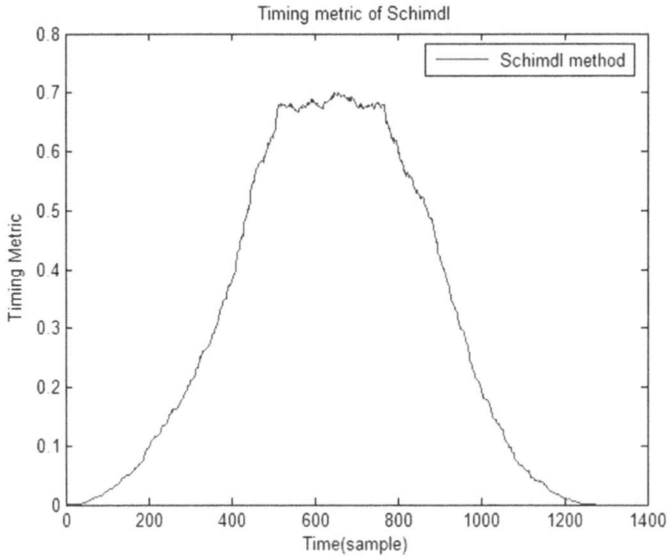

Fig. 6.8 Time-scaling curve of the Schmidl and Cox algorithm [19–21]

the timing-synchronization judgment. For AWGN channels, the length of this flat segment is equal to the length of the cyclic prefix.

6.3.1.2 Minn Algorithm

It can be observed from Fig. 6.8 that, owing to the existence of the cyclic prefix, there is a flat area at the peak value, and the length of the flat area is not fixed. This is related to cyclic-prefix and multipath-channel delays and results in an inaccurate timing estimation. Minn improved Schmidl and Cox's method and used the opposite number for the first half of the training sequence in the second half. The structure of the training sequence is shown in Fig. 6.9.

Let the effective symbol length of the OFDM be L, which is the same as that of the Schmidl and Cox algorithm. A special character is used as the training sequence to find a character with the same sample values in the time domain. After this character passes through the channel, the front and back halves remain the same, except for the phase difference caused by the frequency offset. A pseudo-random sequence can be transmitted at odd points, and a zero value can be uploaded at even-frequency points.

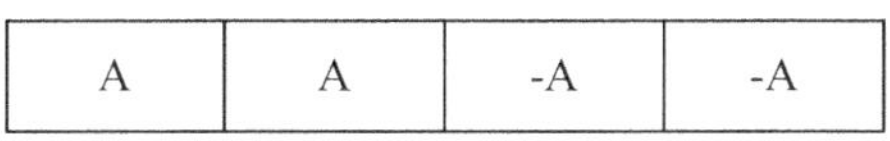

Fig. 6.9 Structure of the Minn-algorithm training sequence

In the Schmidl and Cox algorithm, an IFFT with half the length of the training sequence is used to transform the PN sequence, and then the transformed characters are repeated to obtain the required training characters. The difference between the Minn algorithm and the Schmidl and Cox algorithm is that, in the former, one-fourth of the length of the training sequence is taken for the IFFT transformation, and the result is the value of A in Fig. 6.9. The training sequence is then constructed according to Fig. 6.9. Thus, it is concluded that the training sequence still meets the requirements of the first and second halves, and the timing offset can still be estimated using the training-sequence correlation.

The expression for the timing scale is as follows:

$$M_{Minn}(d) = \frac{\left|P_{Minn}(d)^2\right|}{(R_{Minn}(d))^2}. \tag{6.24}$$

Of which,

$$P_{Minn}(d) = \sum_{m=0}^{1} \sum_{k=0}^{L/4-1} r^*\left(d + \frac{L}{2}m + k\right) r\left(d + \frac{L}{2}m + k + \frac{L}{4}\right) \tag{6.25}$$

$$P_{Minn}(d) = \sum_{m=0}^{1} \sum_{k=0}^{L/4-1} \left| r\left(d + \frac{L}{2}m + k + \frac{L}{4}\right)\right|^2. \tag{6.26}$$

Figure 6.10 shows the time-scaling curve of this algorithm when the number of subcarriers in the AWGN channel is 1024, the time delay is 529, and the carrier frequency offset is 0.25.

Figure 6.11 shows a diagram for calculating the correlation function of the Minn algorithm. Similar to the Schmidl and Cox algorithm, in the correct symbol starting position, each sample value corresponds to another, and a very good correlation is obtained at this time. At the incorrect symbol starting position, some sample values cannot correspond to each other (as shown by the dotted arrows in the figure: $a \leftrightarrow -a,\ -a \leftrightarrow *$), while the rest still correspond to one another (as shown by the solid arrows in the figure: $b \leftrightarrow b,\ c \leftrightarrow c,\ -b \leftrightarrow -b,\ -c \leftrightarrow -c$).

In this manner, the correlation value near the correct starting position is smaller than that at the correct position, but it is still close. Owing to the influence of channel fading and noise, a peak value appears, resulting in time-synchronization errors.

It can be seen from Figures 6.10 and 6.12 that the Minn algorithm has a small peak value; thus, it obtains relatively accurate symbol synchronization and solves the timing-ambiguity problem of the Schmidl and Cox algorithm. At the timing-synchronization position, $M_{Minn}(d)$ reaches its maximum value. This is because the negative sign of the second half of the training symbols reduces the value of $M_{Minn}(d)$ at an incorrect timing point.

When the timing-synchronization position falls within the cyclic prefix ahead of the ideal position, it can be seen from Fig. 6.11 that the negative sign of the second

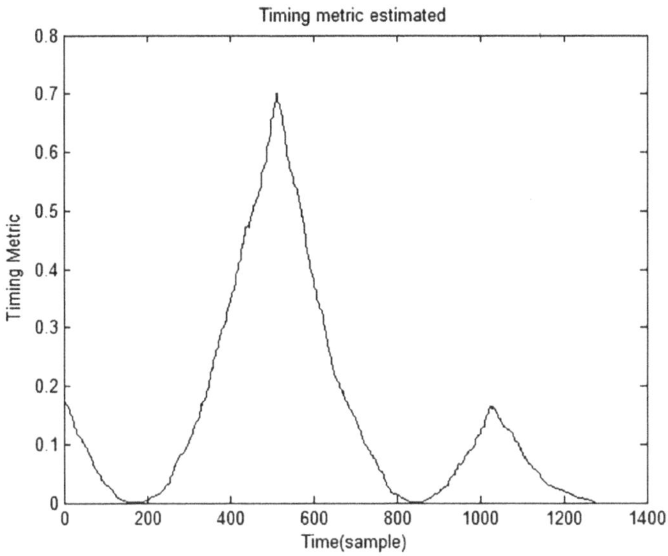

Fig. 6.10 Time-scaling curve of the Minn algorithm [19–21]

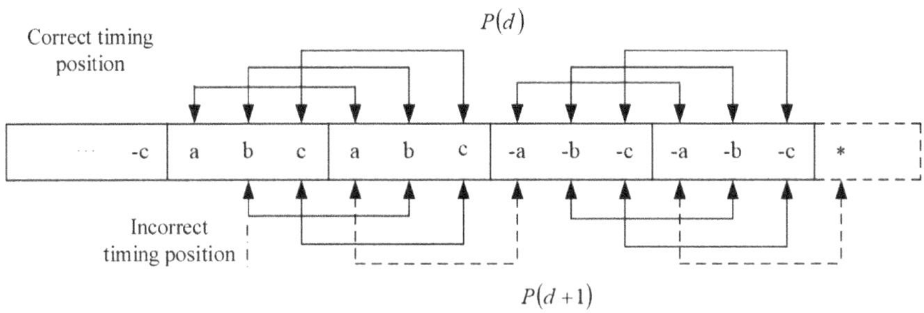

Fig. 6.11 Calculation diagram of the Minn-algorithm correlation function [19–21]

half of the training symbols eliminates the peak-platform phenomenon in the timing estimation of the Schmidl and Cox algorithm. However, from Fig. 6.10, it can also be seen that a large side lobe appears next to the main peak value of the timing-estimation curve, which can easily be misjudged under the influence of channel fading and noise. The Minn algorithm improves the accuracy of the timing estimation to a certain extent by introducing a negative sign on the basis of the training symbol of the Schmidl and Cox algorithm repetition structure; however, it is not the best.

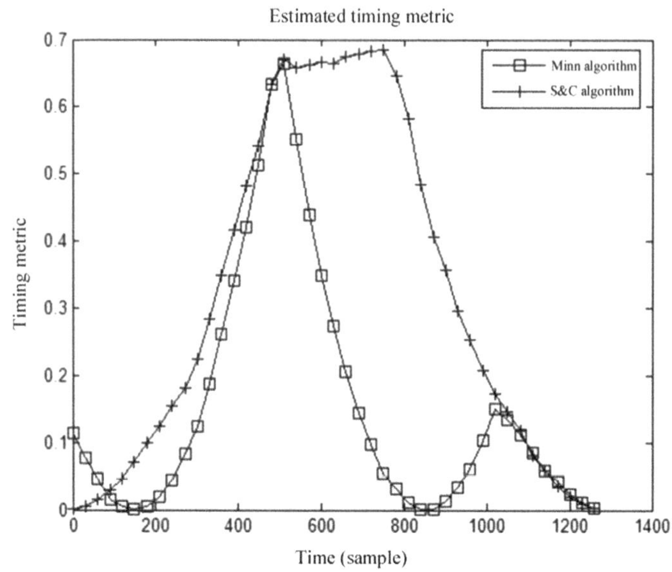

Fig. 6.12 Comparison of the time-scaling curves of the Schmidl and Cox algorithm and the Minn algorithm [19–21]

6.3.2 *Improved Synchronization Algorithms*

Generally, symbol synchronization first obtains the synchronization signal using a correlation method, and then obtains the starting position of the symbol using a certain detection method. The main differences among the various methods are as follows: first, the sequences used to generate correlation signals are different; second, the structure used to generate correlation signals is different; and third, there are different correlations for generating correlation signals. We introduce three improved synchronization algorithms.

6.3.2.1 Symmetric FH-Sequence Symbol Time-Synchronization Algorithm

In the first improved algorithm, the synchronous data block includes two FH sequences of equal lengths. The latter sequence is a reverse copy of the data in the previous sequence. The structure of the synchronous data block is shown in Fig. 6.13.

Fig. 6.13 Synchronous data-block structure

$s(0)\ s(1)\ s(2)\cdots s(L-1)$	$s(L-1)\ \cdots s(2)\ s(1)\ s(0)$

The symmetry of the synchronous data block is significantly correlated. If the influence of channel noise is not considered, the following relationship exists between the two sequences in the synchronous data block:

$$r(2L-1-k) = r(k)e^{j2\pi\varepsilon(2L-1-k)/L}\ k \in [0, L-1], \tag{6.27}$$

where $r(k)$ represents the kth sample value in the synchronization data block at the receiving end.

We define a time-scaling formula as

$$M(d) = \frac{P(d)}{R(d)}$$
$$P(d) = \left|\sum_{k=0}^{L-1} r^*(2L-1-k+d)r(k+d)\right| \quad R(d) = \sum_{k=d}^{d+2L-1} |r(k)|^2, \tag{6.28}$$

where d is the position of the first sample value of the synchronized data block. When $M(d)$ reaches its maximum value, the corresponding $d_{\max}$ is the starting synchronization position of the OFDM symbol.

Figure 6.14a shows the time scaling curve when the number of subcarriers in an AWGN channel is 1024, the time delay is 529, and the carrier-frequency offset is 0.25, and Fig. 6.14b shows that the attenuation coefficient in a multipath channel is [0.2 0.3].

Figure 6.15 shows a comparison of the mean squared error of the time estimation between the Schmidl and Cox (S&C) algorithm and the first improved algorithm in a Gaussian channel.

As can be seen from Fig. 6.14, the peak value of the time-scaling curve of the improved algorithm is very sharp in both the Gaussian and multipath channels. It overcomes the flat-slope problem of the time-scaling curve of the S&C algorithm, and can overcome the multipath interference caused by the wireless channel. The sharper

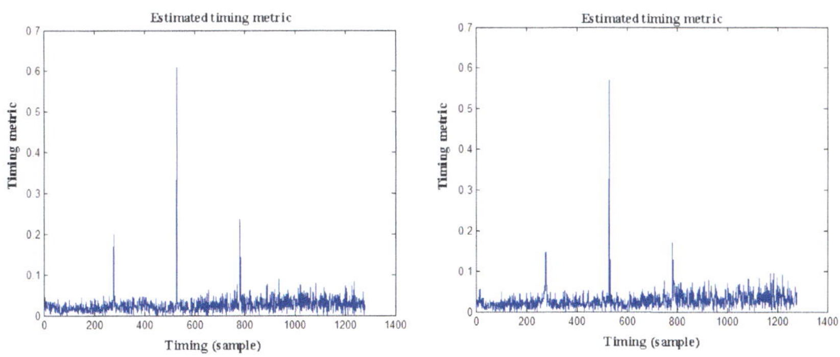

(a) Time-scaling curve in a Gaussian channel. (b) Time-scaling curve in a multipath channel.

Fig. 6.14 Time-scaling curve of the first improved algorithm [19–21]

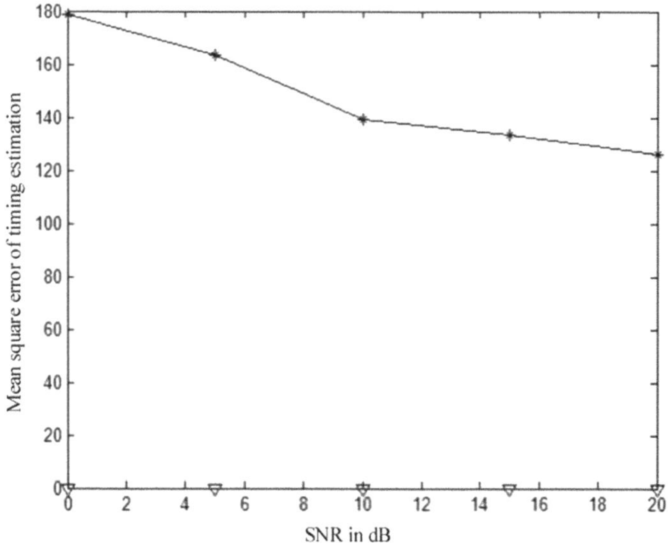

Fig. 6.15 Comparison of mean squared error [19–21]

the peak value of the time-scaling curve, the more accurate is the peak detection. Moreover, the peak value is the starting position of the first sample value of the OFDM symbol, which is an effective symbol timing–synchronization scheme.

From the simulation results in Fig. 6.15, it can be observed that the timing mean squared error of the first improved algorithm is significantly smaller than that of the S&C algorithm, and the time-estimation performance is significantly better than that of the S&C algorithm. Therefore, the algorithm using a training sequence with a symmetrical structure constructed by the FH sequence for time synchronization has a higher accuracy than the S&C algorithm. The timing can be completely and accurately synchronized in a wireless channel with multipath fading.

6.3.2.2 Conjugate-Symmetric FH-Sequence Symbol Time Synchronization Algorithm 1

The structure of the training sequence is shown in Fig. 6.16. A is the time-domain representation of the FH sequence with length $L/4$, and B is the conjugate symmetric sequence of A.

According to the conjugate symmetry of the training sequence, the symbol timing offset estimation function is defined at the receiving end as

Fig. 6.16 Training-sequence structure

A	B	A	B

$$M(d) = \frac{|P(d)|^2}{(R(d))^2}, \tag{6.29}$$

$$P(d) = 4\sum_{k=0}^{L/4-1} r(k+d)r(L+d-k-1) + r\big(L/4+d+k\big)r\big(3L/4+d-k-1\big) \tag{6.30}$$

$$R(d) = \sum_{k=0}^{L/4-1} |r(k+d)|^2 + \big|L/4+k+d\big|^2 + \big|L/2+k+d\big|^2 + \big|3L/4+k+d\big|^2. \tag{6.31}$$

Then, the timing offset is estimated as

$$\theta = \arg\max_d (M(d)). \tag{6.32}$$

Figure 6.17 shows a timing-offset estimation function comparison of the second improved algorithm, Schmidl and Cox algorithm, and Minn algorithm, when the number of subcarriers in the AWGN channel is 1024, the time delay is 500, and the carrier-frequency offset is 0.25.

Figure 6.18 shows the timing-offset estimation function of the second improved algorithm when the attenuation coefficient in the multipath channel is [0 0 0.2 0.3].

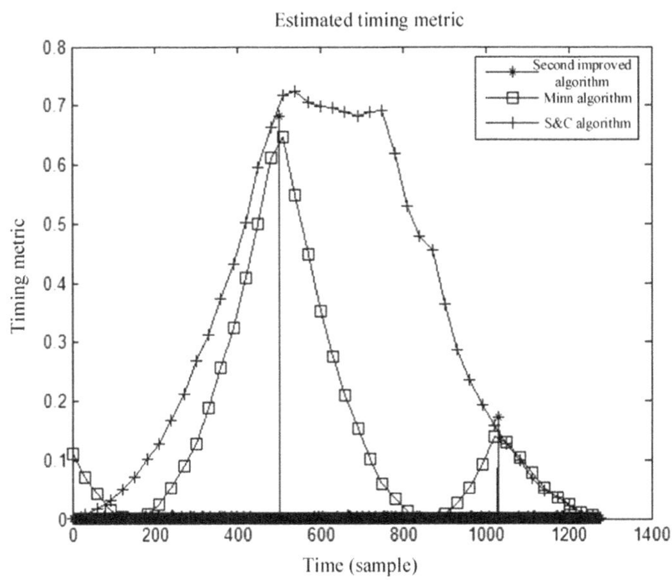

Fig. 6.17 Timing-offset estimation function comparison [19–21]

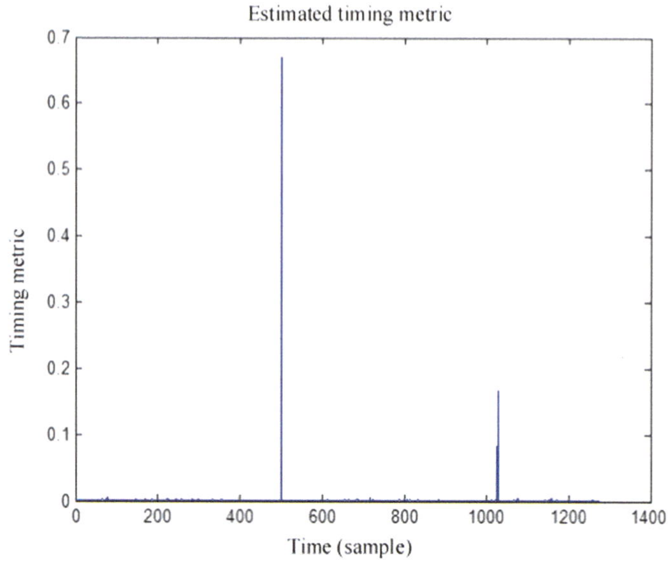

Fig. 6.18 Timing-offset function of the second improved algorithm in a multipath channel

As can be seen from Fig. 6.17, the Minn algorithm avoids the platform problem of the symbol-timing measurement in the Schmidl and Cox algorithm; however, the value around the correct timing point decreases slowly, and the performance is not ideal. The timing curve of the second improved algorithm reaches the maximum value at the correct timing point and exhibits impulse-response characteristics. However, the estimated value of the timing offset corresponding to the other times was very small. Therefore, using this training-sequence structure, the accurate starting point of an FFT window can be determined by searching the maximum value in a certain time window.

As can be seen from Fig. 6.18, the second algorithm overcomes the multipath interference caused by wireless channels in multipath channels and can still obtain very sharp peaks and find the position of the symbol starting point. Therefore, the training sequence with conjugate symmetry constructed using the FH sequence can accurately estimate the starting position of the OFDM symbols at the receiver. This algorithm is an effective symbol-timing synchronization scheme for optical wireless communication.

Figure 6.19 shows a calculation diagram of the correlation function $P(d)$ of the second improved algorithm. From Fig. 6.19, we can intuitively see that the main reason for the unsatisfactory performance of the symbol timing measure is that in the Minn algorithm, when calculating the timing measure, most of the summation terms in the timing-measure calculation formula of two adjacent points are the same; only two terms are different. This will inevitably lead to a close value of the timing-measure function of adjacent points, which makes the timing-measure curve insufficiently sharp, resulting in an error in symbol timing synchronization.

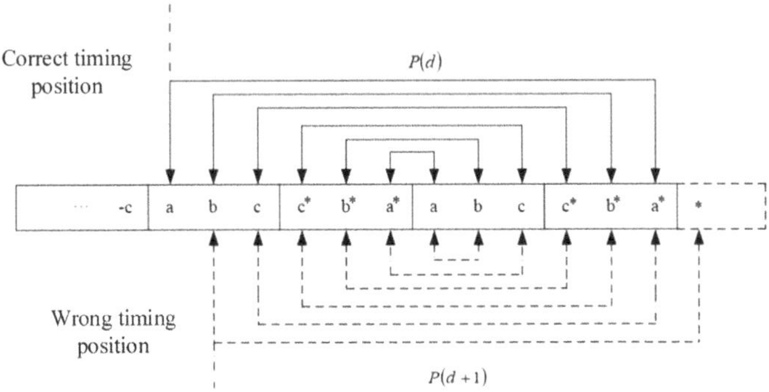

Fig. 6.19 Calculation diagram of the correlation function of the second improved algorithm [19–21]

The second improved algorithm avoids this problem. At the correct starting position of the symbol, each sample value corresponds to another, and a very good correlation is obtained. At the incorrect symbol starting position, the sample values cannot correspond to each other (as shown by the dotted arrows in the figure); thus, the timing measurement curve has a very sharp peak and can be timed accurately.

6.3.2.3 Conjugate-Symmetric FH-Sequence Symbol Timing Synchronization Algorithm 2

Taking advantage of the fact that the FH sequence is less affected by frequency deviation, in the third improved algorithm, the training sequence is composed of an FH sequence, and its structure is shown in Fig. 6.16. *A* is the FH sequence with a length of *L*/4 in the time domain, and sequence *B* and sequence *A* are conjugate symmetric; that is, $A(k) = B^*(L/4 + 1 - k)$, where $k \in [1, L/4]$. We define P(d) and R(d):

$$P(d) = \sum_{k=0}^{L/2-1} r_{d+k} r_{L+d-k-1} \tag{6.33}$$

$$R(d) = \sum_{k=0}^{L/2} |r_{d+k}|^2 \tag{6.34}$$

$$M_1(d) = P^2(d) \big/ R^2(d). \tag{6.35}$$

We define the time-scaling formula as

$$M_2(d + 1) = 4 \cdot (M_1(d + L/4 + 1) - M_1(d)) \tag{6.36}$$

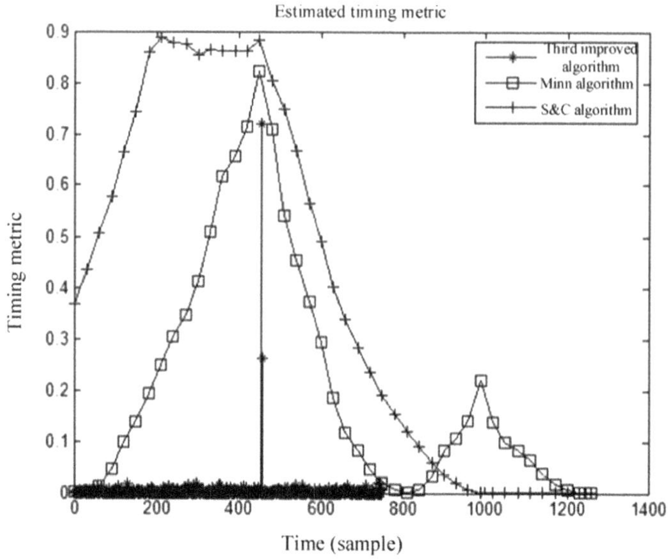

Fig. 6.20 Time-scaling curves of three algorithms in a Gaussian channel

$$\theta = \arg\max_{d}(M_2(d) > 0) \tag{6.37}$$

where d is the time-search variable, and θ is the correct starting point of the OFDM symbol.

The simulation system parameters were set as follows: the number of subcarriers was 1024, data-modulation mode was quadrature phase-shift keying (QPSK), cyclic prefix was 256 sampling points, time delay was 200 sampling points, relative frequency deviation was 0.25, and attenuation coefficient in the multipath channel was [0 0 0.2 0.3]. The timing-performance simulation curves of the different timing algorithms are shown in Figs. 6.20 and 6.21.

As can be seen from Figs. 6.20 and 6.21, in the Gaussian channel, owing to the influence of the cyclic prefix, a platform with the same length as the cyclic prefix appears in the time-scaling curve in the Schmidl & Cox algorithm. In the multipath channel, the length of the platform is less than that of the cyclic prefix because some intervals of the cyclic prefix are damaged by intersymbol interference, which causes uncertainty in the timing estimation.

The Minn algorithm eliminates the platform problem of Schmidl's algorithm, and a peak appears at the corresponding time. However, the time-scaling curve changes slowly at both ends of the correct starting point, and there are also subpeaks, resulting in symbol timing errors. The third improved algorithm can effectively eliminate symbol timing errors and improve the accuracy of the timing estimation. Its performance is significantly better than that of the Schmidl and Cox and Minn algorithms. This is a more effective symbol timing scheme.

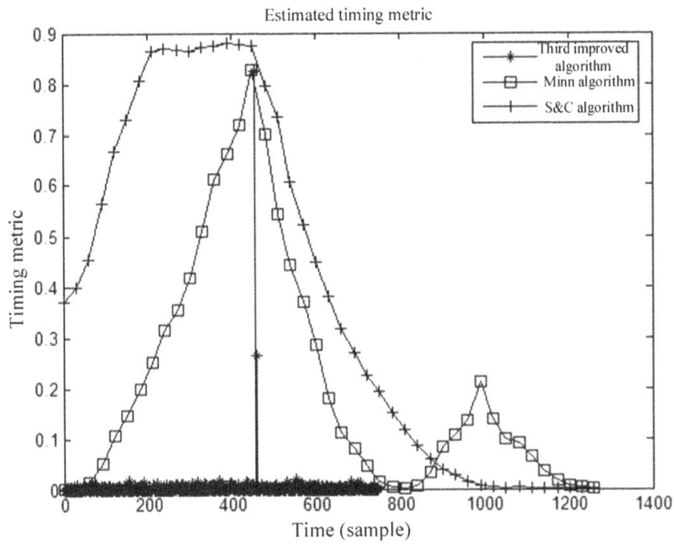

Fig. 6.21 Time-scaling curves of three algorithms in a multipath channel

As can be seen from Figs. 6.22 and 6.23, with an increase in the signal-to-noise ratio, the mean value of the timing-estimation deviation of the Schmidl and Cox and Minn algorithms gradually decreases; however, there are still timing-estimation deviations when the signal-to-noise ratio is low. As a result, the integer multiple-frequency deviation estimation after an FFT is affected by the phase noise and random channel noise caused by timing uncertainty. The mean value of the timing estimation deviation of the third improved algorithm is significantly smaller than those of the Schmidl and Cox and Minn algorithms, and the estimation deviation is 0.

Therefore, compared with the other two algorithms, the performance of the third improved algorithm is significantly improved, and is suitable for a synchronization system in the case of a low signal-to-noise ratio in optical wireless communication.

6.4 Frequency-Synchronization Algorithms

6.4.1 Carrier-Frequency Offset Estimation Algorithms

Owing to the small bandwidth of each subchannel and the high sensitivity to the carrier-frequency offset, OFDM requires highly accurate carrier synchronization. To effectively use limited data, capture parameters over a wide range, and realize high-precision synchronization, the frequency-synchronization process can be divided into several implementation processes.

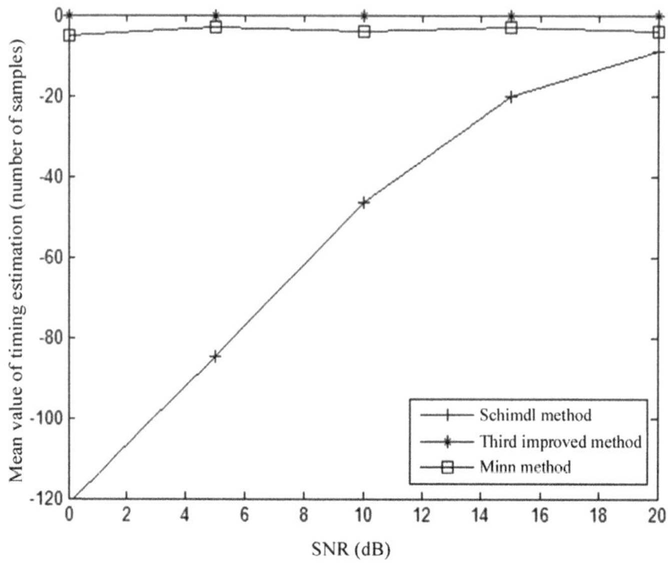

Fig. 6.22 Mean value of the timing estimation of three algorithms in a Gaussian channel

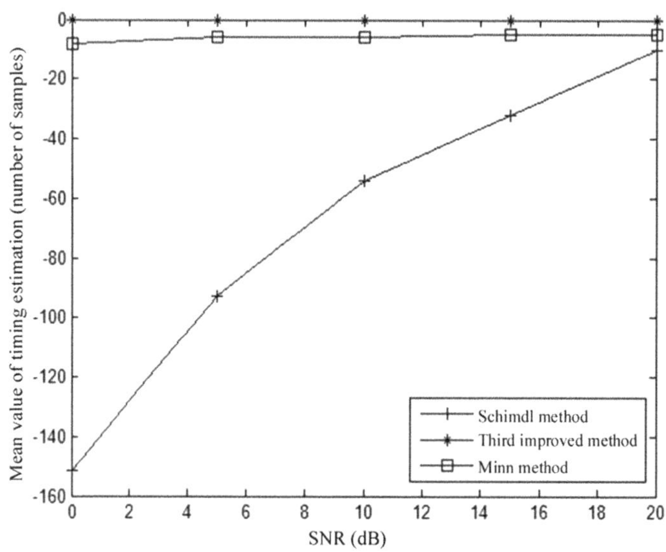

Fig. 6.23 Mean value of the timing estimation of three algorithms in a multipath channel

In coarse-synchronization mode, the synchronizer reduces the large initial deviation of the parameters to a small range, but it does not require complete and accurate synchronization. In precise-synchronization mode, the synchronizer further reduces the residual error of the parameters and improves the estimation accuracy. When the frequency offset is accurately estimated, the frequency drifts, owing to the Doppler effect and other factors; therefore, it is also necessary to track the frequency.

The frequency offset can be decomposed into integer and decimal parts, which are relative to the subcarrier spacing. The integer part is the frequency-division offset equal to the integer multiple of the subcarrier interval, and the fractional part refers to the frequency-division offset that is smaller than the subcarrier interval. The fractional frequency-doubling offset can be estimated using the redundancy introduced by the data-protection interval.

The integer part only shifts the information symbols on the subchannel. It does not destroy the orthogonality between subcarriers, but leads to a complete error of the entire demodulation result. The fractional part causes subchannel interference, destroys the orthogonality between subcarriers, and increase the bit-error rate of the system. Generally speaking, the coarse frequency synchronization in an OFDM system mainly estimates and compensates for the integer-frequency offset, whereas the fine frequency synchronization estimates and compensates for the fractional-frequency offset.

The maximum-likelihood algorithm using a cyclic prefix proposed by Jan [22] has the advantages of little computation and a simple implementation and can be used for symbol timing. However, the frequency-estimation range of this algorithm is small, the timing estimation is rough, and it is easily affected by a multipath channel; therefore, it is difficult to use directly in an actual system.

Moose's frequency-domain correlation algorithm [11] estimates the frequency offset by comparing the phase difference between the symbol subcarriers of two repeated OFDM training sequences. It is simple to implement but its cost is high. The frequency-offset estimation range can reach only half of the carrier interval. Classen [23] used the autocorrelation function to jointly synchronize the frequency offset and time; however, the frequency offset adopts the step-search method, which requires extensive calculation.

Schmidl [13] improved Classen's method by improving its speed without reducing the time and frequency synchronization accuracy. Moreover, it better solved the joint synchronization of time and frequency of the OFDM system. However, its symbol synchronization is ambiguous, which further affects the frequency-offset estimation and is only applicable to AWGN channels.

Hsieh and Wei classified algorithms for estimating the carrier-frequency deviation into three categories [24]:

(1) Data-aided algorithms: The estimate is based on specific training information in the transmitted signal.
(2) Non-data-aided algorithms: The estimate analyzes the received signal in the frequency domain.

(3) Cyclic-prefix-based algorithms: The cyclic prefix is estimated using the structural characteristics of the OFDM symbols.

6.4.2 Maximum-Likelihood (ML) Synchronization Algorithm Based on a Cyclic Prefix

The maximum-likelihood (ML) estimation algorithm based on a cyclic prefix uses the information carried by the cyclic prefix and adopts maximum-likelihood estimation to complete the symbol-timing synchronization and carrier-frequency synchronization. The algorithm was derived based on the assumption that the channel is an additive white Gaussian noise channel.

Considering an OFDM system with N subcarriers and L cyclic-prefix lengths, the length of each OFDM symbol is actually $N + L$ samples. After the channel transmission, the symbol-timing offset at the receiving end is usually expressed as the delay of the received signal, and the frequency offset is usually expressed as the phase distortion of the received signal in the time domain. Then, the received time-domain signal with timing and frequency offsets can be expressed as

$$r(n) = s(n-d)e^{j2\pi\Delta f_c n/N} + \eta(n), \tag{6.38}$$

where $s(n) = \frac{1}{\sqrt{n}}\sum_{k=0}^{N-1} S_k e^{j2\pi nk/N}$, d is the symbol-timing synchronization point—that is, the starting position of the OFDM symbol, but the system does not know it—Δf_c represents the relative frequency offset, and $\eta(n)$ is Gaussian white noise with a zero mean value.

Assuming that complex data S_k after constellation mapping are independent, $s(n)$ is a linear combination of independent and identically distributed random variables after the IFFT. If the number of subcarriers is sufficiently large, it can be seen from the central-limit theorem that $s(n)$ is approximately a complex Gaussian process, and its real and imaginary parts are independent of each other.

However, after a cyclic prefix is introduced into $s(n)$, there is a correlation between each pair of samples with an interval of N; hence, the received signal $r(n)$ contains information on timing deviation d and frequency offset Δf_c, which makes it possible to jointly estimate d and Δf_c.

As shown in Fig. 6.24, 2N + L consecutive samples $r(n)$ are observed, where these samples contain a complete $N + L$-sample OFDM symbol. We define two sets:

$$\begin{aligned} I &= \{d, \wedge, d+L-1\} \\ I' &= \{d+N, \wedge, d+N+L-1\}, \end{aligned} \tag{6.39}$$

where set I is the cyclic prefix of the ith symbol, which contains the same elements as in I'. We take $2N + L$ observation points as vector r:

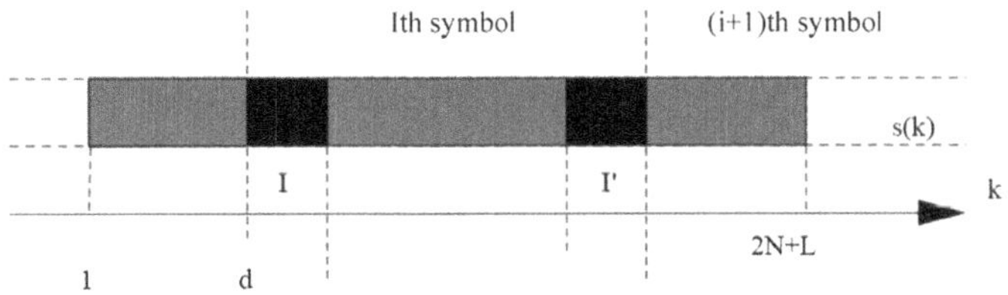

Fig. 6.24 OFDM symbol structure with a cyclic prefix

$$r = (r(1), \wedge, r(2\mathrm{N} + L))^T. \quad (6.40)$$

The elements in sets I' and I correspond to each other. Therefore, the following related characteristics exist:

$$E\left[r(n)r^*(n+m)\right] = \begin{cases} \sigma_s^2 + \sigma_n^2 & m = 0 \\ \sigma_s^2 e^{-j2\pi \Delta f_c} & m = N \\ 0 & otherwise \end{cases}, \quad (6.41)$$

where $\sigma_s^2 = E\left[\left|s(n)^2\right|\right] and\ \sigma_n^2 = E\left[\left|\eta(n)^2\right|\right]$ respectively represent the energy of the useful signal and additive white Gaussian noise, and d and Δf_c are, respectively, the symbol timing-synchronization position and carrier-frequency deviation to be estimated.

The log-likelihood function $\Lambda(d, \Delta f_c)$ with respect to d and Δf_c is defined as the logarithm of the joint probability-density function $f\left(\vec{r}\big/d, \varepsilon\right)$ of 2N + L observed samples in $\vec{r}$, given d and Δf_c. In 2N + L sample values, only the corresponding elements in set I and set I' are correlated. The other points can be regarded as independent of each other.

For simplicity, the conditional probability-density function f(r(k)) is directly expressed by $f(r(k)/d, \Delta f_c)$; thus, the logarithmic-likelihood function $\Lambda(d, \Delta f_c)$ can be written as

$$\begin{aligned} A(d, \Delta f_c) = \log f\left(\vec{r}\big/d, \Delta f_c\right) &= \log\left\{\prod_{n\in I} f[r(n), r(n+N)] \prod_{n\in I\cup I'} [f(r(n)]\right\} \\ &= \log\left\{\prod_{n\in I} \frac{f(r(n), r(n+N))}{f(r(n))f(r(n+N))} \prod_{n} [f(r(n)]\right\}. \end{aligned} \quad (6.42)$$

In the above formula, both one-dimensional and two-dimensional probability-density distribution functions were used. The product term $\prod_n f(r(n))$ is the product of all 2N + L points, so its value is independent of symbol starting point d. Assuming that the information source is an independent equipotential distribution, the real and imaginary parts of r(n) are independent of each other, so the value of $\prod_n f(r(n))$ is also independent of the frequency offset Δf_c. Maximum-likelihood estimation is required to estimate the values of d and Δf_c to maximize $\Lambda(d, \Delta f_c)$. Therefore, omitting $\prod_n f(r(n))$ does not affect the maximum-likelihood estimation of d and Δf_c. Therefore, the above formula can be simplified as

$$A(d, \Delta f_c) = \log \prod_{n \in I} \frac{f(r(n), r(n+N))}{f(r(n))f(r(n+N))} = \sum_{n=d}^{d+L-1} \log \frac{f(r(n), r(n+N))}{f(r(n))f(r(n+N))}. \quad (6.43)$$

Because $r(n)$ is a complex Gaussian random signal, $f(r(n), r(n+N))$ is a two-dimensional complex Gaussian partial probability-density function. Using the correlation characteristics in Eq. (6.40), we obtain

$$f(r(n), r(n+N)) = \frac{\exp\left(-\frac{|r(n)|^2 - 2\rho \mathrm{Re}\left\{e^{j2\pi \Delta f_c} r(n) r^*(n+N)\right\} + |r(n+N)|^2}{\left(\sigma_s^2+\sigma_n^2\right)\left(1-\rho^2\right)}\right)}{\pi^2 \left(\sigma_s^2 + \sigma_n^2\right)^2 \left(1-\rho^2\right)}, \quad (6.44)$$

where ρ is the amplitude of the $r(n)$ and $r(n+N)$ correlation coefficient:

$$\rho = \left| \frac{E\{r(k) r^*(k+N)\}}{\sqrt{E\left\{|r(k)|^2\right\} E\left\{|r(k+N)|^2\right\}}} \right| = \left| \frac{\sigma_s^2 e^{-j2\pi \Delta f_c}}{\sigma_s^2 + \sigma_n^2} \right| = \frac{\sigma_s^2}{\sigma_s^2 + \sigma_n^2} = \frac{SNR}{SNR+1}, \quad (6.45)$$

where SNR is the signal-to-noise ratio. When $f(r(n))$ is the probability-density function of a one-dimensional complex Gaussian distribution, the following is obtained:

$$f(r(n)) = \frac{1}{\pi\left(\sigma_s^2 + \sigma_n^2\right)} \exp\left(-\frac{|r(n)|^2}{\sigma_s^2 + \sigma_n^2}\right). \quad (6.46)$$

Substituting Eqs. (6.46) and (6.48) into Eq. (6.45), after some algebraic operations, we obtain

$$\Lambda(d, \Delta f_c) = c_1 + c_2 |\gamma(d)| \cos\left[2\pi \Delta f_c + \angle \gamma(d)\right] - \rho \phi(d), \quad (6.47)$$

where c_1 and c_2 are constants, and $c_2 > 0$, so it does not affect the maximum-likelihood decision.

The simplified Eq. (6.49) can be expressed as

$$\Lambda(d, \Delta f_c) = |\gamma(d)| \cos\left[2\pi \Delta f_c + \angle \gamma(d)\right] - \rho \Phi(d), \quad (6.48)$$

where

$$\gamma(m) = \sum_{n=m}^{m+L-1} r(n)r^*(n+N) \tag{6.49}$$

$$\Phi(m) = \frac{1}{2}\sum_{n=m}^{m+L-1} |r(n)|^2 + |r(n+N)|^2, \tag{6.50}$$

where ρ represents the amplitude of the correlation coefficient between r(n) and r(n + N) and $\angle\gamma(d)$ represents the angle of the complex number $\gamma(d)$. $\gamma(m)$ represents the sum of the correlation values between L consecutive pairs of N sample values. The first term of Eq. (6.50) is the weighted modulus of $\gamma(m)$, where the weight is determined by the frequency deviation, and the second term is the energy term, independent of the frequency deviation, which depends on the correlation coefficient ρ.

The maximum-likelihood algorithm needs to simultaneously estimate the symbol-timing synchronization position and carrier-frequency deviation. Therefore, the maximization process of the log-likelihood function above should be completed in two steps.

$$\max_{(d,\Delta f_c)} \Lambda(d, \Delta f_c) = \max_{(d)} \max_{(\Delta f_c)} \Lambda(d, \Delta f_c) = \max_{(d)} \Lambda\left[d, \Delta \hat{f}_{ML}(d)\right]. \tag{6.51}$$

In terms of the frequency deviation Δf_c, to maximize Eq. (6.53), the cosine term should be 1, and $2\pi\Delta f_c + \angle\gamma(d) = 2n\pi$, where n is an integer.

In this manner, the maximum-likelihood estimation of frequency deviation Δf_c is obtained:

$$\Delta \hat{f}_{ML}(d) = -\frac{1}{2\pi}\angle\gamma(d) + n. \tag{6.52}$$

In general, the carrier-frequency deviation should be within a small range; therefore, it is taken as $n = 0$:

$$\Delta \hat{f}_{ML}(d) = -\frac{1}{2\pi}\angle\gamma(d), \tag{6.53}$$

The maximum-likelihood function of timing deviation d is

$$\Lambda\left[d, \Delta \hat{f}_{ML}(d)\right] = |\gamma(d)| - \rho\Phi(d). \tag{6.54}$$

Equation (6.56) is only related to d, and the maximum-likelihood estimation $\hat{d}_{ML}$ of d can be obtained by maximizing $\Lambda\left[d, \Delta \hat{f}_{ML}(d)\right]$, and then substituting $\hat{d}_{ML}$ into $\Delta \hat{f}_{ML}(d) = -\frac{1}{2\pi}\angle\gamma(d)$. The maximum-likelihood estimation $\Delta \hat{f}_{ML}$ of the frequency

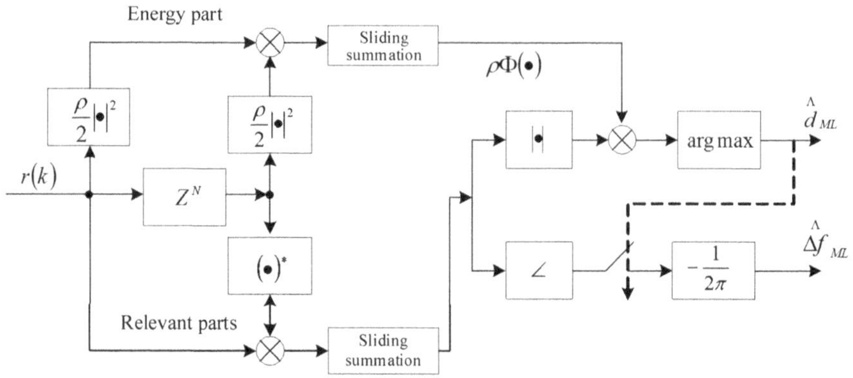

Fig. 6.25 Block diagram of maximum-likelihood estimation

deviation Δf_c can likewise be obtained. Therefore, the joint maximum-likelihood estimation of d and Δf_c becomes

$$\hat{d}_{ML} = \arg\max_{d}\left[|\gamma(d)| - \rho\Phi(d)\right] \tag{6.55}$$

$$\Delta \hat{f}_{ML}(d) = -\frac{1}{2\pi}\angle\gamma(d). \tag{6.56}$$

From Eqs. (6.57) and (6.58), it can be seen that timing estimation is the key to the maximum-likelihood joint estimation algorithm, and the accuracy of the timing estimation determines the accuracy of the frequency-offset estimation.

The block diagram of the maximum-likelihood symbol synchronization and carrier-frequency synchronization is shown in Fig. 6.25. Each time 2N + L sampling points of the received signal are stored in the cache, $\gamma(m)$ and $\Phi(m)$ are calculated according to Eqs. (6.31) and (6.32), respectively, and finally estimated according to Eqs. (6.37) and (6.38), respectively.

Two factors can affect the above judgment results: the number of cyclic-prefix samples and the correlation coefficient ρ determined by the SNR. The former receiver is known, whereas the latter can be fixed. $\gamma(m)$ can provide estimates of symbol timing d and Δf_c. When the modulus of $\gamma(m)$ is weighted and added to the energy term $\Phi(m)$, the peak value will appear at $\hat{d}_{ML}$; at this time, the phase of $r\left(\hat{d}_{ML}\right)$ is proportional to $\Delta \hat{f}_{ML}$.

Figure 6.25 shows the principle block diagram of maximum-likelihood estimation. The signal schematic diagram for jointly determining the symbol timing and carrier-frequency deviation in the AWGN channel includes N = 1992 subcarriers, protection-interval length L = 56, carrier-frequency deviation $\Delta f = 0.25$, and SNR = 15 dB. Figure 6.26a shows a graph for determining the symbol-timing position $\hat{d}_{ML}$ according to Eq. (6.57), and Fig. 6.26b shows that at this determined position,

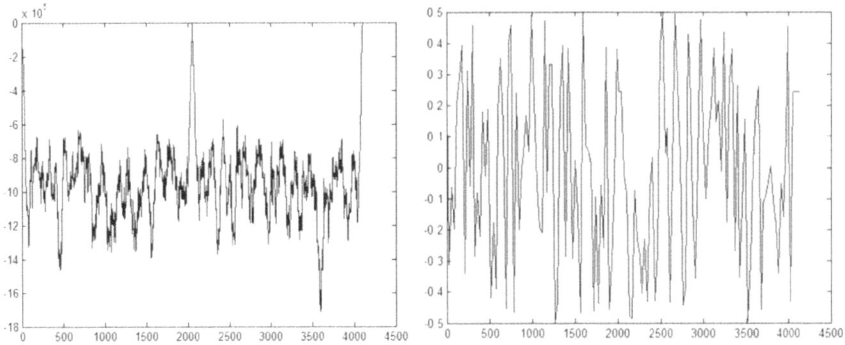

(a) Maximum-likelihood symbol-timing position estimation. (b) Maximum-likelihood carrier-frequency offset estimation.

Fig. 6.26 Graph of the joint determination of symbol timing and carrier-frequency deviation using maximum-likelihood estimation [19–21]

the maximum-likelihood estimation $\Delta \hat{f}_{ML}$ of the carrier-frequency deviation can be obtained according to Eq. (6.58).

It can be observed from Fig. 6.26 that the symbol-timing synchronization position can be obtained at the maximum value of the $\Lambda\left[d, \Delta \hat{f}_{ML}(d)\right]$ curve of the timing-decision function. The correct estimated value of the frequency offset can be obtained, based on this correct symbol timing point.

From Fig. 6.27, we can see that the deviation between the estimated frequency offset and the actual frequency offset is very small, which shows that the maximum-likelihood joint estimation algorithm has an ideal estimation performance for a frequency offset in a Gaussian channel.

Timing estimation is key to the maximum-likelihood estimation algorithm. The number of samples in the cyclic prefix affects the judgment results. Figure 6.28 shows $L = 56$ and 512. From the figure, we can see that the cyclic prefix is too small, and the side lobe has too much impact on the main peak value at the starting point of the symbol timing. However, the length of the cyclic prefix should not be too large, which will affect the signal-transmission rate, and the floor effect will occur when the cyclic-prefix length is too large to a certain limit.

As can be seen from Fig. 6.29, the position of the symbol timing has different degrees of error with the starting position of each OFDM symbol. This is because the algorithm uses the correlation between the cyclic prefix and the copied part in the OFDM symbol. In the AWGN channel, there is a good correlation between the data pairs in sets I and I'. After the multipath channel transmission, the correlation between the data pairs is destroyed; thus, the performance of the ML algorithm is affected, and its performance in the fading channel is degraded.

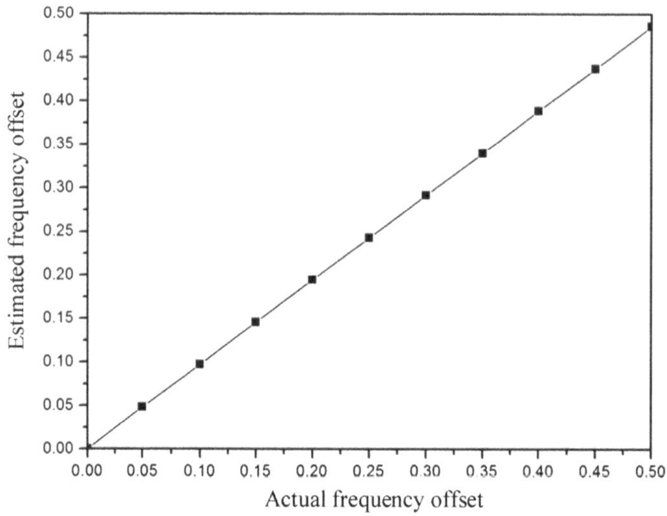

Fig. 6.27 Comparison of the actual frequency offset and estimated frequency offset of the maximum-likelihood algorithm [19–21]

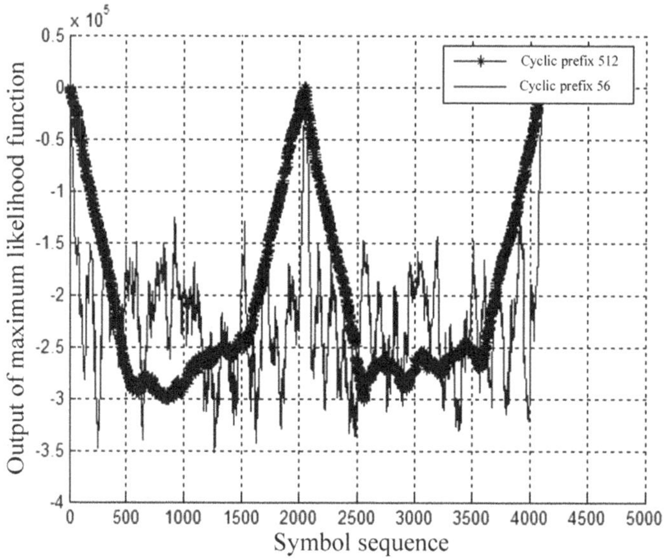

Fig. 6.28 Timing-position estimation of symbols with different cyclic prefixes [19–21]

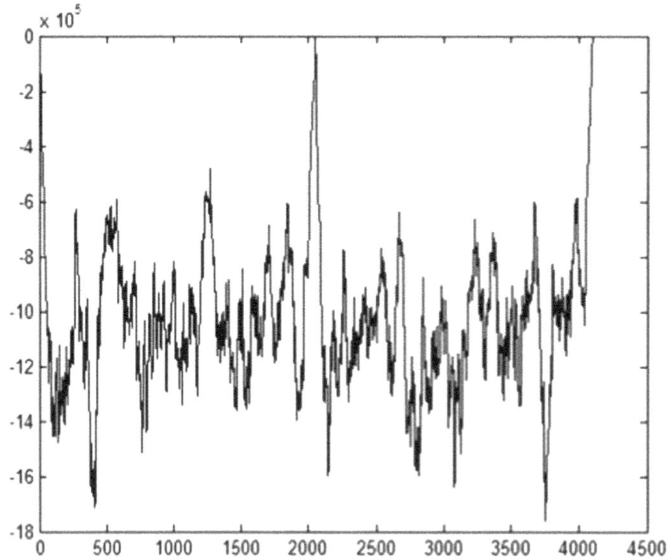

Fig. 6.29 Timing-position estimation of symbols in multipath channels [19–21]

The maximum-likelihood (ML) algorithm has low complexity and can simultaneously realize symbol-timing synchronization and carrier-frequency offset estimation; however, ML also has some disadvantages.

(1) In essence, the ML algorithm uses the correlation of the cyclic prefix (CP) to estimate the timing and frequency deviation. The calculation accuracy also depends directly on the length of the CP. However, the CP is a simple copy of the transmission signal and does not exhibit sharp autocorrelation characteristics. Therefore, the autocorrelation values at timing point d and several points on both sides are relatively large, as calculated according to the ML algorithm. This results in rough and inaccurate timing when noise is present. In addition, from Eq. (6.58), it can be seen that the frequency-offset estimation of the ML algorithm depends on the time offset. If the timing is misjudged, it is difficult to accurately estimate the frequency offset.
(2) A significant problem of the ML algorithm is that it can only be used in an additive white Gaussian noise (AWGN) channel. The function of the CP is to overcome the influence of time dispersion caused by the multipath effect. The estimation of the time and frequency offsets in the ML algorithm assumes an AWGN channel as the premise. In a multipath time-dispersive channel, the signal in the CP is interfered with by the previous symbol signal, which significantly affects the performance of the algorithm. Owing to the existence of intersymbol interference, the ML algorithm is not practical for multipath channels.
(3) Because we cannot determine n in the derivation process and consider $n = 0$, the frequency-offset acquisition range is too small. This is only half of the

subcarrier interval. However, it cannot be used directly in an actual system. This can only be used to estimate the fractional frequency offset. It must cooperate with other integer frequency-offset estimation algorithms to complete the entire frequency-offset estimation.

In summary, from the perspective of accuracy and estimation range, the maximum-likelihood algorithm using a cyclic prefix has little computation, a simple implementation, and can be used for frequency and symbol timing synchronization. However, the frequency-estimation range of the algorithm is too small, and the timing estimation is rough; therefore, it is generally difficult to use directly in actual systems.

6.4.3 ML Algorithm Improvement

6.4.3.1 ML Algorithm for Continuous Symbols

In traditional ML estimation, only one OFDM symbol or $n + 1$ samples are sent; in other words, the resulting estimator only generates the maximum-likelihood estimation of the symbol timing and frequency deviation of each symbol, according to the information carried by the current symbol.

Garcia [25] proposed a continuous symbol-averaging algorithm based on an ML estimation algorithm. Its main idea is to calculate the correlation value of each symbol at the corresponding position; then, average the M symbols, and finally, determine the symbol-synchronization position according to the maximum value of the modular sum of its part and the imaginary part. This algorithm uses the information of multiple OFDM symbols to generate a stable symbol clock and oscillator frequency, and improves the performance of the estimator.

Suppose that m consecutive OFDM symbols are considered, as shown in Fig. 6.30.

The received signal model can be defined as

$$r(k) = \tilde{s}(k - d)\exp\left(j2\pi\,\Delta fk\big/N\right) + n(k) \qquad -\infty < k < \infty, \tag{6.57}$$

where d, Δf, and $n(k)$ are the same as described above. It is assumed that the observation period contains M complete OFDM symbols. I_m and $I_m^{'}$ are the cyclic prefix and its copy of the mth ($m = 0, 1,\ldots, M - 1$) OFDM symbol, respectively:

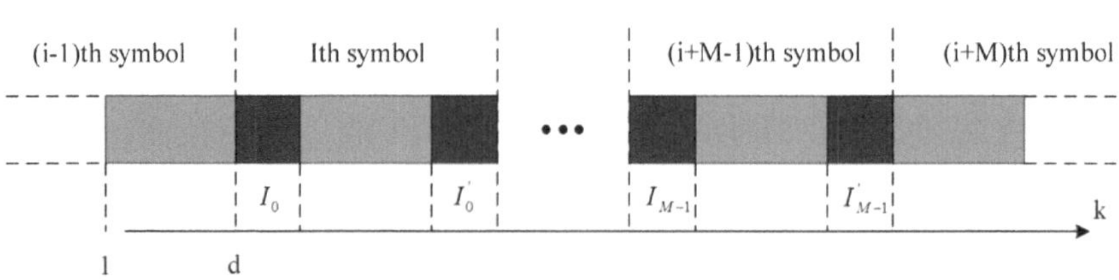

Fig. 6.30 *M* consecutive OFDM symbols [19–21]

$$\begin{aligned} I_m &= \{m(N+L)+d, \cdots, m(N+L)+d+L-1\} \\ I'_m &= \{m(N+L)+d+N, \cdots, m(N+L)+d+N+L-1\}. \end{aligned} \tag{6.58}$$

We define two sets:

$$I \equiv \prod_{m=0}^{M-1} I_m \qquad I' \equiv \prod_{m=0}^{M-1} I'_m \tag{6.59}$$

$$\prod_{k \in I} \frac{f(r(k), r(k+N))}{f(r(k))f(r(k+N))} = \prod_{m=0}^{M-1} \left(\prod_{k \in I_m} \frac{f(r(k), r(k+N))}{f(r(k))f(r(k+N))} \right). \tag{6.60}$$

Therefore, the log-likelihood function can be obtained as

$$\begin{aligned} \Lambda(d, \Delta f) = & \sum_{m=0}^{M-1} \sum_{k=d}^{d+L-1} \mathrm{Re}\left\{e^{-j2\pi \Delta f} r^*(k+m(N+L)) r(k+m(N+L)+n)\right\} \\ & - \frac{\rho}{2} \sum_{m=0}^{M-1} \sum_{k=d}^{d+L-1} \left\{|r(k+m(N+L))|^2 + |r(k+m(N+L)+N)|^2\right\} \\ = & \ |\Gamma_N(d)| \cos\{2\pi \Delta f + \angle\Gamma_N(d)\} + \Gamma_0(d). \end{aligned} \tag{6.61}$$

Of which,

$$\Gamma_N(d) = \sum_{m=0}^{M-1} \gamma_N(d+m(N+L)) \tag{6.62}$$

$$\Gamma_0(d) = \sum_{m=0}^{M-1} \gamma_0(d+m(N+L)). \tag{6.63}$$

The maximum-likelihood estimate is

$$\hat{d}_{serial} = \arg \max_{d} \{|\Gamma_N(d)| + \Gamma_0(d)\} \tag{6.64}$$

$$\Delta \hat{f}_{serial} = -\frac{1}{2\pi} \angle\Gamma_N\left(\hat{d}_{serial}\right). \tag{6.65}$$

Figure 6.31 shows the block diagram of this estimator. As can be seen from Fig. 6.32, using 12 symbols to average the symbol-timing synchronization can reduce the length of the cyclic prefix, achieve the same precise timing synchronization effect as using a cyclic prefix of 512, and improve the signal transmission and spectral efficiencies; however, the algorithm has high computational complexity.

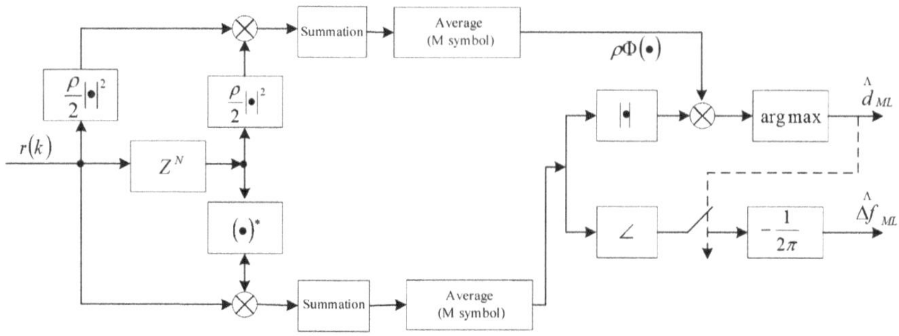

Fig. 6.31 Block diagram of carrier and symbol timing estimation for M consecutive OFDM symbols

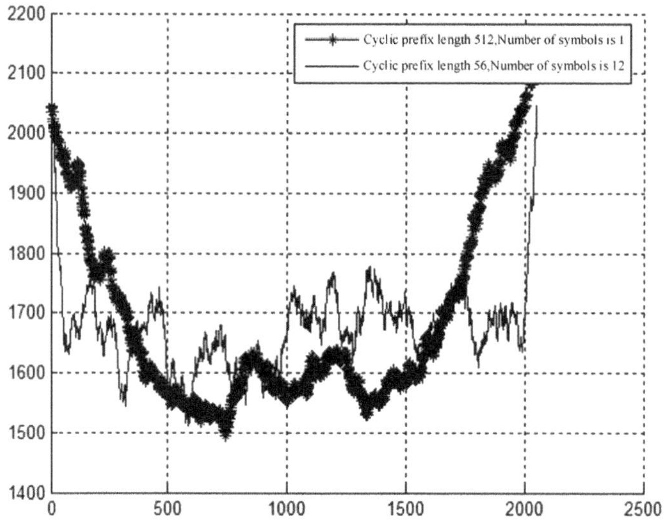

Fig. 6.32 Symbol-timing position estimation of the maximum-likelihood estimation algorithm with different numbers of symbols and cyclic prefixes [19–21]

6.4.3.2 Partial-CP ML Algorithm

Based on the traditional ML algorithm, an estimation algorithm based on a partial CP is adopted [25]. This algorithm discards heavily polluted cyclic prefixes, and adopts a correlation algorithm. Its performance in a multipath channel is better than that of the traditional ML algorithm. The timing estimation and frequency offset are changed to:

$$\hat{d}_{ML} = \arg\max\left[|\gamma(d)| - \rho\phi(d)\right] \tag{6.66}$$

$$\Delta\hat{f}_{ML} = -\frac{1}{2\pi}\angle\gamma\left(\hat{d}_{ML}\right). \tag{6.67}$$

Of which,

$$\gamma(m) = \sum_{n=m}^{m+L'-1} r(n)r^*(n+N) \tag{6.68}$$

$$\phi(m) = \frac{1}{2}\sum_{n=m}^{m+L'-1} |r(n)|^2 + \left|r(n+N)^2\right|. \tag{6.69}$$

It can be seen that the partial-CP correlation algorithm selects a correlation window with a fixed length to perform the partial CP correlation. A continuous peak is obtained during each CP period, and the midpoint of each continuous peak is selected as the starting position of the FFT window to synchronize the symbols.

The partial-CP correlation algorithm has two characteristics: It uses only part of the cyclic prefix for operation, and the autocorrelation operation is selected as the algorithm. It is evident that this algorithm has two advantages. First, the number of related points is reduced, and thus, the amount of computation is reduced. Second, the midpoint of the continuous peak of the correlation value is taken as the starting position of the FFT window, and the step of searching for the maximum value in the ML algorithm is omitted.

6.4.3.3 Frequency-Offset Estimation Algorithm Based on a Training Sequence

Frequency-offset estimation is typically divided into two steps: integer and fractional subcarrier-interval frequency-offset estimation. In the time domain, the fractional frequency offset is estimated using the partial-CP correlation algorithm, and in the frequency domain, the integer frequency offset is estimated using the training sequence of the third improved FH conjugate-symmetry algorithm.

An integer frequency offset will cause the information symbols output after the FFT to shift circularly on the subchannel; that is, the number of integers of the carrier interval will be shifted by the number of bits. It does not destroy the orthogonality of each subcarrier but leads to a complete error of the entire demodulation result. The bit-error rate of the system is approximately 0.5.

The basic concept of integer frequency offset estimation is to estimate the normalized integer frequency offset at the receiver, according to the conjugate symmetry between the first and second parts of the training sequence. The frequency-offset decision function is

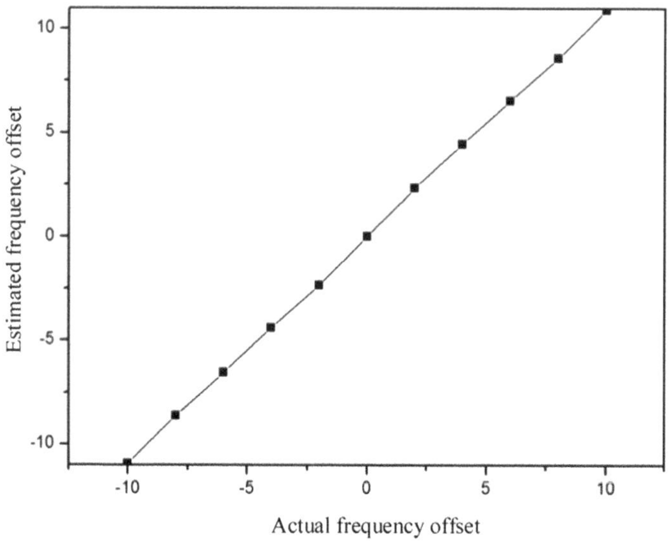

Fig. 6.33 Comparison of the actual frequency offset and estimated frequency offset [19–21]

$$\varepsilon_2 = \frac{angle\left[\sum_{k=0}^{L/4-1} y(k)y\left(\frac{L}{2}-k\right)\right] \times N}{2\pi \times \frac{N}{32}}. \tag{6.70}$$

It provides the frequency domain data after correcting for the decimal frequency offset and FFT.

Figure 6.33 shows a comparison of the actual frequency offset and estimated frequency offset of the algorithm. Figure 6.34 shows the mean squared error of integer multiple-frequency offset estimation when the attenuation coefficient of the improved frequency-offset algorithm is [0 0 0.2 0.3], and the normalized frequency deviation is 2 in the multipath fading channel.

From Fig. 6.33, we can see that the frequency-offset estimation range of the improved frequency-offset scheme is −10 to 10, and the error between the estimated frequency offset and the actual frequency offset is very small. It can be seen from Fig. 6.34 that the mean squared error of the integer frequency-offset estimation is very small. The algorithm performs well in estimating integer multiple-frequency offsets.

When the signal-to-noise ratio increases, the system performance does not improve significantly. Therefore, the algorithm is suitable for a synchronization system with a low signal-to-noise ratio in optical wireless communication.

Table 6.1 shows a comparison of the frequency-offset estimation ranges of several schemes, after normalizing the subcarrier frequency interval for the same training-sequence length. It can be observed that the proposed frequency-offset estimation range is greater than that of the Schmidl and Minn schemes.

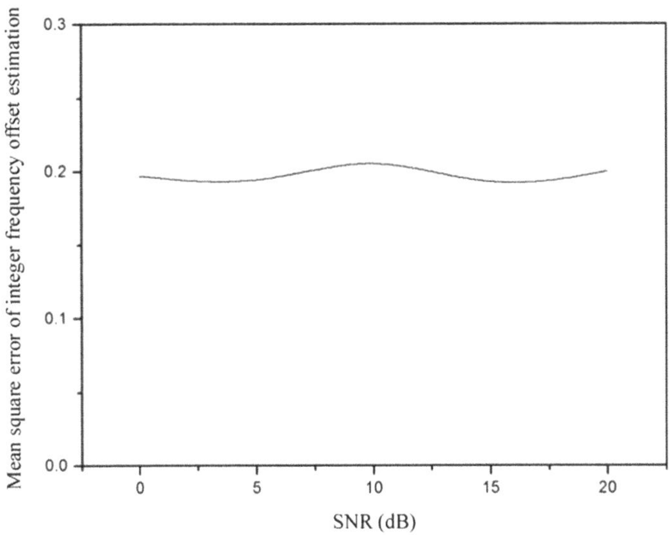

Fig. 6.34 Mean squared error of the integer frequency offset estimation [19–21]

Table 6.1 Comparison of normalized frequency-offset estimation ranges of three schemes

Program	Schmidl	Minn	Improved frequency offset
Frequency-offset estimation range	$\lvert\varepsilon\rvert \le 1$	$\lvert\varepsilon\rvert \le 2$	$\lvert\varepsilon\rvert \le 10$

Chapter Summary

Three symbol-timing synchronization algorithms based on the Frank and Heimiller (FH) sequence were discussed: the symmetric FH-sequence symbol timing synchronization algorithm and conjugate-symmetric FH-sequence symbol timing synchronization algorithms 1 and 2. These three algorithms not only overcame the problem of an autocorrelation platform area in the time-scaling curve of the Schmidl and Cox algorithm but also solved the influence of a slow peak-decline speed and side lobes of the scaling curve on both sides of the timing position of the Minn algorithm.

The time-scaling curve of the third improved conjugate-symmetric FH algorithm only has a peak with impulse response characteristics at the correct timing point, while the corresponding values at other times are very small. Therefore, by searching the maximum value in a certain time window, we can find the exact starting point of the FFT window.

A training sequence with conjugate symmetry was constructed using the FH sequence, which strongly inhibited the frequency offset. A frequency-offset estimation algorithm based on the third improved FH conjugate-symmetry sequence was discussed. The frequency-offset range estimated by the algorithm was related to the length of the training sequence. If the length of the training sequence was properly designed, a relatively large frequency-offset estimation range could be achieved.

References

1. Roberts RD (1996) Qualitative analysis of the impact of clock timing error and/or frequency offsets on an OFDM waveform demodulator 80–85
2. Shelswell P (1995) The COFDM modulation system: the heart of digital audio broadcasting. Electron Commun Eng J 7(3):127–136
3. Gudmundson M, Anderson PO (1996) Adjacent channel interference in an OFDM system. In: Proceedings of vehicular technology conference-VTC, vol 2, pp 918–922
4. Yin CC, Luo T (2004) Multicarrier broadband wireless communication technology. Beijing University of Posts and Telecommunications Press, Beijing, pp 137–140
5. Moose PH (1994) A Technique for orthogonal frequency division multiplexing frequency offset correction. IEEE Trans Commun 42(10):2908–2914
6. Yang B, Letaief K, Cheng RS et al (2000) Timing recovery for OFDM transmission. IEEE J Sel Areas Commun 18(11):2278–2291
7. Fan CX (2002) Communication principle. National Defense Industry Press, Beijing
8. Du MS, Ji HF (2006) The analysis of several training sequences on the performance of correlation. Inf Secur Commun Priv 2:93–94
9. Daffara F, Adami O (1995) A new frequency detector for orthogonal multicarrier transmission techniques. In: IEEE vehicular technology conference, no 2, pp 804–809
10. Neustein SM, Kahn P, Krellenstein DJ et al (1997) Frame synchronization of OFDM systems in frequency selective fading channels. In: IEEE vehicular technology conference, no 3, pp 1807–1811
11. Classen F, Meyr H (1994) Frequency synchronization algorithms for OFDM systems suitable for communication over frequency selective fading channel. In: IEEE vehicular technology conference, no 3, pp 1655–1659
12. Schmidl TM, Cox DC (1997) Robust frequency and timing synchronization for OFDM. IEEE Trans Commun 45(12):1613–1621
13. Minn H, Zeng M, Bhargava VK (2000) On timing offset estimation for OFDM systems. IEEE Commun Lett 4(7):242–244
14. Park B, Cheon H, Kang C et al (2003) A novel timing estimation method for OFDM systems. IEEE Commun Lett 7(5):239–241
15. Sandell M, van de Beek J, Börjesson PO (1998) Timing and frequency synchronzation in OFDM system using the cyclic prefix. In: Proceedings of the international symposium on synchronization, pp 16–19
16. Tureli U, Liu H, Zoltowski MDOFDM (2000) Blind carrier offset estimation: ESPRIT. IEEE Trans Commun 48(9):1459–1461
17. Tureli U, Liu H, Zoltowski MD (1997) A high efficiency carrier estimator for OFDM communications. IEEE Commun Lett (2):104–106
18. Beek JVD, Sandell M, Isaksson M et al (1995) Low-complex frame synchronization in OFDM systems. IEEE
19. Classen F, Meyr H, Classen F et al (1994) Synchronization algorithms for an OFDM system for mobile communication. ITG Fachbericht 18(1):245–257
20. Wu R (2009) Studies on synchronization technique in FSO-OFDM. Xi'an University of technology, Xi'an
21. Ke XZ, Rui WU, Zhao L (2009) FSO-OFDM symbol synchronization based on FH sequence. Semicond Optoelectron 30(4):586–589
22. Ke XZ, Luo WL, Zhao L et al (2009) An improved timing and frequency synchronization scheme for FSO-OFDM 4(2):44–46

23. Hsieh MH, Wei CH (1998) Channel estimation for OFDM systems based on comb-type pilot arrangement in frequency selective fading channels. IEEE Trans Consum Electron 44(1):217–225
24. Fernandez-Getino Garcia MJ, Edfors O, Paez-Borrallo JM (2001) Frequency offset correction for coherent OFDM in wireless systems. IEEE Trans Consum Electron 47(1):187–193
25. Li XC (2007) Research on synchronization in OFDM transmission system. Guizhou University, Guiyang

Chapter 7
Peak-to-Average Power Ratio in OWC-OFDM Modulation Systems

The main disadvantage of an orthogonal frequency-division multiplexing (OFDM) system is that the peak-to-average power ratio (PAPRj) is high. If the PAPR is too large, the OFDM signal is prone to nonlinear distortion when it passes through the amplifier, which destroys the orthogonality between the subcarriers and increases the bit-error rate. This chapter studies the high-PAPR problem in optical wireless communication (OWC)-OFDM systems.

7.1 Definition and Statistical Properties of the Peak-to-Average Power Ratio

7.1.1 Definition of the Peak-to-Average Power Ratio

In wireless communication, the frequency of the data that must be transmitted is usually low. If it is transmitted according to the data frequency itself, it is not conducive to reception and synchronization. Using carrier transmission, a data signal can be loaded and modulated onto the signal of a high-frequency carrier. The receiver receives the data signal according to the carrier frequency and extracts the required data signal through the difference in the wave amplitude.

Similar to other multicarrier modulation (MCM) systems, OFDM systems face the problem of an excessive peak-to-average power ratio, which is the ratio of the maximum peak power to the average power in a period of time. Because the OFDM symbol is composed of multiple independently modulated subcarrier signals, when the phase of each subcarrier is the same or similar, the superimposed signal is modulated by the same initial phase signal, resulting in a large instantaneous power peak. This leads to a higher PAPR.

X. Ke, *Principles and Applications of Optical Wireless Orthogonal Frequency-Division Multiplexing*, Optical Wireless Communication Theory and Technology,
https://doi.org/10.1007/978-981-97-7973-4_7

Because the dynamic range of general power amplifiers is limited, a signal with a larger PAPR easily enters the nonlinear region of the power amplifier, resulting in nonlinear signal distortion, obvious spectrum-spread interference, and in-band signal distortion. This severely degrades the performance of the entire system. High PAPRs have become a major technical bottleneck for OFDM.

Considering the ideal situation, operations such as pilot insertion and channel coding are not performed here. The input data are first converted into a serial-to-parallel signal and then mapped into frequency-domain symbols using four-way quadrature amplitude modulation (4-QAM). The frequency-domain signal is represented as

$$X = [X_0, X_1 \cdots X_K \cdots X_N], \tag{7.1}$$

where N is the number of subcarriers, and the frequency-domain information of the Kth subcarrier is set to $X_K = a_K + jb_K$. Then, after an inverse fast Fourier transform (IFFT), the output OFDM time-domain signal is expressed as

$$x(n) = \sum_{k=0}^{N/2-1} 2\left(a_k \cos\frac{2\pi kn}{N} - b_k \sin\frac{2\pi kn}{N}\right), \quad n = 0,\ 1 \ \ldots \ N-1. \tag{7.2}$$

In an OWC system, because the OFDM signal is a positive real signal, its peak-to-average ratio can be defined as [1]

$$PAPR = 10\lg\frac{\max\left[|x(n)|^2\right]}{E\left[|x(n)|^2\right]}, \tag{7.3}$$

where E[·] represents the mathematical expectation. Assuming that the channel is in a very special case at a certain moment, that is, all N subcarriers are in the same phase, then,

$$PAPR(dB) = 10\log_{10}N. \tag{7.4}$$

When the N subchannels are summed with the same phase, the obtained signal peak power will be N times the average power; thus, the peak-to-average power ratio of the baseband signal is $PAPR = 10\log_{10}N$. For example, in the case of $N = 256$, the system's peak power will be $PAPR = 24\ dB$, which is a very extreme case. The peak-to-average power ratio in an OFDM system usually does not reach this value.

7.1.2 Statistical Properties of the Peak-to-Average Power Ratio

In general, the peak value of a signal $x(n)$ is the maximum value of its envelope $|x(n)|$. However, the probability of a peak value is usually very small, as shown in Fig. 7.1; thus, it is not meaningful to use $|x(n)|$ to define the peak value of the signal.

The PAPR performance of an OFDM system is generally described by the statistical characteristics of the ratio of the peak power to the average power of the OFDM signal. Two main functions can effectively measure the PAPR distribution in OFDM systems: the cumulative distribution function (CDF) and the complementary cumulative distribution function (CCDF) [2].

The CDF is defined as the cumulative function in which the amplitude of the signal is less than a certain peak-to-average power ratio. It can also be defined as the probability that the peak-to-average power ratio is less than a certain threshold value, while the CCDF is defined as the probability that the signal's peak-to-average power ratio is greater than a certain threshold value. The two have the following relationship: CCDF = 1 − CDF. The probability distribution that the peak value of the OWC-OFDM system is greater than a certain threshold value z is [3]

$$P(power > z) = 1 - P(power \leq z). \tag{7.5}$$

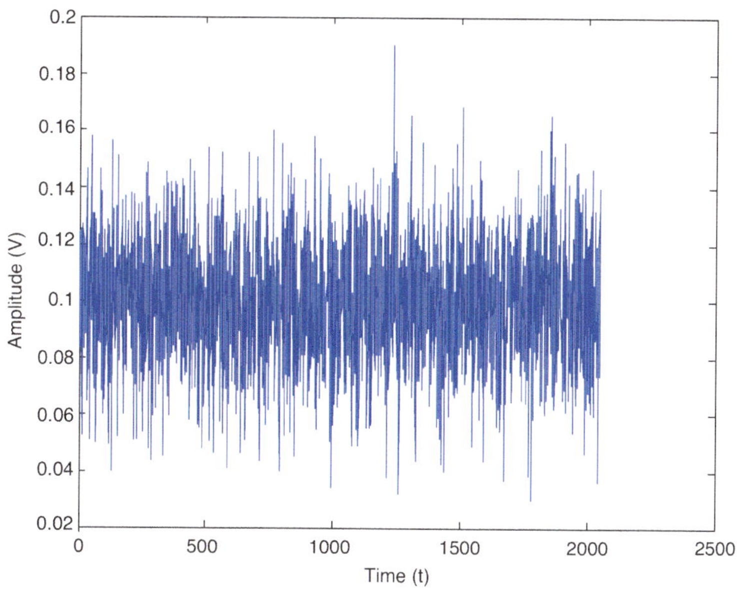

Fig. 7.1 Amplitude of an OFDM signal

For an OFDM system containing N subcarriers, it can be known from the central-limit theorem that for a large number of subcarriers N, the real and imaginary samples of $x(n)$ obey a Gaussian distribution with a mean 0 and variance 0.5. Therefore, the amplitude of the OFDM symbol follows a Rayleigh distribution, and the power obeys a centered χ^2 distribution with two degrees of freedom (the mean is 0 and the variance is 1). Hence, the CDF can be obtained as

$$P(power \le z) = F_{power}(z) = \int_0^z \exp(-y)dy = 1 - \exp(-z). \tag{7.6}$$

In fact, the PAPR performance is usually measured by the CCDF; that is, the probability that the PAPR exceeds a certain threshold $PAPR_0$ is calculated.

When Nyquist sampling is used on the OFDM signal, the sampling points are uncorrelated; therefore, the probability distribution of the peak-to-average power ratio of the OFDM signal exceeding the threshold z is

$$CCDF(PAPR) = P(PAPR > PAPR_0) = 1 - \left(1 - e^{-PAPR_0}\right)^N. \tag{7.7}$$

According to Eq. (7.7), as the number of subcarriers N increases, the probability of an occurrence higher than the threshold value z increases; that is, the probability of a high PAPR also increases.

7.1.3 Causes and Consequences of Peak-to-Average Power Ratios

The peak-to-average power ratio directly affects the operating cost and efficiency of a system. In an OWC-OFDM system, the main reason for a high PAPR is that the OFDM signal is superimposed by N orthogonal subcarriers in the time domain. When the number of subcarriers N reaches a certain level, it can be determined from the central-limit theorem that the OFDM symbol waveform will be a Gaussian random process, and its envelope is highly unstable.

When the N subcarriers are accumulated at the peak point, the maximum peak value is generated; thus, causing a high PAPR. This phenomenon causes nonlinear distortion when the OFDM signal passes through a semiconductor laser. The distortion destroys the orthogonality between subcarriers, deteriorates the transmission performance, and increases the system bit-error rate.

The laser is expected to work in the linear region as much as possible to ensure the transmission quality of the signal. However, owing to the threshold characteristics of the laser, a loading area that is too low or too high results in nonlinear distortion. A large PAPR is a problem that OWC-OFDM systems must face, and measures should be taken to reduce the probability of a signal with a high peak power, thereby avoiding nonlinear distortion when the signal passes through the laser.

7.2 Methods to Reduce PAPR

7.2.1 Limiting Technology

Slicing technology is the simplest and most direct way to reduce the PAPR in OFDM systems. The basic idea is that the electrical signal receives nonlinear processing before it is sent to the optoelectronic modulator; that is, pre-distortion processing is performed on the large peak signal so it does not exceed the dynamic range of the optoelectronic modulator. This prevents the system from having a large PAPR. The most commonly used clipping-like techniques include clipping filtering, peak windowing, and peak cancellation.

7.2.1.1 Clipping Filters

Figure 7.2 shows a graph of the principle of the limiting method. The abscissa is the amplitude of the signal before clipping and the ordinate is the amplitude of the signal after clipping. As seen in Fig. 7.2, the amplitude of the signal after limiting was controlled within $\pm A_{\max}$. When clipping techniques are used to reduce the PAPR of a signal, the signal is filtered out when its amplitude exceeds the set threshold.

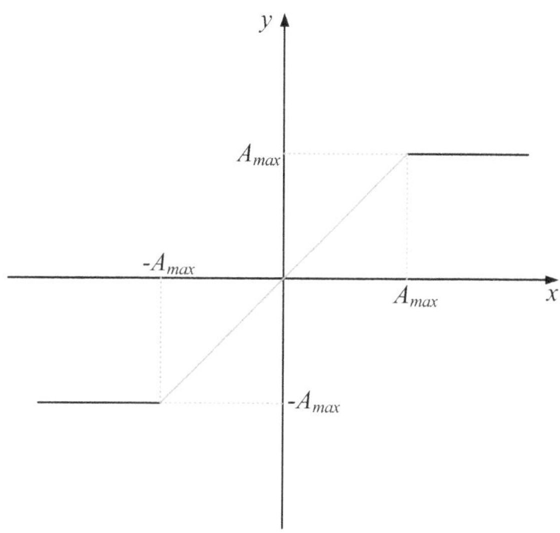

Fig. 7.2 Graph of the limiting principle

In OFDM signals, the probability of a large peak is very low. Although clipping is a very direct and effective method to reduce the peak-to-average power ratio, the clipping process is nonlinear. It can be regarded as allowing the original OFDM signal to pass through a rectangular-window function, where the amplitude of the window function is set as the maximum amplitude of the desired signal. Introducing this window function inevitably leads to serious in-band interference and out-of-band noise, which increases the overall OWC-OFDM system bit-error rate.

7.2.1.2 Peak Windowing Technique

Slicing uses a nonlinear operation to reduce the signal amplitude, resulting in in-band interference and out-of-band radiation. The clipping process can be regarded as the process of multiplying the OFDM signal by a rectangular function. When the amplitude of the OFDM signal is less than the given threshold, the value of the rectangular function is set to 1, and when the amplitude of the OFDM signal is greater than the given threshold, the position of the value of the rectangular function is <1.

The spectrum of the OFDM signal after limiting is the convolution of its spectrum before limiting and the window-function spectrum. Its out-of-band spectrum characteristics are mainly determined by the spectrum of the rectangular-window function. In the time domain, the bandwidth of the window function should not be too wide. If it is, more signal samples will be affected, which increases the bit-error rate of the system and reduces the system performance.

7.2.2 Coding Technology

Coding technology [4] selects different types of codes in the OFDM transmission system, transmits the code group with better PAPR performance as an OFDM symbol, and discards other code combinations with poor performance. This technology is a linear operation of the OFDM signal, and theoretically, it can perfectly solve the high-PAPR problem.

However, the coding methods selected by such methods are generally complex, and when used to process OFDM signals, the coding length will increase significantly with the continuous increase in subcarriers, resulting in a large number of useless codes. When the number of system subcarriers is relatively large, the coding technology selects a set of useful codes for transmission, while other useless code groups are discarded. This reduces the resource utilization and transmission efficiency of the OFDM system. Commonly used coding methods include block coding, Golay complementary sequences (GCS), and Reed–Muller codes [5].

Golay complementary sequences are a class of binary sequences with complementary properties, as proposed by Golay in 1961 [6]. Sequences $x = (x_0, x_1, \ldots, x_{N-1})$ and $y = (y_0, y_1, \ldots, y_{N-1})$ are pairs of sequences of length N, and the shift autocorrelation function is defined as

$$C_x(i) = \sum_{k=0}^{N-1} x_k x_{k+i}. \tag{7.8}$$

If satisfied,

$$C_x(i) + C_y(i) = \begin{cases} 2N, i = 0 \\ 0, i \neq 0 \end{cases}. \tag{7.9}$$

Then, sequence x and sequence y are Golay complementary sequences that form a pair. Taking a discrete Fourier transform (DFT) of both sides of the above equation, we obtain

$$|X(f)|^2 + |Y(f)|^2 = 2N, \tag{7.10}$$

where $|X(f)|^2$ represents the power spectrum of x; that is, the DFT is equal to the autocorrelation function:

$$X(f) = \sum_{k=0}^{N-1} x_n e^{-f2\pi nft}, \tag{7.11}$$

where T is the sampling interval of sequence x. From the spectrum conditions, it can be seen that the upper bound of its power spectrum is $2N$; that is,

$$|X(f)|^2 \leq 2N. \tag{7.12}$$

Assuming that the power of sequence x is 1, the average power of $X(f)$ in Eq. (7.11) is N. Therefore, the PAPR value of OFDM obtained using the Golay complementary sequence satisfies

$$PAPR \leq 2N/N = 2 = 3\,dB. \tag{7.13}$$

However, when the number of subcarriers is large, to ensure that the PAPR of the system can be reduced simultaneously under a certain coding speed, Golay complementary sequence coding can be used to group the subcarriers, which will decrease the error-detection and correction capabilities.

In summary, using a coding method to reduce the PAPR of an OFDM system has the advantages of being relatively simple, stable, easy to implement, and obtains good PAPR-suppression performance; however, it also introduces a large number of useless code groups.

7.2.3 Probabilistic Techniques

Probabilistic technology focuses on reducing the probability of a high PAPR in OFDM signals. The basic concept is to use multiple sequences to represent the same transmitted information. Under a given PAPR threshold, a group that is lower than the given threshold is selected. It has the smallest PAPR signal for the subsequent transmission, thereby reducing the probability of a high-peak-power signal appearing. Figure 7.3 shows a schematic diagram of a probabilistic technique for reducing the peak-to-average power ratio.

Probabilistic technology involves performing correlation mapping on the input data through a linear transformation: $Y_n = A_n \cdot X_n + B_n$ $(0 \le n \le N - 1)$. Y_n is the input vector of the signal after an N-point IFFT and X_n is the original frequency-domain data.

The key to this method is finding the optimal N-point vectors A and B so that the peak value of the time-domain OFDM symbol to be transmitted ($y = IFFT(Y)$) is small. If the CCDF of the original signal is $P\{PAPR > PAPR_0\}$, where $PAPR_0$ is the PAPR threshold, this method generates M optional signals through M parallel inverse discrete Fourier operations, which carry the same information. It then selects the signal with the smallest PAPR that is lower than the given threshold for transmission.

In these M signals, the CCDF of the optimal signal is $[P\{PAPR > PAPR_0\}]^M$, so the probability that PAPR is greater than a given threshold will be as small as possible. In addition, it is necessary to correctly transmit the sideband information to ensure that the original signal can be correctly demodulated at the receiving end.

Probabilistic technology can effectively reduce the probability of a high peak value of the OFDM signal, has better PAPR characteristics, and is suitable for any constellation-modulation type and any number of subcarriers; thus, it reduces the PAPR of an OFDM system and is considered medium PAPR technology.

These technologies include selective mapping (SLM), partial transmit sequences (PTSs), pulse shaping (PS), tone injection (TI), and tone rejection (TR). Among them, SLM and PTS focus on choosing an appropriate vector, A. They both take advantage of the restriction that the elements in the N A-vectors have unit magnitudes; namely,

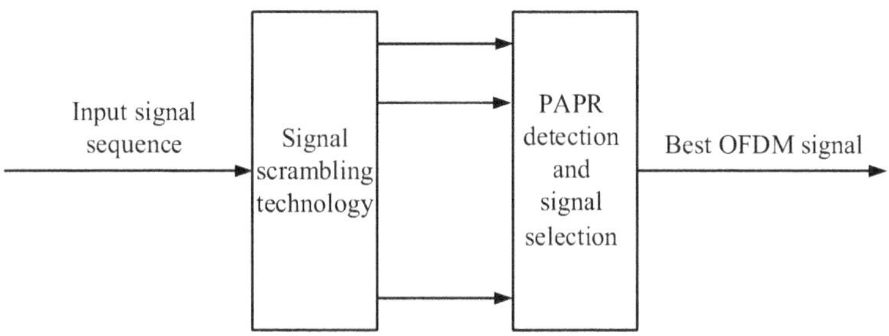

Fig. 7.3 Probabilistic technique to reduce the peak-to-average power ratio

$$A_n = e^{j\theta}, \theta \in [0, 2\pi], 1 \le n \le N. \quad (7.14)$$

This is a pure phase rotation. Relatively speaking, SLM and PTS have low technical complexity, are easy to implement, and are the most typical representatives of probabilistic technologies. The next section discusses these two solutions in detail.

7.3 Probabilistic Approaches to Reducing the Peak-to-Average Power Ratio in OFDM Systems

Probabilistic technology aims to minimize the probability of signal peaks occurring. Because this technology is a linear transformation, it will not cause signal distortion; therefore, this type of technology can effectively reduce the peak-to-average power ratio of the signal.

7.3.1 Partial Transmit-Sequence Method

The main idea of PTS [7] is to decompose data X to be input into V groups of non-overlapping sub-blocks $X = \sum_{v=1}^{V} X_v$, select a suitable phase-factor sequence {bv = exp(jϕv), v = 1, 2, ...,V}, adjust the phase of the V sub-blocks after the above block, and then superimpose them to obtain

$$Y = \sum_{v=1}^{V} b_v X_v. \quad (7.15)$$

Taking the inverse Fourier transform of Y, and according to the linearity of the inverse Fourier transform, we obtain

$$y = IFFT\{Y\} = IFFT\left\{\sum_{v=1}^{V} b_v X_v\right\} = \sum_{v=1}^{V} b_v IFFT\{X_v\}. \quad (7.16)$$

The phase of the sub-data block is appropriately adjusted using the rotation phase factor such that the PAPR of the OFDM signal generated by the superposition of the adjusted sub-block is lower than the PAPR of the original OFDM signal. Figure 7.4 shows the schematic diagram.

The main idea of traditional PTS is to divide the data symbols to be transmitted into several different sub-blocks and select a set of phase-rotation factor sequences that minimize the peak-to-average power ratio of the signal. This set of phase factors is called an optimal phase-rotation factor sequence.

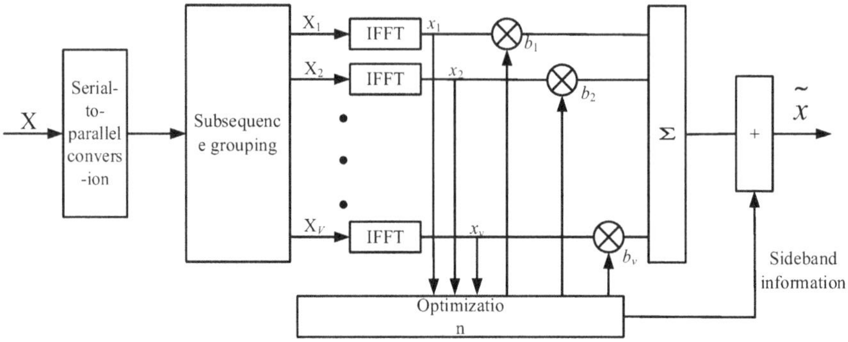

Fig. 7.4 Schematic diagram of the PTS implementation

The traditional PTS algorithm has some disadvantages. Theoretically, $(b_1, b_2, \ldots, b_V)$ can take any value between $\{0, 2\pi\}$; however, generally, $(b_1, b_2, \ldots, b_V)$ can be taken in a discrete phase set. When the PTS method is used to divide the sequence into V groups and the phase value is W, the value of $(b_1, b_2, \ldots, b_V)$ has W^V types. Each time PTS is implemented, VW^V IFFT transforms must be calculated. The amounts of complex multiplication and complex addition required to calculate the IFFT of each N point are

$$\begin{aligned} n_{mul} &= \frac{N}{2}\log_2 N \\ n_{add} &= N\log_2 N. \end{aligned} \tag{7.17}$$

The total computation is

$$\begin{aligned} n_{mul} &= VW^V\frac{N}{2}\log_2 N \\ n_{add} &= VW^V N\log_2 N. \end{aligned} \tag{7.18}$$

High-performance hardware is required to support the OFDM system. Therefore, it is necessary to make new improvements to the traditional PTS algorithm. While suppressing the PAPR, the improved algorithm should also significantly reduce the complexity of the operation and improve the practicability of the algorithm.

The receiver must know the specific twiddle factor used by the transmitter, which is usually communicated to the receiver in the form of sideband information. Therefore, the sideband information must be carefully protected and sent to the receiver without errors to ensure the normal operation of the PTS-OFDM system.

The main purpose of the PTS technique is to search for a sequence of phase twiddle factors that satisfies the following equation:

$$\left\{b_1^*, b_2^*, \ldots, b_v^*\right\} = \underset{(b_1, b_2, \cdots, b_v)}{\arg\min} \left(\max_t \left| \sum_{v=1}^{V} b_v IFFT\{X_v\} \right| \right), \tag{7.19}$$

where arg min(·) represents the decision condition [8] used when the function takes the minimum value. Finally, the OFDM signal with the smallest peak-to-average power ratio is transmitted. To enable the receiving end to identify different phases, it is generally necessary to transmit the phase factor to the receiving end as sideband information.

It can be seen from Eq. (7.19) that, in addition to limiting the value range of (b_1, b_2, ..., b_V) (for example, only taking values in exp{±1, ±j}), the computational complexity can also be reduced by adopting an appropriate segmentation method or an optimal phase-factor search method, as well as a PAPR-calculation optimization technique [9].

The algorithm proposed by Huber and Muller specifically implemented this theory [10]. The algorithm limits the value of the phase-rotation factor to 0 or π, and the value of b_v to $\{\pm 1, \pm \mathrm{j}\}$, and calculates the peak-to-average power ratio of the signals after all the phase twiddle factors are combined. Then, the full search method is used to obtain the set of phase twiddle-factor combinations that minimizes the system PAPR value.

The complexity of the full search method increases exponentially with an increase in the number of sub-blocks, and thus, it is a great burden for an OFDM system with a large number of sub-blocks. Therefore, it is necessary to reasonably select the number of blocks, value range of the phase factor, and size of the number of carriers to reduce the PAPR of the system and reduce the computational complexity to a reasonable range. In general, the phase factor must be transmitted to the receiver as sideband information.

7.3.1.1 Comparison of OFDM Performance Under Different Segmentation Methods

In OFDM systems, vectors are first used to define data symbols $X = (X_1, X_2, \ldots, X_n)$, and then these vectors are divided into V groups, denoted by $\{X_v, v = 1, 2, \ldots, v\}$. There are three types of segmentation: random (computers can only do pseudo-random), adjacent, and interleaved. Figure 7.5 presents a schematic diagram of the three segmentation methods.

Adjacent division divides N/V adjacent subcarriers into one subsequence, random division randomly divides each subcarrier into V sub-sequences, and interleaving division divides a subcarrier into subsequences of interval size V. The above three division methods all follow one principle: Each subcarrier only appears in one subsequence, and the number of subcarriers contained in the V subsequences is equal.

The effect of reducing the PAPR in an OFDM system is different for PTS algorithms that use different partitioning methods. Among the three segmentation

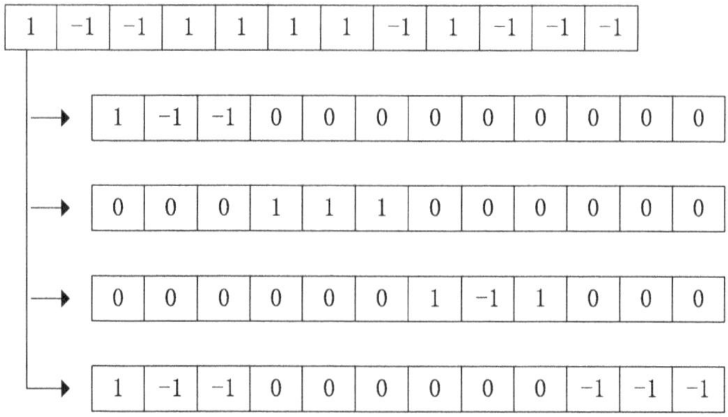

Fig. 7.5 Schematic diagram of adjacent segmentation

methods, the subsequence obtained by the random-segmentation method after the IFFT, has the lowest autocorrelation. Therefore, under the same conditions, the system peak-to-average power ratio obtained by the random-segmentation method was much improved, compared with the other two segmentation methods.

In terms of reducing the peak-to-average power ratio, random partitioning is the best, adjacent partitioning is the second, and interleaving partitioning is the worst; however, interleaving partitioning can effectively reduce the computational complexity of PTS-OFDM systems, while adjacent partitioning has the worst complexity. The split method is the easiest to implement.

Here, we provide a simple example of a subcarrier division. Suppose an OFDM system contains 12 subcarriers and is divided into four subsequences; then, each subsequence contains $12/4 = 3$ subcarriers.

Figures 7.5, 7.6, and 7.7 show the adjacent, interleaving, and random segmentation methods [11], respectively

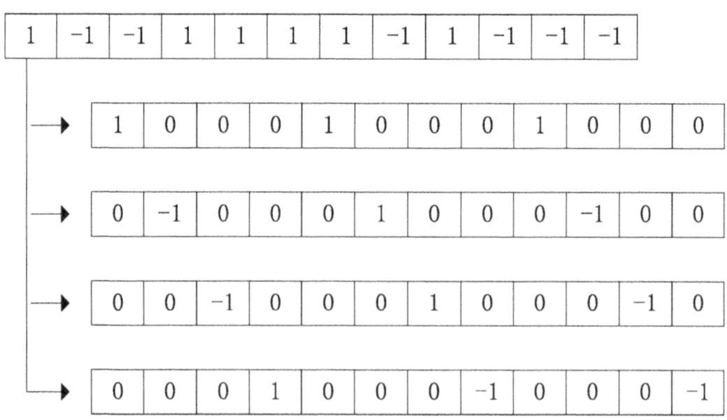

Fig. 7.6 Schematic diagram of interleaving segmentation

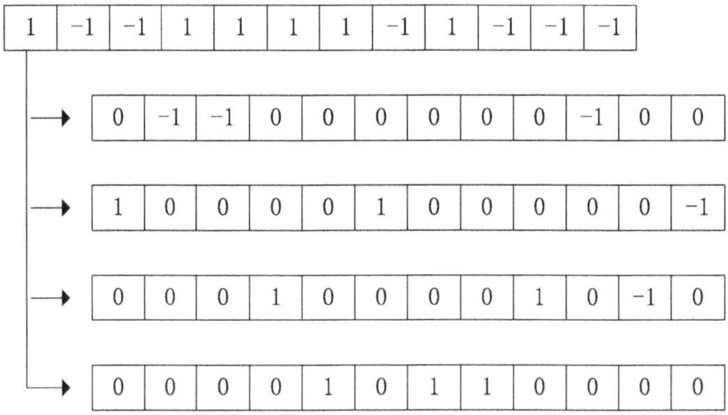

Fig. 7.7 Schematic diagram of random segmentation

Figure 7.8 compares the PAPR performance of different segmentation methods. It can be seen from Fig. 7.8 that for reducing the PAPR, the random-adjacent segmentation method is very close to random segmentation. The random-adjacent segmentation method is better than the adjacent-segmentation method in reducing the PAPR, and is better than the random-segmentation method in reducing the algorithm complexity. Therefore, the random-adjacent segmentation method was used in the simulation. The simulation results also showed that the performance of any PTS method was much better than that of traditional OFDM.

7.3.1.2 Comparison of OFDM Performance Under an Optimized Number of Blocks

Different numbers of blocks V were used to simulate and verify the impact on system performance when $V = 2, 4, 6$, and 8. The parameters are listed in Table 7.1.

Figure 7.9 shows the complementary cumulative-distribution function of the PAPR obtained when the number of subcarriers $N = 128$, the random-plus-adjacent division method was used, and $V = 2, 4, 6$, and 8.

As shown in Fig. 7.9, as the number of divided sub-blocks increases, the PAPR decreases further. At CCDF $= 10^{-2}$ and $V = 8$, the PAPR performance is 4.5 dB better than that of the original OFDM signal. In addition, with the continuous increase in V, the complexity of the system increases, and the rate of change of the curve progressively decreases; that is, the ability to improve the system also decreases. Therefore, considering the compromise between system performance and computational complexity when using the PTS algorithm, it is more appropriate to take $V =$ 4 as the number of sub-blocks.

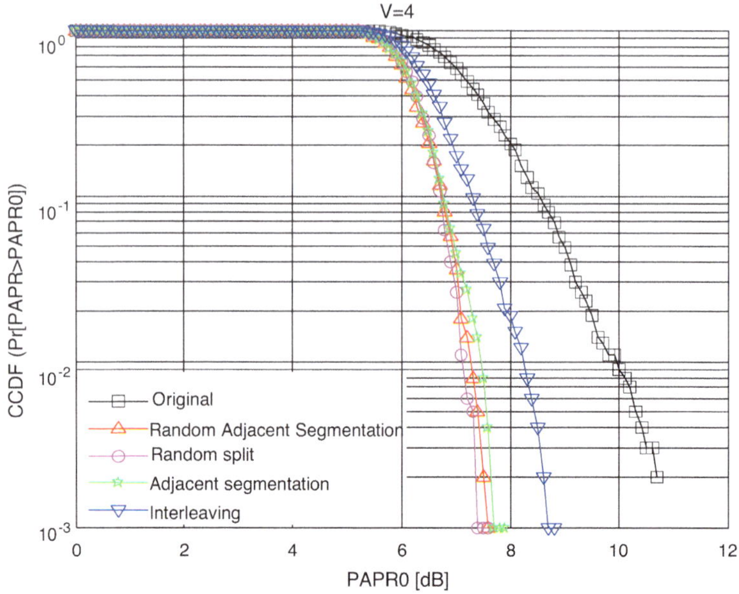

Fig. 7.8 Comparison of PAPR performance under different segmentation methods

Table 7.1 Simulation parameters for different numbers of blocks

Simulation parameters	Type/value
Randomly generated OFDM numbers	10000
Number of subcarriers N	128
Number of blocks V	2, 4, 6, 8
Oversampling rate L	4
Sub-block partitioning method	Random + adjacent
Modulation	4-QAM
Phase factor (b)	1, −1, j, −j

7.3.1.3 Comparison of OFDM Performance Under Different Numbers of Subcarriers

Different numbers of subcarriers were used to simulate and verify the impact on the system performance when $N = 64$, 128, and 256. The parameters are listed in Table 7.2.

Figure 7.10 shows the CCDF of PAPR obtained when the number of sub-blocks is $V = 4$, the random-plus-adjacent segmentation method is used, and N = 64, 128, and 256.

As shown in Fig. 7.10, as the number of subcarriers N increases, the PAPR of the system increases. This is because the larger the number of subcarriers N, the greater

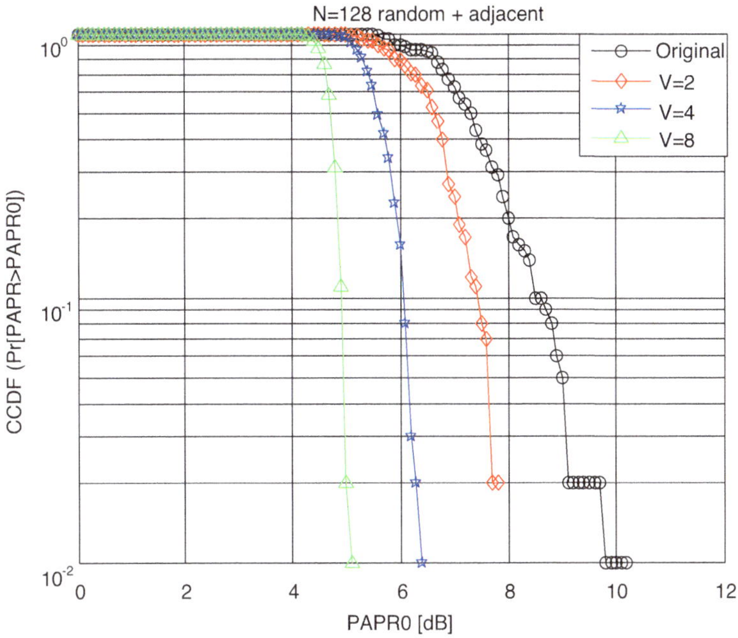

Fig. 7.9 CCDF curve for different numbers of blocks

Table 7.2 Simulation parameters of different carrier numbers

Simulation parameters	Type/value
Randomly generated OFDM numbers	10000
Number of subcarriers N	64, 128, 256
Number of blocks V	4
Oversampling rate L	4
Sub-block partitioning method	Random + adjacent
Modulation	4-QAM
Phase factor (b)	1, −1, j, −j

the probability of the same phase appearing for different numbers of subcarriers. A higher peak-to-average power ratio will be generated when subcarriers of the same phase are superimposed.

At $CCDF = 10^{-2}$ and $N = 64$, the PAPR performance of a PTS-OFDM system was improved by 3 dB at most, compared with the original OFDM signal. In addition, with a continuous increase in the number of subcarriers, the computational complexity of the system continues to increase. The simulation results also demonstrate that the PTS algorithm is suitable for any number of subcarriers.

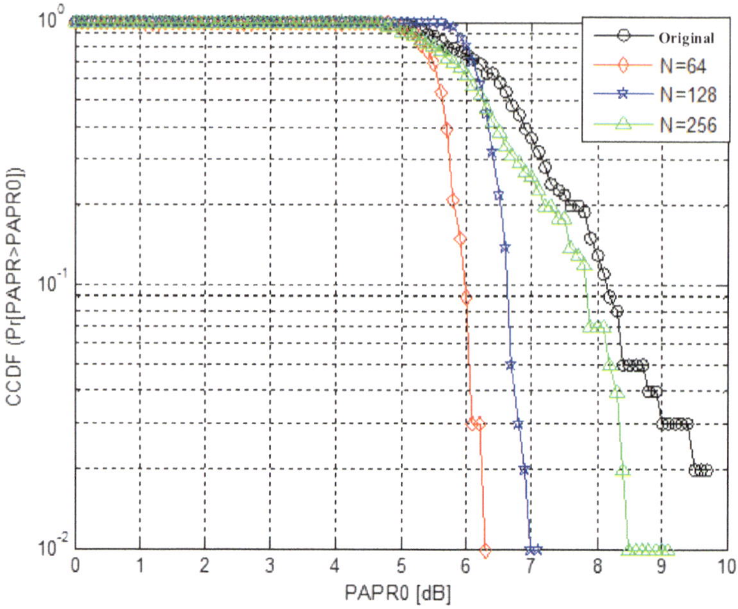

Fig. 7.10 CCDF curve when the number of carriers is different

7.3.1.4 Comparison of OFDM Performance Under Different Phase Factors

PTS technology divides the carrier information into several groups and then adjusts the phase of the information to a certain extent. The PAPR performance of the control system is improved by multiplying the information by a phase factor to rotate its phase.

Different phase-factor numbers P are used to simulate and verify the effects on the system performance when $P = 2, 4$, and 8. The parameters are listed in Table 7.3.

Figure 7.11 shows the CCDF of the PAPR obtained with the number of subcarriers $N = 128$, using the random-plus-adjacent division method, and $P = 2, 4$, and 8.

Table 7.3 Simulation parameters for different phase factors

Simulation parameters	Type/value
Randomly generated OFDM numbers	10000
Number of subcarriers N	128
Number of blocks V	4
Oversampling rate L	4
Sub-block partitioning method	Random + adjacent
Modulation	4-QAM
Phase factor (b)	2, 4, 8

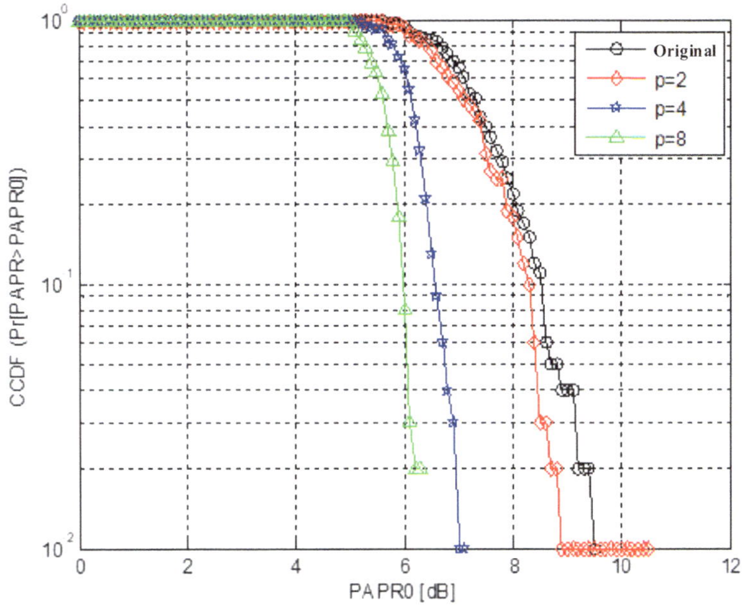

Fig. 7.11 CCDF curve with different numbers of phase factors

It can be seen from Fig. 7.11 that with the increase in the phase factor, the PAPR increasingly decreases. Because the larger P is, the more types of auxiliary phase factors are selected, and the better the performance of the PTS-OFDM system is. However, adding more phase rotations increases the computational complexity of the system. Theoretically, the weighted phase can be arbitrarily selected within [0, 2π]; however, if the phase-factor set $W = [\pm 1, \pm j]$, the multiplication of complex numbers can be avoided and the complexity of the system can be greatly reduced. Therefore, the phase factor added to the final experiment was $[\pm 1, \pm j]$.

From the above analysis and summary, it can be concluded that the factors that affect the PAPR of the system are as follows:

(1) The segmentation method adopted by the OFDM signal. Traditional PTS segmentation methods include random, adjacent, and interleaving segmentation. Random segmentation had the best effect in reducing the PAPR performance of the system; however, its system complexity was the largest. The interleaving segmentation had the lowest system complexity, but reduced the PAPR performance. Adjacent segmentation was better than interleaving segmentation in reducing the PAPR performance.
(2) Number of carrier blocks. When the number of sub-blocks V of the carrier is 2, 4, and 8, by adopting the random-adjacent division method, the PAPR performance of the system improves as the number of carrier blocks increases. With an increase in the number of blocks, the types of phase factors also increase; therefore, there are more choices of phases. In this manner, the system can

choose a set of parameters that minimize the PAPR performance of the system with many alternative phase factors. The group data were transmitted.

(3) Number of phase factors. When the number of phase factors is two, four, and eight, using the random-adjacent division method, the PAPR performance of the system improves as the number of phase factors increases. As the number of candidate phase factors increases, the possibility of finding the optimal phase factor that improves the PAPR system performance also increases.

(4) Number of subcarriers. In a system using PTS technology, when the number of subcarriers increases, the possibility of superimposing high peaks with the same phase increases, causing the PAPR of the system to decrease. However, once a multicarrier system adopts the PTS method, the obtained PAPR result is better than that of the traditional OFDM system.

Figure 7.12 shows the CCDF curve comparison of the PAPR distribution of the original OFDM and PTS-OFDM systems. Here, the CCDF performance curve was used to describe the peak-to-average power-ratio suppression. Based on the above analysis, the simulation parameters are set as follows: 10000 random independent OFDM signals, 4-QAM modulation mode, $N = 128$, random phase-sequence values of $\{\pm 1, \pm j\}$, and an oversampling rate of 4.

In Fig. 7.12, the CCDF curves of the original OFDM system and the peak-to-average power ratio of the PTS-OFDM system are compared. The ordinate CCDF refers to the probability of the PAPR value in the system exceeding a certain threshold

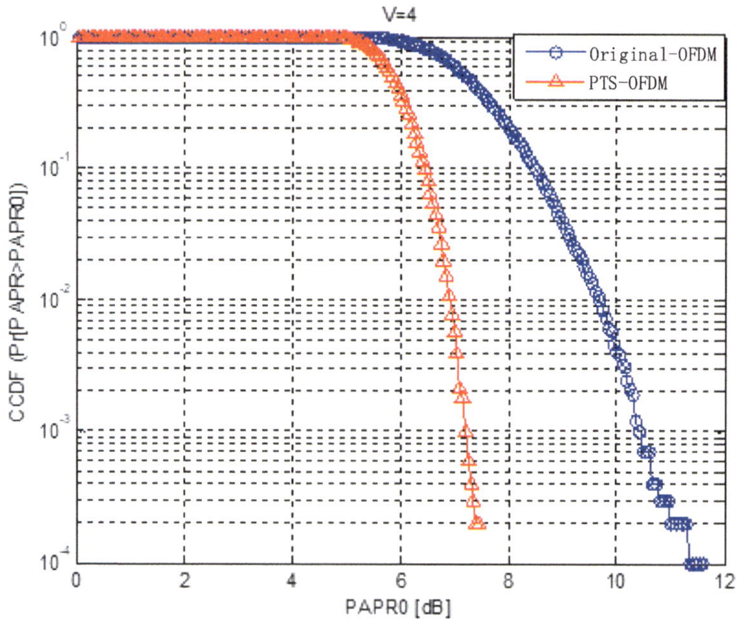

Fig. 7.12 Comparison of CCDF curves for PAPR reduction between OFDM and PTS-OFDM

value. The abscissa represents the threshold $PAPR_0$. The blue line is the initial peak-to-average power ratio of about 11.5 dB, and at the probability of $CCDF = 10^{-4}$, the red line is the PAPR of the system reduced by the optimized PTS algorithm, and the value is about 7 dB. The optimized PTS algorithm is obviously reduced by about 4.5 dB, and the results show that the PTS method can effectively reduce the peak-to-average power ratio of an OWC-OFDM system.

7.3.2 *Selective-Mapping Method (SLM)*

The basic idea of the selective-mapping method (SLM) [12] is to use M statistically independent vectors $X(k)$ ($k = 1, 2, \ldots, M$) to represent the same information, and select the corresponding OFDM time-domain sequence x_k in the sequence when the smallest peak-to-average power ratio is transmitted. Figure 7.13 shows a schematic diagram.

The signal in the transmitter of the OWC-OFDM system can be expressed as

$$x_k = IFFT[X_n],\ (n, k = 0, \ldots, N-1). \tag{7.20}$$

Suppose there are M different random phase-sequence vectors of length N, as shown in Eq. (7.21):

$$P^{(\mu)} = \left(P_0^{(\mu)}, P_1^{(\mu)}, \ldots, P_{N-1}^{(\mu)}\right),\ (\mu = 1, 2, \ldots, M), \tag{7.21}$$

where $P_i^{(\mu)} = \exp\left(j\varphi_i^{\mu}\right)$ and φ_i^{μ} is uniformly distributed within $[0, 2\pi)$. The corresponding dot products of the M phase vectors with the input sequence X of the IFFT are performed, and then M different output sequences $X^{(\mu)}$ are obtained, namely,

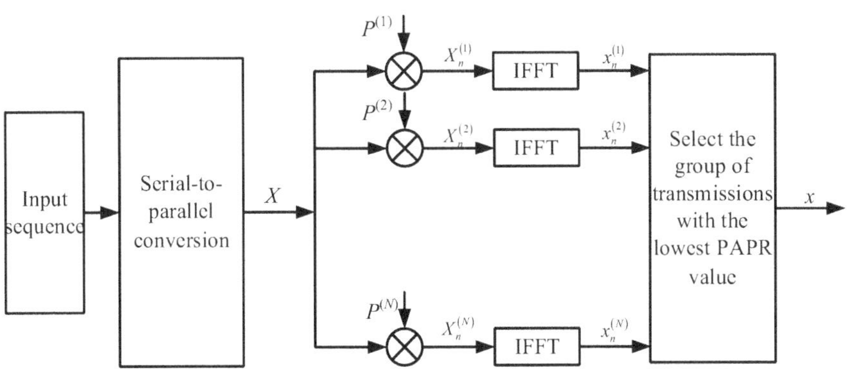

Fig. 7.13 Schematic diagram of the selective-mapping method

$$
\begin{aligned}
X^{(\mu)} &= \left(X_0^{(\mu)}, X_1^{(\mu)}, \ldots, X_{N-1}^{(\mu)}\right) = \langle X \cdot P^{\mu} \rangle \\
&= \left(X_0 P_0^{(\mu)}, X_1 P_1^{(\mu)}, \ldots, X_{N-1} P_{N-1}^{(\mu)}\right).
\end{aligned}
\tag{7.22}
$$

The symbol $\langle\cdot\rangle$ represents a location multiplication between signals. Then, the IFFT calculation is performed on the obtained M different sequences X(μ), and M different output sequences $\left(x_0^{(\mu)}, x_1^{(\mu)}, \ldots, x_{N-1}^{(\mu)}\right)$ are obtained.

Under the condition of a given PAPR threshold value, one channel with the smallest peak-to-average power ratio is selected from the M time-domain signal sequences for transmission. At the receiving end, the received signal is first subjected to a fast Fourier transform, and then, the original signal is restored according to the opposite operation of the transmitting end, according to the received sideband information. The PAPR-suppression performance of the SLM method depends only on the design of the phase factor, which is applicable to various modulation methods, and has no strict requirements on the number of subcarriers in the system.

If the threshold of the PAPR is greater than $PAPR_0$, the probability that the PAPR of the original OFDM sequence exceeds the threshold is defined as $P(PAPR > PAPR_0)$. Therefore, the probability that the PAPRs of the M sequences $X^{\mu}(\mu = 0, 1, \ldots, M-1)$ exceed the threshold will become $[P\,(PAPR > PAPR_0)]^M$. The CCDF according to the PAPR distribution in the OFDM system can be expressed as

$$
[P(PAPR > PAPR_0)] = 1 - \left(1 - e^{-PAPR_0}\right)^K. \tag{7.23}
$$

Subsequently, after using the SLM algorithm, the CCDF of the PAPR distribution of the OFDM system is

$$
[P(PAPR > PAPR_0)]^M = \left(1 - \left(1 - e^{-PAPR_0}\right)^K\right)^M. \tag{7.24}
$$

$M = 1$ is the CCDF of the PAPR distribution of the original OFDM system. The SLM algorithm has an obvious PAPR-suppression ability for the OFDM signal.

Figure 7.14 describes the peak-to-average power ratio when the phase sequence M takes different values. The CCDF performance curve is used to describe the peak-to-average power ratio suppression. The simulation parameters are as follows: 10000 random independent OFDM signals, 4-QAM modulation mode, 128 carriers, and an oversampling rate of 4.

It can be seen from Fig. 7.14 that when $M = 32$ and $CCDF = 10^{-4}$, the peak-to-average power ratio performance of the system, after the SLM algorithm, is reduced by about 5 dB, compared to the original signal. With the increase of the random phase number, the peak-to-average power ratio decreases gradually. It can be seen that with the increase of the value of M, the suppression performance of the peak-to-average power ratio is increasingly improved; however, when the M value increases to a

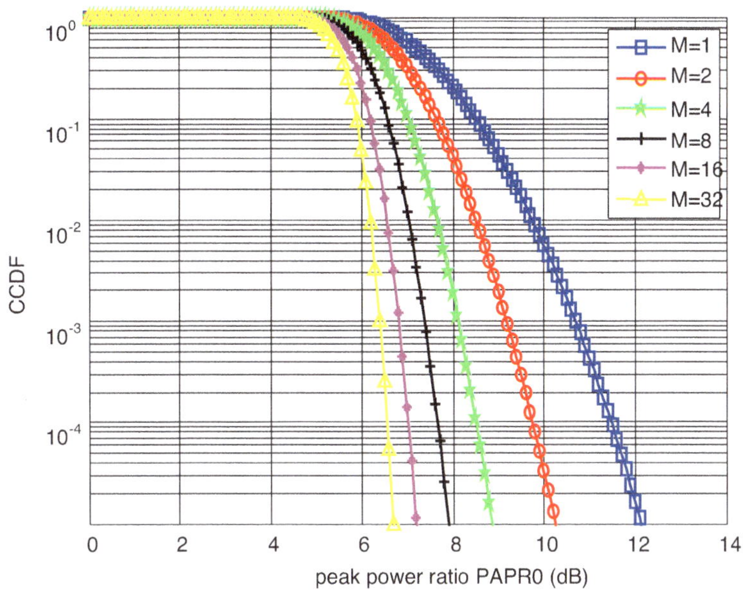

Fig. 7.14 CCDF simulation curve of PAPR under different phase-vector M conditions in an SLM system

certain extent, the effect of suppressing the peak-to-average power ratio gradually weakens.

Theoretically, the selective-mapping method can greatly reduce the peak-to-average power ratio of the system, but at the cost of increasing the system complexity. First, the N-point inverse Fourier transform needs to be calculated $M - 1$ times, which increases the complexity of the calculation. Second, additional sideband information needs to be transmitted to ensure that the receiving end of the system can completely recover the transmitted signal, thus reducing the data-transfer rate of the system.

The selective-mapping method reduces the probability of peaks in the OFDM system by introducing sideband information instead of directly reducing the peak signal. It thereby reduces the bit-error rate of the system, and there is no in-band or out-of-band noise.

7.3.3 Channel Simulation Analysis

Following is a comparative analysis of the constellation diagrams of the OWC-OFDM and PTS-OFDM systems. Moreover, the constellations of the OWC-OFDM and SLM-OFDM systems are analyzed under gamma-gamma channel submultiplicative mixed noise and additive white Gaussian noise.

7.3.3.1 Channel Model

To study the performance of the OWC-OFDM system, we first discuss the effects of noise in the channel.

As shown in Fig. 7.15, the noise model assumes that the signal is $s(t)$ and the noise is $n(t)$. If the mixed superpositioned waveform is in the form of $s(t) + n(t)$, then $n(t)$ is called additive noise. If the superimposed waveform is in the form of $s(t)k(t) + n(t)$, where $k(t)$ is multiplicative noise, and when both $n(t)$ and $k(t)$ are superimposed on the signal, mixed noise acts on the system.

Additive white Gaussian noise (AWGN) is the most basic noise and interference model. Its amplitude distribution follows a Gaussian distribution, whereas the power-spectral density is uniformly distributed. We define multiplicative noise as a channel model that obeys the gamma-gamma distribution [13]. The noise model can be expressed using Eq. (2.56).

(1) **Influence of Gaussian white noise on the system constellation diagram**

Figures 7.16, 7.17 and 7.18 show that at a CCDF = 10^{-4} probability, the peak-to-average power ratio of the original OFDM system is 11.5 dB, and the peak-to-average power ratio of the SLM-OFDM system is reduced to 6.5 dB. The amplitudes of the OFDM and SLM-OFDM signals are normalized to the same interval, and the OFDM and SLM systems are analyzed in the case of additive noise with SNR = 20, SNR = 15, and SNR = 10 under an AWGN channel.

Fig. 7.15 Noise model

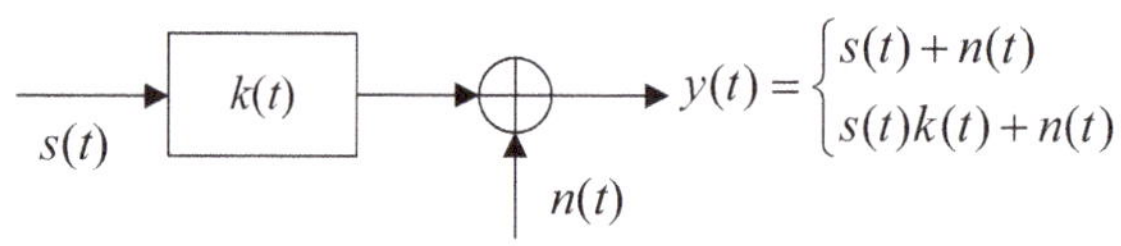

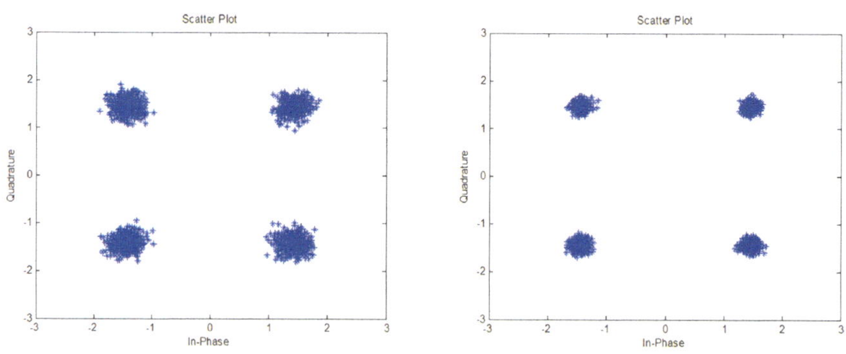

(a)OFDM (b)SLM -OFDM

Fig. 7.16 Constellation diagram of OFDM and SLM-OFDM demodulated signals in the case of SNR = 20 [14, 15]

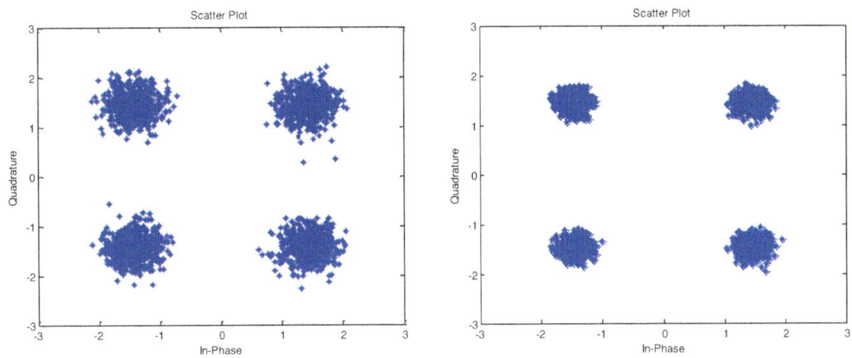

(a) OFDM (b) SLM-OFDM

Fig. 7.17 Constellation diagram of OFDM and SLM-OFDM demodulated signals with SNR = 15 [14, 15]

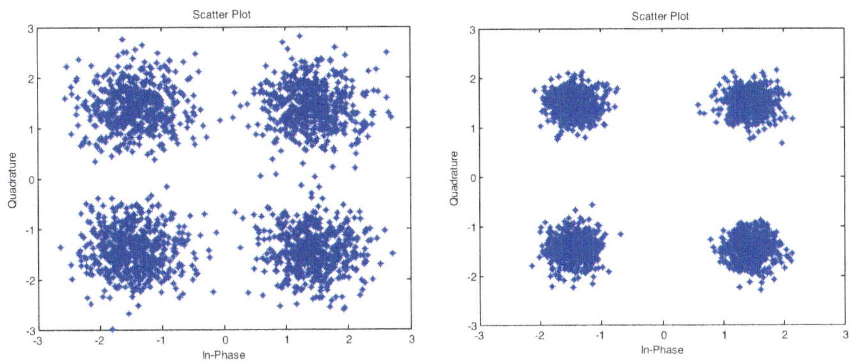

(a) OFDM (b) SLM-OFDM

Fig. 7.18 Constellation diagram of OFDM and SLM-OFDM demodulated signals in the case of SNR = 10 [14, 15]

These figures show the constellation diagrams of the demodulation ends of the OFDM systems, where the abscissa represents the amplitude of the I-channel signal, the ordinate represents the amplitude of the Q-channel signal, and both the abscissa and ordinate are dimensionless quantities.

Figure 7.16 shows the constellation diagram of the demodulated signal at the receiving end of the OFDM and SLM-OFDM systems when the SNR of the additive noise is 20 in an AWGN channel.

Figure 7.17 shows the constellation diagrams of the demodulated signals at the receiving end of the OFDM and SLM-OFDM systems when the SNR of the additive noise is 15 in an AWGN channel.

Comparing Figs. 7.16, 7.17 and 7.18, we can see that after the OFDM and SLM-OFDM systems are affected by white Gaussian noise in the channel, the aggregation degree of each constellation point in the constellation diagram will be affected to a certain extent. The less the system is affected, the higher the degree of aggregation of the constellation points.

In addition, comparing the constellation diagrams of (a) and (b) in Figs. 7.16, 7.17 and 7.18 it is found that in the case of the same SNR value, the aggregation degree of the constellation points in the (b) figures is higher than that shown in the (a) figures. This shows that after adding the SLM algorithm, the resistance to Gaussian white-noise interference increases and the bit-error rate of the system decreases.

(2) **Influence of multiplicative noise on the system constellation diagram**

The OFDM signal passes through an atmospheric channel following the gamma–gamma distribution, and the signal is affected by multiplicative noise. The influence of the OFDM and SLM-OFDM signals on the constellation diagram of the receiver is analyzed under the weak and moderate turbulence characteristics of a gamma–gamma-distribution atmospheric-turbulence channel model.

Figures 7.19, 7.20 and 7.21 show that at a probability of CCDF = 10^{-4}, the peak-to-average power ratio of the original OFDM system is 11.5 dB, and the peak-to-average power ratio of the SLM-OFDM system is reduced to 6.5 dB. The amplitudes of the OFDM and SLM-OFDM signals were normalized to the same interval, and the constellation diagrams of the demodulation ends of the OFDM and SLM-OFDM systems under different light-intensity fluctuation variances were analyzed.

Figures 7.19, 7.20 and 7.21 show the constellation diagrams of the demodulated signals at the receiving ends of the OFDM and SLM-OFDM systems under light-intensity fluctuation variances $\sigma_R = 0.1$, 0.3, and 0.5, respectively, of the multiplicative noise in the gamma–gamma channel.

Comparing Figures 7.19, 7.20 and 7.21, it can be observed that the constellation diagram of the modulated signal is in a better convergence state under weak turbulence, and more scattered in medium turbulence.

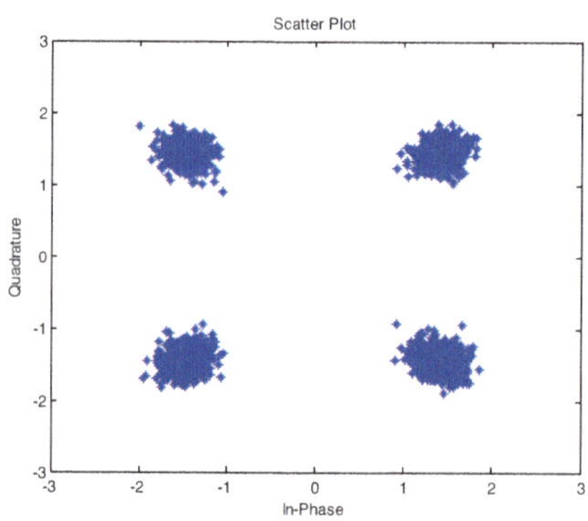

(a) OFDM

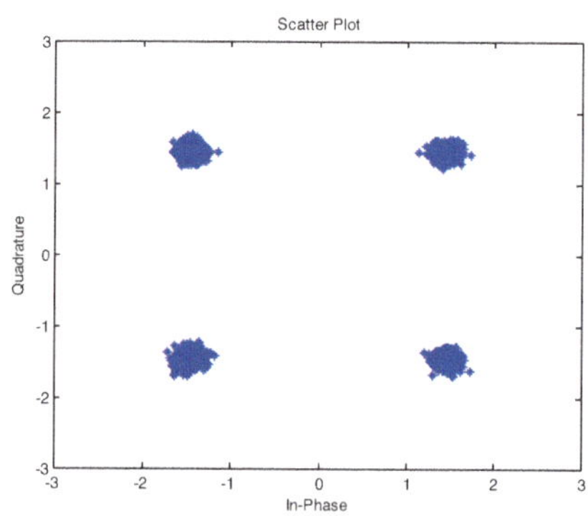

(b) SLM-OFDM

Fig. 7.19 Constellation diagram of the OFDM and SLM-OFDM demodulated signals in case $\sigma_R = 0.1$ [14, 15]

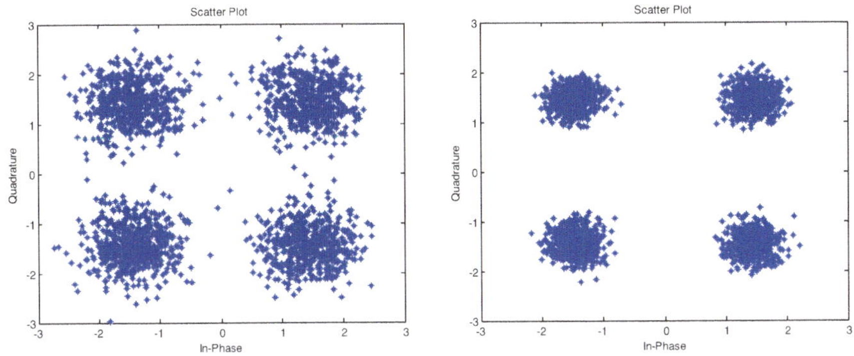

(a) OFDM (b) SLM-OFDM

Fig. 7.20 Constellation diagram of OFDM and SLM-OFDM demodulated signals in case $\sigma_R = 0.3$

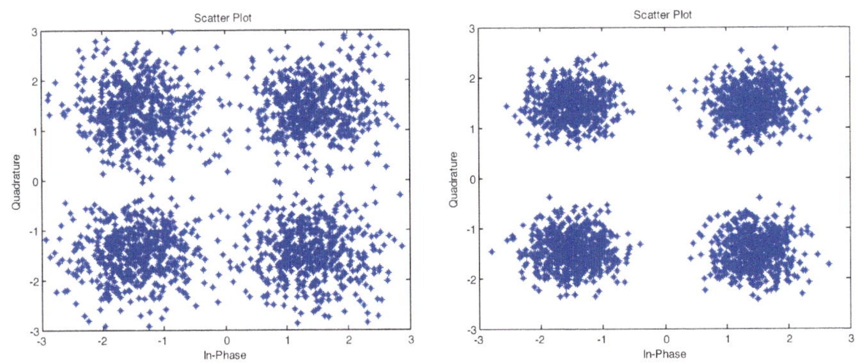

(a) OFDM (b) SLM-OFDM

Fig. 7.21 Constellation diagram of OFDM and SLM-OFDM demodulated signals in case $\sigma_R = 0.5$ [14, 15]

In addition, under weak and moderate turbulence, by analyzing and comparing the constellation diagrams of (a) and (b) in Figures 7.19, 7.20 and 7.21, it is found that under the same light-intensity fluctuation variance, the constellation-point aggregation degrees of graph (b) are better than those of graph (a). This shows that after adding the SLM algorithm, the SLM-OFDM system has a good suppression effect on the atmospheric turbulence of the gamma–gamma channel.

(3) **Influence of mixed noise on the system constellation diagram**

The influence of the OFDM and SLM-OFDM systems on the demodulated signal constellation at the receiver is analyzed under a mixed-noise channel model, considering the additive noise in an AWGN channel and the multiplicative noise in a gamma–gamma channel. Figures 7.22, 7.23, and 7.24 show the conditions under which the peak-to-average power ratio of the original OFDM system is 11.5 dB and the peak-to-average power ratio of the SLM-OFDM system is reduced to 6.5 dB at a probability of CCDF = 10^{-4}.

Moreover, the amplitudes of the OFDM and SLM-OFDM signals are normalized to the same interval, and the effect of different SNR and light-intensity fluctuation variances on the constellation diagrams of the demodulation ends of the OFDM and SLM-OFDM systems are analyzed.

Figure 7.22 shows the constellation diagram of the demodulated signal at the receiving end of the OFDM and SLM-OFDM systems when the SNR value of the additive noise is 25 in the AWGN channel, and the light-intensity fluctuation variance of the multiplicative noise is $\sigma_R^2 = 0.1$ in the gamma–gamma channel.

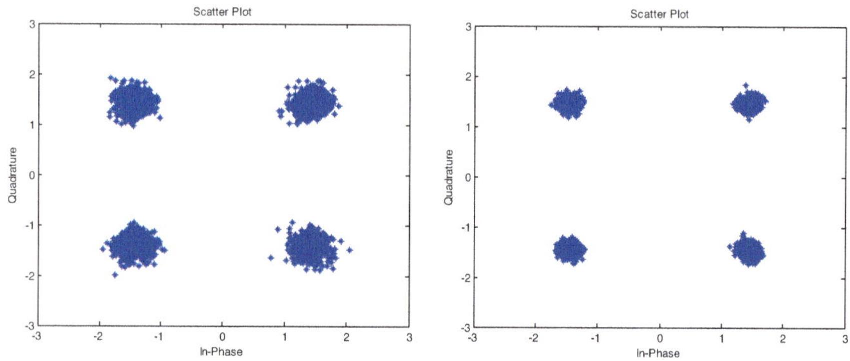

(a) OFDM (b) SLM-OFDM

Fig. 7.22 Constellation diagram of OFDM and SLM-OFDM demodulated signals in case $SNR = 25$, $\sigma_R = 0.1$ [14, 15]

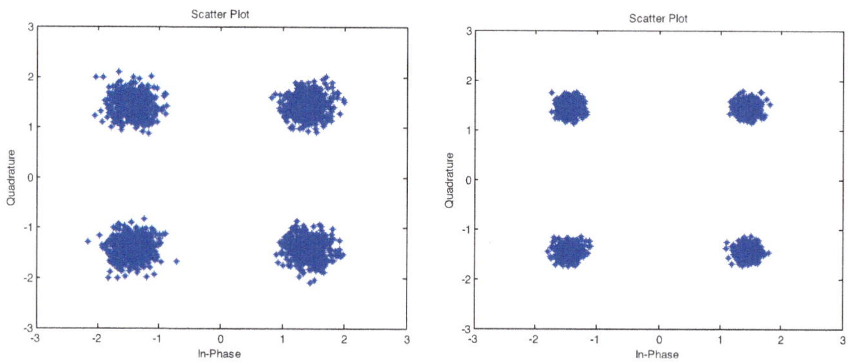

(a) OFDM (b) SLM-OFDM

Fig. 7.23 Constellation diagram of OFDM and SLM-OFDM demodulated signals in case $SNR = 20$, $\sigma_R = 0.1$ [14, 15]

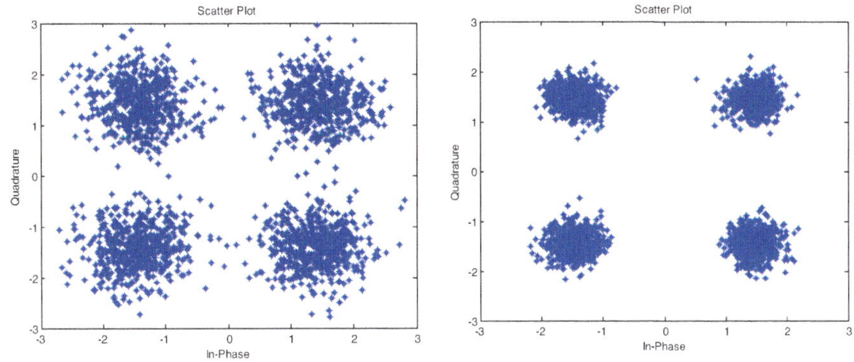

(a) OFDM (b) SLM-OFDM

Fig. 7.24 Constellation diagram of OFDM and SLM-OFDM demodulated signals in case $SNR = 20, \sigma_R = 0.3$ [14, 15]

Figures 7.23 and 7.24 show the constellations of the demodulated signals at the receiving ends of the OFDM and SLM-OFDM systems when the SNR value of the additive noise is 20 in the AWGN channel and the light-intensity fluctuation variance of the multiplicative noise in the gamma–gamma channel is $\sigma_R^2 = 0.1$ and 0.3, respectively.

Figures 7.22, 7.23 and 7.24 show the influence of mixed noise on the constellation diagrams of the demodulator when the SNR is 25, 25, and 20, respectively, under the condition of the channel light-intensity fluctuation variance $\sigma_R^2 = 0.1, \sigma_R^2 = 0.1$, and $\sigma_R^2 = 0.3$, respectively. Comparing Figs. 7.22 and 7.23, it can be seen that when the light-intensity fluctuation variance is equal, the larger the SNR, the higher the aggregation degree of points in the constellation diagram, and the better the system performance. Comparing Figs. 7.23 and 7.24, it can be seen that when the variance value of the light-intensity fluctuation σ_R increases from 0.1 to 0.3, the degree of aggregation of the constellation points progressively decreases. The system performance also worsens.

In addition, under weak and moderate turbulence, by analyzing and comparing the constellation diagrams of (a) and (b) in Figs. 7.23 and 7.24, it is found that the same SNR value and light-intensity fluctuation variance jointly act on the system. In the (b) figures, the aggregation degree of the constellation points is higher than that in the (a) figures. This shows that after adding the SLM algorithm, the mixed-noise resistance is increased, and the bit-error rate of the system is reduced.

7.3.3.2 Eye-Diagram Analysis

Under a mixed-noise channel model considering the combined action of additive noise in the AWGN channel and multiplicative noise in the gamma–gamma channel, the closure of the eye diagrams of the demodulated signals at the receiving ends of the OFDM and SLM-OFDM systems is analyzed to further verify the feasibility of the SLM algorithm to reduce the PAPR.

Figures 7.25, 7.26 and 7.27 show that at the probability of CCDF = 10^{-4}, the peak-to-average power ratio of the original OFDM system is 11.5 dB, and the peak-to-average power ratio of the SLM-OFDM system is reduced to 6.5 dB. The amplitudes of the OFDM and SLM-OFDM signals are normalized to the same interval, and the effects of different SNR and light-intensity fluctuation variances on the demodulation-end eye diagram of the OFDM and SLM-OFDM systems are analyzed.

Figures 7.25 and 7.26 show the eye diagrams of the demodulated signals at the receiving ends of the OFDM and SLM-OFDM systems when the SNR values of the AWGN channel are 25 and 15, respectively, and the light-intensity fluctuation variance $\sigma_R^2 = 0.1$ of the multiplicative noise in the gamma–gamma channel.

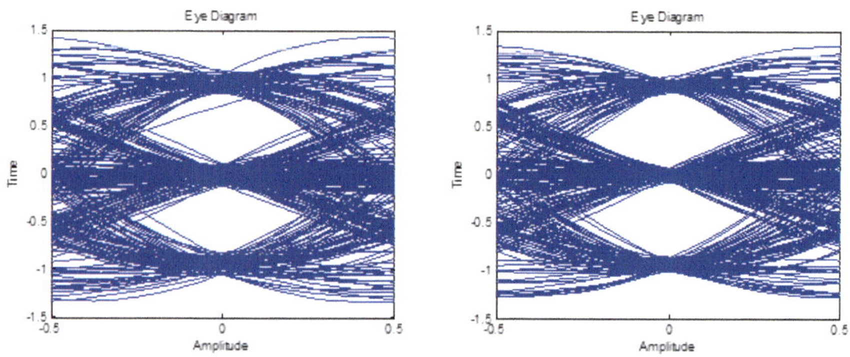

(a) OFDM (b) SLM-OFDM

Fig. 7.25 Eye diagrams of the OFDM and SLM-OFDM demodulated signals in case *SNR* = 25, $\sigma_R = 0.1$ [14, 15]

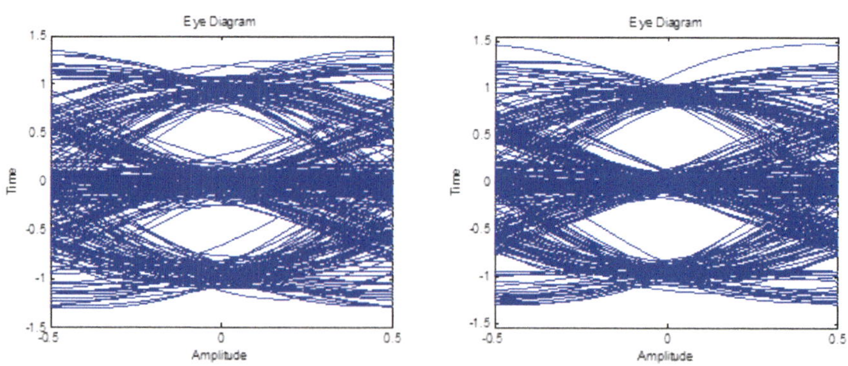

(a) OFDM (b) SLM-OFDM

Fig. 7.26 Eye diagrams of the OFDM and SLM-OFDM demodulated signals in case *SNR* = 15, $\sigma_R = 0.1$ [14, 15]

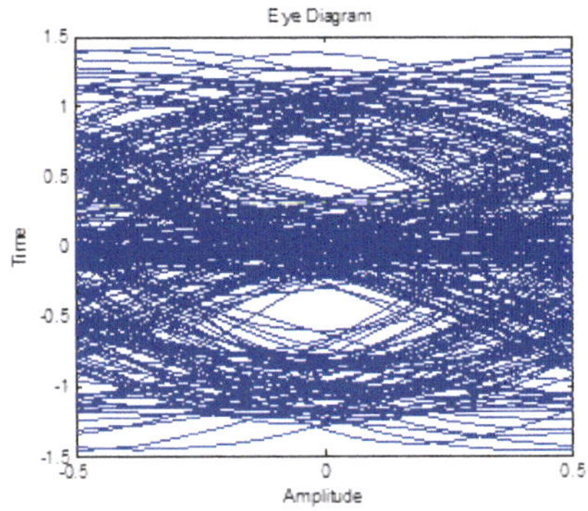

(a) OFDM

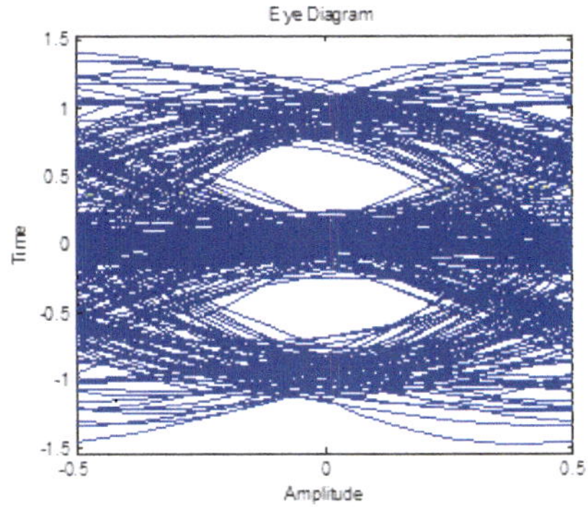

(b) SLM-OFDM

Fig. 7.27 Eye diagrams of the OFDM and SLM-OFDM demodulated signals in case $SNR = 15$, $\sigma_R = 0.3$ [14, 15]

Figure 7.27 shows the eye diagrams of the demodulated signals at the receiving ends of the OFDM and SLM-OFDM systems when the SNR value of the additive noise in the AWGN channel is 15 and the light-intensity fluctuation variance $\sigma_R^2 = 0.3$ of the multiplicative noise in the gamma–gamma channel.

By observing the eye diagrams under different noise conditions, we can analyze the impact of channel noise on the performance of optical wireless communication systems. Comparing Figs. 7.26 and 7.27, it can be seen that when the variance of the light-intensity fluctuation is equal, the larger the SNR, the higher the aggregation degree of points in the constellation diagram, and the better the system performance.

Comparing Figs. 7.26 and 7.27, it can be seen that when SNR = 15 and the σ_R value increases from 0.1 to 0.3, the eye diagram of the OFDM system is very closed, so the system is greatly affected by multiplicative noise. Comparing Figs. 7.26 and 7.27, with weak and medium turbulence, the SLM-OFDM demodulated signal has a larger "eye" opening than the OFDM demodulated signal, and the eye-diagram trace is also clearer. This shows that the OFDM signal after adding the SLM algorithm is better than the original OFDM signal in terms of channel-noise resistance.

In addition, because the amplitudes of the signals are normalized to the same interval, the average power of the SLM-OFDM signal increases under the same noise condition, and the signal-to-noise ratio is also improved; thus, the noise resistance of the SLM-OFDM system increases, which verifies the effectiveness of the algorithm.

7.3.3.3 Bit-Error Rate Analysis

The peak-to-average power ratio of the system is 11.5 dB at a probability of CCDF $= 10^{-4}$ when the original OFDM signal is not controlled by the peak-to-average power ratio. After adding the SLM algorithm, the peak-to-average power ratio of the SLM-OFDM system drops to 6.5 dB. Figure 7.28 shows the comparison curve

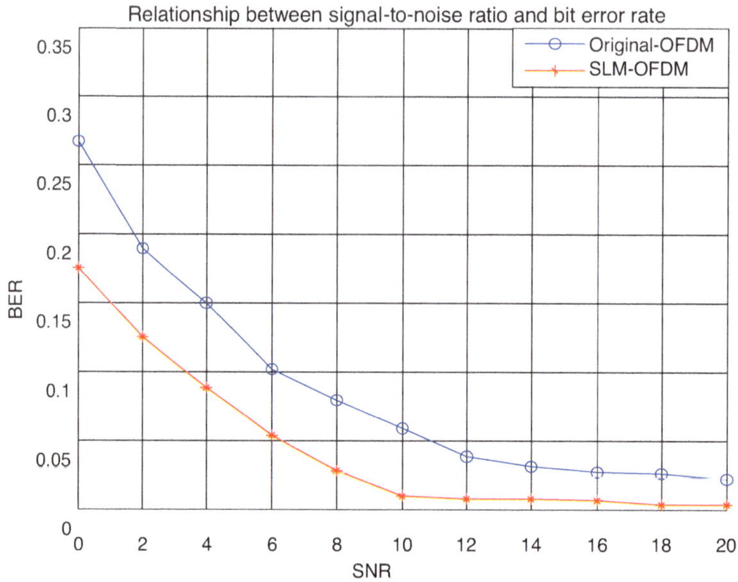

Fig. 7.28 Bit-error rate curve in the case of mixed noise [14, 15]

of the bit-error rate (BER) acting on the original OFDM and SLM-OFDM systems with different SNR values when the light-intensity fluctuation variance $\sigma_R^2 = 0.5$.

As can be seen from Fig. 7.28, when the SNR is high, owing to the influence of atmospheric turbulence, the bit-error rates of the original OFDM and SLM-OFDM systems do not further decrease with an increase in the SNR but converge around a value. For example, when $\sigma_R^2 = 0.5$, the signal-to-noise ratio is higher than 15 dB, the bit-error rate of the original OFDM system reaches 0.02, and the bit-error rate of the SLM-OFDM system reaches 0.003. When the signal-to-noise ratio is lower than 10 dB, both systems are significantly affected by additive noise.

Moreover, it can be clearly observed that the BER curve of the overall SLM-OFDM system is always below the BER curve of the OFDM system. Therefore, the BER of the OFDM system after the SLM algorithm is used is significantly higher than that of the original OFDM system.

7.3.4 Analysis of Results

We conducted experiments at short distances with a test-signal length of 10,000 points. At the receiving end, we used MATLAB to demodulate the collected data and then compared and analyzed the performance of the OWC-OFDM system through constellation and eye diagrams. The measurement principle of the OWC-OFDM wireless-laser communication system is shown in Fig. 7.29.

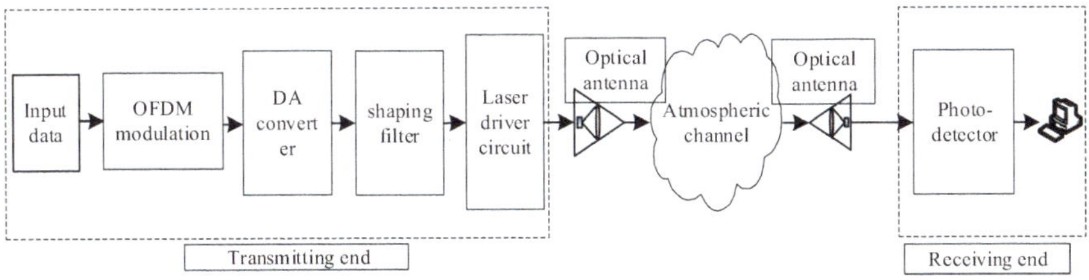

Fig. 7.29 OWC-OFDM measurement-principle diagram

7.3.4.1 Constellation Analysis

Figure 7.30 shows that at a probability of CCDF = 10^{-4}, the peak-to-average power ratio of the original OFDM system is 11.5 dB, and the peak-to-average power ratio of the SLM-OFDM system is reduced to 6.5 dB. The amplitudes of the signals were all normalized to the same interval, and then the constellation diagrams of the demodulation ends of the OFDM and SLM-OFDM systems for the short-distance experiment were created.

Subfigures (a) and (b) in Fig. 7.30 are the demodulation constellation diagrams of the OFDM and SLM-OFDM signals after the experiment, respectively. From the aggregation degree of the constellation diagrams, it can be seen that the experimentally measured SLM-OFDM-signal constellation diagram is more aggregated, compared with the OWC-OFDM-signal constellation diagram. Hence, after the SLM algorithm, the peak-to-average power ratio of the system is reduced.

In addition, the amplitude of the signal is normalized to the same interval under the same noise condition, so that the average power of the SLM-OFDM signal is greater than the average power of the OFDM signal, thereby improving the signal-to-noise ratio and reducing the bit-error rate of the system. The experimental results

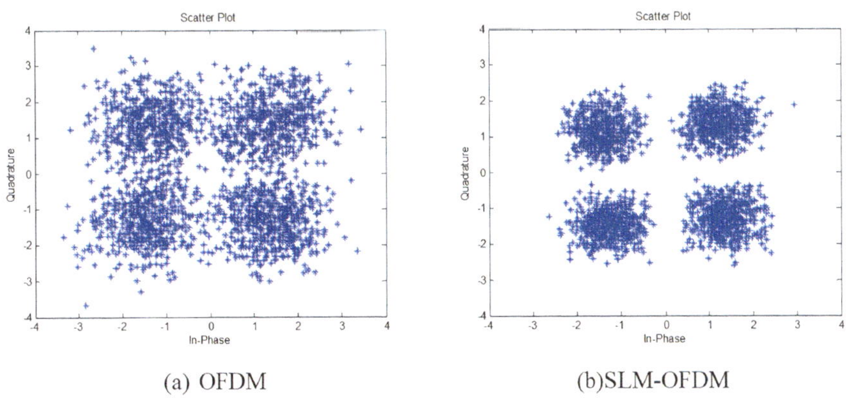

Fig. 7.30 Constellation diagrams of the OFDM and SLM-OFDM experimental demodulated signals [14, 15]

also verify that the quality of the OFDM signal after the SLM algorithm is improved, compared to the original OFDM signal during channel transmission.

7.3.4.2 Eye-Diagram Analysis

Figure 7.31 shows that at a probability of CCDF = 10^{-4}, the peak-to-average power ratio of the original OFDM system is 11.5 dB, and the peak-to-average power ratio of the SLM-OFDM system is reduced to 6.5 dB. The amplitudes of the signals were all normalized to the same interval, and then the demodulation-end eye diagrams of the OFDM and SLM-OFDM systems were created.

Observing Fig. 7.31, it can be clearly seen that the "eyes" of the SLM-OFDM demodulated signal have a larger opening height than those of the OFDM demodulated signal. In addition, they are more correct and the traces are clearer, indicating that the OFDM signal after adding the SLM algorithm is better than the original OFDM signal in channel-noise resistance. Thus, the probability of misjudgments at the receiving end is reduced, as is the bit-error rate of the system. The experiment also verifies the feasibility and effectiveness of the algorithm in practical applications.

7.4 Coding Technology to Reduce the Peak-to-Average Power Ratio

A multicarrier symbol is the modulated product of multiple independent subcarriers. Therefore, a coding method can be used to select a low-peak codeword for transmission, thereby effectively reducing the PAPR of a multi-carrier system. In this

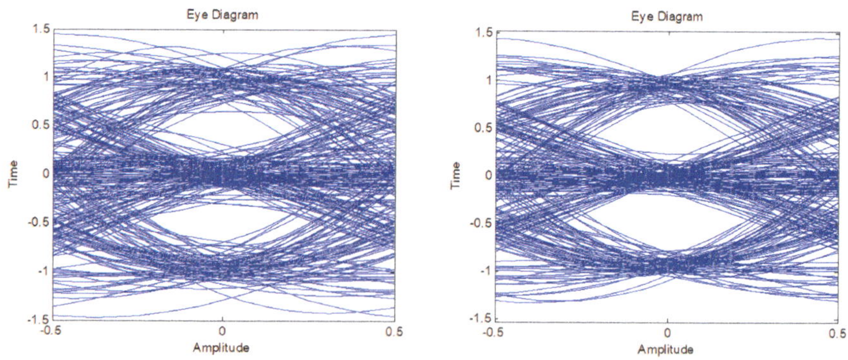

(a)OFDM (b)SLM-OFDM

Fig. 7.31 Eye diagrams of the experimental OFDM and SLM-OFDM demodulated signals [14, 15] section, coding methods are used to reduce the PAPR algorithm of an atmospheric laser multicarrier modulation system, and the system performance is analyzed.

The coding method used to reduce the peak-to-average power ratio selects or constructs codewords with small peak-to-average power ratios as transmission codewords. This concept can also be called peak-reduction coding (PRC). Coding technology is used to improve the PAPR value of the multicarrier signal at the expense of reducing the information-transmission rate.

OFDM technology can not only improve the spectrum utilization but also effectively suppress the multipath effect [16]. However, OFDM signals are composed of multiple independent modulated subcarrier signals. According to the central-limit theorem, the time-domain signal approximately follows a Gaussian distribution.

A higher peak-to-average power ratio naturally requires the analog-to-digital (A/D), D/A, and high-power amplifiers in the system to have a larger linear range; otherwise, it will cause spectrum expansion and in-band distortion of the transmitted signal to increase the bit-error rate, resulting in an increase in the system complexity. Therefore, reducing the PAPR is one of the key technologies in OFDM systems.

Many scholars at home and abroad have conducted extensive research on this problem and have proposed many solutions. According to their principles, these methods can be roughly divided into signal pre-distortion, signal coding, symbol scrambling, and signal-space expansion [17–20]. Signal pre-distortion causes signal distortion and increases the out-of-band radiation [17]; symbol scrambling adds auxiliary sideband detection information, which increases the redundancy of the system [18]. Signal-space expansion technology must obtain the best mapping table; when the number of carriers is large, the amount of computation is huge [19].

The coding algorithm is a distortion-free technology that involves seeking a smaller PAPR in the codeword combination. Usually, the smaller the required PAPR, the fewer the code groups that can be selected. They are specifically related to block coding, Golay complementary sequences, and Reed–Muller codes [20].

7.4.1 Block Coding

7.4.1.1 Block-Coding Method

The scheme for reducing the PAPR of an OFDM system using a block coding (BC) method was originally proposed by Jones and Wilkinson [21]. By introducing an odd parity bit into the codeword, eight types of codewords with the smallest PAPR were selected from the 16 types of codewords that may exist in an OFDM system composed of four subcarriers modulated by binary phase-shift keying (BPSK) for transmission.

The coding rules of the BC method are as follows:

$$b_4 = b_1 \oplus b_2 \oplus b_3 \oplus 1,$$

where $b_i, i = 1, 2, 3$, are information bits, b_4 is odd-parity coded bits, and $\oplus$ is binary addition. Table 7.4 shows the codewords and corresponding PAPRs that may be stored before and after block coding in an OFDM system composed of four subcarriers modulated by BPSK, where the oversampling factor $L = 8$. After block coding, among the 16 possible codewords in the original system, only eight codewords with a PAPR of 2.48 are allowed to transmit; codewords with other PAPR values are discarded.

The block-coding method can be adapted to an OFDM system in which the number of subcarriers is an integer multiple of four. Assume that the number of codewords that may be used is C_{poss} and the number of codewords allowed to be used after block encoding is C_{perm}.

Figure 7.32 shows the PAPR values that can be obtained under different C_{perm}/C_{poss} conditions. These values were obtained by calculating the PAPR of all possible codewords. For example, for an OFDM system with eight subcarriers modulated by BPSK, if the allowed codewords are equal to the number of possible codewords, that is, when block coding is not used, the PAPR of the OFDM system is 9.03 dB. If three-eighths of the possible codewords can be used, which is equivalent to a coding efficiency of 0.82, the resulting PAPR is 4.01 dB. If only one-fourth of the codewords are allowed to be used, that is, the coding efficiency is three-fourths, the resulting PAPR is 3.01 dB.

It can be seen from Fig. 7.32 that, as the number of subcarriers increases, C_{perm}/C_{poss} must be reduced to achieve the desired PAPR value; therefore, the block-coding

Table 7.4 Possible codewords and PAPR (BC) for OFDM systems

Numbers	Unencoded $b_1b_2b_3b_4$	PAPR (dB)	Block coding $b_1b_2b_3b_4$	PAPR (dB)
1	0000	6.02	0001	2.48
2	0001	2.48	0001	2.48
3	0010	2.48	0010	2.48
4	0011	3.73	0010	2.48
5	0100	2.48	0100	2.48
6	0101	6.02	0100	2.48
7	0110	3.73	0111	2.48
8	0111	2.48	0111	2.48
9	1000	2.48	1000	2.48
10	1001	3.73	1000	2.48
11	1010	6.02	1011	2.48
12	1011	2.48	1011	2.48
13	1100	3.73	1101	2.48
14	1101	2.48	1101	2.48
15	1110	2.48	1110	2.48
16	1111	6.02	1110	2.48

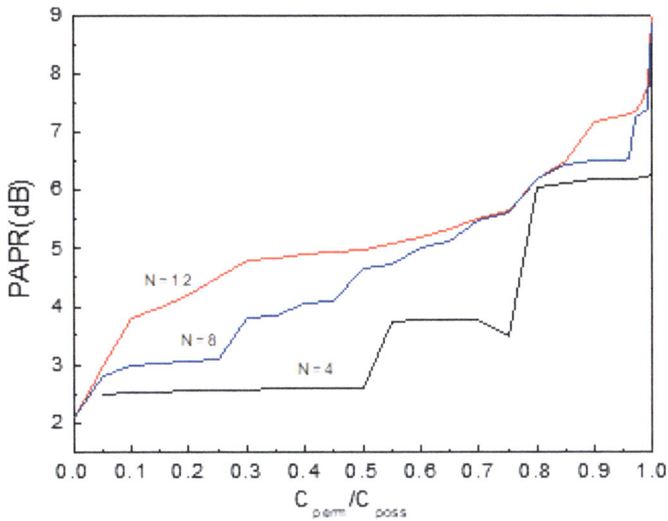

Fig. 7.32 PAPR values for different C_{perm}/C_{poss}

method has certain limitations. If the following formula is used:

$$R = 1 + \frac{1}{N}\log_2\frac{C_{perm}}{C_{poss}},$$

by converting C_{perm}/C_{poss} to coding efficiency R, the results shown in Fig. 7.33 can be obtained. It shows the PAPR values that can be obtained by the OFDM system under different coding efficiencies.

Fig. 7.33 PAPR under different coding efficiencies

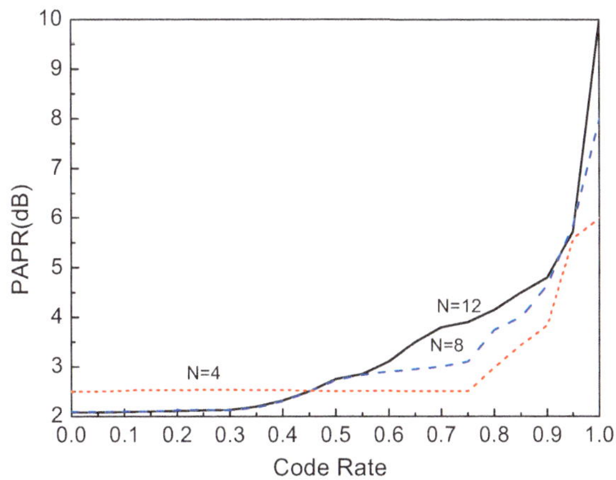

Table 7.5 PAPR performance improvement for the block-coding method

Number of subcarriers N	Unencoded PAPR (dB)	Coding efficiency R	Block coding PAPR (dB)	$PAPR_{rd}$ (dB)
4	6.02	3/4	2.48	3.54
8	9.03	7/8	6.53	2.50
12	10.79	11/12	9.21	1.58
16	12.04	15/16	10.88	1.16
20	13.01	19/20	12.09	0.92

As seen in Fig. 7.33, under the premise of a proper coding efficiency, the PAPR performance of an OFDM system can be significantly improved. In addition, there is a trade-off between the coding efficiency and PAPR; that is, the higher the coding efficiency, the larger the achieved PAPR, and the lower the coding efficiency, the smaller the achievable PAPR.

Table 7.5 shows the PAPR performance improvement obtained using the block-coding method for OFDM systems with different numbers of subcarriers, where $PAPR_{rd}$ represents the PAPR improvement value after coding. As can be seen from the table, with an increase in the number of subcarriers (the corresponding coding efficiency also increases), the performance of the block-coding method worsens the PAPR.

For an OFDM system with a large number of subcarriers, the codeword with the required PAPR value can be selected from all possible codewords using a table lookup; however, with an increase in the number of subcarriers, the table-lookup method becomes complicated and unrealistic.

7.4.1.2 Simple Block-Coding Method

Reference [22] proposed a method called simple block coding (SBC) to reduce the PAPR in OFDM systems with different subcarrier numbers. A codeword of length $N - 1$ is expanded into a codeword of length N by adding a one's complement codeword at the end of the information bits. The encoding rules for the SBC method are as follows:

$$b_N = \overline{b}_{N-2}.$$

That is, the SBC method adds the second-to-last digit of the information bit to the last digit of the information bit, and $\overline{b}_{N-2}$ represents the complement of the $N - 2$nd information bit; thus, the PAPR of an OFDM system composed of N-1 subcarriers can be reduced. The corresponding coding efficiency is $(N - 1)/N$.

For an OFDM system composed of four subcarriers modulated by BPSK, after SBC coding, the codeword length is five and the coding efficiency is 4/5. The

16 possible codeword combinations composed of four information bits become 16 possible codeword combinations with a codeword length of five after SBC encoding.

The difference between SBC and BC is that SBC does not determine the codeword with the smallest PAPR value from all possible codeword combinations for transmission; it reduces the maximum PAPR value of the codeword combination by adding a one's complement bit at the end of the information. Although the PAPR value of the encoded codeword increases for individual codewords, the overall PAPR value of the encoded codeword is extremely low [21].

If the coded bits are also regarded as information bits, for an OFDM system with a BPSK-modulation codeword length of five, there are 32 possible codeword combinations. For an OFDM system formed by four-bit codewords after SBC, there are only 16 codeword combinations, which is equivalent to selecting 16 suitable codewords from 32 codeword combinations for transmission and discarding other codewords.

7.4.1.3 Complement Block-Coding Method

In the block-coding method, the number of complement bits is one. The complement bits are all separated, and there is no case in which the complement bits are connected. If the number of complement bits is increased in the SBC method, the positions of the complement bits are changed, the complement bits are connected together, the PAPR of an OFDM system with any number of subcarriers can be reduced, and the coding efficiency is not greater than $N - 1/N$ [22, 23].

For a codeword of an OFDM system with a certain number of subcarriers modulated by BPSK, the PAPR of the system can be changed by increasing the number of complement bits in the codeword sequence or changing the position of the complement bits. For convenience of analysis, in the following description, the codeword sequence consists of two parts: the information-bit sequence and the complement-bit sequence.

If the length of the codeword sequence is N, the number of bits is N, and the number of complement bits is m, then the number of information bits is $N - m$. When the number of complement bits in the codeword sequence increases, the PAPR of the system can be changed. For example, for an unencoded original input information sequence {11111111}, when the oversampling factor $L = 8$, the corresponding PAPR is 9.03 dB. When the number of complement bits is 1, 2, and 3, the encoded codeword sequences are {11111110}, {11111100}, and {11111000}, respectively, and the corresponding PAPRs are 6.53 dB, 5.22 dB, and 5.99 dB, respectively.

The peaks before and after encoding are shown in Fig. 7.34 [22]. When the codeword length is constant, changing the number of complement bits in the codeword sequence can change the number of complement bits in the information sequence and the PAPR of the system. However, the ability to reduce the PAPR does not increase as the number of complement bits increases. As shown in Fig. 7.34, the PAPR of a codeword with three's complement bits is greater than that of a codeword with two's complement bits.

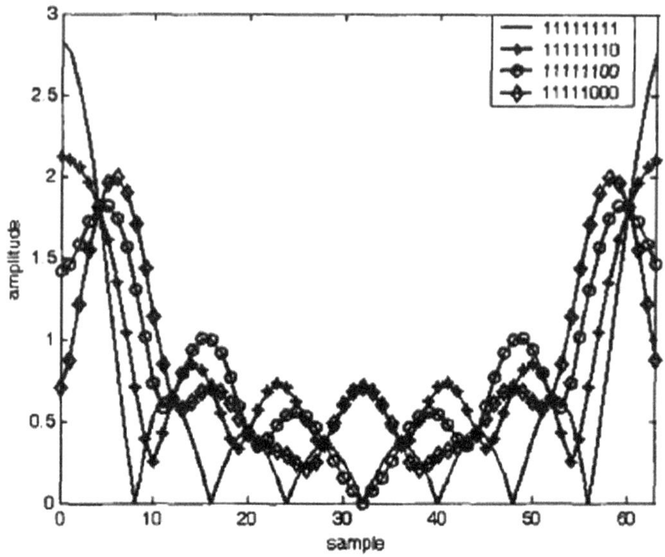

Fig. 7.34 OFDM signal peaks with different numbers of complement bits [23]

When the number of complement bits is constant, different PAPR performance improvements can be obtained by changing the position of the complement sequence. The position of the complement sequence is defined as the starting position of the first bit of the complement sequence in the codeword sequence.

For example, when the number of complement bits is two, changing the position of the complement sequence can yield different codeword sequences. When the codeword sequence is {11111111}, {11111100}, {11111001}, {1111001100}, and {11100111}, the corresponding PAPR values are divided into 9.03 dB, 5.22 dB, 3.01 dB, 4.15 dB, and 5.36 dB, respectively.

Figure 7.35 shows the peak situation when the number of complement bits is two, but the positions are different. It can be seen from Fig. 7.35 that changing the position of the complement sequence can change the PAPR of the system when the other conditions remain unchanged. It can be seen from the above analysis that the PAPR performance can be improved by changing the number of complement bits or the position of the complement sequence.

7.4.2 Coding and Decoding Algorithms Based on Golay Complementary Sequences and Reed–Muller Codes

Golay proposed a binary sequence with complementary properties [25]: a pair of equal-length finite sequences $a = [a_0, a_1,..., a_i,..., a_{N-1}]$ and $b = [b_0, b_1, ...,b_i, ..., b_{N-1}]$. If the autocorrelation characteristics of the two satisfy

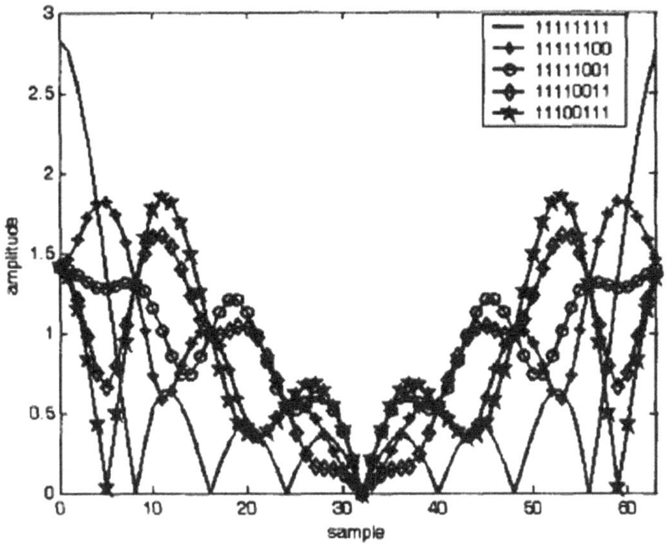

Fig. 7.35 OFDM signal peaks at different positions in the complement sequence [24]

$$\sum_{i=0}^{N-1-n} a_i a_{i+n}^* + \sum_{i=0}^{N-1-n} b_i b_{i+n}^* = 2N\delta(n) = \begin{cases} 2N; n = 0 \\ 0; \textit{otherwise} \end{cases}, \tag{7.25}$$

then, a and b are said to be complementary sequences to each other. N is the sequence length, $\delta(n)$ is the impulse function, and n is the interval length. Golay complementary sequences have the following properties.

Property 1. The sequences obtained by satisfying any one of the following conditions are a pair of Golay complementary sequences:

(a) a and b after the exchange are also a pair of Golay complementary sequences.
(b) The reverse sequence of a and b are a pair of Golay complementary sequences.
(c) The reverse sequence of b and a are also a pair of Golay complementary sequences.
(d) After inversion, a and b are a pair of Golay complementary sequences.
(e) After inversion, b and a are also a pair of Golay complementary sequences.

According to the first property, for a given Golay complementary-sequence pair, $a = \{01100101\}$ and $b = \{00101010\}$, the following new Golay complementary-sequence pairs can be generated:

$$\{00101010\} \text{ and } \{01100101\}; \quad \{10100110\} \text{ and } \{00101010\};$$

$$\{01010100\} \text{ and } \{01100101\}; \qquad \{10011010\} \text{ and } \{00101010\};$$

$$\{11010101\} \text{ and } \{01100101\}.$$

Property 2. The sum of the squares of the length of the complementary sequence is two.

Property 3. If $a = [a_0, a_1,\ldots, a_i,\ldots, a_{N\text{-}1}]$ and $b = [b_0, b_1, \ldots, b_i, \ldots, b_{N\text{-}1}]$ are a Golay complementary-sequence pair, then sequences P and Q are also a pair of Golay complementary sequences, where $P = ab = \{a0, a1, \ldots, ai, \ldots, aN - 1\ b0, b1, \ldots, bi, \ldots, bN - 1\}$ and $Q = a\overline{b} = \left\{a_0 a_1 \cdots a_i \cdots a_{N-1}\overline{b_0 b_1} \cdots \overline{b_i} \cdots \overline{b}_{N-1}\right\}$. $\overline{b}$ represents the inversion of b.

Property 4. If $a = [a_0, a_1, \ldots, a_i, \ldots, a_{N\text{-}1}]$ and $b = [b_0, b_1, \ldots, b_i, \ldots, b_{N-1}]$ are a pair of Golay complementary sequences, then U after interleaving and V are also pairs of Golay complementary sequences, where $U = \{a_0 b_0, a_1 b_1 \ldots a_i b_i \ldots a_{N-1} b_{N-1}\ b_0, b_1, \ldots, b_i, \ldots, b_{N-1}\}$ and $V = \left\{a_0\overline{b}_0 a_1\overline{b}_1 \ldots a_i\overline{b}_i \ldots a_{N-1}\overline{b}_{N-1}\right\}$.

These properties play an extremely important role in reducing the PAPR using Golay complementary sequences. The instantaneous powers of the complementary-sequence pairs can be expressed as

$$p_a(t) = \sum_{k=0}^{N-1} a_k e^{j2\pi k\Delta ft}\left(\sum_{m=0}^{N-1} a_k e^{j2\pi m\Delta ft}\right)^* = \sum_{k=0}^{N-1}\sum_{m=0}^{N-1} a_k a_m e^{j2\pi(k-m)\Delta ft}$$

$$p_b(t) = N + 2\sum_{u=1}^{N-1} R\{A_a(n)\}e^{j2\pi u\Delta ft} \tag{7.26}$$

$$p_b(t) = N + 2\sum_{u=1}^{N-1} R\{A_a(n)\}e^{j2\pi u\Delta ft} \tag{7.27}$$

Adding Eqs. (7.26) and (7.27) gives

$$p_a(t) + p_b(t) = 2N + 2\sum_{u=1}^{N-1} (R\{A_b(n)\} + R\{A_a(n)\})e^{j2\pi u\Delta ft} \le 2N, \tag{7.28}$$

where $A_a(n)$ is the autocorrelation function of a and $A_b(n)$ is the autocorrelation function of b.

Because the instantaneous power is $0 \le p_b(t) \le 2N$ and $0 \le p_a(t) \le 2N$ and its average power $P_{ave} = N$, the peak-to-average ratio of a Golay complementary sequence is $PAPR \le \frac{2N}{N} = 2(3dB)$.

Reference [26] proved that the complementary sequence of a Golay code can be constructed using the Reed–Muller (RM) encoding method. The RM code method is highly efficient. It divides the non-binary RM code into several cosets to separate codewords with large peak-to-average power ratios, thereby reducing the peak-to-average ratio.

Reference [25] introduced two inferences, according to which, the input binary sequence is encoded to form a Golay complementary sequence [23–25]. The following inferences were made.

Inference 1. In $ZRM_{2^h}(2, m)$, each RM (1, m) has $\frac{m!}{2}$ cosets and a common coset expression: $2^{h-1}\sum_{k=1}^{m-1} x_{\pi(k)}x_{\pi(k+1)}$, which forms $2^{h(m+l)}$ Golay complementary sequences of length 2^m. $RM_{2^h}(r, m)$ represents a Reed–Muller code of r-order 2^h base and a length of 2^m, π is a permutation of {1, 2, …, m}, m is the code length, h is the modulation depth, and $h \geq 1$.

Inference 2. For any permutation π of symbols {1, 2, …, m}, select c, $c_k \in Z_{2^k}$; then, $a(x_1, x_2, \ldots, x_m) = 2^{h-1}\sum_{k=1}^{m-1} x_\pi(k)x_\pi(k+1) + \sum_{k=1}^{m} c_k x_k + c$ is a Golay complementary sequence of length 2^m on Z_{2^h}.

7.4.2.1 Encoding Algorithm

Golay complementary sequences have many excellent properties; they have great potential for reducing PAPR, and their use can extend a Golay complementary sequence. The specific encoding algorithm is as follows:

(1) Let the number of subcarriers be 2^m and the number of input binary sequences be $w + h(m+1)$; we take $w = INT\left(\log_2 \frac{m!}{2}\right)$, and use the $M = 2^h$ constellation mapping. We take the first w bits of the input sequence to form a decimal number d, and for the remaining $h(m+1)$ bits of the input sequence, select h bits in turn to form $(m+1)$ groups of decimal information symbols: $u_1, u_2, \ldots, u_{m+1}$, $u \in Z_{2^h}$.

(2) Generate a generator matrix. For $RM_{2^h}(r, m)$, where $0 \leq r \leq m$, for each pair of integers r and m, there is an r-order RM code with a code length of 2^m; then, the corresponding generator matrix is $G = [G_0\ G_1 \ldots G_l \ldots G_r]^T$. Among them, G_0 is an all '1' vector with a code length of $n = 2^m$; G_1 is a matrix of order $m \times 2^m$, and its ith row vector is as follows: 2^{m-i} zeroes are followed by 2^{m-i} ones. This result is repeated 2^{i-1} times in each row (i = 1, 2, …, m). G_l is an $\binom{m}{l} \times 2^m$-order matrix, and the row of G_l is composed of the vectors obtained by the inner product of all l row vectors of G_l. We multiply $u_1, u_2, \ldots, u_{m+1}$ by each row vector of $RM_{2^h}(r, m)$, and then add them modulo 2^h to obtain S, $S = \sum_{i=1}^{m} u_i x_i + u_{m+1} \left(\bmod 2^h\right)$.

(3) According to inference 1, the $d + 1$st row vector in the coset matrix is selected by the decimal value d and added to S, according to modulo 2^h, thereby forming a Golay complementary sequence.

7.4.2.2 Decoding Algorithm

In linear codes, an important basis for measuring the error-correction and error-detection ability of a coding scheme is the Hamming distance [26–29]. In a codeword set composed of multiple equal-length codewords, the minimum Hamming distance is the minimum distance that defines the distance between any two codewords. Let $a = [a_0, a_1, \ldots, a_{N-1}]$ be a sequence of length N, $a \in Z_H$, slightly redefined as $\sum_{i=0}^{N-1} \min(a_i, H - a_i)$, where H represents the length of the Z lattice.

Therefore, the condition for the decoding algorithm to be established is that if $wt_{2^k}(e) < 2^{m+k-2}, k = 0, \ldots, N-1$, is satisfied, then r can be correctly translated into c. Here, e represents the error-code sequence, r represents the Golay complementary sequence after demapping at the receiving end, and c represents the initial coding sequence. w is the weight of the error sequence e modulo 2^k, which is defined as $wt_{2^k}(i) = \min\left(i \bmod 2^k, 2^k - \left(i \bmod 2^k\right)\right)$. For $RM_{2^h}(r, m)$, when $h \geq 1$, the specific decoding algorithm is as follows:

(1) The length received by the receiver is a 2^m sequence, $r \in Z_{2^k}$; let $k = 0$ and $r_0 = r$;
(2) Define sequence y, where yi is $(y)_i = 2^{k-1} - wt_{2^{k+1}}((r_k)_i), i = 0, 1, \ldots, 2^m - 1$;
(3) Let $\hat{y}$ be the fast Hadamard code transform of y. In $j \in Z_{2^m}$, determine the maximum magnitude of $\hat{y}$, that is, $(\hat{y})_j$. According to whether $(\hat{y})_j$ is positive or negative, let w be 0 or 1, and let the binary representation of j be $w_1, w_2, \ldots, w_m$. Use w and j to determine f_k; that is,

$$f_k = \left(\sum_{i=1}^{m} w_i x_i + w\right) \bmod 2^{h-k}. \tag{7.29}$$

(4) If $k = h - 1$, the decoding output is $(2^{h-1} f_{h\text{-}1} + 2^{h-2} f_{h-2} + \ldots f_0) \bmod 2^h$. Otherwise, $r_{k+1} = (r_k - 2^k f_k) \bmod 2^h$. Increase the k value, return to step 2, and continue the calculation until $k = h - 1$.

The decoding algorithm is a hard-decision method that is simple and easy to implement; however, with an increase in the code length, the order of decoding increases, and the computational complexity increases accordingly.

7.4.2.3 System-Performance Analysis

Using quadrature phase-shift keying (QPSK) ($h = 2$) modulation and 16 subcarriers, the coding method is as follows:

First, construct an $RM(1, 4)$ generator matrix:

Set $m = 4, n = 2^m = 16$, and $r = 1$, and construct a one-order RM code. The three matrices of the generating matrix are as follows.

$$G_0 = [1111111111111111] = [x_0]; \tag{7.30}$$

$$G_1 = \begin{bmatrix} 0000000011111111 \\ 0000111100001111 \\ 0011001100110011 \\ 0101010101010101 \end{bmatrix} = \begin{bmatrix} x_1 \\ x_2 \\ x_3 \\ x_4 \end{bmatrix}. \tag{7.31}$$

Next, construct (4!/2 = 12) cosets of $RM(1, 4)$:

$$\begin{aligned}
(0002002000022202) &= 2(x_1x_2 + x_2x_3 + x_3x_4) \\
(0002022000022022) &= 2(x_1x_3 + x_2x_4 + x_3x_4) \\
(0000022000220202) &= 2(x_1x_3 + x_2x_3 + x_2x_4) \\
(0002020000200222) &= 2(x_1x_3 + x_3x_4 + x_2x_4) \\
(0000022002020022) &= 2(x_1x_4 + x_2x_4 + x_2x_3) \\
(0002002002000222) &= 2(x_1x_4 + x_3x_4 + x_2x_3) \\
(0002000200202202) &= 2(x_1x_2 + x_3x_4 + x_1x_3) \\
(0002000202002022) &= 2(x_1x_2 + x_1x_4 + x_3x_4) \\
(0000002202200202) &= 2(x_2x_3 + x_1x_3 + x_1x_4) \\
(0000020202200022) &= 2(x_2x_4 + x_1x_3 + x_1x_4) \\
(0000020200222002) &= 2(x_2x_4 + x_1x_3 + x_1x_2) \\
(0000002202022002) &= 2(x_2x_3 + x_1x_4 + x_1x_2).
\end{aligned} \tag{7.32}$$

Then, determine the amount of input information (bits). Because w = [log2 (4!/2)] = 3, and $h(m + 1) = 10$, the input binary information is 13 bits. The first three bits are used to select the coset of a codeword, and the last 10 bits are used to select a codeword in this coset.

For example, if the input is [0111011010001], [011] is used to select the third coset (0000022000220202), the last 10 digits are used to select the linear combination $2x_1 + 3x_2 + x_3 + 0x_4 + x_0$ = (1122001133002233). Then, S =

$\sum_{i=1}^{m} u_i x_i + u_{m+1}\left(\bmod 2^k\right)$ is used to obtain the codeword (1122023133222031). In this manner, a Gray complementary-sequence codeword in the third coset is obtained. With {0, 1, 2, 3} corresponding to the QPSK-modulated {±1, ±j}, the input signal of the IFFT can be obtained.

Consider a multicarrier system with QPSK ($h = 2$) modulation and 16 ($m = 4$) subcarriers as examples. The number of simulations is 1000, and the oversampling rate is 16. The performance of the system is analyzed before and after encoding, as shown in Figs. 7.36 and 7.37. It can be seen from the figure that before coding, the PAPR value is close to 10.5 dB; after coding, the PAPR value can be effectively reduced to below 3 dB.

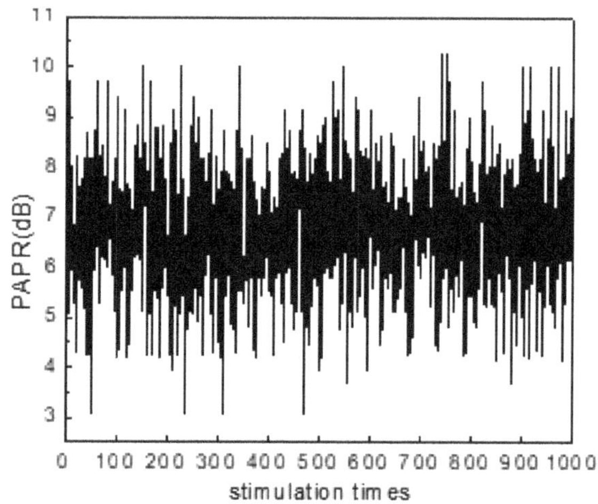

Fig. 7.36 Change curve of unencoded PAPR [30]

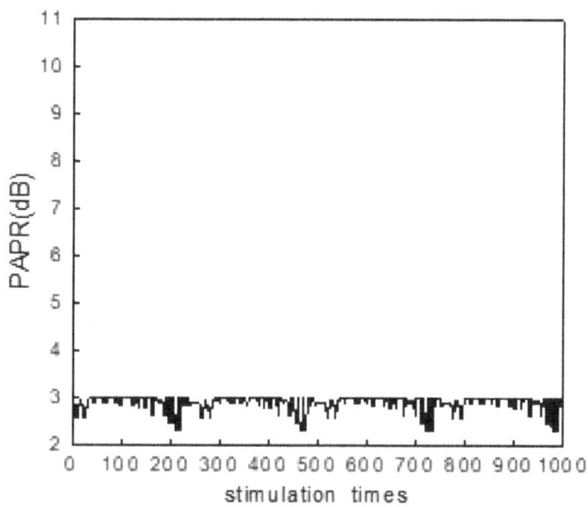

Fig. 7.37 Change curve of PAPR after coding [30]

Golay complementary sequences with Reed–Muller codes can separate codewords with larger PAPR, thereby reducing the PAPR. However, this method has a certain scope of application, which is affected by the type of constellation. It can be determined from the inference that this method is only suitable for MPSK modulation with a constant envelope, not for MQAM modulation with a variable envelope.

Now, consider a multicarrier system using QAM modulation and 16 ($m = 4$) subcarriers as an example. The number of simulations is 1000, and the oversampling rate is 16. The performance of the system before and after encoding is analyzed. It can be seen from Figs. 7.38 and 7.39 that before using the coding method, the PAPR under QAM modulation is concentrated at approximately 9 dB, the maximum is 14 dB, and the minimum is approximately 5 dB, far exceeding 3 dB. After encoding, the performance of PAPR has improved, but it is still concentrated around 4.5 dB.

The modulation depth and code length have an impact on the coding efficiency. Using the Golay complementary-sequence and Reed–Muller coding method, $w = INT\left(\log_2 \frac{m!}{2}\right)$ bits are used to select one of the $\frac{m!}{2}$ cosets, and $(m + 1)$ bits are used to select one of the 2^{m+1} codewords in each coset. The coding efficiency (referred to as the code rate) of the block code is $R = \frac{w+h\cdot(m+1)}{2^m \cdot h}$, where w is the number of Golay complementary sequences, m is the code length, and h is the modulation depth.

When $h = 1$, BPSK is used; when $h = 2$, QPSK is used; and when $h = 3$, 8 PSK is used. Figure 7.40 shows the curve of the code rate with the modulation depth, and Fig. 7.41 shows the curve of the code rate with the code length.

In Fig. 7.40, when the code lengths are 2, 4, and 8, there is a relationship between the code rate and modulation depth. It can be seen from Fig. 7.40 that under any modulation depth, the longer the code length, the lower the code rate; however, under the same code length, the code rate does not change much with the increase in the modulation depth.

Fig. 7.38 Graph of 16 × oversampling using QAM without coding [30]

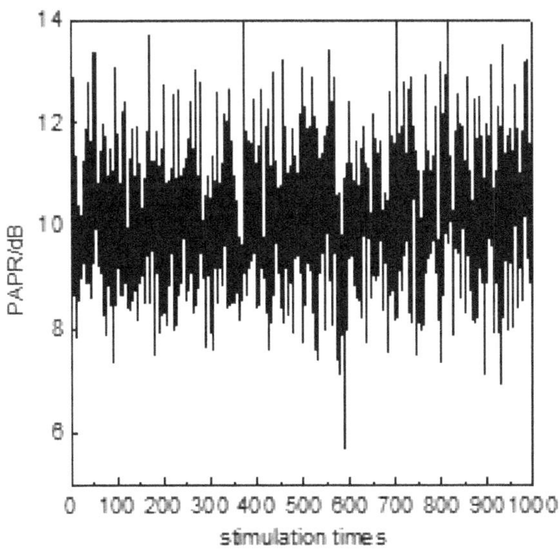

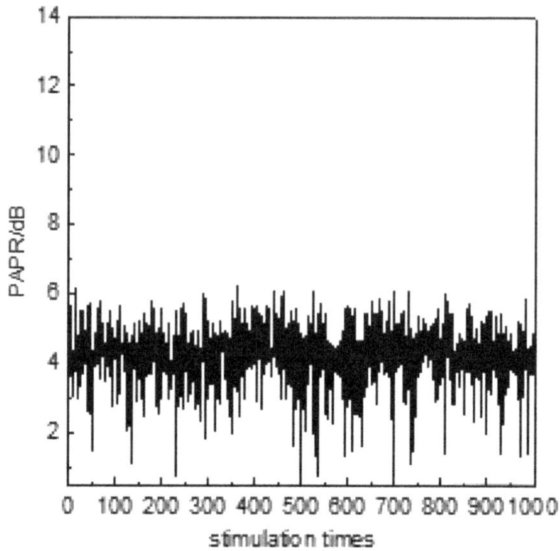

Fig. 7.39 Graph of 16 × oversampling using coded QAM [30]

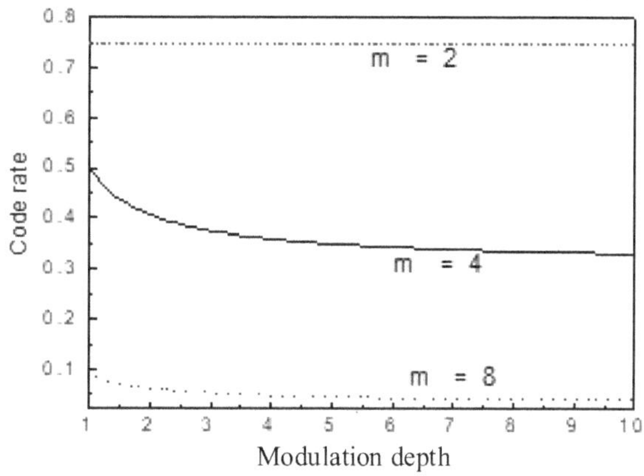

Fig. 7.40 Variation curve of code rate R with modulation depth h under different code lengths m [30]

In Fig. 7.41, when the modulation modes are BPSK, QPSK, and 8PSK, the relationship between the code rate and the code length is shown. It can be seen from the figure that at any code length, the deeper the modulation depth and the lower the code rate. Moreover, at the same modulation depth, when the code length m increases, the code rate decreases exponentially.

The use of a multi-carrier system can not only effectively suppress interference, but also improve spectrum utilization; however, the problem of a high peak-to-average power ratio seriously restricts its application. This method, based on the combination of a Golay complementary sequence and a Reed–Muller code, is very effective in reducing the PAPR of a multicarrier signal. The research results show that [31]

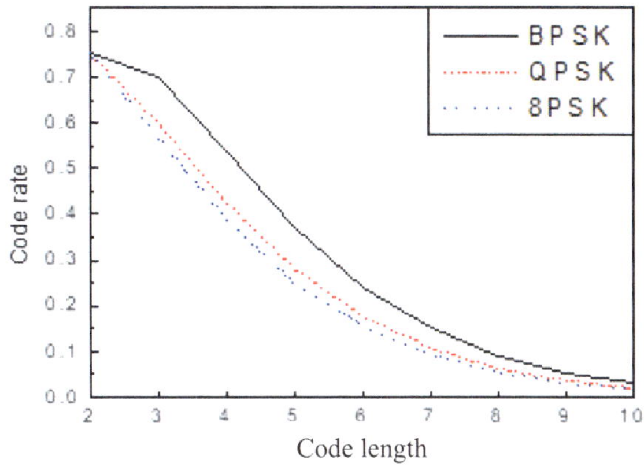

Fig. 7.41 Curve of code rate *R* versus code length *m* under different modulation depths *h* [30]

this method can not only reduce the peak-to-average power ratio to <3 dB but also improve the error-correction and detection ability of the system.

However, this encoding method has certain restrictions on the type of constellation. It can be seen through simulation that this encoding method is only suitable for MPSK modulation with a constant envelope, not for MQAM modulation with a variable envelope. When the number of subcarriers is large, the code rate becomes extremely low. To avoid making the code rate too low, it should be increased by selecting an appropriate modulation depth.

Chapter Summary

The reasons for the PAPR of an OWC-OFDM system and methods to reduce it were analyzed. The selective mapping (SLM) and partial transmit sequence (PTS) algorithms were introduced, and the effectiveness of the two algorithms was verified.

The influence of the SLM and PTS algorithms on the OWC-OFDM system was analyzed using constellation diagrams, and an experiment verified the effectiveness of the SLM algorithm in the OWC-OFDM system.

The results showed that the optical OFDM signal was affected by atmospheric turbulence and atmospheric molecular scattering when transmitted in the atmosphere. An SLM algorithm at the transmitting end could effectively reduce the peak-to-average ratio of the optical OFDM signal. After the amplitude of the signal was normalized to the same interval, the average power of the SLM-OFDM signal was increased, thereby improving the signal-to-noise ratio and reducing the influence of channel noise on the optical OFDM system.

References

1. Song T, You X, Shen L (2002) A method to reduce PAPR in OFDM systems based on BCM. J Appl Sci 20(4):399–403
2. Chen H, Haimovish M (2003) Iterative estimation and cancellation of clipping noise for OFDM signals. IEEE Commun Lett 7(7):305–307
3. Kaiho H, Hattori T (2008) Reduced complexity of spread OFDM for PAPR reduction and its basic performance. In: The 9th Asia-Pacific conference on IEEE conference, pp 311–414
4. Jones AE, Wilkinson TA, Bajrton SK (1994) Block coding scheme for reduction of Peak to mean envelope power ratio of multicarrier transmission schemes. Electron Lett 30(25):2098–2099
5. Yeongeheol Y, Seok C (2003) A novel scheme to reduce peak-to-average power in OFDM based onstandard arrays of linear block codes.In: IEEE international symposium on information theory
6. Golay MJE (1961) Complementary series. IRE Trans Inf Theory IT(7):82–87
7. Ryu H-G, Youn K-J (2002) A new PAPR reduction scheme: SPW (Subblock phase weighting). IEEE Trans Consum Electron 48(1):81–89
8. Lim DW, Heo SJ, No JS (2006) A new PTS OFDM scheme with low complexity for PAPR reduction. IEEE Trans Broadcast 52(1):77–82
9. Muller SH, Huber JB (1997) OFDM with reduced peak-to-averagepower ratio by optimum combination of partial transmit sequences. Electron Lett 33(5):368–369
10. Cimini LJ Jr, Sollenberger NR (2000) Peak-to-average power ratio reduction of an OFDM signal using partial transmit sequences. IEEE Commun Lett 4(3):86–88
11. Kang SG, Kim JG, Joo EK (1999) A novel subblock partition scheme for partial transmit sequence OFDM. IEEE Trans Broadcast 45(3):333–338
12. Robert B, Robert F, Johannes BH (1996) Reducing the peak-to-average power ratio of multicarrier modulation by selected mapping. IEEE Electron Lett 32(22):2056–2057
13. Xie M (2012) Research on the impact of atmospheric dispersion on FSO-OFDM system. Xi'an University of Technology, Xi'an, pp 35–36
14. Ke X, Ye K, Liu J (2017) Experimental research on peak-to-average power ratio control methods in FSO-OFDM systems. Infrared Laser Eng 46(06):180–186.
15. Liu J (2016) Experimental research on peak-to-average power ratio control methods in FSO-OFDM systems. Xi'an University of Technology, Xi'an
16. Yin C, Luo T, Le G (2004) Multicarrier Broadband Wireless Communication Technologies [M]. Beijing University of Posts and Telecommunications Press, Beijing
17. Huang X, Lu J (2001) Reduction of peak-to-average power ratio of OFDM signals with companding transform. Electron Lett 37(8):506–507
18. Muller SH, Huber JB (1997) SLM, peak-to-average power ratio by optimum combination of partial sequences. Electron Lett 33(5)
19. Sumasu A, Ue T, Uesugi M, Kato O, Homma K (2000) A method to reduce the peak power with signal space expansion (ESPAR) for OFDM system.Electron Lett 405–409
20. Olfat M, Liu KJR (2004) Low peak to average power ratio cyclic Golay sequences for OFDM systems. IEEE Commun Soc 993–997
21. Narahashi S, Nojima T (1994) New phasing scheme of N-multiple carriers for reducing peak-to-average power ratio. Electron Lett 30(17):1382–1383
22. Fragicomo S, Matrakidis C, O'Reilly JJ (1998) Multicarrier transmission peak to average power reduction using simple block code. Electron Lett 34(10):953–954
23. Olfat M, Liu KJR (2004) Low peak to average power ratio cyclic golay sequences for OFDM systems.IEEE Commun Soc 993–997
24. Ding M, Liu Y (2003) Performance analysis and simulation of OFDM systems using complementary sequence block coding. J Circ Syst (3) 7–12
25. Davis JA, Jedwab J (1997) Peak-to-mean power control and error correction for OFDM transmission using Golay sequences and reed-muller codes. Electron Lett 33(4):267—268

26. Davis JA, Jedwab J (1999) Peak-to-mean power control in OFDM, Golay sequences and reed-muller codes. IEEE Trans Inf Theory 11(45):2397–2417
27. Wang X, Xiao G (2001) Error Correction Codes—Principles and Applications. Xidian University Press, Xi'an
28. Chen L, Guo Z, Hua J et al (2003) A method to suppress peak-to-average power ratio based on Golay complementary sequences. Digit TV Digit Video 12:25–27
29. Yang M (2006) Research on peak-to-average power ratio reduction techniques in OFDM systems. Lanzhou University, Lanzhou
30. Zhang H. Research and implementation of peak-to-average power ratio reduction in wireless optical wireless communication multicarrier modulation. Xi'an University of Technology, Xi'an
31. Zhao L, Ke X, Sun L (2011) Research on reducing peak-to-average power ratio in FSO-OFDM systems. Infrared Laser Eng 40(09):1749–1753

Chapter 8
Channel Estimation and Assignment

A optical wireless orthogonal frequency-division multiplexing (OFDM) channel is mainly affected by particle scattering, such as dust, rain, snow, and fog. An atmospheric channel is not as fixed and predictable as a wired channel, but has greater randomness, which distorts a signal passing through it. Delay also affects OFDM symbols. Whether increasing the transmit power will suppress the attenuation of an optical signal is also unclear. This chapter introduces common algorithms for channel estimation and assignment.

8.1 Optical OFDM Channel Estimation

Channel estimation is the process of estimating the model parameters of an assumed channel model from the received data. If the channel is linear, the system impulse response is estimated. It should be emphasized that channel estimation is a mathematical representation of the effect of the channel on the input signal, and a "good" channel estimation minimizes some estimation error [1]. Therefore, the main task of channel estimation is to accurately identify or estimate the channel through a certain method to obtain the transmission characteristics in the time or frequency domain.

For an optical OFDM system, channel estimation involves estimating the frequency-response value on each subcarrier, $\hat{H}_k$, where $k = 0, 1, 2 \ldots, N - 1$, is the number of the subcarrier. Figure 8.1 shows a block diagram of a general model for OFDM channel estimation.

Optical OFDM channel estimation is divided into frequency- and time-domain estimation. The time-domain estimation of an atmospheric channel is carried out before the fast Fourier transformation (FFT), and the frequency-domain estimation is carried out after the transformation. It is also possible to combine the two to estimate simultaneously, so that their respective advantages can be brought into full play.

X. Ke, *Principles and Applications of Optical Wireless Orthogonal Frequency-Division Multiplexing*, Optical Wireless Communication Theory and Technology,
https://doi.org/10.1007/978-981-97-7973-4_8

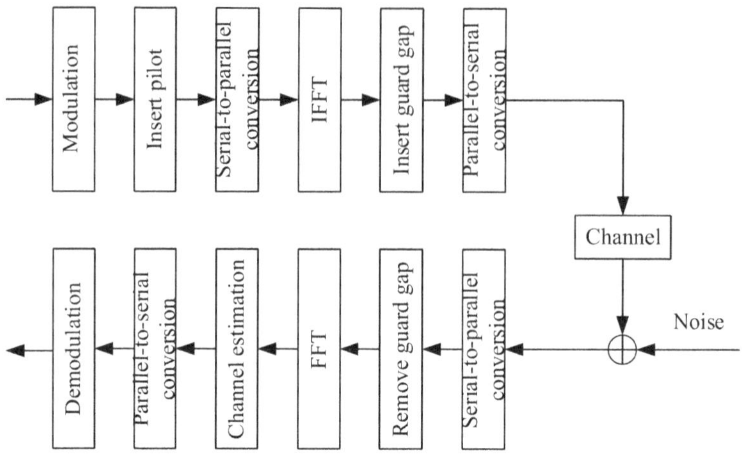

Fig. 8.1 Channel-estimation principle block diagram

8.1.1 Channel-Estimation Classification

Depending on the different OFDM channel resources, OFDM channel estimation is generally divided into the following four categories [2].

(1) Decision estimation, based on the training sequence. To better estimate the time-varying channel, training symbols must be sent periodically. If the channel changes rapidly, the actual channel can be very different from the originally estimated one. This means that bit errors are generated more easily in optical OFDM symbols; thus, the number of training sequences must be increased.

To solve this problem, decision-based channel estimation is proposed. This algorithm demodulates the data, decodes them, and decides using the last channel-estimation value; that is, it is equivalent to having a level of negative feedback in the receiver. The decision result will also have a significant influence on the receiver's accuracy.

(2) Blind channel estimation. A blind channel-estimation method uses certain characteristics inherent to the modulated signal, which are not related to the specific information-bearing bits. Alternatively, the decision-feedback method can be used to estimate the channel.

(3) Semi-blind channel estimation. This channel-estimation method combines the advantages of blind estimation and training sequence–based estimation. Generally speaking, the estimation method usually involves designing a training sequence or periodically inserting pilot symbols into the data.

Blind and semi-blind channel-estimation algorithms do not need or require short training sequences, and have high spectral efficiency; hence, they have been widely studied. However, the computational complexity of general blind and semi-blind estimation methods is relatively high, and problems may occur, such as phase blur (subspace-based methods), error propagation (decision-feedback methods), slow convergence, or falling into local minima. They require more observational data, which limits their usefulness to some extent.

(4) Estimation based on a reference signal. This type of algorithm determines the parameters to be estimated according to certain estimation criteria, or gradually tracks and adjusts the estimated values of these parameters according to certain criteria. This method type is characterized by the need for reference signals (pilots) or training sequences.

Estimation based on training or a pilot sequence is collectively referred to as estimation based on a reference signal. Channel-estimation algorithms based on a training sequence are suitable for burst-transmission systems. By sending a known training sequence, the channel is initially estimated at the receiving end, and when useful information data are sent, the decision is updated using the initial channel-estimation result to complete a real-time channel estimation.

Channel estimation based on pilot symbols is suitable for continuous-transmission systems. By inserting known pilot symbols into the transmitted useful data, the channel-estimation result of the pilot location can be obtained. Then, using the channel-estimation result of the pilot location, the channel characteristics of the useful data location are obtained by interpolation, and the channel estimation is completed [3–6].

8.1.2 Channel Estimation Based on the Least-Squares Criterion

In pilot-based channel estimation, the channel response $\hat{H}\left(n_p, k_p\right)$ at the position of the pilot should be obtained. The specific position of the pilot is represented by a coordinate $\left(n_p, k_p\right)$. Commonly used estimation criteria include the least-squares (LS) and minimum mean-squared error (MMSE) estimation criteria. The channel-estimation equation of the OFDM LS method is given below. Here, we only consider the reception equation for the pilot subcarrier.

$$Y_m(k) = X_m(k)H_m(k) + W_m(k), \tag{8.1}$$

where $m = 0,\ 1, 2, \ldots, M-1$ is a subcarrier, with M subcarriers in total. The received frequency-domain signal is denoted as $Y_m(k)$, the transmitted frequency-domain signal is denoted as $X_m(k)$, the channel impulse-response function in the frequency domain is denoted as $H_m(k)$, and the noise at the receiving end of the system is denoted as $W_m(k)$. The cost function [7–9] is given according to the least-squares definition:

$$J_{LS} = \left(Y_m - \hat{H}_m X_m\right)^* \left(Y_m - \hat{H}_m X_m\right). \tag{8.2}$$

In Eq. (8.2), our objective cost function is written as J_{LS}, * represents the conjugate operation, and $0 \leq m \leq M-1$, is the number of a pilot. Y_m, $\hat{H}_m$, and X_m represent the signal value at the m pilot position received by the receiver, the estimated channel-response value, and the sent pilot-signal value, respectively. According to the above formula, we can derive the LS frequency-response equation for $\hat{H}_m(k)$:

$$\hat{H}_m(k) = \frac{Y_m(k)}{X_m(k)} = H_m(k) + \frac{W_m(k)}{X_m(k)}, \tag{8.3}$$

where $\hat{H}_m(k)$ represents the estimated channel-response value in the frequency domain, and $Y_m(k)$, $X_m(k)$, and $W_m(k)$ represent the received signal, the transmitted signal, and the noise signal at the pilot position in the frequency domain, respectively. In this manner, the channel response of the pilot subcarrier can be obtained. If the impulse response $\hat{h}_m(k)$ in the time domain is obtained, after an FFT transformation, $\hat{H}_m(k)$ can be written as

$$\hat{h}_m(k) = \frac{1}{M} \sum_{m=0}^{m-M-1} \hat{H}_m(k) \exp(j2\pi m/M). \tag{8.4}$$

The impulse response in the time domain is denoted by $\hat{h}_m(h)$, where m = 0, 1, 2, ..., M − 1. After the channel values of the pilot subcarriers are obtained, the channel-response values for all positions must be obtained by interpolation.

The steps of the pilot-based LS algorithm include the following:

(1) Design the pilot structure;
(2) Obtain the channel value at the pilot using the LS criterion;
(3) Use interpolation or other methods to obtain the channel response of the remaining positions, according to the pilot value.

The commonly used pilot structure is a comb-type pilot. Assuming that each OFDM symbol has a total of N subcarriers, the number of pilots inserted after every fixed number of subcarriers is N_p, which is determined using the following formula:

$$X(k) = X\left(m\frac{N}{N_P} + j\right) = \begin{cases} Pilot & j = 0 \\ Data\ j = 1, 2, \ldots, (N/N_P - 1) \end{cases}, 0 \leq m \leq N_P - 1. \tag{8.5}$$

It can be seen that if the noise is small, the LS-criterion estimation result is ideal.

8.1.3 Channel Estimation Based on the MMSE Criterion

According to the minimum mean-squared error (MMSE) criterion, the cost function is defined as [7]

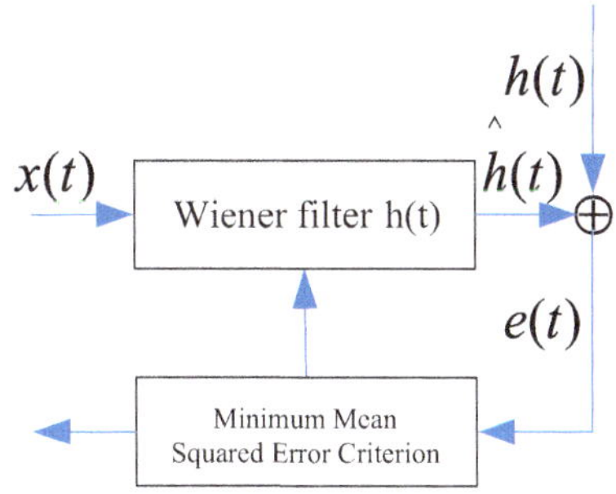

Fig. 8.2 Wiener filter-system block diagram

$$J_{MMSE} = E\left[\left(H_p - \hat{H}_p\right)^*\left(H_p - \hat{H}_p\right)\right]. \tag{8.6}$$

Under certain conditions, the linear MMSE can also be used to estimate the channel parameters, which is called filtering. The basic idea is to find the best transfer function for the MMSE between the output waveform and the input. Figure 8.2 shows the block diagram of a Wiener filter.

The parameter $\hat{H}$ to be estimated can be expressed as

$$\hat{H} = \sum_{p=1}^{M} w_p Y_p = W_p^T Y_p. \tag{8.7}$$

$Y_p = [Y_1, Y_2, \ldots, Y_M]^T$ is the received pilot signal, $W_p = [w_1, w_2, \ldots, w_M]^T$ is the tap coefficient of the filter, and M is the number of pilot symbols. Below, we derive the relation to minimize the cost function C.

Let $e_p = H_p - \hat{H}_p = H_p - W_p^T Y_p$ and $w_p = a_k + jb_k$. Taking the partial derivative of the above cost function J_{MMSE} with respect to w_p, we have

$$\begin{aligned} \frac{\partial J_{MMSE}}{\partial w_k} &= \frac{\partial\left\{E\left[\left(H_p - \hat{H}_p\right)^*\left(H_p - \hat{H}_p\right)\right]\right\}}{\partial w_k} = \frac{\partial\left\{E\left[e_p^* e_p^*\right]\right\}}{\partial w_k} \\ &= \frac{\partial\left\{E\left[e_p^* e_p^*\right]\right\}}{\partial a_k} + j\frac{\partial\left\{E\left[e_p^* e_p^*\right]\right\}}{\partial b_k} \\ &= E\left\{-e_p^*\frac{\partial e_p}{\partial a_k} - e_p\frac{\partial e_p^*}{\partial a_k} + j\left[e_p^*\frac{\partial e_p}{\partial b_k} + e_p\frac{\partial e_p^*}{\partial b_k}\right]\right\} \\ &= E\left\{-e_p^* Y_k - e_p Y_k^* + j\left[e_p^*(-jY_k) + e_p\left(jY_k^*\right)\right]\right\} \\ &= -2E\left[e_p Y_k^*\right] \end{aligned} \tag{8.8}$$

In Eq. (8.8), $\frac{\partial J_{MMSE}}{\partial w_k}$ represents the partial derivative of cost function J_{MMSE} with respect to w_k; the error generated by channel estimation is represented by e_p, a_k and b_k represent the real and imaginary parts of the noise signal, respectively, and e_p^*

represents the conjugate of the error signal. Let $\frac{\partial J_{MMSE}}{\partial w_k} = 0$; then,

$$\begin{aligned} E\left[e_p Y_k^*\right] &= E\left[\left(H_p - W^T Y_p\right) Y_p^*\right] \\ &= E\left[H_p Y_k^*\right] - W^T E\left[Y_p Y_k^*\right]. \\ &= 0 \end{aligned} \tag{8.9}$$

$E[e_p Y_k^*]$, $E\left[Y_p Y_k^*\right]$, and $E\left[H_p Y_k^*\right]$ are the cross-correlation function matrices of the pilot signal, the channel response, and the received pilot signal, respectively. The above formula is expressed as a matrix:

$$WR_{TI} = R_{HI} \Rightarrow W = R_{HI} R_{TY}^{-1}. \tag{8.10}$$

The "⇒" symbol stands for derivation. Substituting the above formula into Eq. (8.8) yields

$$\hat{H}_p = W^T Y_p = R_{HY} R_{YY}^{-1} Y_p, \tag{8.11}$$

where R_{HY} represents the impulse-response function of the pilot signal and the cross-correlation function of the received signal, R_{YY} represents the autocorrelation function of the received pilot signal, R_{YY}^{-1} represents the inverse operation of the autocorrelation function matrix, and W^T represents the tap coefficient of the filter. Then, R_{HY} can be expressed as

$$R_{HY} = E\left[H_p Y_p^H\right] = E\left[H_p \left(X_p H_p + N\right)^H\right] = R_{YY} X_p^H \tag{8.12}$$

$$\begin{aligned} R_{TY} &= E\left[Y_p Y_p^H\right] = E\left[\left(X_p H_p + N\right)\left(X_p H_p + N\right)^H\right] \\ &= X_p R_{HH} X_p^H + \sigma^2 I \end{aligned} \tag{8.13}$$

In Eqs. (8.12) and (8.13), $X_p = \operatorname{diag}(X_1, X_2, \ldots, X_M)^T$ represents a diagonal matrix. The diagonal-matrix symbol is composed of M pilot symbols $X_1, X_2, \ldots, X_M$, which means that the conjugate transpose of the transmitted pilot signal matrix is represented by X_p^H, and σ^2 is the variance value of the noise. I is the identity matrix of order M. Substituting Eqs. (8.12) and (8.13) into Eq. (8.11) yields

$$\begin{aligned} \hat{H}_p &= R_{HY} R_{YY}^{-1} Y_p = R_{HH} X_p^H \left(X_p R_{HH} X_p^H + \sigma^2 I\right) Y_p \\ &= R_{HH}\left[X_p^{-1}\left(X_p R_{HH} X_p^H + \sigma^2 I\right) X_p^{H-1}\right]^{-1} X_p^{-1} Y_p \\ &= R_{HH}\left[R_{HH} + \sigma^2\left(X_p^H X_p\right)^{-1}\right] X_p^{-1} Y_p \\ &= R_{HH}\left[R_{HH} + \sigma^2\left(X_p^H X_p\right)^{-1}\right] \hat{H}_{LS} \end{aligned} \tag{8.14}$$

To simplify the operation process, we replace $\left(X_p^H X_p\right)^{-1}$ with the expectation $E\left[\left(X_p^H X_p\right)^{-1}\right]$. Assuming that the transmission signal is transmitted with equal probability, it can be written as

$$E\left[\left(X_p^H X_p\right)^{-1}\right] = E\left[|1/x_k|^2 I\right] \tag{8.15}$$

The average signal-to-noise ratio satisfies the formula. If it is expressed as $SNR = E\left[|x_p|^2\right]/\sigma^2$, then Eq. (8.13) can be written as

$$\hat{H}_p = R_{HH}\left(R_{HH} + \frac{\beta}{SNR}I\right)^{-1}\hat{H}_{LS} \tag{8.16}$$

In Eq. (8.16), the channel-estimation value $\hat{H}_p$ is obtained using the MMSE algorithm, $\beta = E|x_p|^2 E|1/X_p|^2$ is a quantity that varies with different modulation methods, and $\hat{H}_{LS}$ represents the channel value obtained by the LS algorithm.

If Q is used to replace $R_{HH}(R_{HH} + \beta I/SNR)^{-1}$ in the above formula, the MMSE criterion can be understood as Wiener channel estimation obtained using matrix filtering based on LS. In fact, Eq. (8.16) is related to the mathematical and statistical properties of the selected training sequence, but has little to do with the specific value. Even if the training sequence changes, it only needs to be calculated once. This significantly reduces the complexity of MMSE channel estimation.

8.2 Particle-Filter Algorithms

For linear and Gaussian-distribution filtering problems, Kalman filtering can achieve an optimal solution under a Bayesian algorithm. For weakly nonlinear systems, Kalman filtering cannot directly meet the requirements; however, extended Kalman filtering (EKF) can be used to linearize nonlinear systems. Many require filtering to solve; however, for strong nonlinear systems, EKF will lead to divergence.

In recent years, Julier et al. proposed the unscented Kalman filter (UKF) [8], which avoids the errors introduced in the process of linearizing nonlinear systems, and is more efficient than the extended Kalman filter. This method has better performance, but it still requires the desired probability-density function to satisfy a Gaussian distribution.

8.2.1 Bayesian Estimation Method

8.2.1.1 Bayesian Principle

Bayesian analysis provides a method to calculate the probability of a hypothesis, based on the prior probability of the hypothesis, the probability of observing different data given the hypothesis, and the observed data. This method integrates prior information about unknown parameters with sample information, obtains the posterior information according to the Bayesian formula, and then infers the unknown parameters using the posterior information. The essence of Bayesian theory is to use known measurement information to estimate the posterior probability density of a system-state variable [9], and then to correct the new observation value of the system-state variable, which can be expressed as

$$p(b|a) = \frac{p(a|b)p(b)}{p(a)}. \tag{8.17}$$

In Eq. (8.17), $p(b|a)$ is the conditional probability of b given a, also known as the posterior probability. $p(b|a)$ depends on the specific value of b, $p(a|b)$ is the conditional probability of given a, and $p(b)$ and $p(a)$ are called the marginal probability densities of b and a.

The distribution of the parameters to be estimated in a Bayesian formula generally needs to be known in advance; this is also known as the prior distribution. Bayesian theory holds that when no parameter information is known, the parameter changes within its range with equal probability; that is, the prior distribution of the parameter is uniformly distributed within its range of values. This is also known as the Bayesian hypothesis [10].

In practice, the system needs to be modeled first; that is, two equations need to be known: the state-space model and the measurement model. The state-space model describes how the system state changes over time, and the measurement equation describes a certain relationship between the system state and an observed value. In optical wireless OFDM channel estimation, the state equation is how the state value of the atmospheric channel is updated with time, and the observation equation describes the relationship between the received signal value and the response value of the atmospheric channel at the current moment.

We use x_t *and* z_t to respectively, represent the state and measurement values of the system at time t, $X_t = \{x_0, x_1, \ldots x_t\}$ to represent all the system-state quantities from the start time to time t, and $Z_t = \{z_0, z_1, \ldots, z_t\}$ to represent all the measured values from the start time to time t. Then, the system-state and measurement equations are as follows:

State equation:

$$X_t = F_t(t, X_{t-1}, u_t, V_t). \tag{8.18}$$

Measurement equation:

$$Z_t = H_t(t, X_{t-1}, u_t, W_t). \tag{8.19}$$

V_t and W_t represent the system noise and measurement noise, respectively, and the external input variable of the system can be expressed as u_t. In general, if the influence of external inputs is not considered, the state equation and measurement equation can be described as

$$x_{t+1} = f_t(u_t, v_t) \tag{8.20}$$

$$z_t = g_t(x_t, w_t). \tag{8.21}$$

In Eqs. (8.20) and (8.21), the state value at time $t + 1$ is represented by x_{t+1}, z_t represents the measurement value at time t, and f and g represent the system-state equation and the measurement equation, respectively. The conditional probability-density function model is used to express Eqs. (8.20) and (8.21) as $p(X_t|X_{t-1})$ and $p(Z_t|X_t)$; thus, the general state-estimation model is established.

8.2.1.2 Recursive Bayesian Estimation and Importance Mining

Bayesian estimation theory uses the measurement value $Z_t = \{z_0, z_1, \ldots, z_{t-1}\}$ before time t to estimate the current state; that is, to calculate the posterior probability-density function x_t. The next step is to consider how to obtain the posterior probability density $p(x_t|Z_t)$. If the prior probability density $p(x_{t-1}|Z_{t-1})$ at the current moment can be obtained recursively from the posterior probability $p(x_t|Z_t)$ at the previous moment, then $p(x_t|Z_t)$ is obtained through $p(x_t|Z_{t-1})$ and the entire recursion process can be completed.

Generally, only two steps—prediction and update—constitute the Bayesian-algorithm process. The steps of the prediction process are as follows.

Assuming that the posterior probability at time $t - 1$ is $p(x_{t-1}|Z_{t-1})$, initialize $p(x_0|Z_0) = p(x_0)$, according to the Chapman-Kolmogorov formula. The state prior probability density at time t is

$$p(x_t|Z_{t-1}) = \int p(x_t|x_{t-1})p(x_{t-1}|Z_{t-1})dx_{t-1}. \tag{8.22}$$

In Eq. (8.22), $\int(\cdot)$ represents the integral operation. Equation (8.22) means that the prior probability $p(x_t|Z_{t-1})$, excluding the measured value at time $t - 1$, can be obtained by integrating the one-step transition probability $p(x_{t-1}|x_{t-1})$ and the posterior probability $t - 1$ at time $p(x_{t-1}|Z_{t-1})$.

The a posteriori estimation step of the update process is as follows: First, $p(x_t|Z_t)$ is derived from the observation value z_k and the prior probability $p(x_t|Z_{t-1})$ at time

t. First, by the Bayesian conditional-probability formula,

$$p(b|a) = \frac{p(a|b)p(b)}{p(a)}. \tag{8.23}$$

It can then be deduced that

$$p(x_t|Z_t) = \frac{p(Z_t|x_t)p(x_t)}{p(Z_t)}, \tag{8.24}$$

where $Z_t = \{Z_{t-1}, z_t\}$ represents all observations before time t; therefore,

$$p(Z_t|x_t) = p(Z_{t-1}, z_t|x_t). \tag{8.25}$$

Substituting Eq. (8.25) into Eq. (8.24) yields

$$p(x_t|Z_t) = \frac{p(Z_{t-1}, z_t|x_t)p(x_t)}{p(Z_{t-1}, z_t)}. \tag{8.26}$$

From the conditional-probability formula $p(a, b) = p(a|b)p(b)$ and the joint-distribution probability formula $p(a, b|c) = p(a|b, c)p(b|c)$ in probability theory,

$$p(Z_{t-1}, z_t) = p(z_t|Z_{t-1})p(Z_{t-1}) \tag{8.27}$$

$$p(Z_{t-1}, z_t|x_t) = p(z_t|Z_{t-1}, x_t)p(Z_{t-1}|x_t). \tag{8.28}$$

In Eq. (8.27), $p(Z_{t-1})$ represents the probability of all observations before time t-1, which is obtained from the Bayesian formula:

$$p(Z_{t-1}|x_t) = \frac{p(x_t|Z_{t-1})p(Z_{t-1})}{p(x_t)}. \tag{8.29}$$

Substituting Eq. (8.29) into (8.28), we obtain

$$p(Z_{t-1}, z_t|x_t) = \frac{p(z_t|Z_{t-1}, x_t)p(x_t|Z_{t-1})p(Z_{t-1})}{p(x_t)}. \tag{8.30}$$

Substituting (8.30) and (8.31) into (8.30) yields

$$\begin{aligned} p(x_t|Z_t) &= \frac{p(z_t|Z_{t-1}, x_t)p(x_t|Z_{t-1})p(Z_{t-1})p(x_t)}{p(z_t|Z_{t-1})p(Z_{t-1})p(x_t)} \\ &= \frac{p(z_t|Z_{t-1}, x_t)p(x_t|Z_{t-1})}{p(z_t|Z_{t-1})}. \end{aligned} \tag{8.31}$$

Assuming that the measured values are independent of each other,

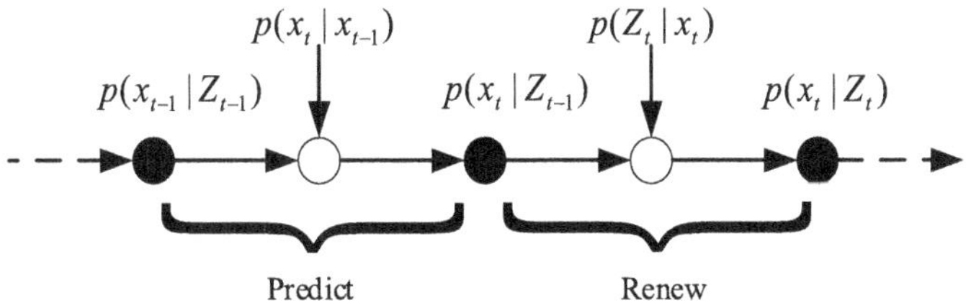

Fig. 8.3 Principle of the Bayesian recurrence process [11]

$$p(z_t|Z_{t-1}, x_t) = p(z_t|x_t). \tag{8.32}$$

Substituting Eq. (8.32) into (8.31) yields

$$p(x_t|Z_{t-1}) = \frac{p(z_t|x_t)p(x_t|Z_{t-1})}{p(z_t|Z_{t-1})}. \tag{8.33}$$

In Eq. (8.33), $p(z_t|x_t)$ is the likelihood function and $p(x_t|Z_{t-1})$ is the prior probability. The entire derivation process of Eq. (8.33) constitutes the posterior $p(x_t|Z_t)$ at time t obtained from the posterior probability $p(x_{t-1}|Z_{t-1})$ at time t-1. The recursive process can obtain the posterior probability-density function at time G from $p(x_t|Z_t)$, according to this analogy.

As shown in Fig. 8.3, the prediction process uses the prior probability $p(x_{t-1}|Z_{t-1})$ at time t, that is, the posterior probability-density function at time $t-1$, and the state-transition model of the system, namely, $p(x_t|Z_{t-1})$, to predict the prior probability $p(x_t|Z_{t-1})$ at time t. After obtaining the observation value at time t, we modify $p(x_t|Z_{t-1})$ to obtain the posterior density $p(x_t|Z_t)$ at time t, and solve the posterior probability at time t + 1 using recursive Bayesian estimation, and so forth.

$p(x_{t-1}|Z_{t-1})$ is usually impossible to obtain directly, and the Bayesian importance sampling (BIS) theorem provides the possibility for hardware implementations of Bayesian algorithms [12].

First, Bayesian importance sampling selects a known easy-to-sample reference distribution $q(X_t|Z_{t-1})$ to replace $p(X_t|Z_t)$, and then weights the reference distribution $q(X_t|Z_t)$ to approximate the posterior probability $p(X_t|Z_t)$. Assuming that $\Psi(X_t)$ represents a function, the desired expression can be written as

$$\begin{aligned} E[\Psi(X_t)] &= \int \Psi(X_t)p(X_t|Z_t)dX_t \\ &= \int \frac{\Psi(X_t)p(X_t|Z_t)q(X_t|Z_t)}{q(X_t|Z_t)}dX_t. \end{aligned} \tag{8.34}$$

In Eq. (8.34), $E[\Psi(X_t)]$ represents the mathematical expectation of function $\Psi(X_t)$, which is given by the Bayesian formula:

$$p(X_t|Z_t) = \frac{p(Z_t|X_t)p(X_t)}{p(Z_t)}. \tag{8.35}$$

Substituting Eq. (8.35) into (8.34) and substituting $w_t(X_t)$ for $\frac{p(Z_t|X_t)p(X_t)}{p(Z_t)}$ yields

$$\begin{aligned} E[\psi(X_t)] &= \int \frac{\psi(X_t)p(Z_t|X_t)p(X_t)q(X_t|Z_t)}{p(Z_t)q(X_t|Z_t)}dX_t \\ &= \int \frac{\psi(X_t)w_t(X_t)q(X_t|Z_t)}{p(Z_t)}dX_t \\ &= \frac{\int \psi(X_t)w_t(X_t)q(X_t|Z_t)dX_t}{p(Z_t)} \end{aligned} \tag{8.36}$$

In Eq. (8.36), $p(Z_t)$ can be written as

$$\begin{aligned} p(Z_t) &= \int p(Z_t, X_t)dX_t \\ &= \frac{\int p(Z_t|X_t)p(X_t)q(X_t|Z_t)}{q(X_t|Z_t)}dX_t. \\ &= \int w_t(X_t)q(X_t|Z_t)dX_t \end{aligned} \tag{8.37}$$

Substituting Eq. (8.37) into Eq. (8.36), we obtain

$$E[\psi(X_t)] = \frac{\int \psi(X_t)w_t(X_t)q(X_t|Z_t)dX_t}{\int w_t(X_t)q(X_t|Z_t)dX_t}. \tag{8.38}$$

Taking independent random samples from the reference distribution $q(X_t|Z_t)$, and calculating the integral using the Monte Carlo method, the approximate expectation of $\Psi(X_t)$ can be obtained:

$$\begin{aligned} \overline{E}[\psi(X_t)] &= \frac{\frac{1}{N_p}\sum_{i=1}^{N_p}\psi(X_t^i)w_t(X_t^i)}{\frac{1}{N_p}\sum_{i=1}^{N_p}w_t(X_t^i)}. \\ &= \sum_{i=1}^{N_p}\psi(X_t^i)\tilde{w}_t(X_t^i) \end{aligned} \tag{8.39}$$

In the formula, $\tilde{w}_t(X_t^i) = \frac{w_t(X_t^i)}{\sum_{i=1}^{N_p} w_t(X_t^i)}$ represents the normalized weight coefficient of the ith particle, $\overline{E}[\psi(X_t)]$ represents the expected approximation of $\psi(X_t)$, X_t^i is extracted from $q(X_t|Z_t)$, and the symbol Σ represents the weighted summation.

8.2.2 Monte Carlo Method

A Monte Carlo method (MCM) [13], also known as a statistical simulation method, is a very important numerical-calculation method guided by probability and statistics theory; it was proposed in the mid-1940s. MCM can also be called a random-sampling or statistical testing method, the basic idea is to randomly take a set of independent samples of the same distribution from a more complex or non-analytical calculation of the function of the true probability distribution, the more the number of samples is drawn, the closer to the true distribution. The detailed mathematical derivation process of the Monte Carlo method is introduced below.

A detailed mathematical derivation of the Monte Carlo method is introduced as follows. Suppose we want to compute an integral such as the following:

$$I = \int_R g(x)dx, \tag{8.40}$$

where $x \in R^n$ is a uniformly distributed random variable. The mean of $g(x)$ is defined by Eq. (8.41) as

$$\overline{g(x)} = \frac{I}{|R|}\int_R g(x)dx = \frac{I}{|R|}, \tag{8.41}$$

where $|R| = \int_R dx$. Assuming that $x_1, x_2, \ldots, x_N$ are the same distribution as x, when N is large, the mean value of $g(x)$ can be approximated as

$$I_N = \frac{|R|}{N}\sum_{i=1}^{N} g(x_i). \tag{8.42}$$

If $g(x)$ has finite variance and I_N converges and obeys the following,

$$\lim_{N\to\infty} \sqrt{N}(I_N - I) \sim N(0, \mathrm{Var}(g(x))), \tag{8.43}$$

the core formula of the Monte Carlo method is

$$p_n(x) = \frac{1}{N}\sum_{i=1}^{N} \delta(x - x_i). \tag{8.44}$$

When $x = x_i$, $\delta(x - x_i) = 1$; when $x \neq x_i$, $\delta(x - x_i) = 0$. $\{x_i, i = 1, \ldots, n\}$ are independent and identically distributed particles drawn from $p_n(x)$, and the probability of each particle is set to $1/n$. We can decompose Eq. (8.39) into the form of $g(x) = f(x)\pi(x)$; then, the integral of Eq. (8.44) is written as

$$I = \int_R f(x)\pi(x)dx, \tag{8.45}$$

where $\pi(x) \geq 0$, and satisfies $\int_R \pi(x)dx = 1$. If N points $\{x_i\}$ can be sampled from distribution $\pi(x)$, then

$$I_N = \frac{1}{N}\sum_{i=1}^{N} f(x_i). \tag{8.46}$$

$\pi(x)$ does not exist in Eq. (8.46); it has been replaced by sampling point $\{x_i\}$. In fact, $\pi(x)$ is the posterior probability density in the Bayesian estimation described earlier.

Generally, the required integral is approximated by sampling a set of weighted particles from the posterior-probability function. The posterior probability can be described as

$$p(X_t|Z_t) = \frac{1}{N_p}\sum_{i=1}^{N_P} \pi_t^i \delta\left(X_t - X_t^i\right). \tag{8.47}$$

The number of random sampling points is N_p, and $\left\{X_t^i, \pi_t^i\right\}_{i=1}^{N_p}$ represents each sampling-point particle and its corresponding weight, and satisfies the condition of $\sum_{i=1}^{N_p} \pi_t^i = 1$. The system-state vector can be expressed as $X_t = \{x_0, x_1, \ldots, x_t\}$, the measured value $Z_t = \{z_0, z_1, \ldots, z_t\}$, and $\delta(\cdot)$ is the Dirac delta function. Thus, the expected value of any function V, according to the Monte Carlo method, can be written as

$$E[\psi(X_t)] = \int \psi(X_t)P(X_t|Z_t)dX_t. \tag{8.48}$$

The discretized Monte Carlo method is approximated as

$$E[\psi(X_t)] = \frac{1}{N_p}\sum_{i=1}^{N_P} \psi(X_t), \tag{8.49}$$

where $E[\psi(X_t)]$ represents the statistical average of $\psi(X_t)$. $P(X_t|Z_t)$ does not appear in Eq. (8.49); its distribution is shown in the distribution of sampling particle $\left\{X_t^i\right\}$, which is independent and identically distributed. It is conceivable that when N_p tends to infinity, the following conclusions can be drawn:

$$\overline{E}[\psi(X_t)] \xrightarrow[N_p\to\infty]{a.s} E[\psi(X_t)]. \tag{8.50}$$

In Eq. (8.50), $\xrightarrow[N_p\to\infty]{a.s}$ means that it always holds when N_p tends to infinity.

The integral approximation of a function can be obtained using the Monte Carlo method with the expected solution of continuous and discrete functions. The importance-sampling method is a common Monte Carlo sampling algorithm, and it is also the core concept of using a particle-filter algorithm.

8.2.3 *Principle of the Particle-Filter Algorithm*

Particle filtering (PF) is based on the sequential Monte Carlo method and recursive Bayesian estimation and was developed in the 1990s. It has been successfully used to solve nonlinear and non-Gaussian systems in engineering.

The filtering problem has received increasing attention. The basic concept is to use a set of particles propagating in the state space with corresponding weights to approximate the posterior probability-density function of the quantity to be estimated through a non-parametric Monte Carlo method. When the number of samples is large, particle filtering can achieve the effect of optimal Bayesian estimation.

However, it is unrealistic to obtain samples directly from the posterior distribution, so sequential importance sampling (SIS) is used to extract samples; that is, a set of random samples and their corresponding weights is used to represent the desired posterior probability-density function. The estimated value of the state is approximated by a weighted average.

From the above discussion, we know that the posterior probability-density function of a system can be represented in the form of particles and corresponding weights; namely,

$$p(X_t|Z_t) \approx \sum_{i=1}^{N_p} \tilde{w}_t^i \delta\left(X_t - X_t^i\right), \tag{8.51}$$

where $\tilde{w}_t^i$ represents the weight of the ith particle.

Assuming that the posterior probability of the system at time t − 1 is $p(X_{t-1}|Z_{t-1})$, it can be expressed in the form of the sum of weighted particles $\left\{X_{t-1}^i, w_{t-1}^i\right\}$, where $i = 1, 2, \ldots, N$ and $\left\{X_{t-1}^i\right\}$ is a set of samples of $q(X_{t-1}|Z_{t-1})$. To obtain $\left\{X_{t-1}^i, w_{t-1}^i\right\}$, the importance density function $q(X_t|Z_t)$ must first be obtained by sampling from $p(X_t|Z_t)$. First, a decomposition transformation must be conducted on $q(X_t|Z_t)$:

$$q(X_t|Z_t) = q(x_t, X_{t-1}|Z_t) = q(x_t, X_{t-1}|Z_t)q(X_{t-1}|Z_{t-1}). \tag{8.52}$$

In this manner, we can first sample $\left\{X_t^i\right\}$ from $q(x_t, X_{t-1}|Z_t)$, and then multiply by the sample in $q(X_{t-1}|Z_{t-1})$ to obtain a sample of $q(X_t|Z_t)$.

Next, a new particle set can be obtained according to Eqs. (8.34) and (8.35), and then the weight-update process is further deduced. Then, Eq. (8.35) is substituted into the Bayesian conditional probability-distribution function (8.53). Then,

$$w_t(X_t) = \frac{p(X_t|Z_t)p(X_t)}{q(X_t|Z_t)}. \tag{8.53}$$

Abbreviating $w_t(X_t)$ to w_t gives

$$w_t = \frac{p(Z_t|X_t)p(X_t)}{q(x_t|X_{t-1}, Z_t)q(X_{t-1}|Z_{t-1})}. \tag{8.54}$$

From Eq. (8.53), the following can be deduced:

$$w_t = \frac{p(Z_t|X_t)p(X_t)}{q(X_{t-1}|Z_{t-1})}. \tag{8.55}$$

From Eqs. (8.54) and (8.55), we can obtain

$$\begin{aligned} w_t &= w_{t-1}\frac{p(Z_t|X_t)p(X_t)}{q(x_t|X_{t-1}, Z_t)p(X_{t-1}|Z_{t-1})p(X_{t-1})} \\ &= w_{t-1}\frac{p(z_t|x_t)p(x_t|x_{t-1})}{q(x_t|X_{t-1}, Z_t)}. \end{aligned} \tag{8.56}$$

Because we are estimating $p(x_t|z_t)$, we first assume that the state to be estimated satisfies the Markov property; namely,

$$q(x_t|X_{t-1}, Z_t) = q(x_t|x_{t-1}, Z_t). \tag{8.57}$$

It can be seen that the two parameters $x_{\{t-1\}}$ and Z_t affect the value of the weight, so the posterior probability-density function can be written as

$$p(X_t|Z_t) \approx \sum_{i=1}^{N_p} \tilde{w}_t^i \delta\left(x_t - x_t^i\right). \tag{8.58}$$

We substitute Eq. (8.57) into (8.56) to calculate the weight w_t^i of each particle; that is,

$$w_t^i = w_{t-1}^i \frac{p\left(z_t|x_t^i\right)p\left(x_t^i|x_{t-1}^i\right)}{q\left(x_t|x_{t-1}^i, Z_t\right)}. \tag{8.59}$$

If $q\left(x_t|x_{t-1}^i, Z_t\right) = q\left(x_t|x_{t-1}^i, z_t\right)$, the importance function is only affected by parameters x_{t-1}^i and z_t and the writable weight-update formula is

$$w_t^i = w_{t-1}^i \frac{p(z_t|x_t^i)p(x_t^i|x_{t-1}^i)}{q(x_t|x_{t-1}^i, z_t)}. \tag{8.60}$$

It can be seen that only x_{t-1} needs to be stored in the recursive process, which is beneficial for a hardware implementation of the algorithm.

The derivation process of Eq. (8.60) is the importance-weight recursive algorithm. Each recursion must sample from the reference distribution. It can be seen from the weight-update Eq. (8.60) that the weight update also considers the likelihood function $p(z_t|x_t^i)$ and state-transition probability $p(z_t|x_{t-1}^i)$. We express the above concept in pseudocode, as shown in Fig. 8.4.

The basic idea of particle filtering is to use weighted-sum particles to represent the posterior probability density of the parameters to be estimated in each iteration process. The particle weights are repeatedly redistributed at the moment, which also constitutes the basis of the particle-filter algorithm and its improved algorithm.

Figure 8.5 shows the state estimation of the particle filter (PF) and extended Kalman filter (EKF) applied to a nonlinear system. It can be seen that the particle-filter method has good nonlinear system-estimation performance and a small mean-squared error.

Although the particle-filter method has great advantages in solving nonlinear non-Gaussian systems, in the sequential importance-sampling process, after many iterations, only a few particle weights are non-zero values. The weights of most particles tend to be zero. This degradation phenomenon is also known as particle starvation.

8.2.4 Limitations of the Particle-Filter Algorithm

8.2.4.1 Particle Degradation

Reference [14] observed that the sequential importance-sampling method increases the variance of the weight as the number of iterations increases. That is, after a certain number of iterations, the weights of many particles tend to be zero, whereas only a few particles have large weights. Thus, extensive computational time is wasted to no effect on the posterior probability.

Reference [15] proposed the idea of an effective sampling scale, using N_{eff} to represent the degree of degradation, which is defined as

$$N_{eff} = \frac{N}{1 + \mathrm{Var}(w_t^{*i})}. \tag{8.61}$$

The true weights of the particles are represented by w_t^{*i}, where $w_t^{*i} = p(x_t^i|Z_t)/q(x_t^i|x_{t-1}^i, z_t)$; however, Eq. (8.61) is generally replaced by Eq. (8.62) in actual use; namely,

Initialization: Set $t = 0$ and extract N_p sampling points from $p(x_0)$

$\{x_t^i, w_t^i\}_{i=1}^{N_p} = SIS(\{x_{t-1}^i, w_{t-1}^i\}_{i=1}^{N_p}, z_t)$

Input: $\{x_{t-1}^i, w_{t-1}^i\}_{i=1}^{N_p}, z_t$

For $i = 1 : N_p$

Draw $x_t^i \sim q(x_t \mid x_{t-1}^i, z_k)$

Update weights using formula

$$w_t^i = w_{t-1}^i \frac{p(z_t \mid x_t^i) p(x_t^i \mid x_{t-1}^i)}{q(x_t \mid x_{t-1}^i, z_t)}$$

End for

Calculate the sum of the total weights $TW = \sum_{i=1}^{N_p} w_t^i$

For $i = 1 : N_p$

Weight normalization: $w_t^i = \frac{w_t^i}{TW}$

End for

Output: $\{x_t^i, w_t^i\}_{i=1}^{N_p}$; that is, the state particle and its weight at time t

Let $t = t + 1$. When the next measurement is obtained, repeat the above process.

Fig. 8.4 Pseudocode of the PF algorithm

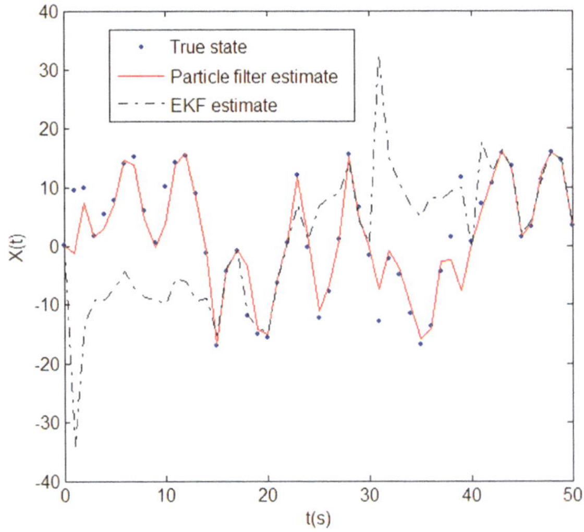

Fig. 8.5 Comparison of the estimation performance of the particle-filter and EKF methods

$$\overline{N_{eff}} = \frac{1}{\sum_{i=1}^{N}\left(w_t^i\right)^2}, \tag{8.62}$$

where w_t^j is the normalized form of the weight. It can be seen that $\overline{N_{eff}}$ satisfies the following relationship: when $1 \le \overline{N_{eff}} \le N$, $w_t^j = 1/N$; when $\exists j \in \{1, 2, \ldots N\}$ that satisfies $w_k^j = 1$, $\overline{N_{eff}} = N$. When $w_k^i = 0, \forall i \neq j, N_{eff} = 1$. It can be seen that the smaller $\overline{N_{eff}}$ is, the more serious the degradation phenomenon is.

The particle-degradation phenomenon can be eliminated by using a sufficient number of particles; however, when the number of sampling points increases, the amount of calculation increases accordingly. Generally speaking, the selection of the number of particles mainly depends on the dimension of the system-state equation, the similarity of the prior-probability and the importance-density functions, and the number of iterations [12].

The current methods to eliminate particle degradation mainly start from two aspects:

(1) Selecting a good importance-density function and
(2) Implementing a resampling strategy.

8.2.4.2 Selecting an Importance-Density Function

The optimal importance function is chosen as [16]

$$\begin{aligned} q\left(x_t|x_{t-1}^i, z_t\right)_{opt} &= p\left(x_t^i|x_{t-1}^i\right) \\ &= \frac{p\left(x_t|x_t^i\right)p\left(z_t|x_{t-1}^i\right)}{p\left(z_t|x_{t-1}^i\right)}. \end{aligned} \tag{8.63}$$

Substituting Eq. (8.63) into Eq. (8.61) yields

$$w_t^i \propto w_{t-1}^i p\left(z_t^i | x_{t-1}^i\right) = w_{t-1}^i \int p(z_t | x_t) p\left(x_t^i | x_{t-1}^i\right) dx_t. \tag{8.64}$$

The biggest disadvantage of the optimal importance function of Eq. (8.64) is that it needs to calculate the integral, which is generally impossible to solve. In practical applications, the usual practice is to replace the importance-density function with the prior probability-density function; namely,

$$q\left(x_t | x_{t-1}^i, z_t\right) = p\left(x_t | x_{t-1}^i\right), \tag{8.65}$$

and derive a new weight-update equation:

$$w_t^i \propto w_{t-1}^i p\left(z_t | x_t^i\right). \tag{8.66}$$

The biggest advantage of this method is that it is simple and easy to implement; however, the observation value is not taken into account, and such an importance function is the most commonly used. In actual use, an appropriate importance function can be selected according to needs and specific situations.

8.2.4.3 Resampling Strategy

The purpose of the resampling strategy is to reduce the number of particles with small weights that have no effect on the posterior probability. These particles with almost zero weights are the main cause of particle degradation. Some type of update processing must be performed on these small-weight particles; this is the particle-resampling method to solve the particle-degradation problem.

The main strategy of resampling is to maintain the total number of particles during the particle-update process, sample the particles with large weights multiple times, and then redistribute the particle weights. The schematic diagram of resampling is shown in Fig. 8.6.

Suppose the posterior probability at time $t - 1$ can be expressed as follows:

$$p(X_{t-1} | Z_{t-1}) \approx \sum_{i=1}^{N_p} \tilde{w}_{t-1}^i \delta\left(x_{t-1} - x_{t-1}^i\right). \tag{8.67}$$

After sampling, Np new particles are obtained. Assuming that $x_k^{i^*}$ is sampled from x_k^i in the obtained Np particles, $p\left(x_k^{i^*} = x_k^i\right) = w_k^j$. After resampling, the weight of each particle is set to 1/Np. After systematic observations, the particles are redistributed; that is, the particles with large weights are assigned to the peak, and the smaller weights are assigned to the flatter positions. After resampling, particles with large weights are allowed to generate multiple new particles, particles with smaller

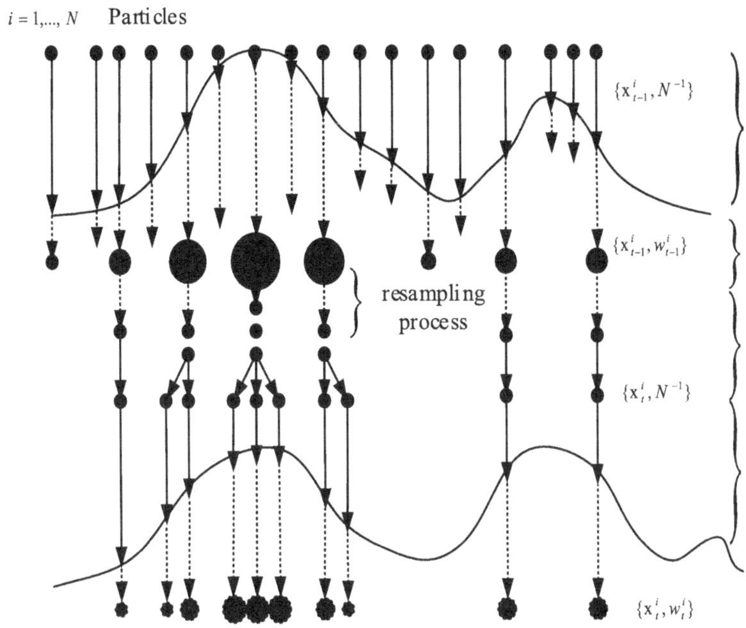

Fig. 8.6 Schematic diagram of the resampling process [11]

weights can be deleted, or a smaller number of new equal-weighted particles can be generated.

The process of resampling is actually a process of constructing a cumulative distribution function. That is, it extracts an integer that satisfies the condition of the random variable u_i from the uniform distribution $U(0, 1)$:

$$\sum_{j=1}^{m-1} w_t^i \le u_i \le \sum_{j=1}^{m} w_k^i. \tag{8.68}$$

Then, x_k^m is recorded and saved as the resampled particle. Figure 8.7 shows a schematic diagram of the CDF algorithm. Figure 8.8 provides a more detailed pseudocode description.

The above process is equivalent to decomposing the [0, 1] interval into N subintervals according to $\lambda_i = \sum^i w_t^j$. When ui falls in the mth interval, particle x_t^m is replicated, as shown in Fig. 8.8.

In practice, whether the resampling process is required is generally determined using an adaptive method. That is, the judgment is made using the effective sampling-scale formula in Eq. (8.72). Threshold value N_{eff} is set according to the requirements. Once $\overline{N_{eff}} \le N_{eff}$, the particle-resampling algorithm must be executed.

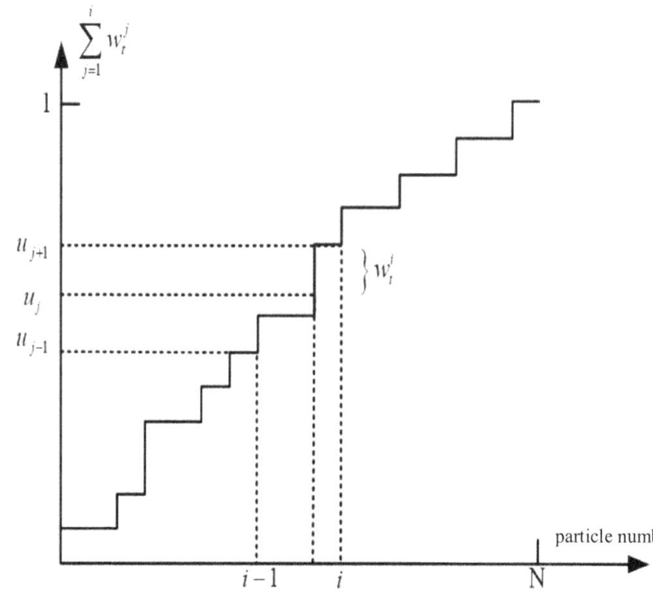

Fig. 8.7 Schematic diagram of the CDF algorithm

Although the particle-resampling algorithm improves the particle-degradation phenomenon to a certain extent, particle depletion occurs; that is, after several resamplings, many particles no longer have the diversity they had at the beginning, and gradually cannot represent the changing trend of the real system state. The general approach is to predict the changing system trend by re-initializing a new set of particle sets, and discarding the invalid previous particle set.

To verify the performance-estimation simulation results of the particle-filter algorithm, our simulation uses a nonlinear model. The state and measurement equations of the system are shown in Eq. (8.69):

$$\begin{cases} x_k = 0.6x_{k-1} + 25\frac{x_{k-1}}{1+x_{k-1}^2} + 10\cos(1.2k) + v_k \\ y_k = \frac{x_k^2}{10} + u_k \end{cases}, \tag{8.69}$$

where $v_k \sim N(0, 2)$, $u_k \sim N(0, 1)$ *and* $N(0.2, 3)$. In the MATLAB simulation, we selected 50 and 100 particles for simulation, and the random resampling method was used in the resampling process. The simulation results are shown in Figs. 8.9 and 8.10.

The abscissa in Figs. 8.9 and 8.10 refers to the simulation time step, and the ordinate refers to the magnitude of the system-state change. It can be seen from the simulation results that when the number of particles is large, the prediction performance of the particle filter is relatively good. However, in only a few moments,

Fig. 8.8 Pseudocode for the resampling process

$\{x_t^{j*}, w_t^j\}_{j=1}^{N_p}$ = RESAMPLE

$\{x_{t-1}^i, w_{t-1}^i\}_{i=1}^{N_p}$

Input: $\{x_{t-1}^i, w_{t-1}^i\}_{i=1}^{N_p}$

Initialize the CDF: $c_1 = 0$

For i=2: N_p

Construct the cumulative distribution function, CDF: $c_i = c_{i-1} + w_k^i$

End for

Set $i = 1$

For j=1: N_p

Generate random numbers u_j from uniform distribution $U(0, 1)$

While $u_j > c_i$

$i = i + 1$

End while

Let $x_t^{j*} = x_t^l$

Let $w_t^{j*} = 1 / N_t$

End for

Output: $\{x_t^i, w_t^i\}_{i=1}^{N_p}$

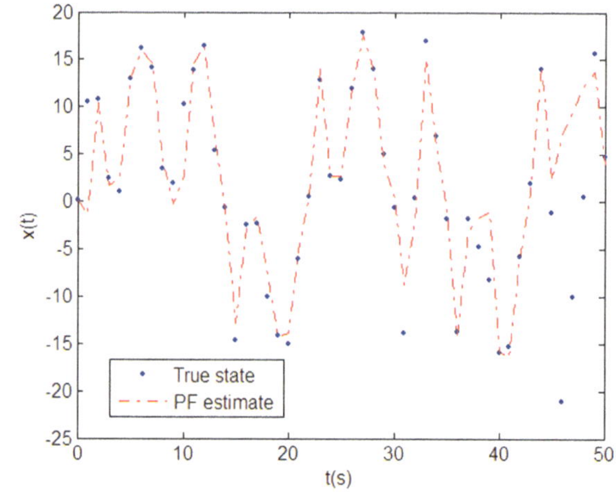

Fig. 8.9 PF-estimated performance curve ($N = 50$)

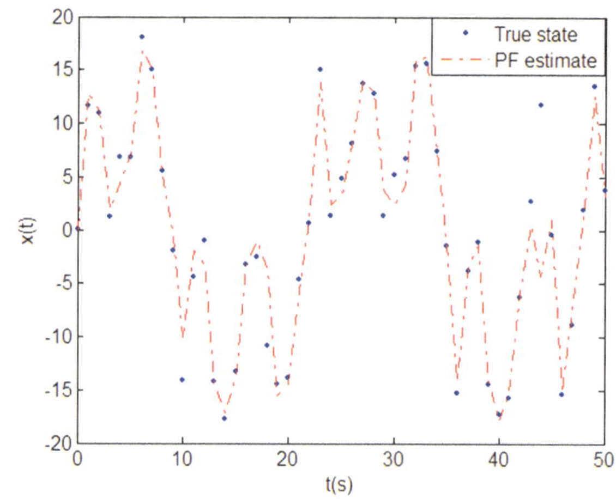

Fig. 8.10 PF-estimated performance curve ($N = 100$)

the estimation effect of the particle-filter algorithm diverges. Most of the time, it can basically reflect the changes of the system trend.

8.2.5 Channel Estimation of an Atmospheric Laser OFDM System Based on Particle Filtering

The particle-filter algorithm is applied to the channel estimation of an atmospheric laser OFDM communication system. The particle-filter method is used to recursively deduce the channel impulse-response function at a given moment, according to the state equation and the measurement equation. The channel estimation is verified by simulation.

8.2.5.1 System-Model Construction

We use the normalized mean-squared error (NMSE) to verify the accuracy of the channel estimation. Figure 8.11 shows the system signal-to-noise ratio (SNR) and the equivalent signal when the NMSE is 0.1, 0.01, and 0.001. The corresponding relationship between the SNR and the signal-to-interference ratio (SIR) is investigated and the performance-simulation analysis results are shown in Fig. 8.11.

In Fig. 8.11, the abscissa represents the signal-to-noise ratio, and the ordinate represents the signal-to-interference ratio. It can be seen from the figure that when the NMSE of the channel estimation is large, the SNR of the entire system decreases very rapidly, whereas under a large NMSE, the signal-to-interference ratio does not increase with the increase of SNR. In contrast, the signal-to-interference ratio increases with the increase of SNR. The system clearly has good performance.

According to the characteristics of the atmospheric channel and the conditions to be met when using particle filtering, the system model is described as follows [13, 16]:

$$X_n = f(X_{n-1}) + V_{n-1} \tag{8.70}$$

$$Z_n = h(X_{n-1}, S_n) + W_n, \tag{8.71}$$

where X_n represents the system channel-state value at time n, Z_n is the observed value of the system, $V_{n\text{-}1}$ and W_n are an independent and identically distributed process

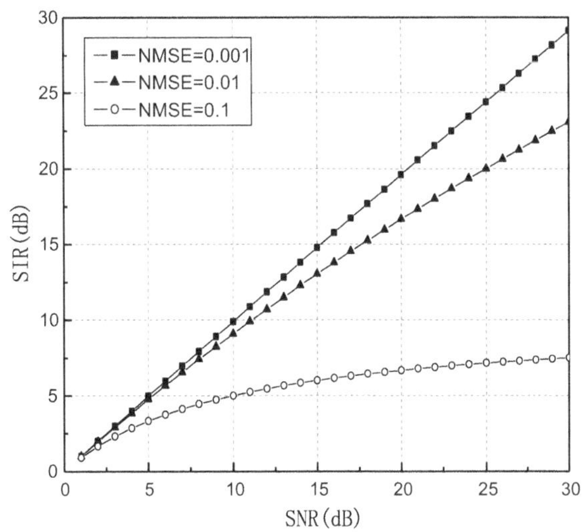

Fig. 8.11 Influence of the channel-estimation error on the system

and observation noise, respectively, $f(\cdot)$ and $h(\cdot)$ can be nonlinear or linear functions, and $f : \Re^{n_x} \times \Re^{n_v} \rightarrow \Re^{n_x}$ and $h : \Re^{n_x} \times \Re^{n_x} \rightarrow \Re^{n_z}$ represent the state-transition function and measurement function of the system, respectively.

The time-varying channel is modeled as a first-order autoregression (AR) model to model an OFDM frame interval; therefore, [16]

the state equation of the system is

$$H_n = \alpha H_{n-1} + (1 - \alpha)V_{n-1}. \tag{8.72}$$

The measurement equation is

$$Y_n = S_n H_n + W_n. \tag{8.73}$$

Among them, $\alpha = J_0(2\pi f_d T)$ reflects the change speed of the time-varying channel, $J_0(\cdot)$ is the zero-order Bessel function, fd represents the maximum Doppler frequency shift, and T represents the duration of one OFDM symbol. Sn is an $N \times N$ diagonal matrix, and the diagonal elements are the Nth OFDM subcarrier transmission symbols. Hn is an $N \times 1$ vector and Vn is an AR model of noise with a mean of 0 and variance σ_V^2. Wn is also an $N \times 1$ vector and follows a distribution with a mean of 0 and variance $N \times 1$.

8.2.5.2 Optical Wireless OFDM Channel Estimation Based on Particle Filtering

A semi-blind estimation method assisted by a training sequence is used; it is carried out in two stages: a training stage and a time-varying channel-estimation stage.

First, the system transmitter sends a training sequence known to the receiver within an optical OFDM symbol time, and the receiver uses the training sequence to complete the initial channel estimation. In the training phase, our detection method is the maximum-likelihood (ML) estimation method; at the next moment, according to the initial channel-estimation value (that is, the channel-estimation value from the previous moment), the system receiver uses particle filtering to blindly estimate the atmospheric channel state at the current moment (without any pilot or training sequence).

The main idea of particle filtering for OFDM channel estimation is to obtain a set of particles and corresponding weights of the initial atmospheric-channel probability distribution by some means in the initial particle stage. Then, in the time-varying channel-estimation stage, within an OFDM symbol block, considering the estimated value of the atmospheric channel at the previous moment, the current observation information, and the limited-character characteristic of the transmitted signal, the terminal calculates the weight of the new atmospheric channel-response particle, according to maximum-likelihood estimation, and then estimates the current atmospheric channel-state information. The atmospheric channel-estimation flowchart of the particle-filter algorithm is shown in Fig. 8.12.

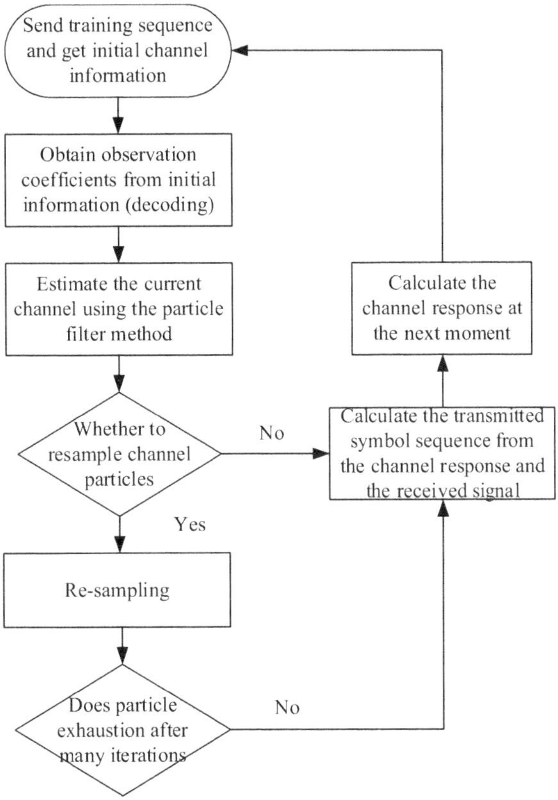

Fig. 8.12 Flowchart of channel estimation using the particle-filter algorithm

8.2.5.3 Simulation Results and Performance Analysis

The particle-filter algorithm is applied to optical wireless OFDM time-varying channel estimation, and the application of the algorithm is simulated. The atmospheric laser OFDM time-varying channel model uses the first-order AR model built above.

The simulation parameters are as follows: quadrature phase-shift keying (QPSK) modulation is used to simulate 100 OFDM symbols, the delay of the four scattering multipaths to the signal is the time length of the subcarriers, and the number of particles is selected as $N = 50$, 100, and 200. The random-resampling algorithm is used for resampling. Figure 5.3 shows the estimated effect of particle filtering on 512 subcarriers.

When using the LS and MMSE algorithms, we use a comb-type pilot structure and an interpolation algorithm at the useful symbols to complete the channel estimation. The channel estimation on the ith OFDM symbol, pth subcarrier is

$$\hat{H}(i,k) = \hat{H}_i(p) + \left(\frac{\hat{H}_i(p+1) - \hat{H}_i(p)}{D_f}\right)(k - k_p) \quad p < k < p+1, \tag{8.74}$$

where p and $p + 1$ represent adjacent subcarriers.

The simulation process is as follows: First, generate a set of binary bit streams through serial-parallel conversion, constellation mapping, and other processes. Then, adopt QPSK modulation, where every two binary bits are a group, corresponding to one of the four complex numbers. Insert a cyclic prefix after the correlation transformation. After the signal is transmitted through our atmospheric-channel model, different algorithms are used to estimate the channel, and then the original transmitted symbol is recovered.

Under different signal-to-noise ratios, the particle-filter method and the current commonly used channel-estimation algorithms (LS and MMSE) are simulated to restore the signal and compare it with the original transmitted binary sequence. The simulation results are shown in Figs. 8.13 and 8.14.

Under the same initial parameter settings, we used the random-resampling algorithm, and the number of particles was selected as $N = 100$ and $N = 200$. The simulation results arc shown in Figs. 8.15 and 8.16, respectively.

Under the same initial parameter settings, we use random and stratified resampling algorithms, and the number of particles is uniformly selected as $N = 100$. The simulation results are shown in Figs. 8.17 and 8.18, respectively.

It can be seen from Fig. 8.16 that the particle-filter algorithm has good estimation performance. Figure 8.15 shows renderings of estimating 512 subcarriers. The performance of the MMSE and LS algorithms is not very different from that of the particle filter when the signal-to-noise ratio is small.

With the increase of SNR, the bit-error rate of the particle-filter algorithm is analyzed using the simulation method. The other two algorithms gain nearly one

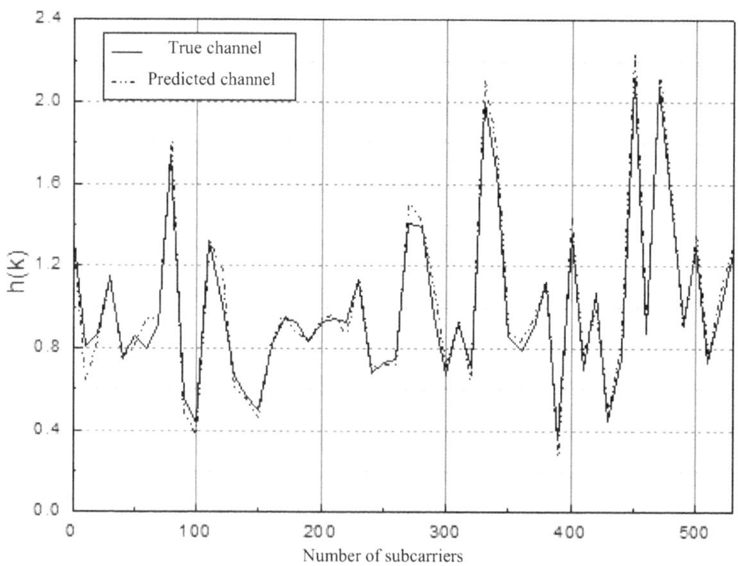

Fig. 8.13 Channel tracking curve

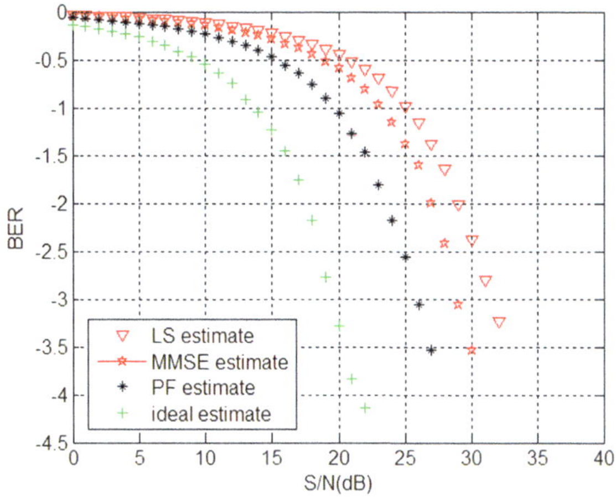

Fig. 8.14 Comparison of BER for OFDM channel estimation using different algorithms ($N = 50$)

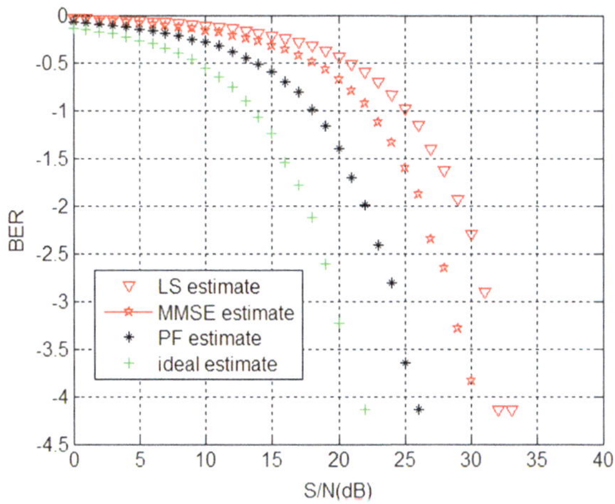

Fig. 8.15 Comparison of bit-error rates for OFDM channel estimation using different algorithms ($N = 100$)

dB. Although it is still distant from the bit-error rate of the ideal channel, this does not affect the excellent estimation performance of the particle-filter algorithm. Research on this strategy will greatly improve the performance of particle-filter channel estimation.

From a comparison of Figs. 8.16, 8.17, and 8.18, it can be seen that different particle numbers have an impact on the channel-estimation performance. From a comparison of Figs. 8.17 and 8.18, it can be seen that the performance of the two resampling methods is similar.

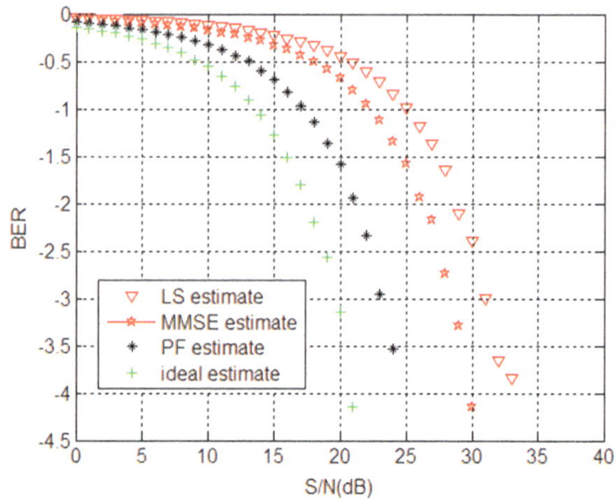

Fig. 8.16 Comparison of bit-error rates for OFDM channel estimation using different algorithms ($N = 200$)

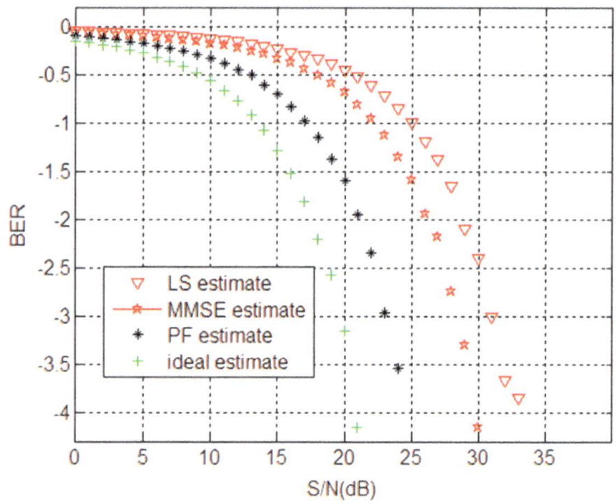

Fig. 8.17 Random resampling algorithm (N = 200)

8.3 Pilot-Assisted Channel-Estimation Algorithms

Because a optical wireless communication (OWC)-OFDM system is affected by the scattering of particles, such as dust, raindrops, and fog, increasing the transmit power does not effectively reduce the effect of scattering on the system [17]. Although a cyclic guard interval is introduced in OFDM, simple frequency-domain equalization can be used at the receiving end to eliminate multipath interference. However, the exact channel-frequency response on each subcarrier must be known before frequency-domain equalization.

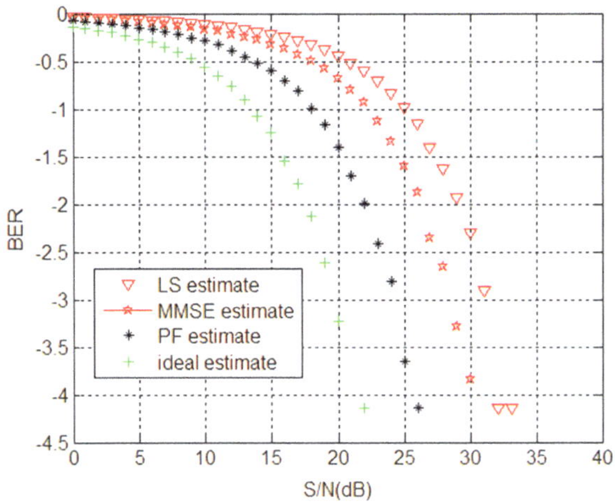

Fig. 8.18 Stratified resampling algorithm (N = 200)

8.3.1 Effects of Fading Channels on OFDM Signals

For the ith OFDM symbol, the received time-domain signal is the convolution of the transmitted signal $x(i, k)$ and the channel impulse response $h(i, k)$:

$$y(i,k) = \sum_{k'=0}^{\tau_{\max}-1} h\big(i,k'\big)x\big(i,k-k'\big) + \eta(i,k) \quad 0 \le k < N + N_G, \tag{8.75}$$

where $\tau_{\max}$ represents the maximum delay spread of the channel, N and NG represent the number of OFDM subcarriers and the number of guard intervals, respectively, $\eta(i, k)$ is Gaussian white noise, and $h(i, k)$ is the time-domain impulse response of the fading channel.

Assuming that the channel impulse response is constant in one OFDM-symbol period (including the guard interval), the system time-domain transmission equation can be expressed as

$$\tilde{y}(i) = \begin{bmatrix} y(i,N_G) \\ \vdots \\ y(i,-1) \\ y(i,0) \\ y(i,1) \\ \vdots \\ y(i,N-1) \end{bmatrix} = \tilde{h}_M(i)\tilde{x}(i) + \tilde{h}_{ISI}(i-1)\tilde{x}(i-1) + \tilde{\eta}$$

$$\tilde{h}_M(i) = \begin{bmatrix} h(i,0) & 0 & 0 & \cdots & \cdots & 0 & 0 \\ h(i,1) & h(i,0) & 0 & 0 & \cdots & 0 & 0 \\ \vdots & \vdots & \ddots & \ddots & \ddots & \vdots & \vdots \\ 0 & \cdots & 0 & h(i,\tau_{\max}-1) & \cdots & h(i,1) & h(i,0) \end{bmatrix}$$

$$\tilde{h}_{ISI}(i-1) = \begin{bmatrix} 0 & \cdots & 0 & h(i-1,\tau_{\max}-1) & \cdots & h(i-1,2) & h(i-1,1) \\ \vdots & 0 & \vdots & 0 & h(i-1,\tau_{\max}-1) & \cdots & h(i-1,2) \\ \vdots & \vdots & \vdots & \ddots & 0 & \ddots & \vdots \\ \vdots & \vdots & \vdots & \vdots & \ddots & \ddots & h(i-1,\tau_{\max}-1) \\ \vdots & \vdots & \vdots & \vdots & \vdots & \ddots & 0 \\ \vdots & \vdots & \vdots & \vdots & \vdots & \vdots & \vdots \\ 0 & 0 & 0 & 0 & 0 & \cdots & 0 \end{bmatrix}$$

$$\tilde{x}(i) = \begin{bmatrix} x(i,N-N_G) \\ \vdots \\ y(i,N-1) \\ y(i,0) \\ y(i,1) \\ \vdots \\ y(i,N-1) \end{bmatrix}, \quad \tilde{x}(i-1) = \begin{bmatrix} 0 \\ \vdots \\ 0 \\ x(i-1,N-\tau_{\max}) \\ \vdots \\ x(i-1,N-2) \\ x(i-1,N-1) \end{bmatrix} \tag{8.76}$$

Assuming $\tau_{\max} < N_G$, after removing the guard interval, the above formula can be simplified to

$$y(i) = \begin{bmatrix} y(i,0) \\ y(i,1) \\ \vdots \\ y(i,N-1) \end{bmatrix} =$$

$$\begin{bmatrix} h(i,0) & 0 & \cdots & 0 & h(i,\tau_{\max}-1) & h(i,\tau_{\max}-2) & \cdots & h(i,1) \\ h(i,1) & h(i,0) & 0 & \cdots & 0 & h(i,\tau_{\max}-1) & \cdots & h(i,2) \\ \vdots & \vdots & \ddots & \ddots & \ddots & \ddots & \vdots & \vdots \\ h(i,\tau_{\max}-1) & h(i,\tau_{\max}-2) & \cdots & h(i,1) & h(i,0) & 0 & \cdots & 0 \\ \vdots & \vdots & \ddots & \ddots & \ddots & \ddots & \vdots & \vdots \\ 0 & \cdots & 0 & h(i,\tau_{\max}-1) & \cdots & h(i,1) & h(i,0) & 0 \\ 0 & \cdots & 0 & 0 & h(i,\tau_{\max}-1) & \cdots & h(i,1) & h(i,0) \end{bmatrix}.$$

$$\begin{bmatrix} x(i,0) \\ x(i,1) \\ \vdots \\ x(i,N-1) \end{bmatrix} + \eta = h_M(i)x(i) + \eta \tag{8.77}$$

The received frequency-domain signal can be expressed as

$$\begin{aligned}
Y(i) &= FFT(y(i)) = FFT(h_M(i)x(i)) + FFT(\eta) \\
&= H_M(i)X(i) + FFT(\eta) \\
&= \begin{bmatrix} H(i,0) & 0 & 0 & 0 \\ 0 & H(i,1) & 0 & 0 \\ 0 & 0 & \ddots & 0 \\ 0 & 0 & 0 & H(i,N-1) \end{bmatrix} \begin{bmatrix} X(i,0) \\ X(i,1) \\ \vdots \\ X(i,N-1) \end{bmatrix} + FFT(\eta). \quad (8.78)
\end{aligned}$$

It can be seen from the above formula that the received OFDM signal in the frequency domain can be expressed as the product of the transmitted signal and the frequency-domain response of the channel. Then, at the receiving end, only single-tap equalization for each subcarrier can eliminate the channel fading; that is,

$$\begin{aligned}
Z(i) = Y(i)./H(i) &= \begin{bmatrix} Y(i,0)/\hat{H}(i,0) + FFT(\eta(i,0))/\hat{H}(i,0) \\ Y(i,1)/\hat{H}(i,1) + FFT(\eta(i,1))/\hat{H}(i,1) \\ \vdots \\ Y(i,N-1)/\hat{H}(i,N-1) + FFT(\eta(i,N-1))/\hat{H}(i,N-1) \end{bmatrix}, \\
&= \begin{bmatrix} \frac{H(i,0)}{\hat{H}(i,0)} X(i,0) \\ \frac{H(i,1)}{\hat{H}(i,1)} X(i,1) \\ \vdots \\ \frac{H(i,N-1)}{\hat{H}(i,N-1)} X(i,N-1) \end{bmatrix} + FFT(\eta) \quad (8.79)
\end{aligned}$$

where $\hat{H}(i,k)$ is the channel-gain estimate. It can be seen that compared with the equalization technology in a single-carrier system, in a multicarrier system, the channel response must be estimated first, and then the estimated channel response is used to equalize the received signal. Therefore, the accuracy of the channel estimation directly affects the entire system performance.

8.3.2 Channel-Estimation Algorithm Based on a Frequency-Domain Pilot

According to the structure of an OFDM symbol, the frequency-domain and time-domain signals can be easily divided. When training data are loaded in the frequency domain, such training sequences are often referred to as pilots.

Pilot-based channel-estimation algorithms generally require three steps to obtain a reliable channel estimation [18–21]:

(1) Reasonably design the pilot structure,
(2) Accurately obtain the channel response at the pilot position, and
(3) Recover the channel responses from all locations, according to the channel responses at the pilot locations.

8.3.2.1 Pilot Structure

Generally speaking, pilot structures are divided into block, comb, and scattered pilots [22–24], as shown in Fig. 8.19.

Because the frequency response of the channel can be regarded as a two-dimensional random signal, the insertion of a pilot frequency can actually be regarded as two-dimensional sampling. To use the inserted pilots to obtain channel estimates for all carriers in all time-frequency spaces, the interval between the inserted pilots must satisfy the Nyquist sampling theorem.

An OFDM frequency-domain signal can be represented as a two-dimensional plane diagram of the time and frequency domains. Therefore, when pilots are inserted into OFDM signals, the minimum interval D_f of the pilot symbols inserted in the frequency direction (in units of subcarrier intervals) should be considered simultaneously with the smallest interval D_t in the time direction (normalized in units of OFDM symbol intervals). To determine D_f and D_t, it is necessary to know the channel changes in the time and frequency domains; namely, the maximum Doppler frequency shift $f_{D_{\max}}$ and the maximum delay spread $\tau_{\max}$ of the channel fading, respectively.

According to the frequency-domain sampling theorem, sampling a signal in the frequency domain corresponds to expanding a period in the time domain. To restore the frequency-domain signal without distortion, the corresponding expanded time-domain signal should not have aliasing distortion. Moreover, the expansion period $1/(D_f \Delta f)$ in the time domain should not be less than the maximum delay spread $\tau_{\max}$:

$$1/(D_f \Delta f) \geq \tau_{\max} \Rightarrow D_f \leq \frac{1}{\tau_{\max} \Delta f}. \tag{8.80}$$

According to the time-domain sampling theorem, to restore the time-domain signal without distortion, the sampling frequency should not be less than twice the

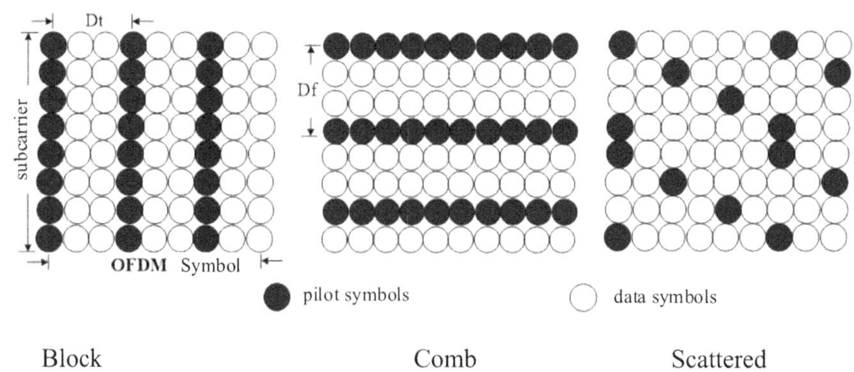

Fig. 8.19 Pilot-structure diagram

signal bandwidth; that is,

$$1/(D_t T) \geq 2f_{D_{\max}} \Rightarrow D_t \leq \frac{1}{2f_{D_{\max}} T}, \tag{8.81}$$

where T is the OFDM symbol period and $\Delta f = 1/T$ is the subcarrier spacing.

D_f and D_t can only take integers in the actual system; that is, $D_f \leq \left[\frac{1}{\tau_{\max}\Delta f}\right]$ and $D_t \leq \left[\frac{1}{2f_{D\max} T}\right]$. Therefore, the total number of pilots contained in a frame is

$$N_{grid} = \left[\frac{N_c}{D_f}\right]\left[\frac{N_s}{D_t}\right], \tag{8.82}$$

where N_c is the number of carriers included in one OFDM symbol, and N_s is the number of OFDM symbols included in one frame.

Because the sampling theorem is satisfied in both the frequency and time domains, if the frequency-response value of the channel at the pilot position can be determined, all the frequency-response values of the entire channel can be obtained.

In addition, because no valid information is transmitted at the position where the pilot symbol is inserted, inserting the pilot symbol will result in a waste of transmission resources. The overhead caused by a pilot insertion is

$$\Lambda = \frac{N_{grid}}{N_c N_s}. \tag{8.83}$$

The signal-to-noise ratio loss is

$$V_{pilot} = 10\log_{10}\left(\frac{1}{1-\Lambda}\right). \tag{8.84}$$

If the subcarrier spacing of the system has been determined, the larger the Doppler frequency shift of the channel, the larger the maximum multipath delay spread, and the larger the required pilot-symbol overhead. Generally, when arranging pilot symbols in an actual system, the first and last OFDM symbols in a frame and the first and last subcarriers of OFDM symbols should contain pilot symbols, to ensure that the estimated value of each frame edge is more accurate.

Assuming that N_p pilot sequences $X_i = \left\{X_i(0)\ X_i(1)\ \ldots\ X_i(N_p-1)\right\}^T$ with the same power are inserted into one OFDM symbol, the power of each pilot signal is σ_s^2, and the position value of the pilot signal is $\left\{k_1\ k_2\ ,\ \ldots,\ k_{N_p}\right\}$. After the receiver undergoes an FFT transformation, the received signal at the pilot position can be expressed as

$$Y_i = \text{diag}(X_i)H_i + N_i, \tag{8.85}$$

where N_i is the noise component with a zero mean and variance σ_n^2. We divide by X_i on both sides of the above formula to obtain

$$\hat{H}_i = W_i^+ h + \tilde{N}_i, \tag{8.86}$$

where W_i is the FFT transformation matrix of $N_p \times L$, and L is the number of multipaths caused by scattering. Then, the variance of the second term characterizing the noise is σ_n^2/σ_s^2. When $N_p \geq L$, h can be estimated as

$$\hat{h} = h + W_i^+ \tilde{N}_i, \tag{8.87}$$

where $(\cdot)^+$ represents the pseudo-inverse of the matrix. The mean-squared error of the channel impulse-response estimation can be expressed as

$$\begin{aligned} E\left\{\left\|\hat{h} - h\right\|^2\right\} &= tr\left\{W_i^+ E\{\tilde{N}_i \tilde{N}_i^H\}\left(W_i^+\right)^H\right\} \\ &= \frac{\sigma_n^2}{\sigma_s^2} tr\left\{W_i^+\left(W_i^+\right)^H\right\} \\ &= \frac{\sigma_n^2}{\sigma_s^2} tr\left\{W_i^H (W_i)^{-1}\right\}, \end{aligned} \tag{8.88}$$

where $tr\{\cdot\}$ represents the trace of the matrix and $\|\cdot\|^2$ represents the second norm of the matrix. Because the selection of each element in W_i depends on the set $\left\{k_1\ k_2\ , \ldots, \ k_{N_p}\right\}$ of pilot-signal positions, the channel-estimation mean-squared error can be minimized by selecting the optimal set $\left\{k_1\ k_2\ , \ldots, \ k_{N_p}\right\}$.

According to the particularity of the FFT transformation matrix, $tr\left\{W_i^H W_i\right\} = N_p L$. Assuming that the L eigenvalues of $\left(W_i^H W\right)^{-1}$ are $\{\lambda_1, \lambda_2, \ldots \lambda_L\}$, and the L eigenvalues of $W_i^H W$ are $\{1/\lambda_1, 1/\lambda_2, \ldots 1/\lambda_L\}$, the channel-estimation mean-squared-error minimization problem can be transformed into the following minimization with constraints problem:

$$\min\left[\sum_{i=1}^{L} \lambda_i\right]$$

$$\text{Restrictions: } \sum_{i=1}^{L} \frac{1}{\lambda_i} = N_p L. \tag{8.89}$$

Because matrix $\left(W_i^H W\right)^{-1}$ is non-negative definite, $\sum^L \lambda_i$ can achieve the minimum value if and only if all its eigenvalues are equal. When matrix Wi satisfies the column orthogonality, $W_i^H W$ has L equal eigenvalues. To ensure the column orthogonality of the matrix W_i, the pilot signal position set must satisfy the following formula:

$$\left\{i, i+\frac{N}{N_p}, \ldots \frac{(N_p-1)N}{N_p}\right\}, i=0,1,\ldots,\frac{N}{N_p}-1. \tag{8.90}$$

That is, when N_p subcarriers are selected as pilot signals at equal intervals, the mean-squared error of the channel estimation can be minimized:

$$E\left\{\left\|\hat{h}-h\right\|^2\right\}=\frac{\sigma_n^2 L}{\sigma_s^2 N_p}. \tag{8.91}$$

8.3.2.2 Estimation of Channel Response at Pilot Locations

After the structure of the pilot frequency is determined, the next step is obtaining the channel response $\hat{H}\left(n_p, k_p\right)$ of the pilot-frequency position, which is represented by the coordinate $\left(n_p, k_p\right)$. On the one hand, the accuracy of $\hat{H}\left(n_p, k_p\right)$ must be guaranteed, and on the other hand, the validity of the estimation method must be considered. The most commonly used estimation criteria are the least-squares (LS) and minimum mean-squared error (MMSE) criteria.

(1) **Estimation algorithm based on the LS criterion**

According to the least-squares criterion, the cost function is defined as [22, 25, 26]

$$J_{LS}=\left(Y_p, \hat{H}_p X_p\right)^*\left(Y_p, \hat{H}_p X_p\right), \tag{8.92}$$

where Y_p, $\hat{H}_p$, and X_p respectively represent the received signal, the estimated value of the channel response, and the transmitted pilot value at the pilot position; $1 \leq p \leq M$ and M is the number of pilots. Let $J_{LS}=0$, and we obtain

$$\hat{H}_p=\frac{Y_p}{X_p}=\frac{H_p X_p}{X_p}+\frac{W_p}{X_p}=H_p+\frac{W_p}{X_p}, \tag{8.93}$$

where W_p is the noise interference at the pilot location.

All operations in this channel-estimation method are performed in the frequency domain, and it is widely used, owing to its simple structure. However, because the LS criterion does not consider noise cancellation, the channel-estimation results will be seriously affected by noise.

The LS algorithm is based on training. When it is applied to an OWC-OFDM system, it usually takes three steps to obtain the channel-estimation value:

(1) Design the pilot structure;
(2) Obtain the channel response at the pilot position through the LS algorithm; and
(3) According to the channel responses at the pilot positions, recover the channel responses at all positions using linear interpolation.

Because the comb-type pilot has excellent performance in a fast time-varying frequency-selective mobile OFDM system, it is selected in this section.

Let the number of subcarriers in each OFDM symbol be N, and insert N_p pilot signals at equal intervals in the frequency domain for each OFDM symbol; then, we have

$$X(k) = X\left(m\frac{N}{N_P} + j\right) = \begin{cases} Pilot & j = 0 \\ Data\ j = 1, 2, \cdots, (N/N_P - 1) \end{cases}, \quad 0 \leq m \leq N_P - 1. \tag{8.94}$$

To reduce the computational complexity, we assume that the pilot signals in each OFDM symbol are the same.

Suppose that in the ith OFDM symbol, the pilot sequence is represented by vector $X_i = \{X_i(0)\ X_i(1) \ldots X_i(N_p - 1)\}^T$, and the received signal on the pilot subchannel is represented by vector $Y_i = \{Y_i(0)\ Y_i(1) \ldots Y_i(N_p - 1)\}^T$.

$$Y_i = diag(X_i)H_i + N_i, \tag{8.95}$$

where N_i is a white Gaussian noise vector on the pilot subchannel, and the frequency-domain characteristics of the pilot subchannel are represented by a vector:

$$\begin{aligned} H_i &= \{H_i(0)\ H_i(1) \ldots H_i(N_p - 1)\}^T \\ &\approx \left\{\tfrac{Y_i}{X_i}(0)\ \tfrac{Y_i}{X_i}(1) \ldots \tfrac{Y_i}{X_i}(N_p - 1)\right\}^T = \hat{H}_i. \end{aligned} \tag{8.96}$$

$\hat{H}_i$ is the channel response at the pilot position in the ith OFDM symbol estimated by the LS algorithm.

(2) **Channel estimation with wiener filtering based on the MMSE criterion**

The mean-squared error is an effective performance measure of random variable estimation; it is easy to calculate and process, and can reflect the error objectively. The minimum mean-squared error (MMSE) algorithm is a statistical estimation based on the Bayesian algorithm.

If signals $x(n)$ and X(n) and observed signals $y(n)$ and Y(n) are generalized stationary, and their autocorrelation function or power-spectral density is known or available, we use linear MMSE. The process of estimating the criterion is called Wiener filtering. The basic concept of Wiener filtering is to find the optimal impulse or transfer function of a linear filter that minimizes the mean-squared error between the output waveform and the input waveform. Figure 8.20 shows the system structure of a Wiener filter.

According to the MMSE criterion, the cost function is defined as

$$J_{MMSE} = E\left[(H_p - \hat{H}_p)^*(H_p - \hat{H}_p)\right]. \tag{8.97}$$

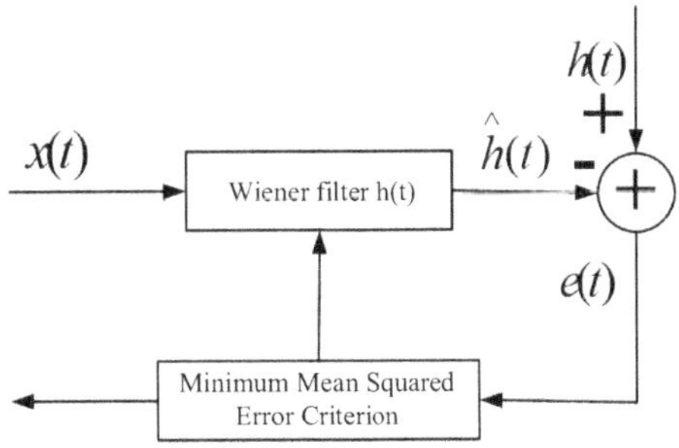

Fig. 8.20 Wiener-filter block diagram

According to the Wiener filter theory [27], it is assumed that

$$\hat{H} = \sum_{p=1}^{M} w_p Y_p = W_p^T Y_p, \tag{8.98}$$

where $Y_p = [Y_1, Y_2, \ldots, Y_M]^T$ represents the received signal at the pilot position, $W_p = [w_1, w_2, \ldots, w_M]^T$ represents the tap coefficient of the filter, and M is the number of pilots.

Definition: $e_p = H_p - \hat{H}_p = H_p - W_p^T Y_p$, $w_k = a_k + jb_k$.

Taking the derivative of Eq. (8.97), we obtain

$$\begin{aligned}
\frac{\partial J_{MMSE}}{\partial w_k} &= \frac{\partial\left\{E\left[(H_p - \hat{H}_p)^*(H_p - \hat{H}_p)\right]\right\}}{\partial w_k} = \frac{\partial\left\{E\left[e_p^* e_p^*\right]\right\}}{\partial w_k} \\
&= \frac{\partial\left\{E\left[e_p^* e_p^*\right]\right\}}{\partial a_k} + j\frac{\partial\left\{E\left[e_p^* e_p^*\right]\right\}}{\partial b_k} \\
&= E\left\{-e_p^* \frac{\partial e_p}{\partial a_k} - e_p \frac{\partial e_p^*}{\partial a_k} + j\left[e_p^* \frac{\partial e_p}{\partial b_k} + e_p \frac{\partial e_p^*}{\partial b_k}\right]\right\} \\
&= E\left\{-e_p^* Y_k - e_p Y_k^* + j\left[e_p^*(-jY_k) + e_p\left(jY_k^*\right)\right]\right\} \\
&= -2E\left[e_p Y_k^*\right].
\end{aligned} \tag{8.99}$$

To minimize J_{MMSE}, $\frac{\partial J_{MMSE}}{\partial w_k} = 0$ must be satisfied; then,

$$\begin{aligned}
E\left[e_p Y_k^*\right] &= E\left[\left(H_p - W^T Y_p\right) Y_k^*\right] \\
&= E\left[H_p Y_k^*\right] - W^T E\left[Y_p Y_k^*\right]. \\
&= 0
\end{aligned} \tag{8.100}$$

We express the above formula in matrix form:

$$W R_{YY} = R_{HY} \Rightarrow W = R_{HY} R_{YY}^{-1}, \tag{8.101}$$

where R_{YY} represents the autocorrelation matrix between the signals received at the pilot position, and R_{HY} represents the cross-correlation matrix between the signal received at the pilot position and the channel response.

Equation (8.98) can be written as

$$\hat{H}_p = W^T Y_p = W = R_{HY} R_{YY}^{-1} Y_p, \tag{8.102}$$

where [28]

$$R_{HT} = E\left[H_p Y_p^H\right] = E\left[H_p (X_p H_p + N)^H\right] = R_{HX} X_p^H \tag{8.103}$$

$$\begin{aligned} R_{YY} &= E\left[Y_p Y_p^H\right] = E\left[(X_p H_p + N)(X_p H_p + N)^H\right] \\ &= X_p R_{HH} X_p^H + \sigma^2 I, \end{aligned} \tag{8.104}$$

where $X_p = \mathrm{diag}(X_1, X_2, \ldots, X_M)^T$ is a diagonal matrix composed of M pilots. I is the identity matrix and σ^2 is the noise variance. Then, the Wiener filter channel-estimation algorithm based on the MMSE criterion can be expressed as

$$\begin{aligned} \hat{H}_p &= R_{HY} R_{YY}^{-1} Y_p = R_{HH} X_p^H \left(X_p R_{HH} X_p^H + \sigma^2 I\right) Y_p \\ &= R_{HH}\left[X_p^{-1}\left(X_p R_{HH} X_p^H + \sigma^2 I\right) X_p^{H^{-1}}\right]^{-1} X_p^{-1} Y_p \\ &= R_{HH}\left[R_{HH} + \sigma^2 \left(X_p^H X_p\right)^{-1}\right] X_p^{-1} Y_p \\ &= R_{HH}\left[R_{HH} + \sigma^2 \left(X_p^H X_p\right)^{-1}\right] \hat{H}_{LS} n, \end{aligned} \tag{8.105}$$

where R_{HH} represents the autocorrelation matrix corresponding to the channel impulse. When the pilot signal X_p changes, the calculation of $X^H X^{-1}$ becomes more complex. To further reduce the complexity, $\left(X_p^H X_p\right)^{-1}$ can be replaced by its expected value $E\left[\left(X_p^H X_p\right)^{-1}\right]$.

In the case of the equal-probability transmission of signals,

$$E\left[\left(X_p^H X_p\right)^{-1}\right] = E\left[|1/x_k|^2 I\right]. \tag{8.106}$$

The average signal-to-noise ratio satisfies $SNR = E\left[|1/x_k|^2 I\right]$; further simplifying Eq. (8.102), we can obtain

$$\hat{H}_p = R_{HH}\left(R_{HH} + \frac{\beta}{SNR} I\right)^{-1} \hat{H}_{LS}. \tag{8.107}$$

$\beta = E|X_p|^2 E|1/X_p|^2$ is a constant that depends on the modulation method. If the autocorrelation matrix R_{HH} and SNR are known a priori, $R_{HH}\left(R_{HH} + \frac{\beta}{SNR} I\right)^{-1}$ only needs to be calculated once.

Definition: $Q = R_{HH}\left(R_{HH} + \frac{\beta}{SNR} I\right)^{-1}$. That is, the Wiener-filter channel-estimation algorithm based on the MMSE criterion can be regarded as a filtering process using the correlation matrix Q based on LS estimation.

(3) **Singular-value decomposition (SVD) estimation algorithm based on the MMSE criterion**

Although the linear MMSE estimator only needs to use the frequency-domain correlation characteristics, and its complexity is lower than that of an estimator that uses both time- and frequency-domain correlations, it still requires a significant number of complex calculations. Moreover, if the channel matrix is singular, the inversion will be impossible to compute or will produce the wrong value. The singular-value decomposition (SVD) algorithm can be inverted and will not find the wrong value, regardless of whether the channel matrix is singular; thus, it reduces the implementation complexity [29, 30].

On the basis of Eq. (8.107), we can first perform singular value decomposition on Q to obtain

$$Q = U \Lambda U^H = U \begin{bmatrix} \delta_1 & 0 & 0 & 0 \\ 0 & \delta_1 & \vdots & 0 \\ \vdots & \vdots & \ddots & \vdots \\ 0 & 0 & \dots & \delta_M \end{bmatrix} U^H, \tag{8.108}$$

where U is an eigenmatrix composed of the eigenvectors of Q, Λ is a diagonal matrix, and δ_k are the M eigenvalues. Substituting the above formula into Eq. (8.107), a channel-estimation formula based on singular-value decomposition can be obtained:

$$\hat{H}_{SID} = U \Lambda U^H \hat{H}_{LS}. \tag{8.109}$$

If only the larger Λ eigenvalues in the L matrix are retained, the above formula can be expressed as

$$\begin{aligned} \hat{H}_{SVD} &= U \Lambda U^H \hat{H}_{LS} \\ &\approx U \begin{bmatrix} \Delta_L & 0 \\ 0 & 0 \end{bmatrix} U^H \hat{H}_{LS} \\ &= \left(\sum_{k=1}^{L} \delta_k u_k u_k^H \right) \hat{H}_{LS} \end{aligned}$$

$$= \left(\sum_{k=1}^{L} u_k^H \hat{H}_{LS} \right) \delta_k u_k. \quad (8.110)$$

u_k is the corresponding eigenvector, Δ_L is the $L \times L$ matrix at the upper-left corner of Λ, and changing the size of L enables a tradeoff between complexity and performance.

8.3.2.3 Interpolation Algorithm in OWC-OFDM Channel Estimation

Figure 8.19 shows a two-dimensional distribution of pilots. Processing can be spatially classified into one-dimensional and two-dimensional processing methods. One-dimensional processing methods can be further divided into channel recovery in the frequency domain (between various subcarriers in one OFDM symbol) and in the time domain (between different OFDM symbols on the same subcarrier).

Frequency-domain channel estimation mainly relies on interpolation filtering. Different interpolation algorithms have a greater impact on the channel-estimation performance. Three main types of channel-estimation recovery algorithms exist in the time domain: interpolation-based algorithms, decision-feedback-based tracking estimation algorithms, and decision-feedback-based and LMS-based tracking methods [21].

Interpolation methods typically include linear and discrete Fourier transform (DFT) interpolation.

(1) **Linear interpolation**

The linear-interpolation algorithm is the simplest interpolation method. This method is based on the frequency-domain response of two adjacent pilot subcarriers, the positional relationship between the data subcarrier and the pilot subcarrier, and the frequency-domain response of data subcarriers between subcarriers [22].

Let $N_p D_f \times 1$ denote the LS estimation value at the pth pilot position of the ith OFDM symbol. First, it is interpolated in the frequency-domain direction, and the channel estimation on the subcarrier of the ith OFDM symbol is

$$\hat{H}(i,k) = \hat{H}_i(p) + \left(\frac{\hat{H}_i(p+1) - \hat{H}_i(p)}{D_f} \right) (k - k_p) \quad p < k < p+1. \quad (8.111)$$

After completing the interpolation in the frequency-domain direction, the same interpolation can be performed in the time-domain direction:

$$\hat{H}(i,k) = \hat{H}_i(p) + \left(\frac{\hat{H}_i(p+1) - \hat{H}_i(p)}{D_f} \right) (i - i_p) \quad p < k < p+1. \quad (8.112)$$

In this manner, channel estimates for all time–frequency points are obtained. The advantage of linear interpolation is that the complexity is low and it is easy to implement. The disadvantage is that when the channel exhibits strong frequency selectivity, the number of required pilots is relatively large.

(2) **DFT interpolation**

Assuming that the LS estimates of the frequency-domain responses on all pilot subcarriers of the ith OFDM symbol are $\hat{H}_i = \{\hat{H}_i(0) \quad \hat{H}_i(1) \quad \ldots \quad \hat{H}_i(N_p - 1)\}^T$, and Np is the number of pilots, this vector represents a sample of the channel frequency-domain response. An estimate of the equivalent time-domain impulse response can be obtained by an inverse DFT (IDFT) transform at point N_p:

$$\hat{h}_i = IDFT(\hat{H}_i). \tag{8.113}$$

In the time domain, a column vector $\hat{h}_i$ of $N_p D_f \times 1$ is obtained by zero-padding $\hat{h}_i$:

$$\hat{h}_i = [\hat{h}_i^T \ 0]^T, \tag{8.114}$$

where $\tilde{h}_i$ represents the zero vector of $1 \times (N_p D_f - N_p)$. Then, $\tilde{h}_i$ is converted to the frequency domain through a DFT transform at point $N_p D_f$ to interpolate in the frequency domain:

$$\hat{H}_i = DFT(\tilde{h}_i). \tag{8.115}$$

After completing the interpolation in the frequency domain, the channel estimates at all time–frequency points can be obtained by linear interpolation in the time domain. DFT interpolation has the advantage of relatively low complexity, owing to the existence of fast DFT and IDFT algorithms.

8.3.2.4 Performance Verification

Because we use comb-type pilots in LS and MMSE channel estimation, we only need to interpolate in the frequency-domain direction to recover the channel-estimation values at all time–frequency points. The channel estimate on the kth subcarrier of the ith OFDM symbol is

$$\hat{H}(i,k) = \hat{H}_i(p) + \left(\frac{\hat{H}_i(p+1) - \hat{H}_i(p)}{D_f}\right)(k - k_p) \quad p < k < p+1. \tag{8.116}$$

Assuming that the system is fully synchronized, the simulation parameters are set as follows: QPSK modulation, 512-point IFFT, 270 useful subcarriers, eight cyclic prefixes, and seven pilot intervals to simulate 100 OFDM symbols.

The simulated channel is a scattering channel. When the number of scattering paths $L = 4$ and 6, the normalized intensity of each path relative to the transmitted

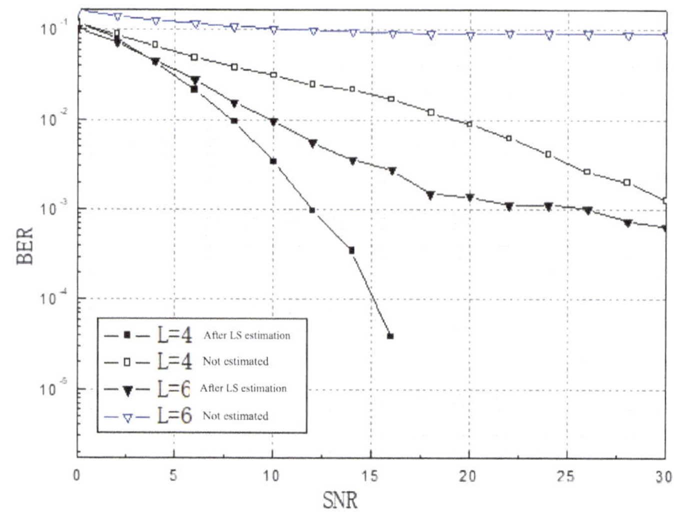

Fig. 8.21 LS channel-estimation bit-error rate diagram

signal is [0.223 0.303 0.334 0.245] and [0.223 0.303 0.334 0.245 0.347 0.458], respectively, and the delay of each path relative to the transmitted signal is [2 3 5 7] and [2 3 5 7 9 13], respectively. When $L = 4$, the bit-error rate can only reach 10^{-3} when the signal-to-noise ratio is 30 dB. When the pilot interval satisfies $L < N_p$, the system performance is significantly improved. Figures 8.21, 8.22, and 8.23 show the system-performance comparison before and after loading the LS, Weiner-MMSE, and SVD-MMSE channel-estimation modules, respectively.

When the pilot interval satisfies $L = 8 > N_p$, the delay of each path is [2 3 5 7 8 9 10 10], and the strength of each path is [0.223 0.303 0.334 0.245 0.347 0.458 0.258 0.426], as shown in Fig. 8.24. Comparing the system performance before and after loading the LS channel-estimation module, because $L > N_p$, the system bit-error characteristics are relatively poor.

8.3.3 *Channel-Estimation Algorithm Based on a Time-Domain Training Sequence*

An OFDM system based on time-domain training-sequence channel estimation is generally referred to as a time-domain synchronous (TDS)-OFDM system. Figure 8.25 shows the frame structure of TDS-OFDM. The time-domain training sequence is placed before the cyclic guard interval and will be used for receiver synchronization and channel estimation.

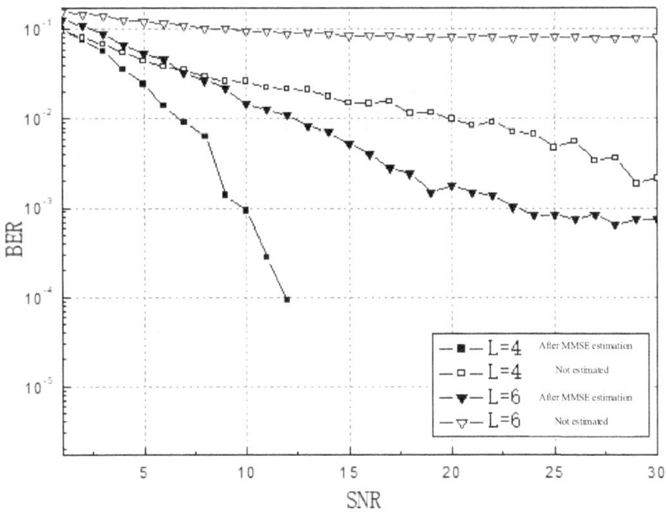

Fig. 8.22 Wiener-filter channel-estimation bit-error rate diagram based on the MMSE criterion

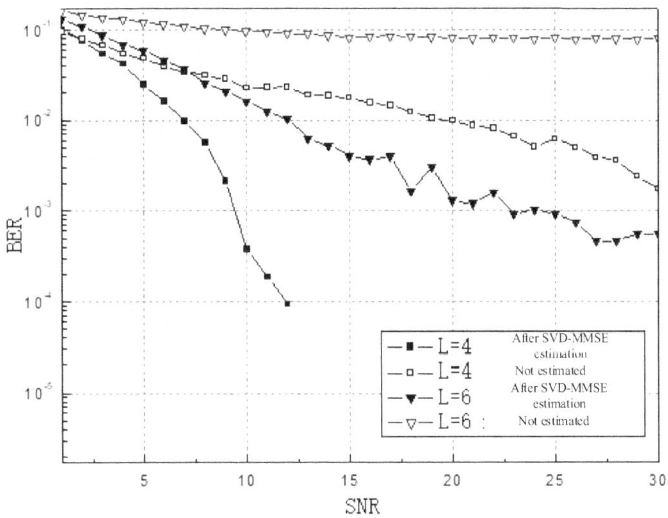

Fig. 8.23 SVD-MMSE channel estimation bit-error rate diagram

8.3.3.1 Channel-Estimation Algorithm Based on a DFT Transform

Assuming that the system is fully synchronized, the time-domain training signal received by the receiver is the linear convolution of the time-domain training sequence $t_1 = \left\{t_1(0) \quad t_1(1) \quad \dots \quad t_1\left(N_p - 1\right)\right\}^T$ and the channel impulse response:

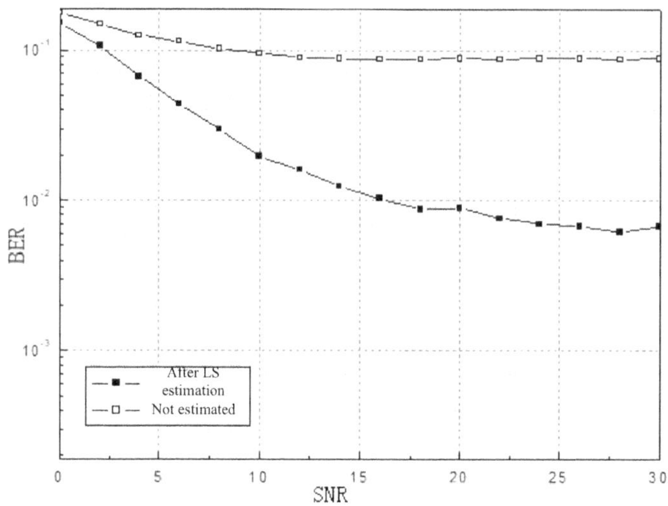

Fig. 8.24 LS channel-estimation bit-error rate when $L > N_\mathrm{p}$

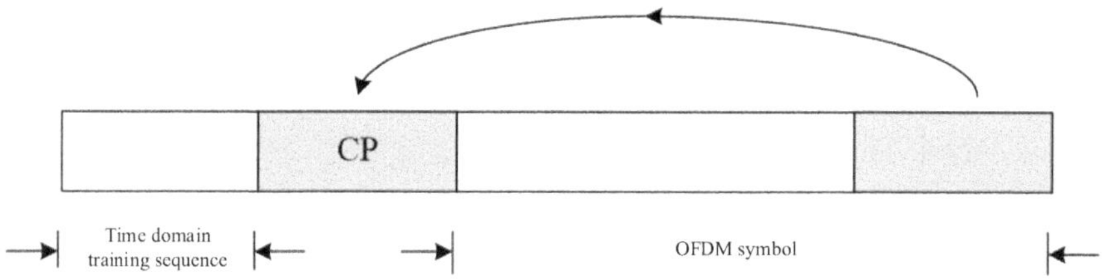

Fig. 8.25 TDS-OFDM frame structure

$$y(k) = \sum_{k'=0}^{L} h(k') \cdot t_1(k - k') + \eta(k) \quad k = 0, \ldots, N_p - 1. \tag{8.117}$$

If the time-domain training sequence is defined as a cyclic repetition, that is, $t_1 = [t, t]^T$,

$$t_1(i) = t_1\left(i + \frac{N_p}{2}\right) \quad i = 0, \ldots, \frac{N_p}{2} - 1. \tag{8.118}$$

If the length of the channel impulse response satisfies $L < N_p/2$, then Eq. (8.117) can be expressed in the form of a circular convolution:

$$y(k) = \sum_{k'=0}^{L} h(k') \cdot t_1(k - k') + \eta(k)$$

$$= \sum_{k'=0}^{L} h(k') \cdot t_1\left(\left(k - k'\right)_{\frac{N_p}{2}}\right) + \eta(k) \quad k' = \frac{N_p}{2}, \ldots, N_p - 1. \tag{8.119}$$

Converted to the frequency domain, it can be expressed as

$$Y(k) = H(k) \cdot T(k) + FFT(\eta(k)) \quad k = 0, \ldots, \frac{N_p}{2} - 1, \tag{8.120}$$

where $T(k)$ is the Fourier transform of $t(k)$, which is a known sequence; hence, the channel response at the time-domain training sequence is

$$\hat{H}_t(k) = \frac{Y(k)}{T(k)} \quad k = 0, \ldots, \frac{N_p}{2} - 1. \tag{8.121}$$

After the channel response $\hat{H}_t(k)$ at the training sequence is obtained, all the channel responses can be further recovered through the interpolation algorithm described above.

The premises of using this estimation algorithm are as follows:

(1) The time-domain training sequence must satisfy the cyclic repetition form defined in Eq. (8.126), and
(2) The length of the training sequence must satisfy $N_p > 2L$.

Compared with pilot-based channel estimation $\left(N_p > L\right)$, this algorithm requires twice the amount of training data for the channel impulse-response length; hence, the spectrum-utilization rate is very low.

8.3.3.2 Prediction Algorithm Based on an Adaptive Filter

An adaptive filter can adjust its own parameters to meet the requirements of some optimal criterion when the statistical characteristics of the input process are unknown or changed. When the statistical properties of the input process are unknown, the process of the adaptive filter adjusting its own parameters is called a "learning" process. When the statistical characteristics of the input process change, the adaptive filter adjusting its own parameters is called a "tracking" process.

The principle of the adaptive filter is shown in Fig. 8.26, where n is the number of iterations, $x(n)$ represents the input signal, $y(n)$ represents the output signal of the adaptive filter, and $d(n)$ is the desired signal (reference signal). The error signal $e(n)$ is calculated from $d(n) - y(n)$. The objective function required by an adaptive algorithm is constructed using the error signal, and an appropriate method for updating the filter coefficients is determined. Minimizing the objective function means that the output signal of the adaptive filter matches the desired signal.

Adaptive-filter applications can be roughly divided into the following four categories: identification, equalization, prediction, and interference cancellation.

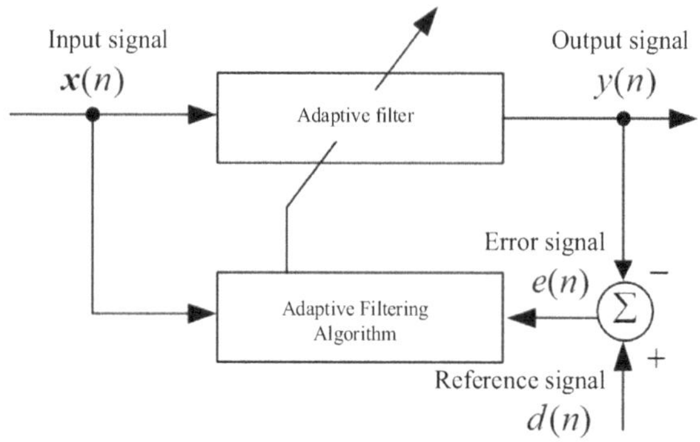

Fig. 8.26 Schematic diagram of an adaptive filter

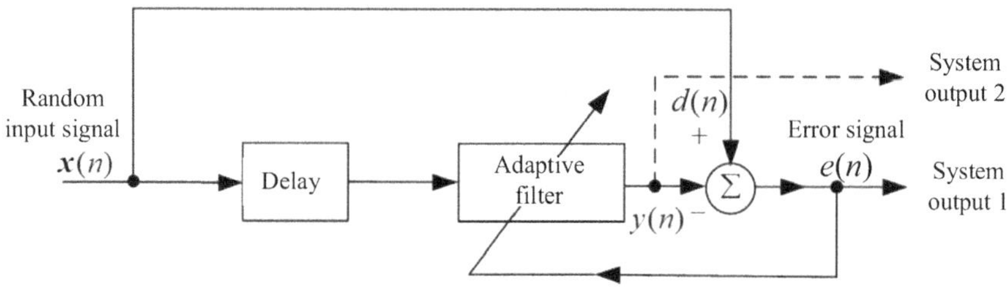

Fig. 8.27 Schematic diagram of an adaptive predictor

Channel estimation based on adaptive filtering utilizes its prediction function. Figure 8.27 shows the principle diagram of adaptive-filter prediction.

8.3.3.3 Semi-blind Channel-Estimation Algorithm Based on LMS

Widrow and Hoff proposed the least mean squares (LMS) algorithm in 1960. The atmospheric channel is a selective-fading channel in the frequency domain, and the value of the channel impulse response of each subcarrier in the coherence time remains correlated. The LMS algorithm is used to predict the initial value of future symbol-position iterations. In this manner, the impulse response of the channel can be directly calculated, and the channel can be compensated using frequency-domain equalization [31].

Because the channel-estimation and tracking algorithm based on decision feedback only needs pilot data in the learning stage, and uses the feedback decision information for channel estimation in the tracking stage, it does not require additional pilot-data information, which can improve the system efficiency [32, 33].

Figure 8.28 shows a schematic diagram of the time-domain LMS channel-estimation algorithm based on an OWC-OFDM system.

$\hat{h}_I(m)$ and $\hat{h}_Q(m)$ are the initial values of the future symbol iterations predicted by the channel impulse response at the time-domain training sequence of horizontally

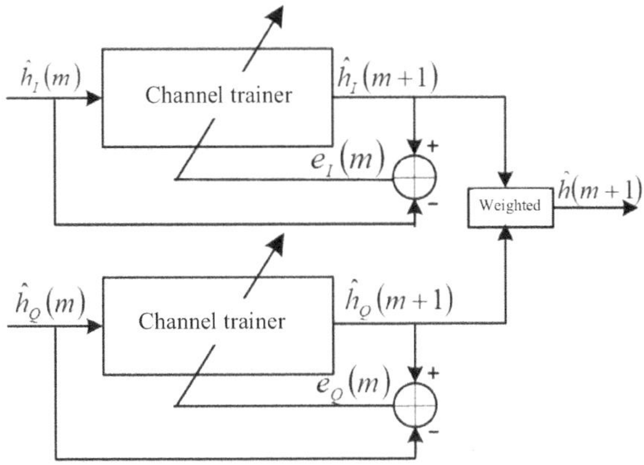

Fig. 8.28 Schematic diagram of LMS channel prediction

and vertically polarized light, respectively. $\hat{h}_I(m+1)$ and $\hat{h}_Q(m+1)$ are the estimated values of the channel impulse response in the horizontal and vertical polarization directions, respectively, of the next cycle in the same subcarrier. $e_I(m)$ and $e_Q(m)$ are error signals. The iterative estimation process is

$$\begin{cases} \hat{h}_I(m+1) = \hat{w}_I^H(m)\hat{h}_I(m) \\ e_I(m) = \hat{h}_I(m+1) - \hat{h}_I(m) \end{cases} \tag{8.122}$$

$$\begin{cases} \hat{h}_Q(m+1) = \hat{w}_Q^H(m)\hat{h}_Q(m) \\ e_Q(m) = \hat{h}_Q(m+1) - \hat{h}_Q(m) \end{cases}. \tag{8.123}$$

The adaptive update process of the weighting coefficient is

$$\hat{w}_I^H(m+1) = \hat{w}_I^H(m) + \mu\hat{h}_I(m)e_I(m) \tag{8.124}$$

$$\hat{w}_Q^H(m+1) = \hat{w}_Q^H(m) + \mu\hat{h}_Q(m)e_Q(m), \tag{8.125}$$

μ is the adaptive step size. Finally, by weighting the channel-prediction values in the two linear-polarization directions, a more accurate channel-estimation value is obtained:

$$\hat{h}(m+1) = \frac{\hat{h}_I(m+1) + \hat{h}_Q(m+1)}{2}. \tag{8.126}$$

After obtaining the channel time-domain response $\hat{h}(i,k)$ for all times, we perform an FFT transformation on it to obtain the estimated vector $\hat{H}(i,k)$ of the frequency-domain channel transfer function (channel-estimation value on the kth subcarrier of the ith OFDM symbol).

Simulation

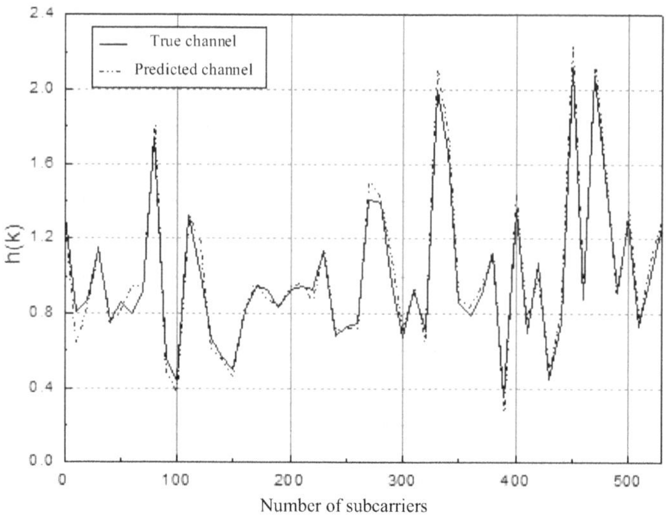

Fig. 8.29 Channel tracking curve

The Monte Carlo method is used to verify the effect of the LMS-based semi-blind channel-estimation algorithm on the system performance. Assuming that the system is completely synchronized, the simulation parameters are set as follows: QPSK modulation, 512-point IFFT, number of subcarriers is 270, cyclic prefix is 18, and 100 OFDM symbols are verified.

The channel is a scattering channel. The number of scattering paths is four, the delay of each path relative to the transmitted signal is [2 3 5 7], the normalized intensity of each path relative to the transmitted signal is [0.123 0.103 0.064 0.05], the Doppler frequency shift is 100 Hz, and the photoelectric efficiency is 0.85.

Finally, the tracking curve of the channel-estimation values of all subcarriers on an OFDM symbol compared to the real channel value is given, as shown in Fig. 8.29. The system bit-error rate is also analyzed. Figure 8.30 shows the bit-error rate curves of the system without channel estimation and with LMS channel estimation. After LMS channel estimation, the system bit-error rate improves by approximately 2 dB.

8.3.3.4 Influence of Nonideal Channel Estimation on a System

The accuracy of the channel estimation will directly affect the receiving performance of the receiver [34–37]. Here, the normalized mean-squared error (NMSE) is used as the standard to measure the quality of the channel estimation, and the influence of the channel-estimation error on the performance of a multicarrier system is analyzed.

$\hat{H}(i, k)$, $H(i, k)$, $X(i, k)$, $Y(i, k)$, and $N(i, k)$ represent the estimated channel response, the true value of the channel response, the transmitted frequency-domain signal, the received frequency-domain signal, and the additive white Gaussian noise

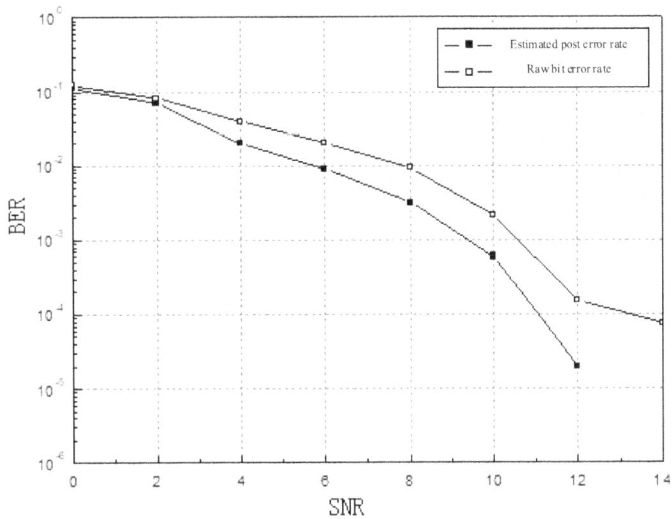

Fig. 8.30 Bit-error rate curve

component on the kth subcarrier of the ith OFDM symbol, respectively. Then, the zero-forcing detection of $X(i, k)$ can be expressed as

$$\begin{aligned}\hat{X}(i,k) &= \frac{Y(i,k)}{\hat{H}(i,k)} = \frac{H(i,k)X(i,k)}{\hat{H}(i,k)} + \frac{N(i,k)}{\hat{H}(i,k)}\\ &= X(i,k)\left(1 + \frac{H(i,k)-\hat{H}(i,k)}{\hat{H}(i,k)}\right) + \frac{N(i,k)}{\hat{H}(i,k)}\\ &= X(i,k) + \frac{H(i,k)-\hat{H}(i,k)}{\hat{H}(i,k)}X(i,k) + \frac{N(i,k)}{\hat{H}(i,k)}\\ &\approx X(i,k) + \frac{H(i,k)-\hat{H}(i,k)}{\hat{H}(i,k)}X(i,k) + N'(i,k).\end{aligned} \tag{8.127}$$

The second term in the above equation represents the extra noise component introduced, owing to the channel-estimation error. Assuming that the power of $N'(i,k)$ is σ_n^2, the channel-estimation error and $N'(i,k)$ are independent of each other, and the power of signal $X(i,k)$ is σ_s^2. The total noise variance is

$$\begin{aligned}\sigma_N^2 &= E\left\{\left|\frac{H(i,k)-\hat{H}(i,k)}{H(i,k)}X(i,k) + N'(i,k)\right|\right\}\\ &= E\left\{\left|\frac{H(i,k)-\hat{H}(i,k)}{H(i,k)}X(i,k)\right|^2 + \left|N'(i,k)\right|^2 + 2\left|\frac{H(i,k)-\hat{H}(i,k)}{H(i,k)}X(i,k)N'(i,k)\right|\right\}\\ &= E\left\{\left|\frac{H(i,k)-\hat{H}(i,k)}{H(i,k)}\right|^2\right\}\sigma_s^2 + \sigma_n^2\end{aligned} \tag{8.128}$$

The NMSE of the channel estimation is defined as

$$\gamma = E\left\{ \left| \frac{H(i,k) - \hat{H}(i,k)}{H(i,k)} \right|^2 \right\}. \tag{8.129}$$

Then, the equivalent signal-to-interference ratio (SIR) can be expressed as

$$\begin{aligned} \overline{\beta} &= \frac{\sigma_s^2}{\sigma_N^2} = \frac{\sigma_s^2}{\gamma\sigma_s^2 + \sigma_n^2} \\ &= \frac{1}{\gamma + \sigma_n^2/\sigma_s^2} \\ &= \frac{\beta}{\gamma\beta + 1}, \end{aligned} \tag{8.130}$$

where $\overline{\beta} = \frac{\sigma_s^2}{\sigma_N^2}$ is the true signal-to-noise ratio of the system.

Figure 8.31 shows the correspondence between the system signal-to-noise ratio and the equivalent signal-to-interference ratio when the NMSE is 0.1, 0.01, and 0.001. It can be seen from the figure that when the NMSE is 0.1, the loss of system SNR caused by inaccurate channel estimation is very serious, especially when the SNR is high. The system SIR hardly increases with the increase of SNR.

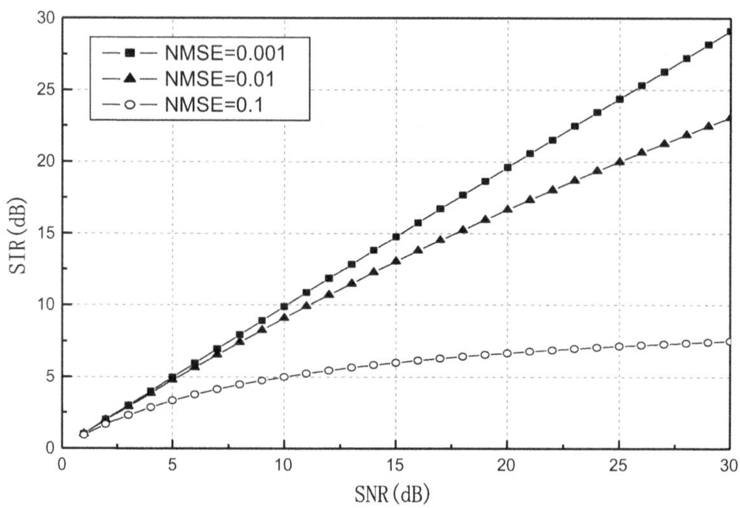

Fig. 8.31 Influence of the channel-estimation error on the system

8.4 Single-User OWC-OFDM Channel Assignment

The system divides an actual channel into several subchannels, and can flexibly allocate transmit power and information bits, according to the actual transmission conditions of each subchannel. For a system where all subchannels use a fixed modulation scheme, the error probability is mainly determined by the subchannel experiencing the worst fading. Therefore, in a frequency-selective fading channel, with the increase of the average signal-to-noise ratio (SNR), the error probability of the system decreases very slowly.

If different modulation schemes are used for different subchannels in the system, and the modulation scheme must be adapted to the SNR of each subchannel, the power allocation to a single subchannel can be optimized.

8.4.1 Water-Injection Algorithm

In the system, by dividing the available channel bandwidth B into N narrower subchannels, the transmission characteristics of each subchannel can be close to the ideal state. Assuming that the transfer function of the channel is $H(f)$, there is additive white Gaussian noise (AWGN) in the channel with a power-spectral density of $N(f)$, the bandwidth of each subchannel is $\Delta f = B/N$, and $|H(f)|^2/N(f)$ is basically unchanged in the subchannel frequency band. The total transmit power of the signal is

$$P_T = \int_B S_x(f)df. \tag{8.131}$$

In Eq. (8.136), $S_x(f)$ is the power-spectral density of the transmitted signal. According to Shannon's theorem, the channel capacity is

$$C = \int_B \log_2\left[1 + \frac{|H(f)|^2 S_x(f)}{N(f)}\right]df. \tag{8.132}$$

If the channel capacity is maximized under the constraints of Eq. (8.136), it can be transformed into

$$\max \int_B \left[\log_2\left(1 + \frac{|H(f)|^2 S_x(f)}{N(f)}\right) + \lambda S_x(f)\right]df. \tag{8.133}$$

In Eq. (8.138), λ is the Lagrange multiplier, which is solved by the variational method:

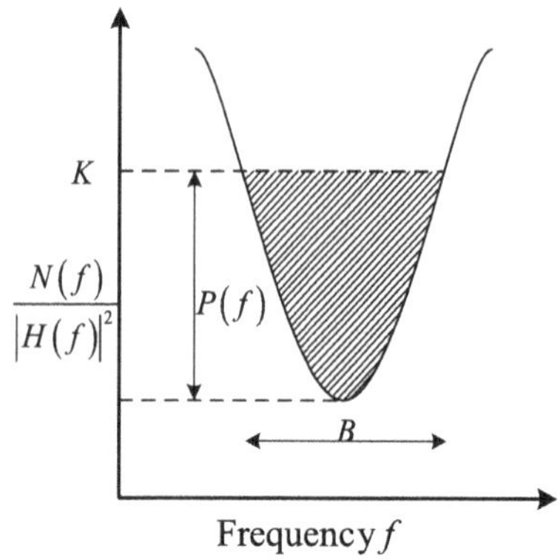

Fig. 8.32 Schematic diagram of the water-injection algorithm

$$S_x(f) = \begin{cases} K - \frac{N(f)}{|H(f)|^2} & f \in B \\ 0 & f \notin B \end{cases} \tag{8.134}$$

$$K = \frac{1}{B}\left(P_T + \int_B \frac{N(f)}{|H(f)|^2} df\right). \tag{8.135}$$

The physical meaning of Eq. (8.133) is as follows: When the signal-to-noise ratio $|H(f)|^2/N(f)$ is large, the allocated power of the corresponding subchannel is also relatively large; otherwise, the allocated power of the subchannel is relatively small or no power is allocated.

Figure 8.32 graphically illustrates the allocation algorithm. The method of maximizing this channel capacity is similar to pouring water into a bowl represented by the solid curve; thus, obtaining $S_x(f)$ can show the maximization of the channel capacity. This is the so-called water-injection allocation algorithm [38].

The water-injection method can achieve the theoretical-maximum channel capacity on a band-limited channel; however, the computational complexity of the algorithm is very large, and the quantization accuracy of the constellation scale must be assumed to be infinitely small, which cannot be achieved in practical applications.

8.4.2 Adaptive Bit-Power Allocation Algorithm

The OWC channel is a time-varying environment, and the channel capacity and signal-to-noise ratio change randomly. Owing to the influence of frequency-selective fading, the channel conditions of different OFDM subchannels differ widely. By allocating different transmit powers and sending different numbers of bits on different OFDM subcarriers, the performance or capacity of the system can be improved.

The idea of adaptive technology is to make the transmission scheme adapt to the channel state as much as possible, by reasonably adjusting and utilizing various parameters and resources when the transmitter obtains some form of channel information.

8.4.2.1 Hughes–Hartogs Algorithm

The Hughes–Hartogs [39] algorithm is a typical iterative-based sequential bit and power allocation algorithm. The basic principle of the algorithm is as follows: First, the number of bits in each subchannel is set to zero, and then all the bits to be allocated are allocated to the corresponding subchannels in turn.

For each allocation, we first find the subchannel with the smallest power increase required to increase one bit, and then increase the number of bits of this subchannel by one. This cycle is repeated until all the bits are allocated, and finally the power required by each subchannel is calculated. The Hughes–Hartogs algorithm can achieve optimal bit and power allocation results. Figure 8.33 shows a flowchart of the algorithm.

The algorithm proceeds as follows:

(1) Set the initial bits of all subchannels to zero; i.e.,

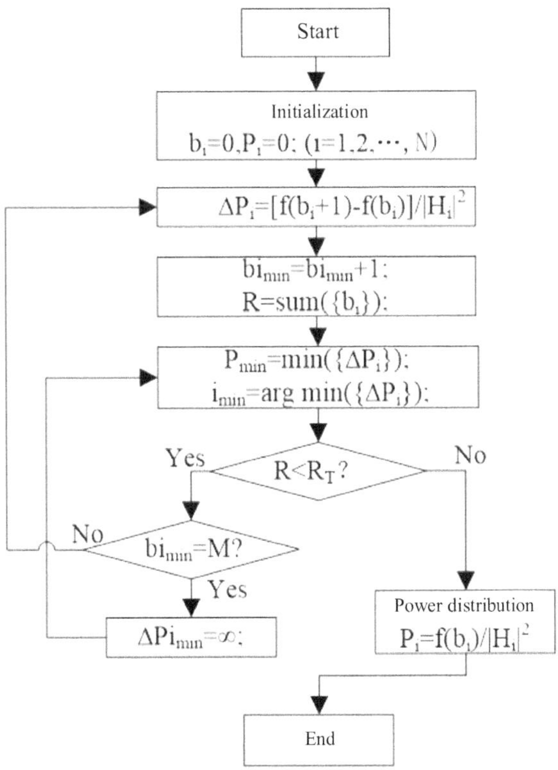

Fig. 8.33 Flowchart of the Hughes–Hartogs algorithm

$$b_i = 0 \quad (i = 1, 2, \ldots, N). \tag{8.136}$$

(2) Calculate the power required to increase each subchannel by one bit; following is the power increment:

$$\Delta P_i = \frac{f(b_i + 1) - f(b_i)}{|H_i|^2} \quad (i = 1, 2, \ldots, N), \tag{8.137}$$

where $f(b_i)$ is the power required to allocate b_i bits.

(3) Find the minimum value in $\{\Delta P_i\}$ and its corresponding subchannel, namely,

$$\begin{cases} P_{\min} = \min(\Delta P_i) \\ i_{\min} = \arg\min_{i=1,2,\ldots,N}(\Delta P_i) \end{cases}. \tag{8.138}$$

(4) Allocate one bit of information to subchannel $i_{\min}$; namely,

$$b_{\min} = b_{i_{\min}} + 1. \tag{8.139}$$

Calculate the total number of bits currently allocated $R = \sum(\{b_i\})$. If $R < R_T$, determine whether $b_{\min} = M$. If so, go to step 5; if $b_{\min} \neq M$, go to step 2. If $R = R_T$, the bit allocation is complete; go to step 6 for power allocation.

(5) Set $\Delta P_{i_{\min}} = \infty$; go to step 3.

(6) Distribute power:

$$P_i = \frac{f(b_i)}{|H_i|^2}, \quad i = 1, 2, \ldots, N. \tag{8.140}$$

At this point, the bit and power allocations are all completed.

The Hughes–Hartogs algorithm requires numerous sorting and searching operations, and the computational complexity is high; thus, it is not very suitable for data transmissions with high real-time requirements.

8.4.2.2 Chow Algorithm

The Chow algorithm [40] allocates bits according to the channel capacity of each subchannel, reducing the number of sorting operations. The algorithm optimizes the spectral efficiency of the system on the premise of ensuring the target bit-error rate.

The bit allocation is completed in three steps. First, the optimal threshold γ_{margin} for system performance is determined; then, the modulation mode of each subchannel is determined, and finally, the bits and power of each subchannel are adjusted. Figure 8.34 shows a flowchart of the Chow algorithm.

The algorithm proceeds as follows:

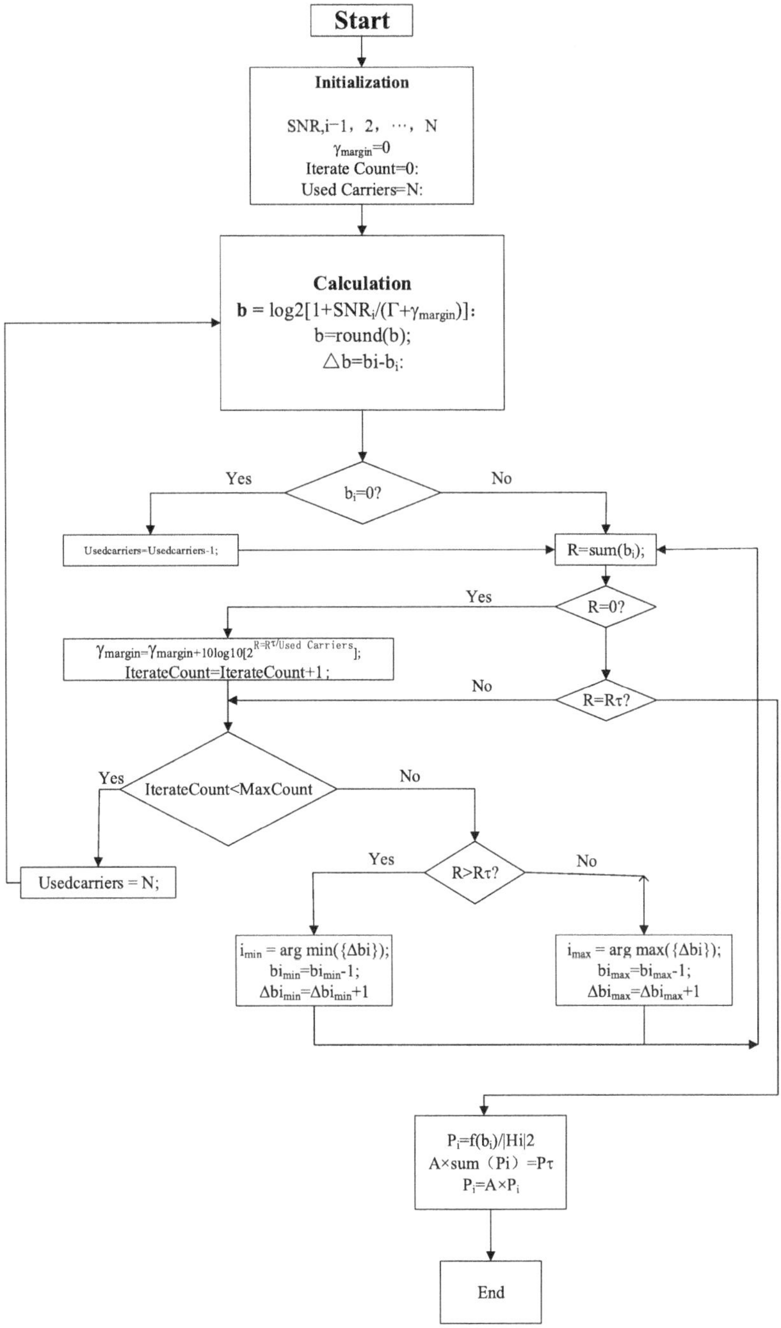

Fig. 8.34 Flowchart of the Chow algorithm

(1) Calculate the signal-to-noise ratio $SNR_i, i = 1, 2, \ldots, N$ of each subchannel, assuming that the signal energy on all subchannels is normalized; $\varepsilon_i = 1, i = 1, 2, \ldots, N$.
(2) Let $\gamma_{m\arg in} = 0(dB)$, the number of iterations IterateCount = 0, the number of used subchannels Usedcarriers = N, where N is the maximum number of available subchannels.
(3) Calculate $b_i,\ \hat{b}_i,\ \Delta b_i$ *and Usedcarriers* sequentially from $i = 1$ to N;

$$b_i,\ \hat{b}_i,\ \Delta b_i\ and\ Usedcarriers$$
$$\begin{cases} b_i = \log_2\left(1 + \dfrac{SNR_i}{\Gamma + \gamma_{m\arg in}}\right) \\ \hat{b}_i = round(b_i) \\ \Delta b_i = b_i - \hat{b}_i \\ Usedcarriers = Usedcarriers - 1\ when\ \hat{b}_i = 0 \end{cases} \tag{8.141}$$

In Eq. (8.149), the *round* (b_i) function is the largest integer not greater than b_i.
(4) Calculate $R = \sum b_i$. If $R = 0$, the channel condition is too poor to be used; go to step 5. Otherwise, go to step 7.
(5) Calculate the new $\gamma_{m\arg in}$,

$$\gamma_{m\arg in} = \gamma_{m\arg in} + 10\lg 2^{\frac{R-R_T}{Usedcarriers}}. \tag{8.142}$$

(6) $IterateCount = IterateCount + 1$.
(7) If $R \neq R_T$ and $IterateCount < MaxCount$, let $Usedcarriers = N$, and go to step 3; otherwise, go to step 8.
(8) If $R > R_T$, find the smallest Δb_i; correspondingly, $\Delta b_i = \Delta b_i + 1$. Repeat this step until $R = R_T$.
    If $R < R_T$, to find the largest Δb_i corresponding to $b_i = b_i + 1$ and $\Delta b_i = \Delta b_i + 1$, repeat this step until $R = R_T$.
(8) Adjust the transmit power on each subchannel so that it corresponds to b_i,

$$P_{e,i} = P_{e,t\arg et}, \forall i. \tag{8.143}$$

(9) Adjust the total transmit power. Multiply all used subchannels by the same scaling factor to make the total signal power $P = P_T$.

Because the Chow algorithm eliminates a considerable amount of searching and sorting, the complexity of the algorithm is reduced. Because it uses the channel capacity as the rate-allocation criterion, it has two shortcomings:

(1) The signal power and transmission rate are directly related, and
(2) The room for optimization is limited.

An actual system does not try to reach the transmission-system capacity, but to make the bit-error rate as low as possible under a given power and rate.

8.4.2.3 Fischer's Algorithm

Fischer's algorithm [40] is one of the most efficient algorithms at present. The algorithm plays a guiding role in both practical applications and theoretical analyses. It optimizes the bit-error rate performance of the system under the constraints of determining the total rate and transmit power of the system. The Fischer algorithm first stores the noise power value $\log_2 N_i$ on each subchannel, and then only needs to perform some integer addition and division operations; thus, its complexity is further reduced, compared to the Chow algorithm. Figure 8.35 shows a flowchart of Fischer's algorithm.

The algorithm proceeds as follows:

(1) It is necessary to know the noise variance Ni on each subchannel; $i = 1, 2, \cdots, N$, N is the total number of subcarriers, RT is the total number of bits, N_{used} is the number of used subchannels, and its initial values $N_{used} = N$. $\boldsymbol{I}$ are the subchannel sets, with initial values $\boldsymbol{I} = \{1, 2, \cdots, N\}$. Then, $LDN_i = \log_2 N_i$ is calculated for $i = 1, 2, \cdots, N$ of each subchannel, and stored for the next use.

(2) We calculate the number of bits that can be allocated to each subchannel in $\boldsymbol{I}$:

$$b_i = \left(R_T + \sum_{j \in \boldsymbol{I}} LDN_j\right) / N_{used} - LDN_j. \tag{8.144}$$

If $b_i < 0$ and $i \in \boldsymbol{I}$, then $N_{used} = N_{used} - 1$; remove the ith subchannel from $\boldsymbol{I}$, and then recalculate b_i until $b_i > 0$, $i \in \boldsymbol{I}$.

(3) Because the calculated b_i is generally not an integer, it must be quantized: $\hat{b}_i = round(b_i)$. The quantization error $\Delta b_i = b_i - \hat{b}_i$, and $R = \mathrm{sum}\left(\left\{\hat{b}_i\right\}\right)$;

(4) If $R = R_T$, go to step 6 for power distribution; otherwise, go to step 5;

(5) If $R > R_T$, then find the smallest Δb_i, with corresponding $\hat{b}_i = \hat{b}_i - 1$, $R = R - 1$, and $\Delta b_i = \Delta b_i + 1$. Repeat this step until $R = R_T$; then, go to step 6.

if $R < R_T$, then find the largest Δb_i, with corresponding $\hat{b}_i = \hat{b}_i + 1$, $R = R + 1$, $\Delta b_i = \Delta b_i - 1$. Repeat this step until $R = R_T$; go to step 6.

(6) Power distribution. Assign the transmit power allocated to each subchannel in $\boldsymbol{I}$, according to Eq. (8.153):

$$p_i = \frac{P_T N_i 2^{\hat{b}_i}}{\sum_{i \in \boldsymbol{I}} N_i 2^{\hat{b}_i}}. \tag{8.145}$$

Fischer's algorithm provides closed-form solutions for bit and power allocation; thus, it has less complexity and is suitable for high-speed wireless data transmissions. Its bit-error rate performance is no worse than that of Chow's algorithm, and in some cases the performance is even better.

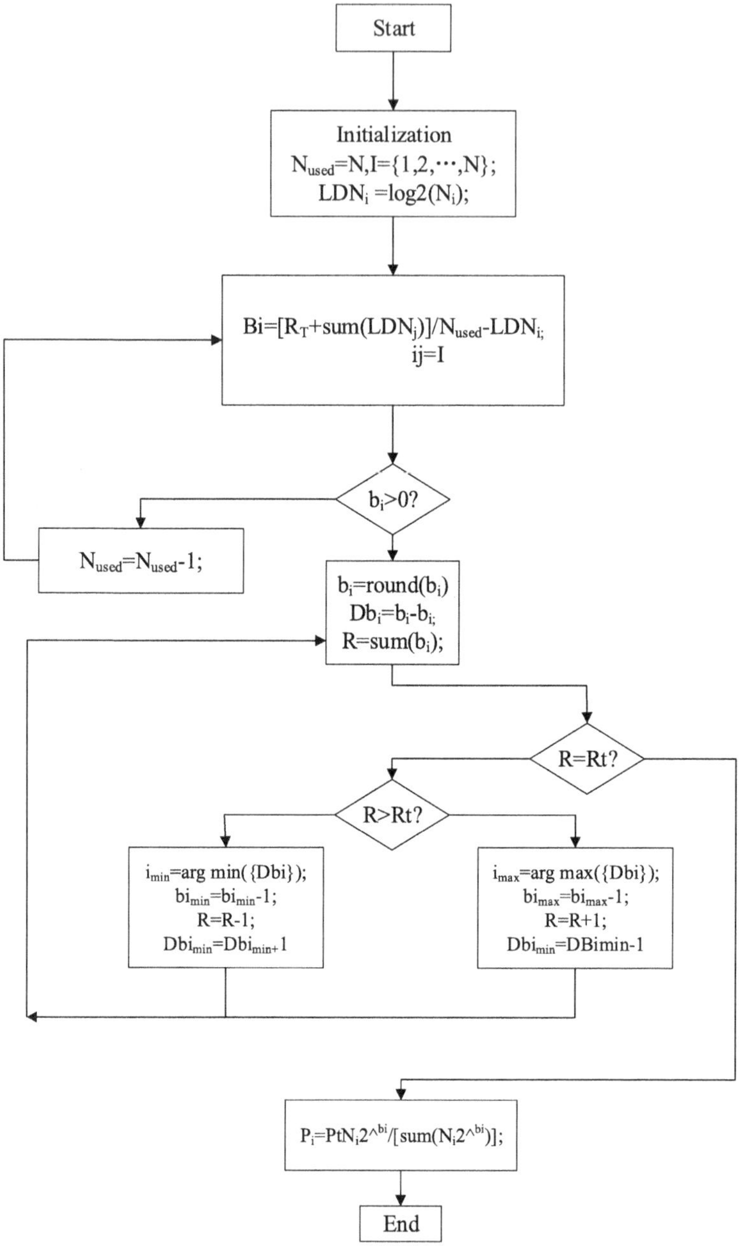

Fig. 8.35 Fischer algorithm flowchart [35]

8.4.3 Improved Adaptive Bit-Power Allocation Algorithm

Most of the current bit-allocation algorithms use M-ary quadrature amplitude modulation (MQAM); however, it has serious shortcomings: First, when the MQAM method is used, the number of bits transmitted on the subchannel is $\{0, 1, 2, 4, 6, \ldots, 2m\}$. Executing 0 to 1 and 1 to 2 requires one bit, but the others must be increased by two bits. Second, the energy of the constellation points in the MQAM modulation mode is not equal, which will introduce a higher PAPR problem to the OFDM system.

Using M-ary phase-shift keying (MPSK) can avoid the above two shortcomings, because it can transmit any number of bits, and it uses equal-energy modulation. This section describes an improved adaptive channel-allocation algorithm for OWC-OFDM systems based on MPSK modulation. It has a fixed bit-transmission rate R_T and bit-error rate (BER), so that the total transmit power P_T of the system can be minimized.

The bit-error rate of MPSK modulation is [36]

$$BER \approx 2Q\left(\sqrt{2\gamma}\sin\left(\frac{\pi}{2^R}\right)\right). \tag{8.146}$$

In Eq. (8.146), the Q function is defined as $Q(\alpha) = \int_\alpha^\infty \frac{1}{\sqrt{2\pi}}\exp\left(\frac{x^2}{2}\right)dx$. Moreover, the signal-to-noise ratio γ and the bit rate R are nonlinear. We simplify Eq. (8.146) to obtain

$$M = 2^R = \frac{\pi}{\sin^{-1}\left(\frac{Q(BER/2)}{\sqrt{2\gamma}}\right)} \tag{8.147}$$

$$R = \log_2^M = \log_2 \pi / \sin^{-1}\left(\frac{Q^{-1}(BER/2)}{\sqrt{2\gamma}}\right)s. \tag{8.148}$$

When x is small, $\sin^{-1}(x) \approx x$; then, Eq. (8.148) can be simplified to

$$R = \log_2\left(\frac{\sqrt{2\gamma}\pi}{Q^{-1}(BER/2)}\right) = \frac{1}{2}\log_2\left[\gamma/\left(\frac{Q^{-1}(BER/2)}{\sqrt{2\pi}}\right)^2\right] = \frac{1}{2}\log_2\left(\frac{\gamma}{\Gamma^*}\right). \tag{8.149}$$

In Eq. (8.149), $\Gamma^* = \left(\frac{Q^{-1}(BER/2)}{\sqrt{2\pi}}\right)^2$; we continue to simplify Eq. (8.149) to obtain

$$\begin{aligned} R &= \frac{1}{2}\log_2\sqrt{\left(\frac{\gamma}{\Gamma^*}\right)} = \log_2\left[1 + \left(\frac{\gamma}{\Gamma^*} - 1\right)\right] \\ &= \log_2\left(1 + \gamma/\left(\frac{\gamma\sqrt{\Gamma^*}}{\sqrt{\gamma} - \sqrt{\Gamma^*}}\right)\right) = \log_2\left(\frac{\gamma}{\Gamma} + 1\right), \end{aligned} \tag{8.150}$$

where $\Gamma = \frac{\gamma\sqrt{\Gamma^*}}{\sqrt{\gamma}-\sqrt{\Gamma^*}}$.

Based on the above equations, the adaptive channel-allocation algorithm is improved. The main idea is to use MPSK modulation to initialize and allocate each subchannel, according to the signal-to-noise ratio and channel gain of each subchannel, and then adjust each subchannel. The bit allocation of each b_i is adjusted so that it satisfies the constraint; that is, $\sum b_i = B$. Figure 8.36 shows a flowchart of the improved algorithm.

The specific allocation process is as follows:

(1) Initialize the allocation. For $i = 1, 2, \ldots, N$, calculate the number of bits on the ith subchannel according to Eq. (8.151):

$$b_i = \log_2\left(1 + \frac{\gamma}{\Gamma}\right). \tag{8.151}$$

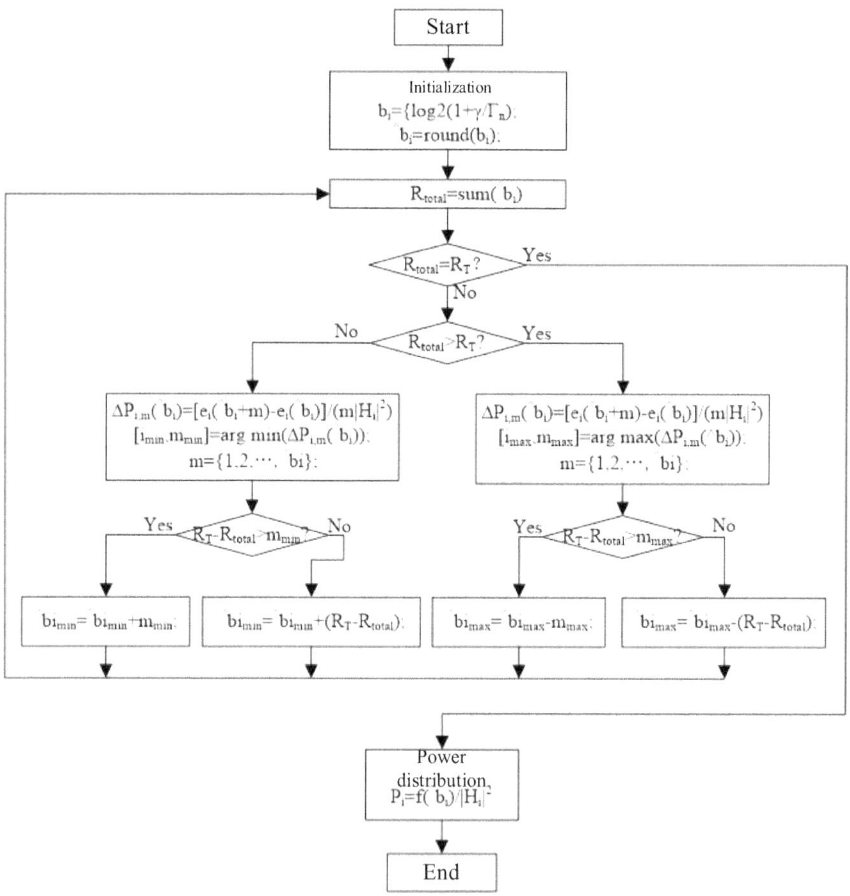

Fig. 8.36 Flowchart of the improved algorithm [41]

b_i is rounded ($b_i = round(b_i)$), where the $round(b_i)$ function produces the largest integer not greater than b_i.

(2) Calculate the total number of bits currently allocated: $R_{total} = \sum_{i=1}^{N} \hat{b}_i$.

If $R_{total} < R_T$, go to step 3 to continue allocating bits to the subchannel.

If $R_{total} > R_T$, go to step 4 to modulate the number of bits already allocated to the subchannel.

If $R_{total} = R_T$, then the bit allocation ends; go to step 5 for power distribution.

(3) For $i = 1, 2, \ldots, N$, calculate the number of bits allocated to the ith subchannel, and increase the average power increment per bit when $1, 2, \ldots, M - \hat{b}_i$ is increased from $\hat{b}_i$:

$$\Delta P_{i,m}\left(\hat{b}_i\right) = \frac{e_i\left(\hat{b}_i + m\right) - e_i\left(\hat{b}_i\right)}{m|H_i|^2} \quad m = 1, 2, \ldots, M - \hat{b}_i. \tag{8.152}$$

Find the subscript corresponding to the minimum value in $\Delta P_{i,m}\left(\hat{b}_i\right)$:

$$[i_{\min}, m_{\min}] = \arg\min\left(\Delta P_{i,m}\left(\hat{b}_i\right)\right). \tag{8.153}$$

In Eq. (8.153), $i_{\min}$ is the subchannel corresponding to the minimum value, and $m_{\min}$ is the number of additional bits corresponding to the minimum value.

If $R_T - R_{total} \geq m_{\min}$, after allocating $m_{\min}$ bits to subchannel $i_{\min}$, go to step 2.

If $R_T - R_{total} < m_{\min}$, find the subchannel with the smallest power increase that can be increased by no more than $R_T - R_{total}$ bits at one time, and obtain a new $[i_{\min}, m_{\min}]$. After making the assignment, go to step 2.

(4) For $i = 1, 2, \ldots$, N, calculate the average power reduction per bit when the number of bits allocated by the ith subchannel is reduced from $\hat{b}_i$ to $1, 2, \ldots, \hat{b}_i$:

$$\Delta P_{i,m}\left(\hat{b}_i\right) = \frac{e_i\left(\hat{b}_i + m\right) - e_i\left(\hat{b}_i\right)}{m|H_i|^2} \quad m = 1, 2, \ldots, \hat{b}_i. \tag{8.154}$$

Find the subscript corresponding to the maximum value in $\Delta P_{i,m}\left(\hat{b}_i\right)$:

$$[i_{\min}, m_{\min}] = \arg\max\left(\Delta P_{i,m}\left(\hat{b}_i\right)\right), \tag{8.155}$$

where $i_{\max}$ is the subchannel corresponding to the maximum value, and $m_{\max}$ is the number of reduced bits corresponding to the maximum value.

If $R_{total} - R_T \geq m_{\max}$, reduce subchannel $i_{\max}$ by $m_{\max}$ bits, then go to step 2.

If $R_{total} - R_T < m_{\max}$, search for the subchannel with the largest power reduction that can be reduced by no more than $R_T - R_{total}$ bits at one time, and obtain a new $[i_{\max}, m_{\max}]$; after allocating, go to step 2.

(5) Power allocation: allocate the minimum transmit power to each subchannel; namely,

$$P_i = \frac{f(\hat{b}_i)}{|H_i|^2}. \tag{8.156}$$

8.4.4 Analysis of Simulation Results

A MATLAB simulation of the improved algorithm was conducted. The simulation parameters were as follows: the number of subchannels $N = 64$, the MPSK modulation method is adopted, the average number of bits carried on each subchannel is four, and the maximum number of bits allowed to be carried is eight. The improved algorithm is compared with the Hughes–Hartogs algorithm and the fixed bit-allocation (FBA) algorithm, and the simulation results are shown in Fig. 8.37.

As shown in Fig. 8.37, the improved algorithm is obviously better than the FBA algorithm. When the SNR is larger, the bit-error rate of the improved algorithm approaches that of Hughes–Hartogs, but the algorithm complexity is reduced. The method of initially allocating partial bits is adopted, which reduces the number of cycles in the bit-allocation process. In the process of allocating the remaining bits, the traditional one-by-one bit allocation method is abandoned, and the method of allocating multiple bits at a time is adopted.

The simulation results show that the bit-error rate of the improved algorithm approaches the bit-error rate of the Hughes–Hartogs algorithm as the signal-to-noise ratio increases; however, the computational complexity of the improved algorithm is greatly reduced.

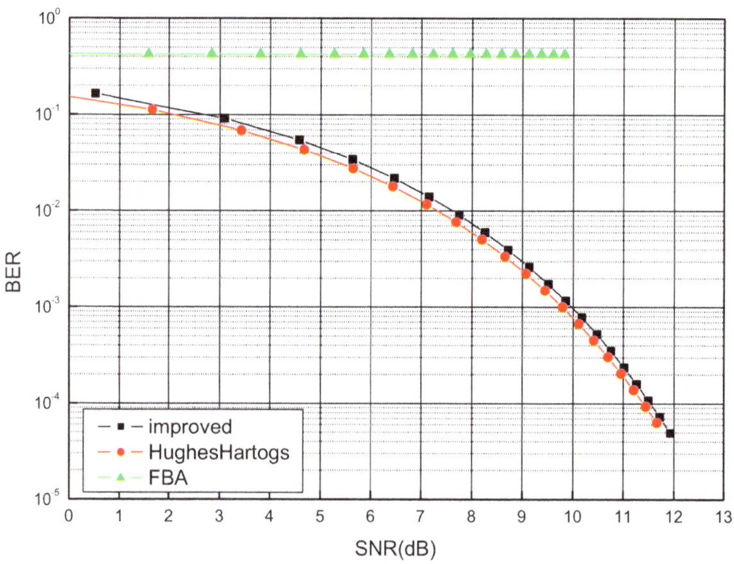

Fig. 8.37 Comparison of bit-allocation algorithms [41]

8.5 Multiuser OWC-OFDM Adaptive Channel Allocation

In a multiuser OWC-OFDM system, all users in the system share all subchannels. Therefore, the adaptive allocation technology in a multiuser system includes two aspects: allocation of subchannels in the system and allocation of bits and power to the subchannels allocated to users.

8.5.1 Principle of a Multiuser Adaptive OWC-OFDM System

We design a multiuser subchannel, bit, and power allocation scheme in which each user can transmit its own signal in any time slot. The scheme determines the subchannel allocated to each user, the number of bits allocated to each subcarrier, and the transmit power according to the fading information of the useful channel. The block diagram of a multiuser FSO-OFDM system is shown in Fig. 8.38.

The transmitter obtains the state information of all subchannels through the multiuser channel estimator. The adaptive subchannel, bit, and power allocator allocates subchannels to K users according to the channel-state information, and allocates bits and transmit power to each subchannel. The system uses a dedicated channel to transmit the allocation information to the receiver in real time.

The serial-to-parallel converter allocates a corresponding number of bits from specific users to each subchannel, according to the adaptive allocation information. The adaptive modulator completes the baseband modulation of each subchannel according to the allocation information, and the adaptively modulated signals are sent in parallel to the IFFT unit. After a parallel/serial conversion, guard-interval

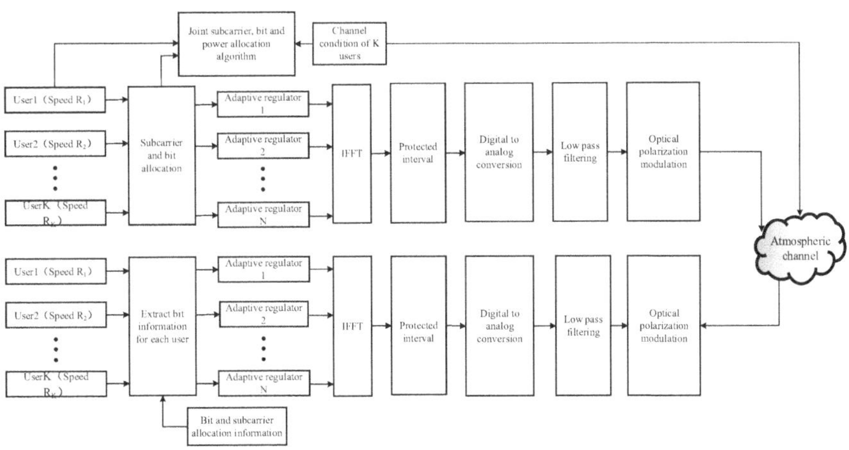

Fig. 8.38 Block diagram of the multiuser adaptive FSO-OFDM system [41]

addition, digital-to-analog conversion, and light-polarization modulation, the signal is emitted.

Each user at the receiving end performs the inverse operation of the transmitting end. First, the received signal undergoes optical-polarization demodulation, low-pass filtering, analog-to-digital conversion, and FFT transformation, after removing the guard interval. The information is distributed; then, each receiver selects its own subchannel for baseband demodulation, and then obtains its own data information through a parallel–serial converter.

The multiuser channel estimator and adaptive subchannel, bit, and power divider are two key parts in the system. The working condition of the multiuser estimator directly determines the performance of the adaptive subchannel, bit, and power divider.

8.5.2 Several Multiuser Adaptive Algorithms

In an OWC-OFDM system with K user subchannels, for user k, each OFDM symbol needs to transmit R_k information bits. Through channel estimation, the transmitter knows the real-time channel gain of each user on each subchannel, and the transmitter can allocate subchannels, bits, and power to each user, according to the channel-state information (CSI).

$R_{k,n}$ and $P_{k,n}$ respectively represent the number of transmitted bits and transmit power of user k on the nth subchannel. The relationship between $R_{k,n}$ and $P_{k,n}$ can be expressed by Eq. (8.157):

$$P_{k,n} = \frac{f_k\left(R_{k,n}\right)}{\left|H_{k,n}\right|^2}, \tag{8.157}$$

where $f_k\left(R_{k,n}\right)$ is the power required to receive $R_{k,n}$ bits when each subcarrier meets a certain bit-error rate when the channel gain is 1. It should be noted that $f_k\left(R_{k,n}\right)$ and k are closely related because different users may have different BER requirements and modulation schemes.

The total transmit power P_T of the OFDM system is the sum of the transmit powers of all users on all subchannels; namely,

$$P_T = \sum_{k=1}^{K} P_k = \sum_{k=1}^{K}\sum_{n=1}^{K} \rho_{k,n} P_{k,n}, \tag{8.158}$$

where P_k represents the sum of the transmit power of user k on all subchannels; that is,

$$P_k = \sum_{n=1}^{K} \rho_{k,n} P_{k,n}, \tag{8.159}$$

where $\rho_{k,n}$ is the probability that the kth user occupies the nth subcarrier.

8.5.2.1 MA-Based Optimal Multiuser-Assignment Algorithm

Cheong Yui Wong et al. produced an optimal joint subchannel and bit allocation algorithm by solving the optimization problem under the margin-adaptive (MA) criterion [41]. The algorithm flowchart is shown in Fig. 8.39, and the algorithm is specifically described as follows:

First, let $R_{k,n} = \rho_{k,n} b_{k,n}$; then, $R_{k,n} \in \left[0, M \rho_{k,n}\right]$ is the maximum modulation order, and the optimization problem is

$$P_T = \min \sum_{k=1}^{K} \sum_{n=1}^{N} \frac{\rho_{k,n}}{a_{k,n}^2} f_k \left(\frac{R_{k,n}}{\rho_{k,n}} \right). \tag{8.160}$$

In Eq. (8.160), $R_{k,n}$ and $\rho_{k,n}$ still satisfy

$$R_k = \sum_{n=1}^{N} R_{k,n} \quad k = 1, 2, \ldots, K \tag{8.161}$$

$$1 = \sum_{k=1}^{K} \rho_{k,n}. \tag{8.162}$$

Using the Lagrange formula,

$$L = \sum_{n=1}^{N} \sum_{k=1}^{K} \frac{\rho_{k,n}}{a_{k,n}^2} f_k \left(\frac{R_{k,n}}{\rho_{k,n}} \right) - \sum_{k=1}^{K} \lambda_k \left(\sum_{n=1}^{N} R_{k,n} - R_k \right) - \sum_{n=1}^{N} \beta_n \left(\sum_{k=1}^{K} \rho_{k,n} - 1 \right). \tag{8.163}$$

Equation (8.163) takes the partial derivative of $R_{k,n}$ and $\rho_{k,n}$ and sets it to zero, yielding

$$\frac{\partial L}{\partial R_{k,n}} = \frac{1}{a_{k,n}^2} f_k' \left(\frac{R_{k,n}}{\rho_{k,n}} \right) - \lambda_k = 0. \tag{8.164}$$

$$\frac{\partial L}{\partial \rho_{k,n}} = \frac{1}{a_{k,n}^2} \left[f_k \left(\frac{R_{k,n}}{\rho_{k,n}} \right) - \frac{R_{k,n}}{\rho_{k,n}} f_k' \left(\frac{R_{k,n}}{\rho_{k,n}} \right) \right] - \beta_n = 0. \tag{8.165}$$

As calculated using Eq. (8.165),

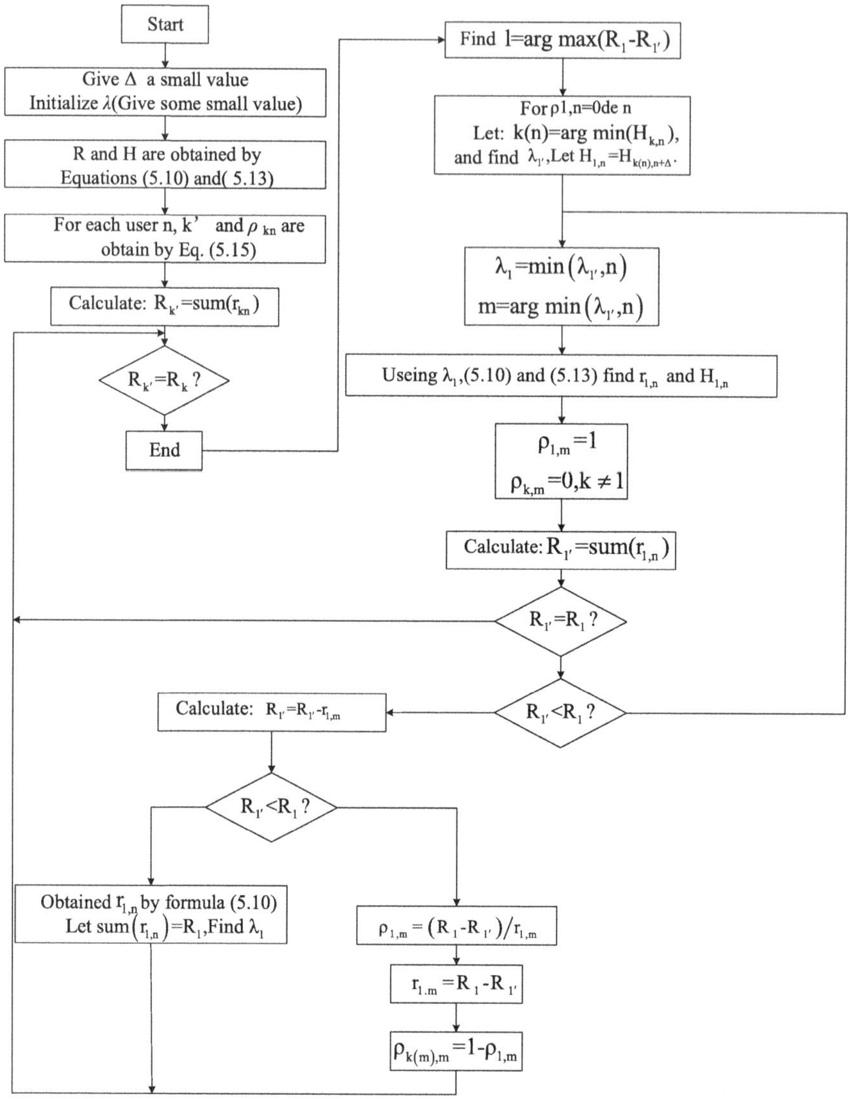

Fig. 8.39 Flowchart of the optimal algorithm [41]

$$R_{k,n} = \rho_{k,n} f_k'^{-1}\left(a_{k,n}^2 \lambda_k\right) \tag{8.166}$$

$$\lambda_k' = \begin{cases} f_k'(0)/a_{k,n}^2, f_k'^{-1}\left(a_{k,n}^2 \lambda_k\right) < 0 \\ \lambda_k, \quad 0 \le f_k'^{-1}\left(a_{k,n}^2 \lambda_k\right) \le M \\ f_k'(m)/a_{k,n}^2, f_k'^{-1}\left(a_{k,n}^2 \lambda_k\right) > M \end{cases} . \tag{8.167}$$

According to Eq. (8.166), we can obtain

$$\rho_{k,n} = \begin{cases} 1 & \beta > H_{k,n}\left(\lambda'_k\right) \\ 0 & \beta \le H_{k,n}\left(\lambda'_k\right) \end{cases}. \tag{8.168}$$

In Eq. (8.168),

$$H_{k,n}(\lambda) = \frac{1}{a_{k,n}^2}\left[f_n\left(f_k^{\prime -1}\left(a_{k,n}^2\lambda_k\right)\right) - a_{k,n}^2\lambda_k f_k^{\prime -1}\left(a_{k,n}^2\lambda_k\right)\right]. \tag{8.169}$$

When a set of $\lambda_k (k = 1, 2, \ldots, K)$ is known, the optimal point of the cost function can be calculated according to Eqs. (8.166) and (8.169). The calculation steps are as follows:

Calculate $H_{k,n}(\lambda)$ according to Eq. (8.169), and then determine $\rho_{k,n}$. For the nth subchannel, select the user with the smallest $H_{k,n}(\lambda)$ to use the subchannel; that is,

$$\rho_{k,n} = \begin{cases} 1 & k = \arg\min\left(H_{k,n}(\lambda)\right) \\ 0 & else \end{cases}. \tag{8.170}$$

From this, $\rho_{k,n}$ can be obtained; however, the constraint condition in Eq. (8.170) is not necessarily established. Thus, it is necessary to find the most appropriate λ_k for each user through an iterative search.

First, $\lambda_k (k = 1, 2, \ldots, K)$ takes a small value. In the iterative process, λ_k is incremented each time for a user k until the transmission-rate requirement of the user is met. The process is repeated until the transmission rates of all users meet the requirement. For user k, with the increase of user λ_k, the $H_{k,n}(\lambda)$ calculated by Eq. (8.177) will decrease accordingly, and more $\rho_{k,n}$ values will become 1. This increases $R_{k,n}$ until it finally meets the transmission-rate requirement of the user.

This scheme has two problems: one is that the calculated $b_{k,n}$ does not necessarily belong to $D = \{0, 1, \ldots, M\}$. The other is that for a subchannel n, there may be multiple users whose $H_{k,n}(\lambda)$ are equal; hence, multiple users might share a subchannel. That is, the value of $\rho_{k,n}$ may be between (0, 1), and simply quantifying $b_{k,n}$ and $\rho_{k,n}$ cannot satisfy the constraints. Therefore, it must be further processed.

For a subchannel N, if multiple users are sharing, let

$$\rho_{k,n} = \begin{cases} \rho_{k',n} = 1 & k' = \arg\ \max\left(\rho_{k,n}\right) \\ \rho_{k,n} = 0 & \forall k \ne k^* \end{cases}. \tag{8.171}$$

After the above subchannel allocation is completed, bit allocation is performed for each user. The algorithm can achieve optimal allocation in theory; however, the problem of shared processing will cause a large performance loss. In addition, the computational complexity of this scheme is quite large, so it is difficult to apply to an actual high-speed data-transmission system.

8.5.2.2 Suboptimal Multiuser Assignment Algorithm

C.Y. Wong et al. proposed a suboptimal algorithm after proposing the optimal multiuser subchannel-allocation algorithm [41]. The algorithm assumes that the number of subchannels required by each user is known, so that subchannel allocation and bit allocation are performed separately, with the subchannel allocation performed first, and the bit allocation performed next. The Wong suboptimal algorithm is described in detail as follows:

(1) Initial allocation. First, the channel-gains $H_{k,n}(k = 1, 2, \ldots K; n = 1, 2, \ldots, N)$ of the N subchannels of the K users are sorted in descending order, and the subchannels with the largest gains are allocated to the users as much as possible.

Subchannel allocation is performed for each of the K users. First, the subchannels with a larger gain relative to user k's channel are processed. Assume that a subchannel is allocated to the user, and the subchannel with the largest channel gain is selected. If the subchannel has not been allocated to another user, the subchannel is allocated to user k, and then the subchannel with the largest channel gain for the next user is processed. Otherwise, no processing is performed, and the next user is directly considered.

The above operations are repeated until the subchannels allocated to all users reach the required number of subchannels or all subchannels are allocated, and the initial subchannel allocation ends.

(2) A subchannel is iteratively optimized, and it is further allocated among the users, attempting to reduce the total transmission power of the system.

A power-reduction factor $P_{i,j}$ is defined, which represents the maximum value of the power reduction caused by removing a certain subchannel from user i and assigning it to user j temporarily. Therefore, the maximum total power saved when exchanging a pair of subchannels for user group (i, j) is $\Delta P_{i,j} = P_{i,j} + P_{j,i}$, and this group of subchannels is denoted as $n_{i,j}$ and $n_{j,i}$.

For all user groups (i, j), $(i, j = 1, 2, \ldots, K)$ and $i \neq j$, we calculate $\Delta P_{i,j}$ and sort $\Delta P_{i,j}$ in descending order. Then, we find the maximum value $\left\{\Delta P_{i,j}\right\}$ in $\Delta P_{i',j'}$, and its corresponding user group $\left(i', j'\right)$ and subchannels $n_{i',j'}$ and $n_{j',i'}$.

If $\Delta P_{i',j'} > 0$, subchannels $n_{i',j'}$ and $n_{j',i'}$ are exchanged between user $\left(i', j'\right)$; that is, the subchannel $n_{j',i'}$ originally allocated to user i' is allocated to user j', and the subchannel $n_{j',i'}$ originally allocated to user j' is allocated to user i'. Then, the distribution matrix ρ is updated so that

$$\begin{cases} \rho_{i',n_{j',\imath}} = 0 \quad \rho_{i',n_{\imath,j'}} = 1 \\ \rho_{j',n_{r_j,j'}} = 0 \quad \rho_{j',n_{j',f}} = 1 \end{cases} \tag{8.172}$$

and $\Delta P_{i,j}$ is recalculated.

The above operations are repeated until all the powers are reduced by the factor $\Delta P_{i',j'} \leq 0$, which means that the total power of the system cannot be reduced any further, and the subchannel allocation ends.

Finally, a single-user-type bit allocation is performed for each user.

The following example illustrates the specific process of Wong's suboptimal algorithm. To illustrate the problem visually and simply, we assume that the number of users $K = 3$ and the number of subchannels $N = 8$ in the OWC-OFDM system. The number of subchannels required by users 1, 2, and 3 are 3, 3, and 2, respectively. The randomly generated channel gain of each subchannel corresponding to each user is shown in Eq. (8.173).

$$CA_{k\times N} = \begin{bmatrix} 1.22\ 1.15\ 1.02\ 0.95\ 0.56\ 0.68\ 0.70\ 1.10 \\ 0.32\ 0.45\ 0.98\ 1.02\ 1.10\ 0.46\ 0.67\ 0.88 \\ 0.99\ 0.95\ 0.85\ 0.50\ 0.55\ 0.67\ 1.00\ 1.32 \end{bmatrix} \tag{8.173}$$

From Eq. (8.173), the sorting matrix $A_{k\times N}$ can be obtained as

$$A_{k\times N} = \begin{bmatrix} 1\ 2\ 4\ 5\ 8\ 7\ 6\ 3 \\ 8\ 7\ 3\ 2\ 1\ 6\ 5\ 4 \\ 3\ 4\ 5\ 8\ 7\ 6\ 2\ 1 \end{bmatrix}. \tag{8.174}$$

In Eq. (8.174), 1 represents the maximum channel gain, and 8 represents the minimum channel gain.

According to the above method, the obtained initial allocation matrix is

$$\rho_{K\times N} = \begin{bmatrix} 1\ 1\ 0\ 0\ 0\ 0\ 0\ 1 \\ 0\ 0\ 1\ 1\ 1\ 0\ 0\ 0 \\ 0\ 0\ 0\ 0\ 0\ 0\ 1\ 1 \end{bmatrix}, \tag{8.175}$$

where 1 means that a channel has been allocated, and 0 means that it is not allocated.

After the above system is iteratively approximated, the final distribution matrix obtained is

$$\rho_{K\times N} = \begin{bmatrix} 1\ 1\ 0\ 0\ 0\ 1\ 0\ 0 \\ 0\ 0\ 1\ 1\ 1\ 0\ 0\ 0 \\ 0\ 0\ 0\ 0\ 0\ 0\ 1\ 1 \end{bmatrix}. \tag{8.176}$$

The premise of this algorithm is that the subchannel requirements of each user are known in advance, so the user's bit requirements must be converted into subchannel requirements. This conversion is performed before the subchannels are allocated, so it is difficult to achieve an ideal match.

The finally obtained subchannel may not be able to meet the user's bit requirement, which leads to a degradation of the algorithm performance. However, the algorithm separates subchannel allocation and bit allocation, which effectively reduces the computational complexity.

8.5.2.3 Real-Time Subchannel-Allocation Algorithm

Guodong Zhang proposed a real-time subchannel-allocation algorithm [41]. The main idea is to first use the single-user water-injection algorithm to allocate the subchannels that each user wants, and then address any subchannel conflicts. If a subchannel is allocated to one of the users, then the bits of other users on this subchannel should be allocated to other subchannels, which will increase the total transmit power.

We select the user with the smallest increase in transmit power. The subchannel is allocated to the user. The above operations are performed on all conflicting subchannels, and the final result is the final allocation scheme. The algorithm is described as follows:

(1) A single-user bit-allocation algorithm is used for each user to obtain the bit–power allocation of each user on each subchannel;
(2) For a conflicting subchannel i, it is assumed that $L(k_1, k_2, \ldots, k_L)$ users need to use it. The total transmit power on subchannel i is $P_i = \sum_{j=1}^{L} P_{i,k_j}$, and all conflicting subchannels are arranged in the descending order of P_i;
(3) These conflicting subchannels are processed sequentially. For subchannel i, assign the subchannel to user k_j; then, for each user $k_h (h \neq j)$, it is necessary to reallocate the bit information allocated on the subchannel to other subchannels.
When user $k_h (h \neq j)$ occupies a subchannel, its transmit power is P_{i,k_j}. The transmit power after reassignment becomes P_{re,k_h}, and the power increment brought by reassignment is $\Delta P_{k_h} = P_{re,k_h} - P_{i,k_h}$. Then, the power increase brought by assigning subchannel to user kj is $\Delta P_{k_j} = \sum_{h=1}^{L} \Delta P_{k_h}$.
(4) Assign subchannel i to the user that brings the smallest power ΔP_{k_j}.
(5) Repeat steps 3 and 4 until all conflicting subchannels are processed.

8.5.2.4 Algorithm Performance Comparison

MATLAB simulations were performed on the above algorithms to compare their performance. The simulation parameters are as follows: the number of users $K = 3$, the number of subchannels $N = 32$, the modulation method is MQAM, and the maximum number of bits allowed to be carried on each subchannel is 8. The simulation results are shown in Fig. 8.40.

As can be seen from Fig. 8.40, the optimal algorithm has the smallest bit-error rate, and the real-time subchannel allocation algorithm has the largest bit-error rate. In theory, the performance of the optimal algorithm should be the best; however, because the algorithm needs to address the problem of users sharing subchannels, this affects the system performance to a certain extent.

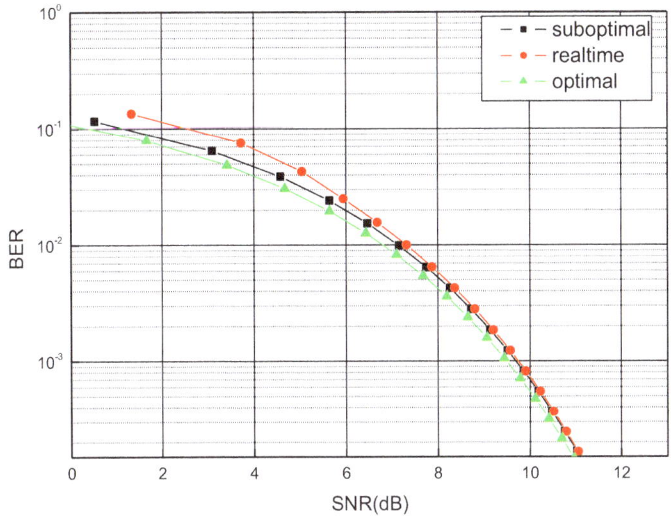

Fig. 8.40 Performance comparison of three subchannel-allocation algorithms [41]

8.5.3 Improving the Algorithm Fairness

Frequently, the multiuser subchannel-allocation strategy only optimizes the overall system performance; whether each user fairly receives the subchannels is ignored. That is to say, if a certain user has the best performance on a certain subchannel, the subchannel is allocated to this user.

When the channel conditions experienced by each user are ideal conditions that are independent and identically distributed, from a statistical point of view, this criterion obtained by optimal performance can also ensure the "fairness" of the number of subchannels obtained by each user; that is, over a long period of time, the number of subchannels allocated to each user is basically the same.

However, in real systems, the situation is often not so simple. Owing to path loss and other reasons, the average received signal-to-noise ratio of users close to the base station is usually higher than that of users far away from the base station. If the allocation is purely based on optimal performance, it is likely that users close to the base station will occupy most of the resources, while users at the edge of the base-station range cannot always be allocated subchannels, and the "fairness" of channel allocation cannot be guaranteed.

To avoid this situation, two functions considering the relative utilization and relative number of bits can be used for allocation; these functions are given below.

8.5.3.1 Subchannel-Allocation Algorithm for Relative Utilization and Bit Numbers

The subchannel-allocation algorithm for relative utilization and bit numbers has high spectrum utilization and real-time requirements. The relative number of bits and the relative utilization are two functions [41] that are used for subcarrier allocation. Combined with adaptive modulation, the algorithm has almost the same effect as the upper bound of frequency performance (the upper bound of performance means that each user selects the subcarrier with the best channel response to meet its own bit-rate requirements, regardless of whether it is occupied by other users).

Assuming that the system has K users and N subcarriers, the quality of service (QoS) requirements of each user are known, the bit requirement of user k is bk, the bit-error rate requirement is BERk, the channel response matrix is $H = \{h_{k,n}\}_{K\times N}$, and $h_{k,n}$ is the subchannel of user k. For a channel response on n, it is assumed that the atmospheric channel is a slow-fading channel; that is, the channel response remains constant during one allocation and transmission.

According to the user's bit-error rate requirement, the highest modulation order of user k on subchannel n can be calculated:

$$c_{k,n} = \{m|m = \max\{f_m(SNR_{k,n})\} \leq BER_{k,n}, m \in M\}. \tag{8.177}$$

In Eq. (8.177), f_m is the bit-error rate function of the m-order modulation mode, M is the set of selectable modulation modes, $M = \{0, 1, 2, 3, 4, 5, 6, 7, 8, 9, 10, 11, 12\}$, corresponding to no modulation, *BPSK*, *QPSK*, 8*QAM*, 16*QAM*, 32*QAM*, 64*QAM*, 128*QAM*, 256*QAM*, 512*QAM*, 1024*QAM*, 2048*QAM*, and 4096*QAM*, respectively. The receiving signal-to-noise ratio $SNR_{k,n}$ is determined by the carrier-to-noise ratio *CNR* of the system and the channel response $H_{k,n}$:

$$SNR_{k,n} = CNR + 20\lg|H_{k,n}|. \tag{8.178}$$

The algorithm is based on traditional beamforming techniques; that is, most of the energy and bits are allocated to the subchannel with the largest eigenvalue. Therefore, $H_{k,n}$ of Eq. (8.178) can calculate the largest eigenvalue, as long as it is transformed by SVD.

The relative number of bits for user k is defined as

$$S_k = \frac{b_k}{\Delta f \sum_{n=1}^{N} c_{k,n}} \quad k \in \{1, 2, \ldots, K\}, \tag{8.179}$$

where Δf represents the bandwidth of the subcarriers, and the denominator represents the total bit rate that can be provided by allocating all subcarriers to a user. S_k represents the degree of satisfaction of the system of user k, and the larger S_k is, the more difficult it is for user k to be satisfied.

The relative utilization of user k on subcarrier n is defined as

$$U_{k,n} = \frac{c_{k,n}}{\sum_{K=1}^{K} c_{k,n}} \quad n \in \{1, 2, \ldots, N\}, \tag{8.180}$$

where $U_{k,n}$ represents the utilization degree of subcarrier n by user k, and the larger $U_{k,n}$ is, the more subcarrier n should be allocated to user k.

The algorithm is based on two functions considering the relative bit-number requirement and relative utilization. This improves the spectrum-utilization rate of the system, meets the requirements of each user, and allocates channels as fairly as possible among the users. The algorithm prioritizes allocating subcarriers to users with relatively large numbers of bits. When specifically assigning a subcarrier to a user, priority should be given to subcarriers with high relative utilization. However, the algorithm is relatively complicated when calculating the highest modulation order on each subcarrier; hence, it is further improved. The improved algorithm is as follows.

8.5.3.2 Improved-Algorithm Implementation Steps

The subcarrier-allocation algorithm based on the relative number of bits and relative utilization adaptively allocates subcarriers according to user needs; however, when the user requirements have been met, it no longer allocates subcarriers, resulting in part of the unfairness. Based on these two functions, the implementation of the algorithm has been improved, mainly by using a relatively simple Chow algorithm when calculating the highest modulation order on the subcarrier. This not only further improves the implementation rate, but more importantly, avoids the resource waste of some subcarriers.

This algorithm is more reasonable and effective. Figure 8.41 shows the flowchart of the improved algorithm. The steps of the algorithm are as follows:

(1) Initialize the subcarrier-allocation matrix $\boldsymbol{A} = (a_{i,j}) = \boldsymbol{0}_{K \times N}$ and the user needs b_k;
(2) Based on the Chow algorithm, calculate $c_{k,n} = \log_2(1 + SNR_{k,n})$;
(3) Calculate the relative bit-count function for all users $S = \{S_k\}_k$;
(4) Select the user k_{user} to which subcarriers are to be allocated. Determine whether S is empty; if so, go to step 8. Otherwise, the user k_{user} that receives subcarriers will be selected. $k_{user} = \arg\max(S)$, and the largest $S_{k_{user}} = \max(S)$ is selected;
(5) Select subcarrier n_{subca} to be allocated to user k_{user}. For all subcarriers n, $n \in \{1, 2, 3, \ldots, N\}$, calculate the relative utilization functions $U_{k_{user}} = \{U_{k_{user},n}\}_K$ and $n_{subca} = \arg\max(U_{k_{user}})$ of user k_{user}, and select the maximum $U_{k_{user},n_{subca}} = \max(U_{k_{user},n})$;
(6) Determine whether $U_{k_{user},n_{subca}}$ is 0. If $\pi(x)$ is 0, delete $S_{k_{user}}$, from S and go to step 3. Otherwise, according to the calculated k_{user} and n_{subca}, set the subcarrier allocation matrix $a_{k_{user},n_{subba}} = 1$, the number of bits of user k_{user} $b_{k_{user}} = b_{k_{user}} - c_{k_{user},n_{subca}}$, and the channel-gain matrix $h_{k_{user},n_{subca}} = 0$, $(k = 1, 2, \ldots, K)$;
(7) Go to step 3 after updating S;

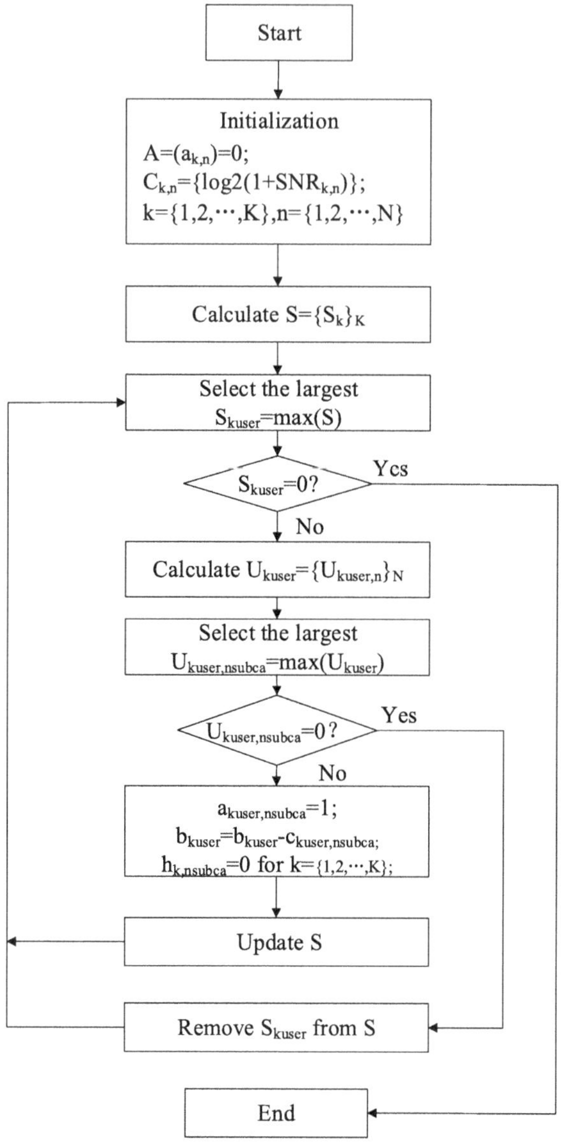

Fig. 8.41 Flowchart of the improved algorithm [41]

(8) End (all subcarriers are assigned to the most suitable users).

8.5.3.3 Performance Simulation

The simulated channel is a Rayleigh channel, the number of users N is 10, the maximum channel delay is 4 μs, the bandwidth of the system is 5 MHz, the number of subcarriers is $K = 1024$, and there are 128 cyclic prefixes (CPs). The modulation method on each subcarrier is determined by the signal-to-noise ratio. If the lowest BPSK modulation is used, the data rate provided by each subcarrier is 4.88 kbits/s, which is the basic rate of the system, denoted as R_b. The user's bit-rate requirement is assumed to be an integer multiple of R_b, and the bit-error rate requirement is 10^{-4}.

(1) **Spectral Efficiency**

To analyze the performance of the improved algorithm in terms of spectral efficiency, the simulation is compared with the subcarrier-allocation algorithm in [41]. Figure 8.42 shows the relationship between the average number of bits per subcarrier *B* and the carrier-to-noise ratio *CNR*. It can be seen from the figure that the improved algorithm has a gain of about 5 dB over the Yu algorithm. Therefore, the improved algorithm has relatively high spectral efficiency.

(2) **Fairness**

The fairness of the subcarrier-allocation algorithm is defined as

$$F = E\left(\min_k\left(\sum_{n=1}^{N} a_{kn}c_{kn}\right)\right)F = E(\min). \tag{8.181}$$

Equation (8.181) represents the minimum number of bits a user can obtain in a subcarrier-allocation algorithm; it can also represent the fairness of the algorithm among users. The larger the value of F, the higher the algorithm efficiency, the closer

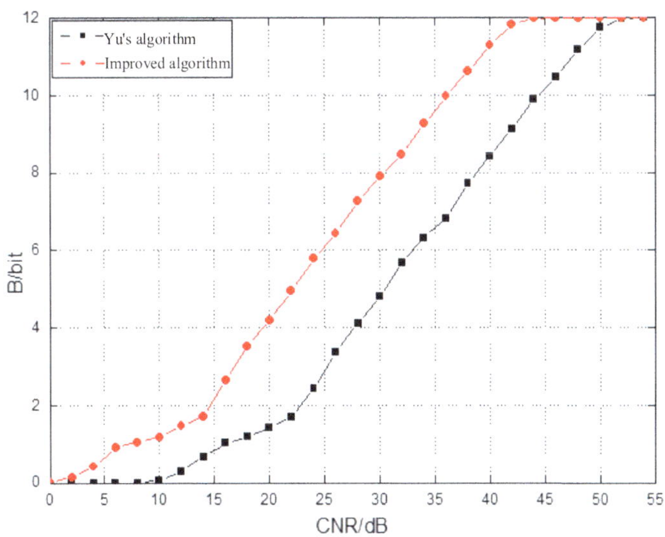

Fig. 8.42 Comparison of the spectral efficiency between Yu's algorithm and the improved algorithm [41]

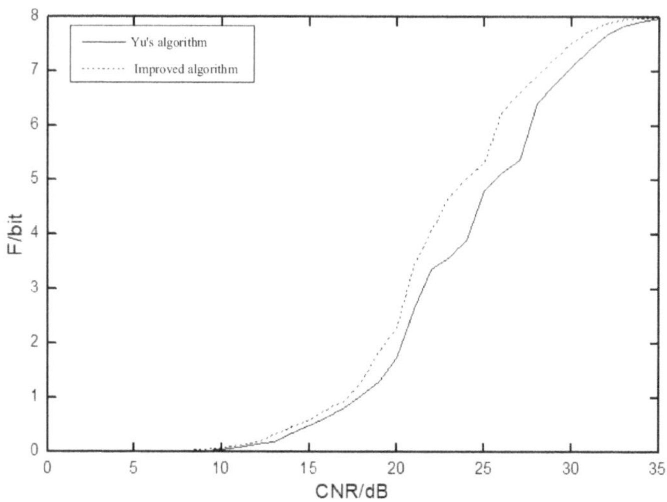

Fig. 8.43 Fairness comparison between Yu's algorithm and the improved algorithm [41]

it is to the average number of bits obtained by users, and the better the fairness. To obtain the fairness-curve relationship, the bit rates of each user are set to be the same, and the total bit-rate requirement is slightly higher than the system capacity.

Figure 8.43 shows the fairness curves for the two algorithms. It can be seen that the improved algorithm is fairer than Yu's algorithm.

The subchannel-allocation algorithm using the relative number of bits and relative utilization can adaptively allocate subchannels according to user needs; however, when the user's requirements have been met, subchannels are no longer allocated; thus, causing part of the unfairness. The improved algorithm uses the Chow algorithm with low computational complexity to calculate the maximum modulation order of the subchannel, which once again increases the implementation rate and reduces the implementation complexity. More importantly, the waste of resources of part of the subcarriers is avoided.

This algorithm is more reasonable and effective. Under the same performance conditions, the improved algorithm has much smaller computational complexity than the original algorithm, and can be used in adaptive OWC-OFDM systems with high real-time requirements.

Chapter Summary

(1) Starting with recursive Bayesian theory and the Monte Carlo sampling method, the particle-filter and resampling algorithms were introduced and derived, the particle-filter algorithm was applied to the model, and a recursive channel-estimation process based on the particle-filter algorithm was deduced.

(2) The particle-filter algorithm and the LS and MMSE algorithms were respectively applied to optical wireless OFDM channel estimation. The results showed that the particle-filter method improved the frequency-band utilization, compared with the traditional channel-estimation method, and improved the system's bit-error rate performance at the same signal-to-noise ratio.
(3) For a single-user OWC-OFDM system, the "water-injecting," Hughes–Hartogs, Chow, and Fischer algorithms were studied, and an adaptive channel-allocation algorithm based on M-ary phase-shift keying (MPSK) was improved. For a multiuser OWC-OFDM system, the optimal multiuser channel-allocation algorithm, sub-optimal multiuser channel-allocation algorithm, and real-time channel-allocation algorithm based on the margin-adaptive (MA) method were studied, and a fairness method based on the relative bit number and relative utilization was proposed. The further improved multiuser adaptive channel-allocation algorithm improved the realization rate and reduced the realization complexity.

References

1. Zhao L, Ke X, Liu J (2009) Research on modeling and LS channel estimation of band atmospheric laser OFDM system. Laser Mag 30(3):38–39
2. Zhang H (2006) Basic principles and key technologies of orthogonal frequency division multiplexing. National Defense Industry Press
3. Negi R, Cioffi J (1998) Pilot tone selection for channel estimation in a mobile OFDM system. IEEE Trans Consum Electron 44(3):1122–1128
4. Coleri S, Ergen M, Puri A, Bahai A (2002a) A study of channel estimation in OFDM systems. IEEE Veh Technol Conf 56(2):894–898
5. Coleri S, Ergen M, Puri A, Bahai A (2002) Channel estimation techniques based on pilot arrangement in OFDM systems. IEEE Trans Broadcast 48(3):223–229
6. Song B (2005) Research on channel estimation and equalization technology in OFDM wireless broadband mobile communication system. Shanghai Jiaotong University, Shanghai
7. Zhou H (2002) Research on key technologies of wireless internet—Research on channel estimation technology of OFDM system. Department of Radio Engineering, Southeast University, Nanjing
8. Julier SJ, Uhlmann JK (2000) Durrant-Whyte H.F.A new method for the nonlinear transformation of means and covariance in filters and estimations. IEEE Trans Autom Control 45(3):477–482
9. Kay SM, Luo P, Zhang W et al (2003) Statistical signal processing basis—Estimation and detection theory, vol 8. Electronics Industry Press, pp 253–259
10. Yang Y (2006) Adaptive dynamic navigation and positioning. Surveying and Mapping Press
11. Zhang H (2008) Target tracking based on particle filter and its experimental research. Xi'an University of Technology, p 6
12. Li H, Wang Y, Jiang M, Yuan D (2007) Doubly fading channel tracking based on particle filter in MIMO-OFDM system. In: 2007 international symposium on microwave antenna, propagation and EMC technologies for wireless communications
13. Doucet A, de Freitas N, Gordon N et al (2001) Sequential Monte Carlo methods in practice. Springer, p 6
14. Doucet A, Godsill S, Andrieu C (2000) On sequential Monte Carlo sampling methods for Bayesian filtering. Stat Comput 10:197–208

15. Liu JS, Chen R (1998) Sequential Monte Carlo methods for dynamical systems. J Naerican Stat Assoc 93:1032–1044
16. Wang H, Chang P (1996) On verifying the first order Markovian assumption for a Rayleigh fading channel. IEEE Trans Commun 45(2):353–357
17. Zhao L, Ke X, Liu J (2009) Research on frequency band FSO-OFDM system modeling and LS channel estimation. Laser J 30(3):38–39
18. Coleri S, Ergen M, Puri A, Bahai. A study of channel estimation in OFDM systems. In: Proceedings Vehicular Technology Conferences, vol 2, pp 894–898
19. Hsieh MH, Wei CH (1998) Channel estimation for OFDM systems based on comb-type pilot arrangement in frequency selective fading channels. IEEE Trans Consum Electron 44(1):217–225
20. Li Y (2000) Pilot-symbol-aided channel estimation for OFDM in wireless system. IEEE Trans Veh Technol 49
21. Lim C-H, Han DS (2006) Robust LS channel estimation with phase rotation for single frequency network in OFDM. IEEE Trans Consum Electron 52(4):1173–1178
22. Lin J-C (2007) LS channel estimation for mobile OFDM communications on time-varying frequency-selective fading channels. In: IEEE communications society subject matter experts for publication in the ICC 2007 proceedings, pp 3016–3023
23. Haykin S (1996) Adaptive filter theory, 3rd edn. Prentice Hall, Englewood Cliffs, New Jersey
24. van de Beek JJ, Edfors O, Sandell M, Wilson SK, Borjesson PO (1995) On channel estimation in OFDM systems. VTC95'fall, vol 2, pp 815–819
25. Edfors O, Sandell M, van de Beek JJ (1996) OFDM channel estimation by singular value decomposition 923–927
26. Edfors O, Sandell M, van de Beek JJ (1998) OFDM channel estimation by singular value decomposition. IEEE Trans Commun 46(7):931–939
27. Zhao L, Ke X (2009) LMS channel estimation algorithm in polarized FSO-OFDM system. Opto-Electron Eng 36(8):80–84
28. Bai BF, Cai YM, Xu YY (2005) A wiener LMS channel tracking algorithm in OFDM system. Chin J Electron Inf 27(11):1699–1703
29. Bai B, Wang M (2006) Wiener LMS channel estimation and tracking algorithm for OFDM systems with transmitter diversity. In: 2006 international conference on communications, circuits and systems proceedings, Guilin, China, pp 1040–1044
30. Speth M, Fechtel SA, Fock G, Meyr H (1999) Optimum receiver design for wireless broad-band systems using OFDM-part I. IEEE Trans Commun 47(11):1668–1677
31. Tang X, Alouini M, Goldsmith AJ (1999) Effect of channel estimation error on M-QAM BER performance in Rayleigh fading. IEEE Trans Commun 47(12):1856–1864
32. Leke A, Cioffi JM (1998) Impact of imperfect channel knowledge on the performance of multicarrier systems. In: Proceeding of the IEEE GLOBECOM'98, vol 2, pp 951–955
33. Tong X, Luo T (2003) The principle and application of OFDM mobile communication technology, vol 6. People's Posts and Telecommunications Press, Beijing, pp 143–144
34. Hughes-Hartogs D (1989) Ensemble modem structure for imperfect transmission media. U.S. Patents 4,833,706
35. Chow PS, Cioffi JM, Bingham JA (1995) A practical discrete multitone transceiver loading algorithm for data transmission over spectrally shaped channels. IEEE Trans Commun 43(2):773–775
36. Prokis JG (2001) Digital communication, English edn., vol 10 Electronic Industry Press, Beijing, pp 441–447
37. Wong CY, Cheng RS, Lataief KB, Murch RD (1999) Multiuser OFDM with adaptive subcarrier, bit, and power allocation. IEEE J Sel Areas Commun 17(10):1747–1758
38. Duan Z (2007) Research on channel allocation technology of FSO-OFDM system. Xi'an University of Technology, Xi'an
39. Wong CY, Tsui CY, Cheng RS, Letaief KB (1999) A real-time sub-carrier allocation scheme for multiple access downlink OFDM transmission. In: IEEE VTS 50th, vol 2, pp 1124–1128

40. Zhang G (2004) Subcarrier and bit allocation for real-time services in multiuser OFDM systems. In: IEEE international conference on communications, vol 5, pp 2985–2989
41. Yu G, Zhang C, Qiu P (2004) A subcarrier allocation algorithm for adaptive orthogonal QDM system. J Zhejiang Univ 38(9):1112–1116

Chapter 9
OFDM System Based on Subcarrier-Modulation Heterodyne Detection

In this chapter, a subcarrier-modulation heterodyne-detection method is used to detect an orthogonal frequency-division multiplexing (OFDM) optical signal. The closed expressions of the average carrier-to-noise ratio and the average bit-error rate of the subcarriers in a turbulent channel are derived, and the OFDM/subcarrier modulation is analyzed. The performance of the heterodyne-detection system and the influence of the Doppler frequency shift and nonlinear-modulation effect of the turbulent atmospheric channel on the system are investigated. Finally, a simulation experiment is conducted on subcarrier-modulation heterodyne detection and the reception of an OFDM optical signal through a turbulent atmospheric channel.

9.1 Basic Principle of OFDM Subcarrier Modulation

The transmitting end of a optical wireless communication system adopts subcarrier modulation. In Fig. 9.1, the baseband signal is modulated to the carrier frequency to generate a subcarrier signal, and then the light intensity of the light source at the transmitting end is modulated. The subcarrier-modulation method can use not only amplitude modulation, but also phase, frequency, or other high-order modulation methods.

It is assumed that the transmitting end of a optical wireless communication system adopts subcarrier-modulation quadrature phase-shift keying (SCM-QPSK) modulation, and uses an optical transmitting antenna to collimate the laser-beam expansion and transmit it to optical wireless.

The receiving end uses non-optical heterodyne detection (NOHD) to receive subcarrier-modulated signal light. Figure 9.2 shows a schematic diagram of the NOHD system. The system includes four parts: an optical receiving antenna, a local-oscillator optical subcarrier-modulation module, a subcarrier heterodyne-detection module, and a difference-frequency signal-demodulation module. The

X. Ke, *Principles and Applications of Optical Wireless Orthogonal Frequency-Division Multiplexing*, Optical Wireless Communication Theory and Technology,
https://doi.org/10.1007/978-981-97-7973-4_9

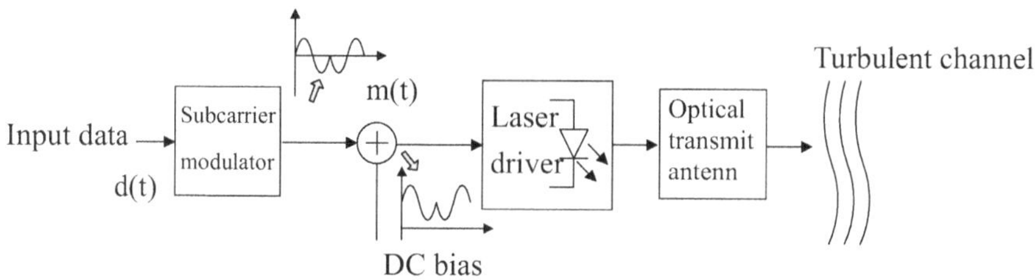

Fig. 9.1 Principle block diagram of the subcarrier modulation at the transmitting end of a optical wireless communication system

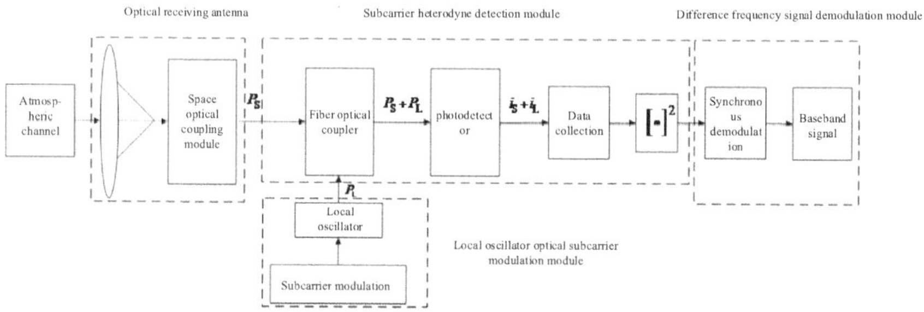

Fig. 9.2 Block diagram of a NOHD system

optical receiving antenna is equipped with a spatial optical-coupling module. The signal light is focused by the optical receiving antenna and forms a focused light field on the surface of the spatial optical-coupling module, which couples the optical wireless signal light into the multimode fiber. The local-oscillator optical subcarrier-modulation module modulates the local-oscillator laser using the subcarrier intensity to output the modulated local-oscillator light.

The subcarrier heterodyne-detection module includes a fiber coupler, a photoelectric detection and collection part, and a square-processing part. The fiber coupler superimposes the signal light field and the modulated local-oscillator light field. The photoelectric detection and collection part detects the surface of the photoelectric detector. The photocurrent is generated by superimposing the light field, and data are collected by an analog-to-digital converter. The square-processing part constructs a cross term through the square process to generate the beat-frequency signal. The beat-frequency signal is the signal optical subcarrier and the local-oscillator optical subcarrier at the receiving end.

The difference-frequency signal-demodulation module mainly performs synchronous demodulation on the difference-frequency signal obtained by the subcarrier heterodyne-detection module, and restores the baseband signal. The NOHD receiving method adds a local-oscillator laser at the receiving end and modulates it with an analog subcarrier. It constructs a cross term through square processing to heterodyne the subcarrier at the transmitting end and the local-oscillator subcarrier at the receiving end. This should improve the system-detection gain and reduce the implementation difficulty.

9.1.1 Mathematical Model of Transmitter Subcarrier Modulation

After adding a DC bias to the SCM-QPSK signal at the transmitting end, the intensity modulates the signal light, and the signal light power can be expressed as

$$P_S(t) = P_s + k_p P_s \cos(\omega_s t + \phi_j), \tag{9.1}$$

where P_s is the average power of the signal light, kp is the optical modulation degree, $0 \leq k_p \leq 1$, ωs is the angular frequency of the subcarrier at the transmitting end, and ϕ_j is the phase of the subcarrier, $\phi_j = j.\pi/4, j = 1, 3, 5, 7$.

9.1.2 NOHD Receiving-End Mathematical Model

The NOHD receiver must perform subcarrier modulation on the local-oscillator laser. The local-oscillator subcarrier signal is composed of a sinusoidal carrier. For simplicity, the optical-modulation degree of the local oscillator is set to be the same as that of the transmitter. After the subcarrier modulation plus DC-bias intensity modulates the local oscillator, the local-oscillator optical power P_{lo} (t) can be expressed as

$$P_{lo} = P_L\left[1 + k_p \cos(\omega_1 t)\right], \tag{9.2}$$

where P_L is the average power of the local-oscillator light, and ω_l is the angular frequency of the local-oscillator light subcarrier. The photodetector detects the signal light field and the light field generated by the local oscillator on the photosensitive surface and then outputs the response current, which is sampled, and the DC component is filtered out using a Butterworth bandpass filter. The signal is expressed as

$$\begin{aligned} u^2 = {} & \alpha^2 g^2 \{P_s^2 k_p^2 \cos 2(\omega_s t) + P_L^2 k_p^2 \cos 2(\omega_L t) \\ & + k_p^2 P_S P_L \cos\left[(\omega_s - \omega_L)t + \phi_j\right] \\ & + k_p^2 P_S P_L \cos\left[(\omega_s + \omega_L)t + \phi_j\right]\} R_L^2, \end{aligned} \tag{9.3}$$

where α is the responsivity of the photodetector, RL is the internal resistance of the photodetector, and g is the amplification gain of the photodetector. The signal in Eq. (9.3) contains four frequency components: $2\omega_s$, $2\omega_L$, subcarrier difference-frequency term $\omega_s - \omega_L$, and subcarrier sum-frequency term $\omega_s + \omega_L$. The subcarrier difference-frequency signal is obtained through a Butterworth bandpass filter:

$$v = \alpha^2 k_{\mathrm{p}}^2 g^2 P_{\mathrm{S}} P_{\mathrm{L}} \cos\left[(\omega_{\mathrm{S}} - \omega_{\mathrm{L}})t + \phi_j\right] R_{\mathrm{L}}^2. \tag{9.4}$$

It can be seen from Eq. (9.4) that the NOHD scheme heterodynes the subcarrier and can effectively amplify the input signal. Because the subcarrier heterodyne is not the heterodyne of the signal light field and the local-oscillator light field, the polarization states of the local-oscillator light and the signal light do not need to be exactly the same. According to Eq. (9.4), the equivalent effective signal power of the difference frequency output is

$$P_c = \frac{\overline{v^2}}{R_L} = \frac{1}{2}\alpha^4 k_p^4 g^4 P_S^2 P_L^2 R_L^3. \tag{9.5}$$

When a optical wireless communication system adopts direct detection, the output effective signal power is [1]

$$P_{\mathrm{o}} = \alpha^2 P_{\mathrm{S}}^2 R_{\mathrm{L}}. \tag{9.6}$$

When the average power of the signal light is the same, the NOHD gain G is expressed as

$$G = \frac{1}{2}\alpha^2 k_p^4 g^4 P_L^2 R_L^2. \tag{9.7}$$

It can be seen from Eq. (9.7) that the NOHD gain is affected by the average power P_L of the local-oscillator light, the internal resistance R_L of the detector, and the amplification gain g of the photodetector. When $\alpha = 1$, $k_p = 1$, and $g = 10$, the average optical power of the local oscillator is at the milliwatt level and a positive–intrinsic–negative (PIN) photodetector is used. It can theoretically obtain a heterodyne-detection gain of 23 dB, which is close to the optical heterodyne-detection gain. However, the actual NOHD gain is limited by the light-modulation degree k_p and the photodetector responsivity α limit.

In an ideal channel, the signal-power spectrum after NOHD squared processing is shown in Fig. 9.3. In Fig. 9.3, it can be seen that the frequency of the subcarrier difference-frequency signal generated after the square processing is 1 MHz, the frequency of the subcarrier sum-frequency signal is 6 MHz, the frequency of the local-oscillator optical-frequency multiplication term is 7 MHz, and the subcarrier of the direct-detection signal is 2.5 MHz. The power is −17 dBm. It can be seen from the NOHD power spectrum that the power of the subcarrier heterodyne signal is 20 dB higher than that of the direct-detection signal, and has a certain amplification gain for the baseband signal.

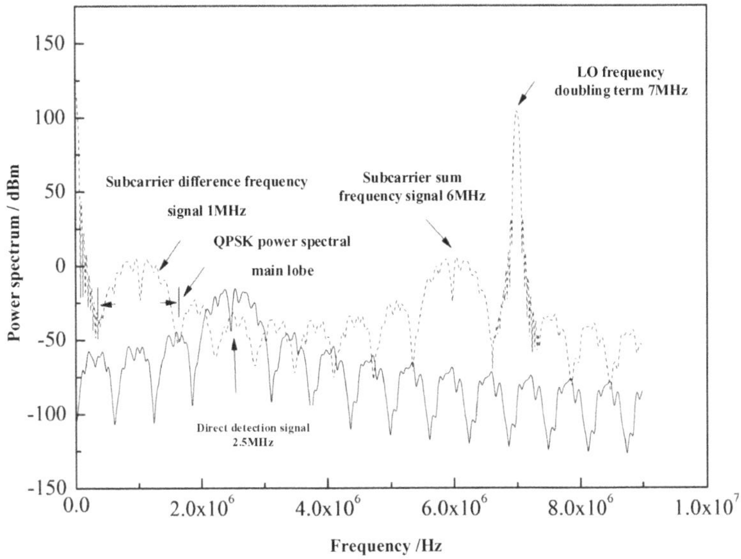

Fig. 9.3 Power spectrum of the NOHD difference-frequency signal and direct-detection signal under ideal conditions

9.2 Basic Principle of a Subcarrier Heterodyne-Detection OFDM System

SCM-OFDM modulation technology refers to modulating the baseband OFDM signal to the carrier-frequency signal and then modulating the intensity to the optical carrier. This technology makes OFDM more compatible with existing wireless systems. This section mainly discusses the working principle and mathematical model of SCM-OFDM/NOHD.

9.2.1 Basic Principle of an OFDM NOHD System

The block diagram of the optical wireless SCM-OFDM/NOHD system is shown in Fig. 9.4. The transmitting end maps the source bit sequence through QPSK modulation to form a modulation information sequence. After a serial-to-parallel conversion, the inverse fast Fourier transform (IFFT) obtains the time-domain sampling sequence from the OFDM signal and inserts a cyclic prefix to obtain a complex OFDM signal. After a parallel–serial conversion with the imaginary part, the modulated carrier frequency is up-converted, and the carrier-frequency signal intensity modulates the optical carrier to generate a signal light and sends it to the optical wireless communication (OWC) link using the optical antenna.

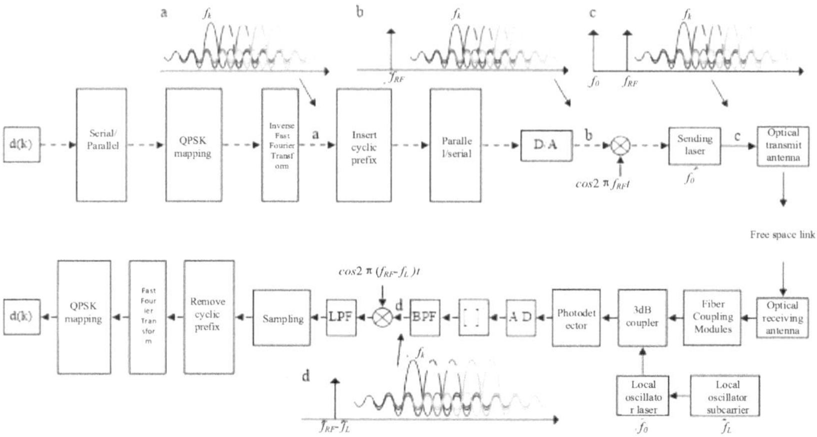

Fig. 9.4 Principle block diagram of a optical wireless SCM-OFDM/NOHD system

The optical antenna at the receiving end receives the signal light and couples it into a multimode fiber through the fiber-coupling module. The local-oscillator laser is modulated by the sine of the subcarrier intensity. The output-modulated local-oscillator light is superimposed with the signal light through the optical coupler. The photocurrent is squared to generate a subcarrier beat-frequency signal, which is synchronously and coherently demodulated to obtain an OFDM-baseband complex signal. The cyclic prefix is removed, and the input bit sequence is demodulated by QPSK mapping after a fast Fourier transform.

9.2.2 Mathematical Model of an SCM-OFDM System

After a serial–parallel conversion of the input bit sequence at the transmitting end, N channels of subcarrier-modulation symbols $d(k)$ are obtained, and the subcarriers are modulated and mapped using QPSK to obtain N complex symbols $X_n = a_n + jb_n$. Each symbol modulates one subcarrier, and after superposition of the modulated signal, N complex signals are obtained. The real part $I(t) = \mathrm{Re}\left[\sum_{k-1}^{N} a_n \exp(j2\pi f_k t)\right]$ and the imaginary part $Q(t) = \mathrm{Im}\left[\sum_{k-1}^{N} b_n \exp(j2\pi f_k t)\right]$ are inserted into the cyclic prefix to obtain the OFDM symbol, where f_k is the kth subcarrier carrier frequency and satisfies $f_k = (k-1)/T_s$, and T_s is the OFDM symbol period.

We convert the imaginary part and the real part of the complex OFDM symbol using a parallel–serial conversion into a signal $G(t)$ and obtain the SCM-OFDM signal using the subcarrier amplitude modulation of $G(t)$. Assuming that the frequency of the subcarrier signal at the transmitting end is f_{RF} and the initial phase is zero, the SCM-OFDM signal can be expressed as

$$S(t) = G(t)\cos(2\pi f_{\mathrm{RF}} t) \tag{9.8}$$

After adding the DC bias to the SCM-OFDM signal, the laser intensity is modulated to generate a double-sideband optical signal. The power of the SCM-OFDM optical signal can be expressed as

$$\begin{aligned} P_S(t) &= P_S[1 + mS(t)]\cos^2(2\pi f_0 t) \\ &= P_S\{1 + m[G(t)\cos(2\pi f_{RF} t)]\}\cos^2(2\pi f_0 t), \end{aligned} \tag{9.9}$$

where f_0 is the frequency of the sending laser, P_{S} is the average power of the sending laser, m is the optical modulation coefficient, $m = \sum_{n-0}^{N-1} m_n$, and m_n is the modulation coefficient of an OFDM signal subcarrier.

The receiving end uses the NOHD method to detect the SCM-OFDM optical signal. As described in Chap. 2, this method requires the modulation of the local oscillator's optical subcarrier; that is, the subcarrier intensity-modulates the local-oscillator laser using the carrier-frequency subcarrier with a frequency of f_{L}. To simplify the analysis, the optical modulation coefficient of the local-oscillator laser is set to be the same as that of the transmitter. Then, the modulated optical power of the local oscillator can be expressed as

$$P_L(t) = P_L[1 + m(\cos 2\pi f_L t)]\cos^2(2\pi f_0 t), \tag{9.10}$$

where P_{L} is the average power of the local-oscillator laser.

When the influence of atmospheric turbulence is not considered, the noise at the NOHD receiving end mainly includes relative intensity noise, thermal noise, and shot noise. The total noise $n(t)$ mean-squared current at the receiving end can be expressed as

$$N_0 = \frac{4K_B T_{abs}\Delta f}{R_L} + 2qI\Delta f + (RIN)I^2\Delta f. \tag{9.11}$$

Assuming that the responsivity of the photodetector is ρ and the total noise of the receiver obeys a stationary Gaussian distribution, the output photocurrent is filtered by a fifth-order Butterworth bandpass filter to remove the DC signal. The Gaussian band-limited white noise passing through the filter is assumed to be $n_1(t)$. The filtered photocurrent is squared and the output signal $v(t)$ can be expressed as

$$\begin{aligned} v(t) &= \{\rho P_S m[G(t)\cos(2\pi f_{RF} t)] + \rho P_L m\cos(2\pi f_L t) + n_1(t)\}^2 \\ &= \rho^2 P_S^2 m^2 G^2(t)\cos^2(2\pi f_{RF} t) + \rho^2 P_L^2 m^2 \cos^2(2\pi f_L t) \\ &+ \rho^2 P_S P_L m^2 G(t)\cos(2\pi f_{RF} t)\cos(2\pi f_L t) \\ &+ \rho P_S m G(t)\cos(2\pi f_{RF} t) n_1(t) + \rho P_L m\cos(2\pi f_L t) n_1(t) + n_1^2(t). \end{aligned} \tag{9.12}$$

Equation (9.12) can be divided into three parts: the interaction of the signal itself, the interaction between the signal and the noise, and the effect of the noise. To obtain the beat-frequency signal centered on $f_{\mathrm{RF}} - f_{\mathrm{L}}$, $v(t)$ passes through the Butterworth bandpass filter, with the subcarrier beat frequency as the center frequency and 2B as the bandwidth. The bandpass filter outputs the subcarrier beat-frequency signal $z(t)$ as

$$z(t) = \rho^2 P_S P_L m^2 G(t) \cos\left[2\pi (f_{RF} - f_L)t\right] + n_2(t), \tag{9.13}$$

where $n_2(t)$ is the band-limited noise after passing through the bandpass filter, and its noise power is $P_L^2 N_0 B/\Delta f + 2(N_0/\Delta f)^2 B^2$.

$z(t)$ is coherently demodulated in the electrical domain. The real part $I(t)$ and imaginary part $Q(t)$ of the complex OFDM signal are obtained by an A/D conversion after sampling and filtering by a low-pass filter. They can be expressed as

$$\hat{v}_I(t) = \rho^2 P_{\mathrm{S}} m^2 P_{\mathrm{L}} \mathrm{Re}\left[\sum_{k=1}^{N} \hat{a}_n \exp(j2\pi f_k t)\right] + \hat{Z}_I(k) \tag{9.14}$$

$$\hat{v}_Q(t) = \rho^2 P_S m^2 P_L \mathrm{Im}\left[\sum_{k=1}^{N} \hat{b}_n \exp(j2\pi f_k t)\right] + \hat{Z}_Q(k), \tag{9.15}$$

where $\hat{Z}_I(K)$ and $\hat{Z}_Q(K)$ are the Fourier transforms of the sampled n2(t). The QPSK demodulated signal is obtained after the Fourier transform of the complex OFDM signal, and the input bit sequence $d(k)$ is recovered by mapping.

It can be seen from Eqs. (9.14) and (9.15) that the NOHD process down-converts the subcarrier at the transmitter and increases the receiver signal-to-noise ratio (SNR), and the heterodyne gain depends on the local-oscillator laser P_{L} and the optical-modulation coefficient m. At the receiving end of the SCM-OFDM optical signal NOHD, it is usually necessary to estimate and compensate the subcarrier phase noise during digital signal processing to ensure the system communication performance.

9.3 BER Model and Numerical Analysis of an SCM-OFDM System

This section analyzes the closed expressions of the subcarrier carrier-to-noise ratio and the average bit-error rate (BER) of the gamma–gamma channel model and finds the appropriate local-oscillator optical power and modulation coefficient through a theoretical numerical analysis to obtain the best carrier-to-noise ratio.

9.3.1 System Carrier-to-Noise Ratio Analysis

The average power of the receiving terminal carrier can be expressed as $(m_n I)^2/2$. Assuming that the modulation coefficients of each subcarrier are equal, the average power of each subcarrier ω_n signal after NOHD heterodyning is

$$C = \frac{1}{2} \cdot (m_n)^4 \cdot \rho^4 P_S^2 P_L^2. \tag{9.16}$$

For atmospheric channels, the refractive-index fluctuations on the transmission path cause amplitude and phase fluctuations, which are manifested as light-intensity fluctuations on the surface of the photodetector and produce parasitic amplitude and phase modulations on the subcarrier. The fluctuation characteristics of the received light intensity can be described using a gamma–gamma light-intensity scintillation model. The influence of atmospheric turbulence on the SCM-OFDM optical signal mainly considers atmospheric attenuation and the intensity fluctuation caused by turbulence.

The optical power received by the photodetector can be expressed as

$$P_{FSO}(t) = P_S(t)LX + n_1(t), \tag{9.17}$$

where L represents the atmospheric-turbulence, optical-fiber coupling-device, and signal-light attenuation due to weather factors, such as rain and fog. X is the random fluctuation caused by atmospheric turbulence, and its probability-density function conforms to the gamma–gamma light-intensity scintillation model. After the photoelectric detector outputs photocurrent A/D sampling, offline signal square processing is performed, and the I channel difference-frequency signal is obtained through a Butterworth bandpass filter whose center frequency is $f_{\mathrm{RF}} - f_{\mathrm{L}}$. It can be expressed as

$$z_I(t) = \rho^2 P_S m^2 P_L LX \sum_{k=1}^{N} d(k)\exp(j2\pi f_k t)\cos(2\pi(f_{RF} - f_L)t + \psi_0) + n_2(t) \tag{9.18}$$

The average signal power on subcarrier f_k after subcarrier heterodyning under the scintillation distribution of a gamma–gamma channel is

$$C = \frac{1}{2} \cdot (m_n)^4 \cdot \rho^4 P_S^2 P_L^2 L^2 X^2. \tag{9.19}$$

The carrier-to-noise ratio (R_{CN}) of the receiving terminal is defined as the ratio of the subcarrier signal power C to the subcarrier noise power $\sigma_{n_2}^2/\mathrm{Ts}$; hence, the average carrier-to-noise ratio under a turbulent flow can be expressed as

$$\langle R_{CN}\rangle = \int_0^{\infty} R_{CN} p_X(x)dx = \frac{(\alpha\beta)^{\frac{\alpha+\beta}{2}}}{\Gamma(\alpha)\Gamma(\beta)} \int_0^{\infty} \frac{CT_s}{\sigma_{n_2}^2} x^{\frac{\alpha+\beta}{2}-1} K_{\alpha-\beta}\left(2\sqrt{\alpha\beta x}\right)dx. \quad (9.20)$$

According to Eq. (9.20), the influence of the subcarrier modulation coefficient and local-oscillator optical power on the average carrier-to-noise ratio is numerically analyzed. The number of subcarriers is $N = 64$, the cyclic prefix is eight, the OFDM subcarrier is modulated by QPSK, and the photodetector responsivity $\rho = 0.8$ A/W. The atmospheric-turbulence intensity uses the strong-turbulence parameter $(\alpha, \beta) = (4, 1)$. When the local-oscillator optical power at the receiving end is 10 dBm, the received optical power P_r is −20, −10, and −5 dBm.

Figure 9.5 shows the relationship between the average carrier-to-noise ratio and the subcarrier-modulation factor. We assume here that the dynamic range of the optical modulation module, A/D and D/A, is large and belongs to the linear range; thus, we ignore the nonlinear distortion effect of the system. It can be seen from Fig. 9.5 that the subcarrier average carrier-to-noise ratio $\langle R_{CN}\rangle$ increases as the subcarrier-modulation factor m_n increases. When $P_r = -10$ dBm, the subcarrier modulation factor m_n increases from 0.1 to 1.6%, and the average carrier-to-noise ratio increases by about 40 dB.

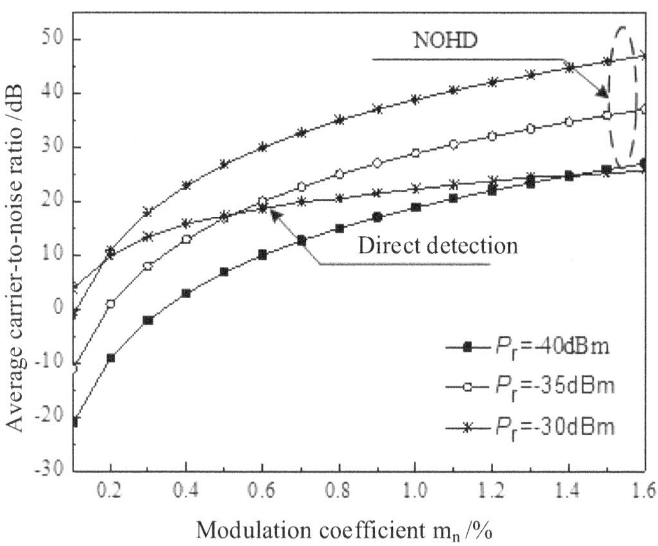

Fig. 9.5 Relationship between the subcarrier average carrier-to-noise ratio and subcarrier modulation coefficient

When the subcarrier-modulation factor is <0.2%, the average carrier-to-noise ratio of direct detection is higher than that of NOHD; however, when the subcarrier-modulation factor is >0.2%, the average carrier-to-noise ratio of the subcarriers obtained by NOHD is obviously better than that of direct detection, and the maximum can be improved by 20 dB. Therefore, using NOHD heterodyne technology can improve the average carrier-to-noise ratio, which is more suitable for complex atmospheric channels characterized by intensity scintillation.

We set the OFDM symbol period T_s to 1280 μs and set the strong-turbulence parameters $(\alpha, \beta) = (4, 1)$. When the received optical power is −45 dBm, −30 dBm, and −15 dBm, respectively, the average carrier-to-noise ratio of the subcarriers varies with the local oscillator. The power-variation curve is shown in Fig. 9.6.

As can be seen from Fig. 9.6, when the received optical power is −15 dBm, the average carrier-to-noise ratio gradually increases as the local-oscillator (LO) optical power increases from −20 to −11 dBm, and then the local-oscillator optical power increases, but the average carrier-to-noise ratio gradually decreases. This is because the excessive optical power of the local oscillator can significantly increase the shot noise of the photodetector and degrade the system performance. When the received optical power is different, the best average carrier-to-noise ratio can be obtained when the local-oscillator optical power is close to −11 dBm.

In Fig. 9.6, when the received optical power is −15 dBm, the local-oscillator optical power can be changed in the range of −20 to −1.25 dBm, and the average carrier-to-noise ratio reduction does not exceed 15% of the maximum value; however, when the received optical power is −45 dBm, the local-oscillator optical power

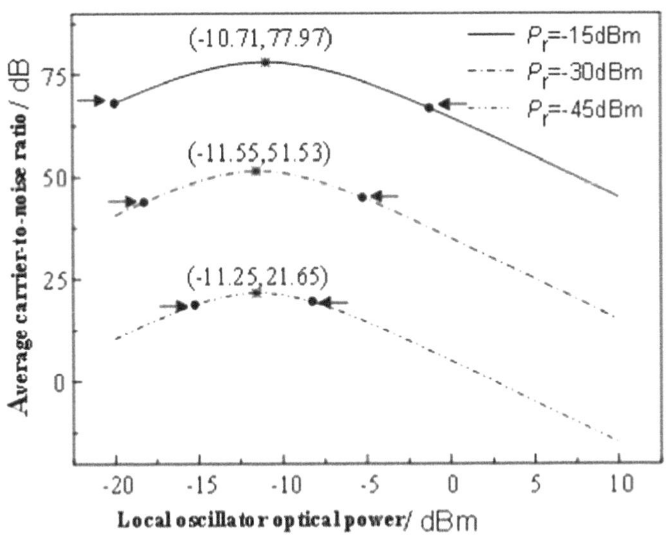

Fig. 9.6 Relationship between the average carrier-to-noise ratio of subcarriers and the modulation factor of LO optical power

only changes in the range of −15.23 to −8.24 dBm to reduce the average carrier-to-noise ratio by no more than 15% of the maximum value. Therefore, it is very important to select the appropriate local-oscillator optical power to obtain a high average carrier-to-noise ratio at the receiving end.

9.3.2 Analysis of System Bit-Error Performance

If the OFDM subcarrier channel is considered to be a linear channel with additive Gaussian noise, and the channel and phase estimation errors are ignored, when the OFDM subcarrier is modulated by QPSK, the bit-error rate of the subcarrier heterodyne-detection system can be given [2]:

$$P_b = \frac{1}{2} erfc\left(\sqrt{\frac{R_{CN}}{2}}\right) = \frac{1}{2} erfc\left(\sqrt{\frac{CT_s}{2\sigma_{n_2}^2}}\right), \tag{9.21}$$

where *erfc* is the complementary error function and $\sigma_{n_2}^2$ is the noise variance of $n_2(t)$. According to the subcarrier carrier-to-noise ratio, the subcarrier average bit-error rate can be expressed as

$$\begin{aligned}\langle P_b \rangle &= \frac{1}{2}\int_0^\infty erfc\left(\sqrt{\frac{R_{CN}}{2}}\right) p_X(x)dx \\ &= \frac{(\alpha\beta)^{\frac{\alpha+\beta}{2}}}{\Gamma(\alpha)\Gamma(\beta)}\int_0^\infty erfc\left(\sqrt{\frac{CT_s}{2\sigma_{n_2}^2}}\right) x^{\frac{\alpha+\beta}{2}-1} K_{\alpha-\beta}\left(2\sqrt{\alpha\beta x}\right)dx.\end{aligned} \tag{9.22}$$

To integrate Eq. (9.22), the Meijer G-function can be used to transform $erfc\left(\sqrt{Z}\right)$ into $\left(1/\sqrt{\pi}\right)G_{1.2}^{2.0}\left[z\middle|\begin{matrix}1\\0.1/2\end{matrix}\right]$, and $K_n(Z)$ into $(1/2)G_{0.2}^{2.0}\left[Z^2/4\middle|\bar{n}/\bar{2}, -n/2\right]$; then, the average bit-error rate can be expressed as

$$\begin{aligned}\langle P_b \rangle \; = \; & \frac{(\alpha\beta)^{\frac{\alpha+\beta}{2}}}{\Gamma(\alpha)\Gamma(\beta)}\int_0^\infty \left(\frac{1}{\sqrt{\pi}}\right) G_{1,2}^{2,0}\left[\frac{(m_n^4\rho^4 P_S^2 P_L^2 L^2)x^2}{4\sigma_{n_2}^2/T_s}\middle|\begin{matrix}1\\0,1/2\end{matrix}\right] \\ & x^{\left(\frac{\alpha+\beta}{2}-1\right)}\left(\frac{1}{2}\right) G_{0,2}^{2,0}\left[\alpha\beta x\middle|\bar{n}/\bar{2}, -n/2\right]dx.\end{aligned} \tag{9.23}$$

After derivation, the subcarrier average bit-error rate is

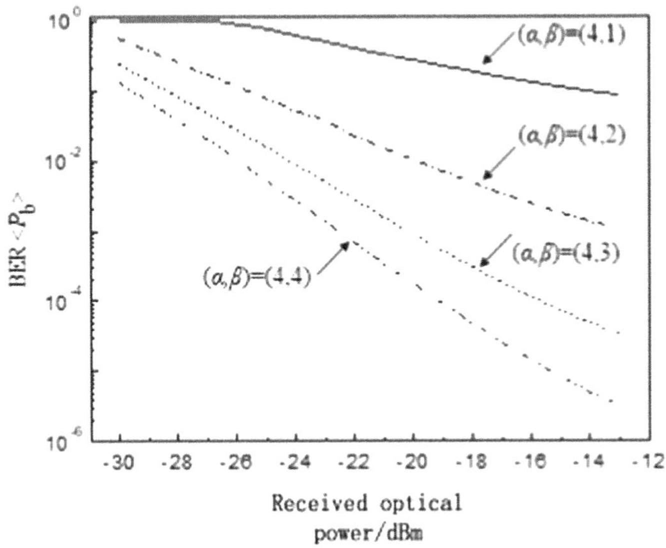

Fig. 9.7 Average bit-error rate curves for different received optical powers

$$\langle P_b\rangle = \frac{2^{(2\alpha+\beta-1)}}{\pi\sqrt{\pi}\Gamma(\alpha)\Gamma(\beta)} G_{5,2}^{2,4}\left[\frac{m_n^4 P_S^2 P_L^2 \rho^4 L^2 T_s}{2^6\sigma_{n_2}^2(\alpha\beta)^2}\middle| \begin{matrix} \frac{1-\alpha}{2}, \frac{2-\alpha}{2}, \frac{1-\beta}{2}, \frac{2-\beta}{2}, 1 \\ 0, 1/2 \end{matrix}\right]. \quad (9.24)$$

Figure 9.7 shows the relationship between the average bit-error rate and the received optical power under different turbulence-intensity conditions when the subcarrier-modulation coefficient $m_n = 8\%$, the local-oscillator optical power $P_L = -10$ dBm, and the OFDM signal is modulated by QPSK. It can be seen from Fig. 9.7 that the light-intensity flicker has a very obvious effect on the average bit-error rate of the receiving end.

When the received optical power is -20 dBm and the turbulence parameters are $(\alpha, \beta) = (4, 1)$ and $(\alpha, \beta) = (4, 4)$, the system average bit-error rates are 10^{-1} and 10^{-4}, respectively. When the turbulence is strong, $(\alpha, \beta) = (4, 1)$, the average bit-error rate can only be improved by one order of magnitude, as the received optical power increases from -30 to -12 dBm. This shows that the NOHD receiving system can improve the detection sensitivity under strong turbulence; however, the intensity scintillation has a great influence on the average bit-error rate.

When the turbulence intensity is weak, $(\alpha, \beta) = (4, 4)$, as the received optical power increases from -30 to -12 dBm, the average bit-error rate can be increased by five orders of magnitude. This indicates that under a weak turbulence intensity, the bit-error rate is significantly reduced when the optical power increases.

9.3.3 SCM-OFDM System Simulation Experiment

To verify the correctness of the theoretical analysis, a simulation experimental platform of the SCM-OFDM subcarrier heterodyne-detection system was built using simulation software, as shown in Fig. 9.8. The parameter settings were as follows: the subcarriers at the sending and receiving ends were 2 MHz and 3.5 MHz, respectively, a semiconductor continuous laser with a 850 nm wavelength was used as the light source at the sending and receiving ends, the optical power of the local-oscillator laser was -10 dBm, the signal light and the local-oscillator light were both carrier-intensity modulated, and the atmospheric channel adopted the gamma–gamma turbulence model.

When the turbulence parameters are medium and strong, $(\alpha, \beta) = (4, 3)$, and the received optical power is -21 dBm, the time-domain waveform and power spectrum of the NOHD difference-frequency signal are shown in Fig. 9.9a, b, and the time-domain waveform and power spectrum of the direct-detection signal are shown in Fig. 9.9c, d, respectively.

Comparing Fig. 9.9a, c, it can be seen that the direct-detection signal is more susceptible to the intensity flicker caused by turbulence; hence, a higher DC offset is required and the high peak-to-average ratio of the OFDM signal will cause nonlinear distortion. Compared with the direct-detection signal, the subcarrier beat frequency can suppress the noise more effectively.

The subcarrier heterodyne detection makes the frequency-band noise near the beat-frequency signal pass into the receiver, while the other frequency-band noises in the subcarrier heterodyne process are filtered out. In Fig. 9.9d, the carrier-to-noise ratio of the power spectrum of the direct-detection signal is 28.2 dB. In Fig. 9.9b, the carrier-to-noise ratio of the power spectrum of the subcarrier heterodyne signal is 36.4 dB, which is 8.2 dB higher than that of direct detection; therefore, NOHD is used. The method enables the useful OFDM signal power to be enhanced, while the noise power is effectively suppressed.

When the medium and strong turbulence parameters are $(\alpha, \beta) = (4, 2)$ and the weak-turbulence parameters are $(\alpha, \beta) = (4, 4)$, the average bit-error rate curves under different received optical-power conditions are shown in Fig. 9.10. The hollow symbols in Fig. 9.10 are the simulated experimental data, and the solid line is the average bit-error rate curve, calculated according to Eq. (9.24).

From Fig. 9.10, it can be seen that the average bit-error rate of the system decreases gradually with the increase of the received optical power. The average bit-error rate

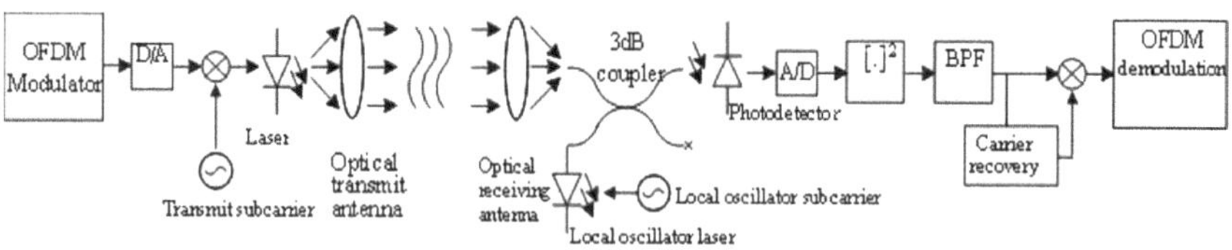

Fig. 9.8 SCM-OFDM/NOHD simulation experiment platform

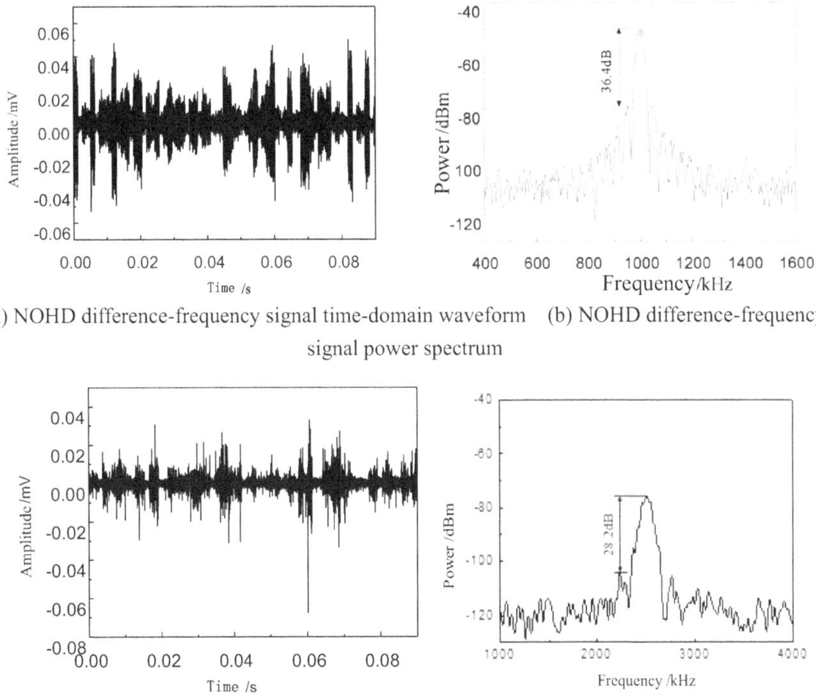

(a) NOHD difference-frequency signal time-domain waveform (b) NOHD difference-frequency signal power spectrum

(c) Direct-detection signal time-domain waveform (d) Direct-detection signal power spectrum

Fig. 9.9 Comparison of the NOHD difference-frequency and direct-detection signals

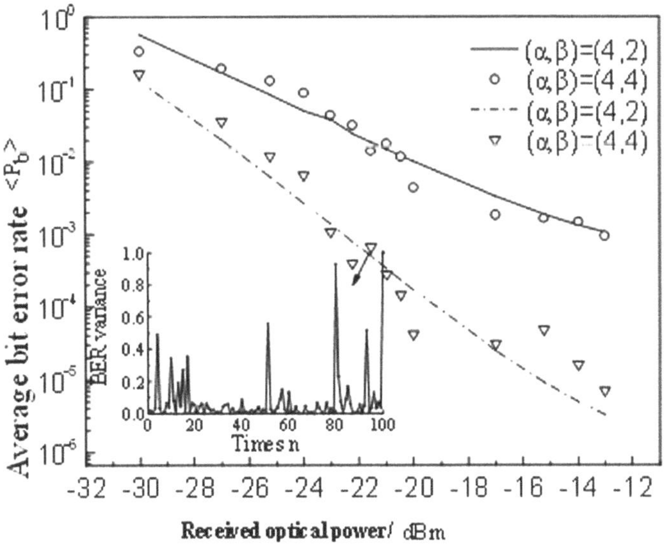

Fig. 9.10 Average bit-error rate curve of the SCM-OFDM/NOHD system transmission

obtained by the experiment is close to the numerical-simulation result of the closed formula derived from the theory.

Under weak-turbulence conditions, when the received optical power is >−17 dBm, the average bit-error rate is above 10^{-5} to ensure the reliability of the optical wireless communication system. During weak turbulence, $(\alpha, \beta) = (4, 4)$, the bit-error rate variance of 100 experiments with a received optical power of −22 dBm is shown in Fig. 9.10. The fluctuation of the system error variance is small, indicating that weak random turbulence has less of an impact on the system.

Figure 9.11 shows the signal constellation diagrams when the received optical power is −20.97 dBm under different turbulence intensities. It can be seen from Fig. 9.11 that the received signal constellation has a certain degree of divergence when the turbulence is strong, $(\alpha, \beta) = (4, 1)$, and the average bit-error rate of the system is 10^{-4}. When the turbulence is weak, $(\alpha, \beta) = (4, 4)$, the received signal constellation is obviously concentrated, and the average bit-error rate can be improved by two orders of magnitude.

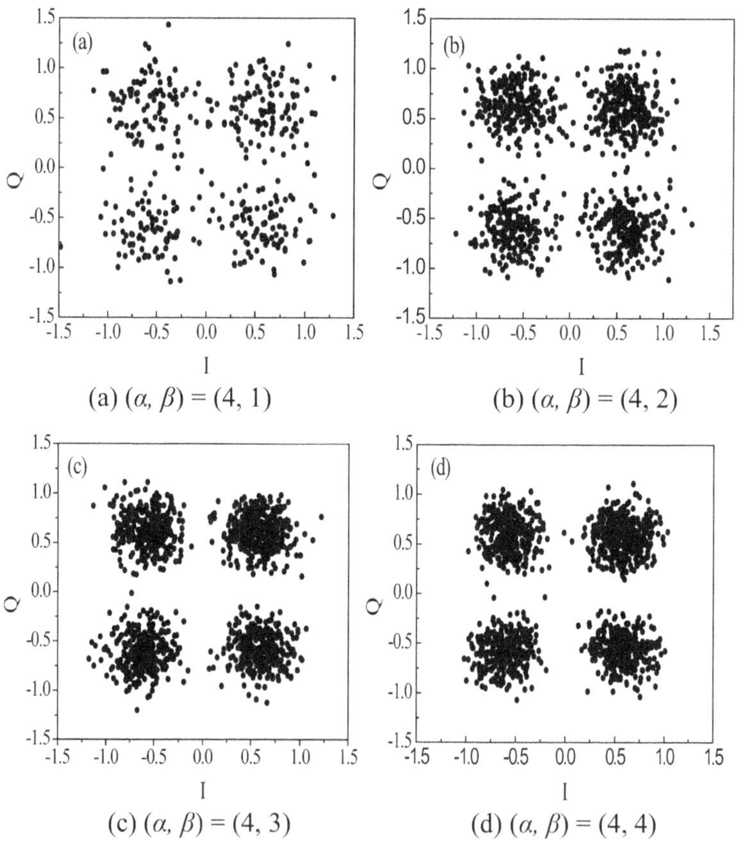

(a) $(\alpha, \beta) = (4, 1)$ (b) $(\alpha, \beta) = (4, 2)$

(c) $(\alpha, \beta) = (4, 3)$ (d) $(\alpha, \beta) = (4, 4)$

Fig. 9.11 Constellation diagrams of different turbulence intensities

In Fig. 9.11, as the turbulence intensity increases, the dispersion-phase noise of the constellation point increases; however, there is no large phase rotation, indicating that the turbulence has little effect on the demodulation of the beat-frequency signal. The atmospheric turbulence mainly causes the amplitude of the NOHD beat-frequency signal to fluctuate, which causes the bit-error rate to increase.

The experimental results show that using subcarrier heterodyne technology can effectively suppress the noise caused by turbulence and improve the overall performance of the system, when the appropriate local-oscillator optical power is selected.

9.3.4 Analysis of the Influence of the Nonlinear Distortion Effect and Frequency Offset on the System

In an SCM-OFDM subcarrier heterodyne-detection system, both the transmitter and the receiver must perform analog modulation on the laser. This causes a nonlinear distortion effect, and the Doppler frequency shift will generate a frequency offset when the atmospheric channel transmits the signal.

This section focuses on the nonlinear distortion and frequency offset in an SCM-OFDM subcarrier heterodyne-detection system. By modeling a detection signal with nonlinear distortion and a frequency offset, the nonlinear modulation and frequency offset are discussed in depth. The orthogonality of the subcarriers decreases the impact on the system.

9.3.4.1 Nonlinear-Effect Influence Analysis

In an SCM-OFDM/NOHD system, both the transmitter and the receiver must perform analog modulation on the laser. During the analog modulation, the electro-optical conversion characteristics of the semiconductor and the corresponding drive circuit have nonlinear distortion effects, which cause the output signal to generate frequencies that do not exist in the input signal.

The components include harmonic products and intermodulation products. Harmonic products can generally be filtered out by filters. Intermodulation products include the intermodulation between the noise components of the device itself and the intermodulation between the noise and the carrier. These intermodulation products can be regarded as interference noise in the output signal. The nonlinear distortion introduced by this non-ideal factor may lead to additional subcarrier interference.

Because the OFDM subcarrier spectra overlap each other, this places strict constraints on the orthogonality between subcarriers. When the orthogonality cannot strictly meet the requirements, the subcarriers will affect each other, resulting in a decrease of the signal-to-noise ratio of the receiver.

The nonlinear distortion of laser modulation mainly comes from the following:

(1) The inherent nonlinearity of the die caused by relaxation oscillation and leakage current in the active region,
(2) The nonlinearity of the output optical-power P-I curve, and
(3) The clipping caused by the accidental swing of the instantaneous driving current of the laser below the threshold current.

An OFDM modulated signal has a large peak-to-average ratio. When the linear working area of the modulator is limited, and an OFDM signal with a high peak-to-average ratio passes through a nonlinear amplifier, the signal will produce nonlinear distortion.

The complex envelope generated by the OFDM signal module at the transmitting end can be expressed as

$$s(t) = \sum_{k=1}^{N_{SC}} c_k \cdot e^{j2\pi f_k t}, \tag{9.25}$$

where c_k is the kth subcarrier information symbol, N_{sc} is the number of subcarriers, f_k is the subcarrier frequency, $f_k = (k-1)/T_s$, and $\mathrm{T_s}$ is the OFDM symbol period.

When the baseband OFDM signal subcarrier modulates, it is necessary to perform a parallel–serial conversion on the real and imaginary parts of the complex OFDM signal. The signal can be expressed as $G(t)$, and the subcarrier signal is obtained by up-converting $G(t)$ with f_{RF} as the carrier frequency. In the laser driving signal, the laser bias current is set to I_{b}; then, the driving current of the laser can be expressed as

$$i(t) = G(t) \cdot \cos(2\pi f_{RF} t) + I_b. \tag{9.26}$$

Owing to the nonlinearity of the P-I characteristic curve of the output optical power of the laser, a nonlinear distortion will be generated when the laser is modulated by the subcarrier. The modulated laser power can be expanded by a Taylor series near the bias point:

$$P(t) = P_0 + \sum_{k=1}^{\infty} \frac{1}{k!} \frac{d^k P}{di^k} (i(t) - I_b)^k. \tag{9.27}$$

The I channel signal can be expressed as

$$G_I(t) = \sum_{k=1}^{N_{SC}} c_k \cos(2\pi f_k t). \tag{9.28}$$

We substitute Eq. (9.28) into Eq. (9.27) to obtain

$$P_I(t) = P_0 + P'\left[G_I(t)\cos(2\pi f_{RF} t)\right] + \frac{1}{2!} P''\left[G_I(t) \cdot \cos(2\pi f_{RF} t)\right]^2$$

$$+ \frac{1}{3!}P'''\left[G_I(t)\cdot\cos(2\pi f_{RF}t)\right]^3 + \cdots,$$

where the first-order term is

$$X(t) = \frac{1}{2}P'\left[\sum_{k=1}^{N_{SC}} c_k\cdot\cos(2\pi f_k t)\cdot\cos(2\pi f_{RF}t)\right]. \tag{9.29}$$

In Eq. (9.28), the second-order distortion term $K(t)$ is expanded as

$$\begin{aligned} K(t) &= \frac{1}{2}P''\left[\sum_{k=1}^{N_{SC}} c_k\cdot\cos(2\pi f_k t)\cdot\cos(2\pi f_{RF}t)\right]^2 \\ &= \frac{1}{8}P''c_k^2(1+\cos(2\pi\cdot 2f_k t)+\cos(2\pi\cdot 2f_{RF}t) \\ &+ \frac{1}{2}\cos(2\pi\cdot 2f_k t + 2\pi\cdot 2f_{RF}t) \\ &+ \frac{1}{2}\cos(2\pi\cdot 2f_k t - 2\pi\cdot 2f_{RF}t)) \end{aligned} \tag{9.30}$$

In Eq. (9.30), $2f_k$ and $2f_{RF}$ are harmonic distortion, and $2f_k + 2f_{RF}$ and $2f_k - 2f_{RF}$ are intermodulation-distortion products of mutual frequencies. It can be seen that the optical power output by the transmitter after subcarrier modulation is affected by the nonlinear effects, resulting in new frequency components.

Next, the influence of the second-order nonlinear distortion generating new frequency components in the NOHD receiver's heterodyne process will be discussed.

Assuming that the frequency of the subcarrier at the receiving end is f_L, and the amplitude of the subcarrier is set to 1, because the subcarrier is a sine wave, it can be considered that there is no nonlinear effect when setting the appropriate bias current and modulation coefficient. Then, the subcarrier time-domain waveform can be expressed as $\cos(2\pi f_L t)$, and the signal obtained by the square of the current sampling after the photoelectric detection at the receiving end can be expressed as

$$\begin{aligned} \{X(t)+K(t)+\cos(2\pi f_L t)\}^2 &= X^2(t)+K^2(t)+\cos^2(2\pi f_L t)+2X(t)K(t) \\ &+ 2X(t)\cos(2\pi f_L t)+2K(t)\cos(2\pi f_L t), \end{aligned} \tag{9.31}$$

where

$$
\begin{aligned}
K^2(t) &= \left[\frac{1}{8}P''\left(\begin{array}{l} 1+\sum_{k=1}^{Nsc} c_k\cos(2\pi\cdot 2f_k t)+\cos(2\pi\cdot 2f_{RF}t) \\ +\frac{1}{2}\sum_{k=1}^{Nsc} c_k\cos(2\pi\cdot 2f_k t+2\pi\cdot 2f_{RF}t) \\ +\frac{1}{2}\sum_{k=1}^{Nsc} c_k\cos(2\pi\cdot 2f_k t-2\pi\cdot 2f_{RF}t)\end{array}\right)\right]^2 \\
&= \left(\frac{1}{8}P''\right)^2\left[\begin{array}{l} 1+\sum_{k=1}^{Nsc} c_k^2\cos^2(2\pi\cdot 2f_k t)+\cos^2(2\pi\cdot 2f_{RF}t) \\ +\frac{1}{4}\sum_{k=1}^{Nsc} c_k^2\cos^2(2\pi\cdot 2f_k t+2\pi\cdot 2f_{RF}t)\end{array}\right. \\
&+\frac{1}{4}\sum_{k=1}^{Nsc} c_k^2\cos^2(2\pi\cdot 2f_k t-2\pi\cdot 2f_{RF}t) \\
&+2\sum_{k=1}^{Nsc} c_k\cos(2\pi\cdot 2f_k t)+2\cos(2\pi\cdot 2f_{RF}t) \\
&+\sum_{k=1}^{Nsc} c_k\cos(2\pi\cdot 2f_k t+2\pi\cdot 2f_{RF}t) \\
&+\sum_{k=1}^{Nsc} c_k\cos(2\pi\cdot 2f_k t-2\pi\cdot 2f_{RF}t) \\
&+2\sum_{k=1}^{Nsc} c_k\cos(2\pi\cdot 2f_k t)\cdot\cos(2\pi\cdot 2f_{RF}t) \\
&+\sum_{k=1}^{Nsc} c_k\cos(2\pi\cdot 2f_k t)\cdot\sum_{k=1}^{Nsc} c_k\cos(2\pi\cdot 2f_k t+2\pi\cdot 2f_{RF}t) \\
&+\sum_{k=1}^{Nsc} c_k\cos(2\pi\cdot 2f_k t)\cdot\sum_{k=1}^{Nsc} c_k\cos(2\pi\cdot 2f_k t-2\pi\cdot 2f_{RF}t) \\
&+\cos(2\pi\cdot 2f_{RF}t)\cdot\sum_{k=1}^{Nsc} c_k\cos(2\pi\cdot 2f_k t+2\pi\cdot 2f_{RF}t) \\
&+\sum_{k=1}^{Nsc} c_k\cos(2\pi\cdot 2f_k t+2\pi\cdot 2f_{RF}t)\cdot\sum_{k=1}^{Nsc} c_k\cos(2\pi\cdot 2f_k t-2\pi\cdot 2f_{RF}t)
\end{aligned}
\tag{9.32}
$$

Equation (9.32) includes input-frequency components $2f_k$, $2f_{RF}$, $2f_k + 2f_{RF}$, and $2f_k - 2f_{RF}$; octave components $4f_k$, $4f_{RF}$, $4f_k + 4f_{RF}$, and $4f_k - 4f_{RF}$; sum-frequency terms $2f_k + 2f_{RF}$ and $4f_K + 2f_{rf}$; and difference-frequency terms $2f_{RF} - 2f_k$ and $4f_{RF} - 2f_k$.

Because the receiving end of the subcarrier heterodyne detection must detect the useful signal of the frequency band $f_{RF} - f_L$, these frequency components will not enter into the useful-signal frequency band; therefore, they can be filtered out by the filter, and will not affect the NOHD difference-frequency signal.

Similarly, the frequency generated by the square of the first-order component of the transmitted signal and the square of the local-oscillator subcarrier signal in the squared signal will not fall within the useful-signal frequency band to affect the NOHD difference-frequency signal. The possible influence of the cross term in Eq. (9.31) is discussed below.

The cross term $2X(t)K(t)$ is expanded as

$$\begin{aligned}2X(t)K(t) &= \frac{1}{2}P'' \cdot P' \sum_{k=1}^{Nsc} c_k^3 \cos^2(2\pi f_k t)\cos(2\pi f_{RF} t) \\ &+ P'' \cdot P' \sum_{k=1}^{Nsc} c_k^3 \cos^2(2\pi f_k t)\left[\frac{1}{2}\cos(2\pi f_{RF} t) + \cos(2\pi \cdot 3f_{RF} t)\right]\end{aligned} \tag{9.33}$$

It can be seen from this item that the signal spectrum of the cross term $2X(t)K(t)$ is centered at f_{RF} and $3f_{RF}$, and the signal is also not in the frequency band of the useful difference-frequency signal $f_{RF} - f_L$; therefore, its influence on the NOHD signal can be eliminated by a filter.

The cross term $2X(t)\cos(2\pi f_L t)$ in Eq. (9.31) is expanded as

$$\begin{aligned}2X(t)\cos(2\pi f_L t) &= P' \sum_{k=1}^{Nsc} c_k \cos(2\pi f_k t)\cos(2\pi(f_{RF} - f_L t)) \\ &+ P' \sum_{k=1}^{Nsc} c_k \cos(2\pi f_k t)\cos(2\pi(f_{RF} + f_L t)\end{aligned} \tag{9.34}$$

Equation (9.34) shows that the sum-frequency term $f_{RF} + f_L$ in the cross term can be filtered out by the Butterworth filter, and a useful NOHD difference-frequency signal $P' \sum c_k \cos(2\pi f_k t)\cos(2\pi(f_{RF} - f_L t))$. can be detected. It can be seen that the NOHD receiving end converts the baseband OFDM signal through the process of the subcarrier difference-frequency $f_{RF} - f_L$. The spectrum is linearly shifted to the carrier with the difference frequency as the center frequency.

The cross term $2K(t)\cos(2\pi f_L t)$ in Eq. (9.31) is expanded as

$$2K(t)\cos(2\pi f_L t) = \frac{1}{2}P'' \sum_{k=1}^{Nsc} c_k^2 \cos^2(2\pi f_k t)\cos(2\pi f_L t)$$

$$+\frac{1}{2}P''\sum_{k=1}^{Nsc}c_k^2\cos^2(2\pi f_k t)\cos(2\pi\cdot 2f_{RF}t)\cos(2\pi f_L t) \quad (9.35)$$

It can be seen from Eq. (9.35) that in the second-order distortion term, the spectrum of the baseband OFDM signal is modulated on the frequency band $f_L - 2f_{RF}$ after being squared, and when f_L, f_{RF}, and the filter bandwidth are not selected properly, $\frac{1}{2}P''\sum_{k=1}^{Nsc}c_k^2\cos^2(2\pi f_k t)\cos(2\pi.2f_{RF}t)\cos(2\pi f_L t)$ may have an impact on the NOHD difference-frequency signal.

It can be seen from the above analysis that the nonlinear-distortion effect generated by the subcarrier modulation of the OFDM signal at the transmitting end will have a non-negligible impact on the NOHD reception, resulting in a decrease in the signal-to-noise ratio. The filter bandwidth can effectively reduce the influence of nonlinear effects on the NOHD receiver.

In the following, an SCM-OFDM/NOHD-system simulation experiment is conducted, considering the influence of nonlinear effects. The channel is a gamma–gamma channel model with weak turbulence, and the subcarrier at the transmitting end is 2.5 MHz.

Figure 9.12 shows the power spectrum of the subcarrier-modulated signal at the transmitter with nonlinear modulation. In Fig. 9.12, the OFDM symbol period is 0.09216 s, the subcarrier frequency interval is 10.9 Hz, and the entire OFDM orthogonal carrier-frequency band is 10.972 = 0.78 kHz.

Figure 9.12 shows the result of a first-order linear modulation; i.e., there is a modulated OFDM signal bandwidth at 2.5 MHz. From Fig. 9.12, it can be seen that

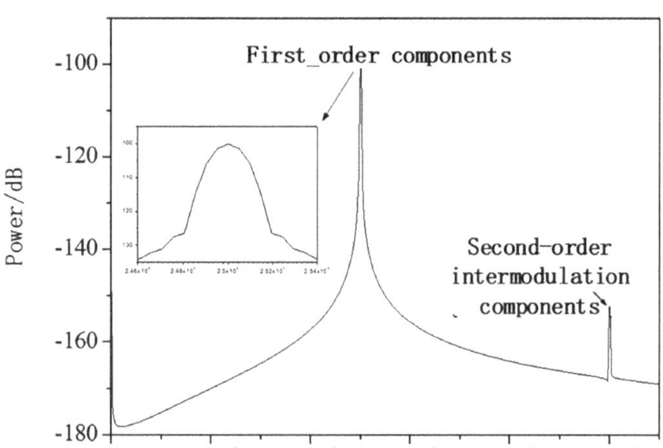

Fig. 9.12 Power spectrum of the subcarrier signal at the transmitter in the case of second-order intermodulation

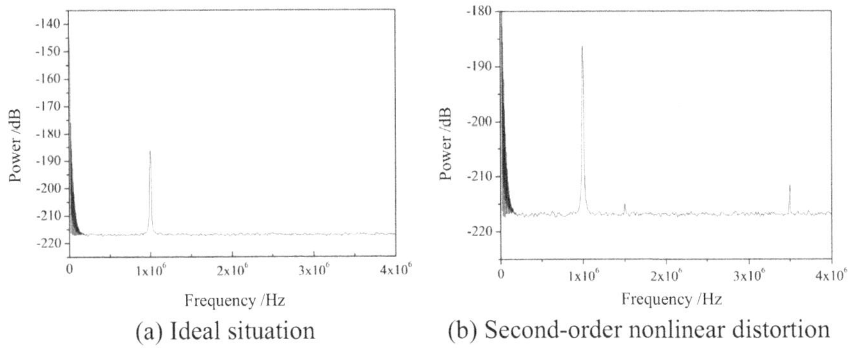

Fig. 9.13 Power spectrum of the squared signal

when there is nonlinear modulation at the transmitting end, the result of the second-order intermodulation in the up-conversion has a carrier-frequency multiplication term of 5 MHz; the multiplication term of the OFDM signal is also modulated to the carrier with the center frequency of 5 MHz. It can be seen from Fig. 9.12 that the output subcarrier signal will not produce spectral aliasing during second-order intermodulation at the transmitter.

The next problem is to study the influence of the subcarrier heterodyne after a nonlinear subcarrier modulation of the OFDM signal.

Figure 9.13 shows the signal power spectrum after square processing in the subcarrier-heterodyne process at the receiving end. Figure 9.13a shows that when there is no nonlinear distortion, the NOHD heterodyne process produces the required difference-frequency term of 1 MHz. However, in Fig. 9.13b, owing to the nonlinear modulation at the transmitting end, in addition to the difference-frequency term of 1 MHz, other frequency components are generated in the subcarrier heterodyne process. The spurious cross terms in the intermediate frequency (IF) bandwidth will produce useful signals and interference, reducing the signal-to-noise ratio.

Figure 9.14 shows the demodulated signal constellation of the SCM-OFDM/NOHD system. Figure 9.14a shows the demodulated signal constellation of a weak turbulent channel without nonlinear distortion, and Fig. 9.14b shows the constellation with nonlinear distortion. To demodulate the signal-constellation diagram of a weak turbulent channel with second-order nonlinear distortion, it can be clearly seen from the comparison of the two diagrams that the nonlinear modulation reduces the signal-to-noise ratio, increases the bit-error rate, and scatters and rotates the constellation diagram at a certain angle. The system needs to be compensated.

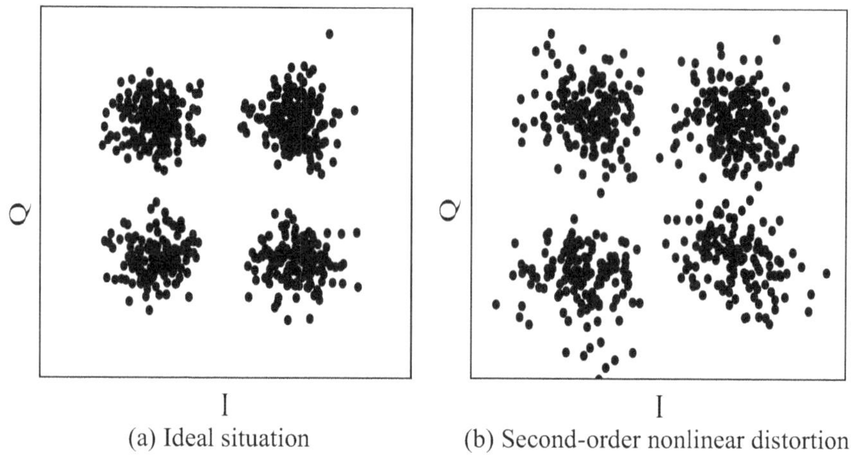

Fig. 9.14 Constellation diagram after demodulation

9.3.4.2 Analysis of the Frequency-Offset Influence on the Signal-to-Noise Ratio

The carrier-frequency offset mainly comes from the frequency offset between the transmitter and receiver local oscillators. The carrier-frequency offset will not only cause the phase rotation of the demodulation constellation point to affect the symbol decision, but also make the OFDM system subcarrier interference more serious. The orthogonality between them is destroyed, causing intercarrier interference (ICI) between subcarriers, resulting in a serious degradation of the system performance.

Assuming that when the SCM-OFDM signal is transmitted in the atmospheric channel, the frequency offset caused by the Doppler frequency shift is Δf, the optical power of the I-channel signal after optical-intensity modulation can be expressed as

$$P(t) = \mathrm{Re}\left[\sum_{k=1}^{N} a_n \exp(j2\pi f_k t) \cdot \exp(j2\pi f_s t) \cdot \exp(j2\pi \Delta f t)\right] \cdot \cos(2\pi f_0 t). \quad (9.36)$$

The optical-intensity modulation of the local-oscillator subcarrier also needs to be performed when the subcarrier is modulated at the receiving end; then, the photocurrent $i(t)$ output by the photodetector can be expressed as

$$i(t) = \sum_{k=1}^{N} a_n \cos(2\pi f_k t + 2\pi f_s t + 2\pi \Delta f) + \cos(2\pi f_L t). \quad (9.37)$$

After squaring, the difference-frequency cross term $v(t)$ is obtained by sampling and passing through the filter, which can be expressed as

$$v(t) = \sum_{k=1}^{N} a_n \cos(2\pi f_k t - 2\pi f_l t + 2\pi \Delta f t). \tag{9.38}$$

Then, the OFDM signal with a frequency offset obtained after the coherent demodulation of the difference-frequency signal at the receiving end can be expressed as

$$r(t) = \sum_{k=1}^{N} a_n e^{j(2\pi f_k t + 2\pi \Delta f)} + N(t), \tag{9.39}$$

where $N(t)$ represents additive white Gaussian noise (AWGN). Then, the demodulated signal at the receiving end can be expressed as

$$c'_{ki} = \frac{1}{T_s} \int_0^{T_s} r(t) e^{-j(2\pi f_k t + 2\pi \Delta f t)} dt. \tag{9.40}$$

Substituting $r(t)$ in Eq. (9.39) into (9.40), we obtain

$$\begin{aligned} c_{li}{}' &= \frac{1}{T_s} \int_0^{T_s} \left[\sum_{l=0}^{N} a_n r(t) e^{-j(2\pi f_k t + 2\pi \Delta f t)} + N(t) \right] e^{-j2\pi f_k t} dt \\ &= \sum_{l=0}^{N} a_n \frac{1}{T_s} \int_0^{T_s} e^{j(2\pi (f_l - f_k)t) + j2\pi \Delta t t} dt + n(t) \\ &= \eta_0 c_k + \sum_{l \neq k} \eta_{k-l} c_l + n_k. \end{aligned} \tag{9.41}$$

The first, second, and third terms correspond to the signal, intercarrier interference (ICI), and additive white Gaussian noise, respectively. Let $\eta_0 = |\eta_0| + \delta$, *and* $= a$; then, ac_k is the useful signal part, of which δc_k belongs to the additional noise. The ICI coefficient is defined as

$$\eta_m = \frac{1}{T_s} \int_0^{T_s} e^{j(2\pi f'_m t + 2\pi \Delta f t)} dt, f'_m \frac{m}{T_s}, m = -(N_{sc} - 1), \ldots, 0, 1, \ldots, (N_{sc} - 1). \tag{9.42}$$

Then, by integrating Eq. (9.42), we can obtain

$$\eta_m = \frac{\sin(\pi(m + \delta))}{\pi(m + \delta)} e^{-j\pi(m+\delta)}, \tag{9.43}$$

where $\delta = \Delta f.T_s$ is the normalized frequency offset. The signal-to-noise ratio $|\eta_0 = \sin(\pi \Delta fT)/\pi \Delta fT|$ can be defined as

$$\gamma\prime = \frac{a^2}{\gamma^{-1} + V_0}, \tag{9.44}$$

where a^2 is the signal power of the useful signal $c_{ki}\prime$, assuming $\gamma = \sigma_k^2/n_k^2$ is the signal-to-noise ratio in the absence of subcarrier interference and γ^{-1} is the thermal noise variance. V_0 is the remaining noise variance and can be expressed as

$$V_0 = E\left[|\eta_0|^2\right] + E\left[\sum_{l \neq k} |\eta_{k-l}|^2\right]. \tag{9.45}$$

Then, the SNR penalty due to the frequency offset can be expressed as

$$\begin{aligned}\Delta\gamma &= 10\log(\gamma') - 10\log(\gamma) = 10\log\left(\frac{a^2}{1 + V_0 \cdot \gamma}\right) \\ &= -10\log a^2 + 10\log(1 + V_0 \cdot \gamma) \\ &\approx \frac{10}{\ln 10}\left(\left(1 - a^2\right) + V_0 \cdot \gamma\right)\end{aligned} \tag{9.46}$$

Assume $E\left[|\eta_0|^2\right] = a_1$:

$$E\left[\xi^2\right] = E\left[|\eta_0|^2\right] - (E[|\eta_0|])^2 = a_1 - a^2. \tag{9.47}$$

Otherwise,

$$E\left[\sum_{l \neq k} |\eta_{k-l}|^2\right] = 1 - E\left[|\eta_0|^2\right] = 1 - a_1. \tag{9.48}$$

Then, the SNR penalty due to the frequency offset can be expressed as

$$\Delta\gamma \approx \frac{10}{\ln(10)} \frac{1}{3} (\pi N \Delta f T_s)^2 \cdot \gamma. \tag{9.49}$$

When the subcarrier adopts QPSK, if the bit-error rate is required to be 10^{-7}, the signal-to-noise ratio requirement is about $15/\sin^2(\pi/4)$ [3]. According to Eq. (9.49), it can be observed that the 64-subcarrier QPSK is normalized. The relationship between the frequency offset and signal-to-noise ratio cost is shown in Fig. 9.15. The figure shows the frequency-offset function of the SNR when the BER is 10^{-7}. It can be seen that to reduce the SNR to <15 dB, the frequency offset δ should be <0.0028.

Fig. 9.15 Relationship between the 64-subcarrier QPSK normalized frequency offset and SNR cost

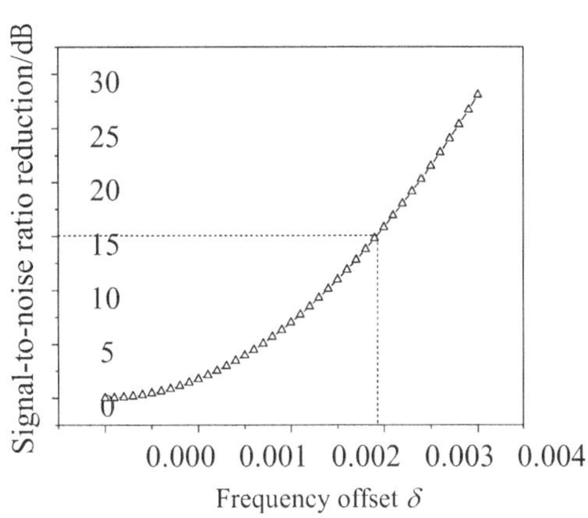

The bit-error rate of a subcarrier QPSK-modulated OFDM signal can be expressed as

$$P_e \approx \frac{1}{2} erfc\left(\sqrt{\frac{\gamma'}{2}}\right). \tag{9.50}$$

According to Eq. (9.50), the bit-error rate of the 64-subcarrier QPSK-modulation SCM-OFDM subcarrier heterodyne-detection system is numerically analyzed, and the results are shown in Fig. 9.16. The figure shows the bit-error rate curve when the normalized frequency offset is 0, 0.002, and 0.004, respectively.

When the normalized frequency offset is 0.002, the bit-error rate curve is similar to that without a frequency offset. The influence of the frequency offset can be ignored during demodulation at the receiving end. However, when the normalized frequency offset increases to 0.004, the system bit-error rate increases greatly, and it is necessary to compensate for the frequency offset to reduce the system bit-error rate.

9.4 Experimental Study

NOHD detection data under different weather conditions were collected and analyzed. The NOHD detection signals of different subcarrier-modulation signal-light modes were collected, and the system performance was analyzed and compared. The NOHD theoretical model was validated.

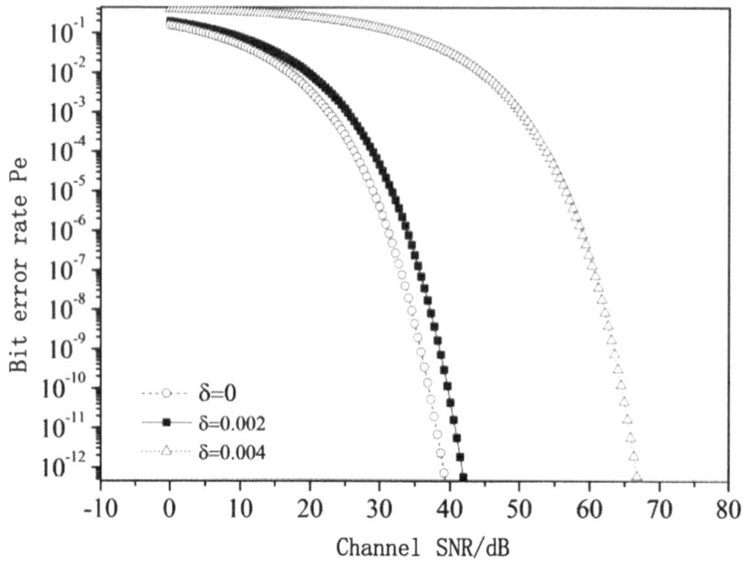

Fig. 9.16 Sixty-four-subcarrier QPSK-modulation SCM-OFDM/NOHD bit-error rate curve

9.4.1 NOHD Experimental-System Design

9.4.1.1 Subcarrier-Modulation Transmitter Design

The working principle of the sending end is as follows: First, a digital signal is generated by the PC and sent to the modulation module. The modulation module modulates the binary digital signal in different manners, and the subcarrier frequency is set to 2.5 MHz. The modulated-subcarrier electrical signal is connected to the laser-drive module, which drives the laser, and the laser emits modulated light waves to convert the electrical signals to optical signals. The signal light is sent through the atmospheric channel using the optical transmitting antenna, and is collimated.

The performance and parameters of each component of the sender are described below.

(1) **Laser-light source and internal-modulation characteristics**

The selectable wavelength ranges of wireless-excitation communication mainly include 810–960, 980–1060, and 1550–1600 nm, of which the 850 nm band belongs to the "atmospheric window"; therefore, the attenuation is small when transmitting in the atmospheric channel. Through the calculation of Eq. (9.51), the atmospheric transmittance can be obtained according to the visibility value.

$$\tau(\lambda) = \exp\left[-\frac{3.912}{V}\left(\frac{\lambda}{0.55}\right)^{-q} \times z\right], \quad (9.51)$$

where V is the visibility (unit: km), z is the transmission distance (unit: km), and λ is the light wavelength. The value of the infrared band q is

$$q = \begin{cases} 1.6 & V \text{ larger} \\ 13 & \text{Medium visibility} \\ 0.585V^{1/3} & V \leq 6 \text{ km} \end{cases}. \tag{9.52}$$

From Eq. (9.52), the attenuation coefficient and visibility table of an 850 nm laser in different weather conditions can be obtained, as shown in Table 9.1.

In the experiment, a 120 mW 850 nm internally modulated semiconductor laser was used as the signal source at the transmitting end. The laser fully met the experimental requirements for signal transmission in a 1 km turbulent atmospheric channel. In this experiment, an analog laser (LQx/RLx Series, model LQA850-120E, Newport Company, USA) was used. The maximum output power was 120 mW, and the maximum analog-modulation frequency was 20 MHz (Table 9.2).

(2) **Subcarrier-modulation module**

In the experiment, the optical wireless communication transmitter adopts subcarrier modulation and uses a field-programmable gate array (FPGA) to perform the symbol conversion, serial-to-parallel conversion, and digital-information modulation. A high-speed A/D-conversion circuit board (AD9788EVB) is used for the digital-to-analog conversion to generate an analog subcarrier signal drive.

Table 9.1 Attenuation coefficient and visibility of an 850-nm laser in different weather conditions

Weather conditions	Visibility (km)	Attenuation coefficient (dB/km)
Very clear	50–20	0.20–0.52
Clear	20–10	0.52–1.0
Light haze	10–4	1.0–2.9
Haze	4–2	2.9–5.8
Thin fog	2–1	5.8–14.0
Light fog	1–0.5	14.0–34.0
Moderate fog	0.5–0.2	34.0–84.9
Thick fog	0.2–0.05	84.9–339.6

Table. 9.2 Transmitting laser parameters

Output wavelength	850 nm ± 15 nm
Output power	120 mW
Power stability	0.5%
Maximum modulation frequency	20 MHz
Operating voltage	4.8–12 VDC
Operating temperature	0–40 °C

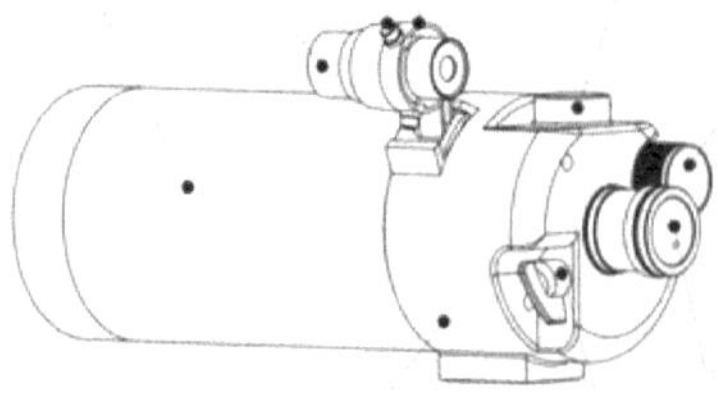

Fig. 9.17 Schematic diagram of the optical transmit/receive antenna structure

Figure 6.4 shows a physical diagram of the laser and the subcarrier-modulation module at the transmitter. The FPGA can be seamlessly connected with the AD9788EVB. Through the AD9788EVB serial peripheral interface, the FPGA can program and read its internal parameters. The FPGA output signals are orthogonal to each other, and the data are input to a DAC9788 data-acquisition module; thus, the FPGA and AD9788EVB work together to output the subcarrier modulation signal.

(3) **Optical transmit/receive antenna**

In the experiment, a Maksutov–Cassegrain astronomical telescope was used for the light-transmitting and light-receiving antennas. It has a catadioptric optical antenna, which is very small, light, adjustable, and easy to operate. The antenna diameter is 105 mm. The telescope is equipped with two sets of eyepieces, including one direct-view eyepiece, which are convenient for collimating and focusing the beacon light and the 850 nm signal laser. The schematic diagram of the optical transmit/receive antenna is shown in Fig. 9.17.

9.4.1.2 NOHD Receiver

In the experiment, the NOHD receiving end mainly includes an optical receiving antenna, a local-oscillator optical subcarrier-modulation module, a subcarrier heterodyne-detection module, and a difference-frequency signal-demodulation module. A block diagram of the NOHD terminal of the optical wireless communication system is shown in Fig. 9.18.

The optical receiving antenna is embedded with a spatial optical-coupling module. First, the spatial optical-coupling efficiency is improved by adjusting the focal length, azimuth, and pitch of the receiving antenna until the photoelectric detector can detect the signal light. Secondly, a subcarrier is generated by the signal source to modulate the local-oscillator laser. The maximum modulation voltage of the local-oscillator light is 2 V.

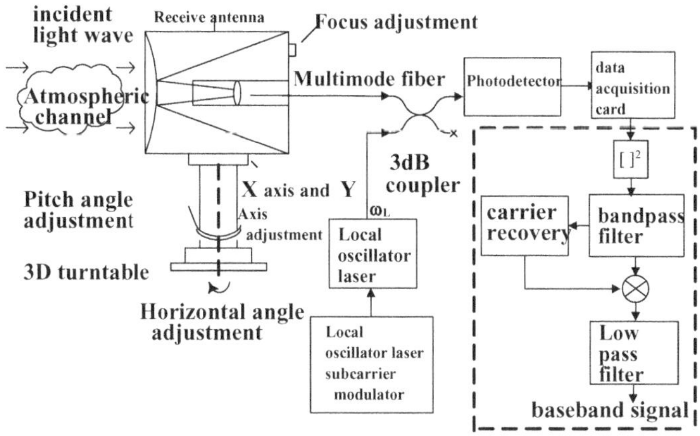

Fig. 9.18 Block diagram of the NOHD terminal in a optical wireless communication system

In the experiment, a sine-wave subcarrier with a 200 mV bias and a 200 mV peak-to-peak value is used to modulate the local-oscillator light, and the subcarrier frequency is set to 3.5 MHz. The local-oscillator laser output is connected to the other input end of the 3 dB coupler. Finally, the output signal of the 3 dB coupler converges on the surface of the PIN detector for detection, and its output voltage is sampled by the A/D data-acquisition card at a 40-MHz sampling frequency. The computer records the output data from the acquisition card in real time, and then squares the signal. The demodulated signal is processed and subjected to carrier recovery.

(1) **Spatial optical-coupling module**

A new spatial optical-coupling structure based on an optical-fiber array is adopted, the spatial light is coupled into multiple multimode fibers through a microlens array, and finally, the multiple fibers are fused together. This structure improves the spatial optical-coupling efficiency. The spatial optical-coupling module is shown in Fig. 9.19. This module solves the reduced coupling efficiency and signal interruptions caused by optical spot damage, and the jitter and drift caused by atmospheric turbulence.

(2) **Local-oscillator laser**

In this experiment, the local-oscillator light source adopts an 850-nm multimode fiber-coupled high-speed modulated laser (AUT-PCL-850-020THS, Shanghai Haoliang Optoelectronics Equipment Co., Ltd.). Its main parameters are shown in Table 9.3.

In the experiment, the local-oscillator (LO) laser does not have specific requirements for the line width; however, its modulation linearity is greatly dependent on the modulation depth. The bias current should be adjusted so that its operating point is in the linear region of the semiconductor power-current characteristic curve.

(3) **Coupler**

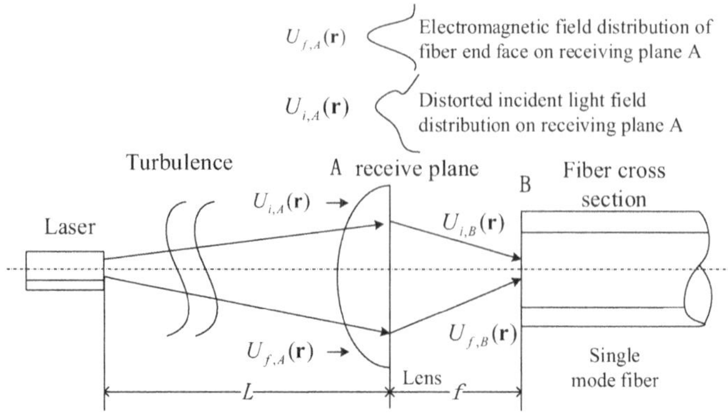

Fig. 9.19 Spatial optical-coupling module

Table 9.3 LO laser parameters

Fiber-coupled output power	≤20 mW
Output wavelength	850 nm ± 5 nm
Fiber-optic connector	FC/PC
Power stability	3%
Operating voltage	5 V
Analog-modulation frequency	≤20 MHz

The NOHD receiving end adopts a 3 dB coupler to couple the signal light into the multimode fiber and the local oscillator beam. The 3 dB coupler selected in the experiment is a multimode fiber and router produced by Shanghai Suyou Communication Devices Co., Ltd. (model MMC-1 × 2–850). The splitting ratio is 50%.

(4) **Photodetector**

A photodetector is a semiconductor device that converts optical signals into electrical signals. The performance of various types of photodetectors is determined by their quantum efficiency, frequency response and corresponding sensitivity. The fixed-gain silicon semiconductor detector of Thorlabs Company in the United States is selected, the model is PDA10A-EC, the detection laser wavelength range is 200–1100 nm, and the output voltage signal.

(5) **Data processing and signal demodulation**

The PIN detector converts the superimposed signal from the signal light and the local-oscillator light into an electrical signal through a photoelectric conversion, and the data are measured by a synchronous high-speed A/D acquisition card (PCI8002A4, Beijing Altay Technology Development Co., Ltd.). Its main characteristic parameters are shown in Table 9.4.

Table 9.4 Synchronous high-speed A/D acquisition-card characteristic parameters

12-bit AD sampling accuracy; sampling rate of each channel is 40 MHz/s Four single-ended synchronous analog inputs
AD 512 K RAM memory per channel
AD programmable measurement range: ±5 V, ±2.5 V, ±1 V, ±0.5 V
Supports analog trigger and digital trigger
One channel 12-bit, DA analog output
Eight digital inputs and outputs

9.4.1.3 One-Kilometer Atmospheric-Turbulence Channel

As shown in Fig. 9.20, the sending end of the experimental link is located in a teaching building of Xi'an University of Technology near the East Second Ring Road in Xi'an, China. It is on the eighth floor, about 30 m from the ground. The receiving end is located on the 16th floor of a building at the intersection of the South Second Ring Road and the East Second Ring Road. The 16th floor is about 60 m above the ground. From the satellite map, the straight-line distance between the two places is about 1.3 km. The positions of the sending and receiving ends of the link are not at the same level, which means this is a slope transmission.

The time is from September 2012 to February 2016, and experimental data from two typical weather conditions were selected for description, analysis, and comparison.

On January 23, 2013, the visibility was hazy, with a southwesterly wind less than level 3, and the temperature varied from −3 to 10 °C. The PM2.5 concentration was 299 $\mu g/m^3$ and the PM10 concentration was 199 $\mu g/m^3$.

On March 20, 2013, the weather was cloudy, the southeast wind was less than level 3, and the temperature varied from 10 to 24 °C. The PM2.5 concentration was 108 $\mu g/m^3$ and the PM10 concentration was 153 $\mu g/m^3$.

9.4.2 Analysis of the Results of Experiments in Different Weather Conditions

To study the performance of the NOHD system, experimental data from a turbulent atmospheric channel in cloudy and hazy weather are compared and analyzed below.

9.4.2.1 Measured Data and Analysis in Cloudy Weather

March 20, 2013 had cloudy weather. SCM-QPSK/NOHD and SCM-QPSK/direct-detection experiments were carried out at 10:00 pm and midnight. The communication channel was a 1.3-km turbulent atmospheric channel.

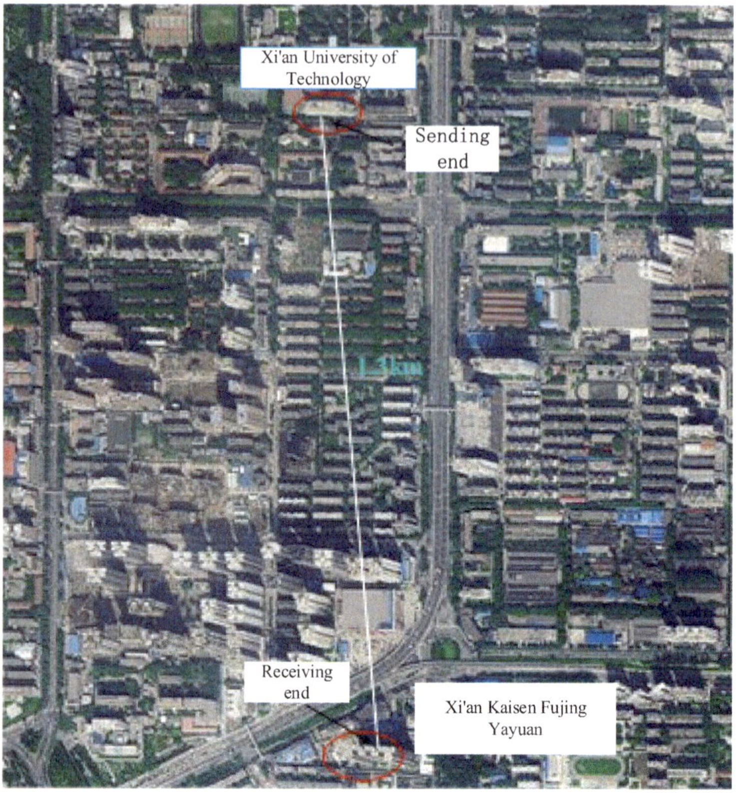

Fig. 9.20 Schematic diagram of the 1 km atmospheric-channel link

Figure 9.21 shows the output signal of the 3-dB coupler, where (a) and (b) show the data measured at 10:00 pm and midnight, respectively. Figure 9.21 shows that the signal-amplitude fluctuation from the signal light-intensity flickering at midnight is greater than that at 10:00 pm.

After squaring, the subcarrier beat-frequency signal obtained by a fifth-order Butterworth bandpass filter is shown in Fig. 9.22, in which (a) and (b) are obtained after data processing the difference-frequency signals at 10:00 pm and midnight, respectively. The subcarrier difference-frequency signal generated in the experiment is 1 MHz. According to the measured weather conditions, the visibility at midnight is lower than that at 10:00 pm; hence, the signal attenuation is more severe. The subcarrier difference-frequency signal obtained at midnight is weaker than that at 10:00 pm.

The *I*-channel signal, obtained by synchronous demodulation after the carrier recovery of the subcarrier difference-frequency signal, is filtered by a low-pass filter. The time-domain waveform of the signal before the decision is shown in Fig. 9.23, where (a) and (b) are the time-domain signal waveforms demodulated at 10:00 pm

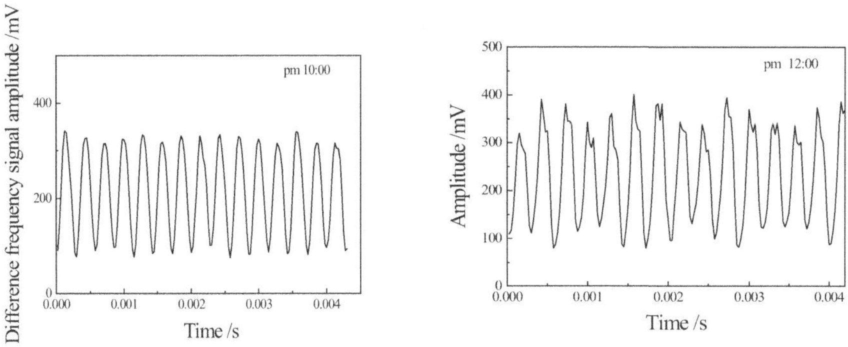

(a) 10:00 pm (b) Midnight

Fig. 9.21 Three-decibel-coupler output signal in cloudy weather [4–6]

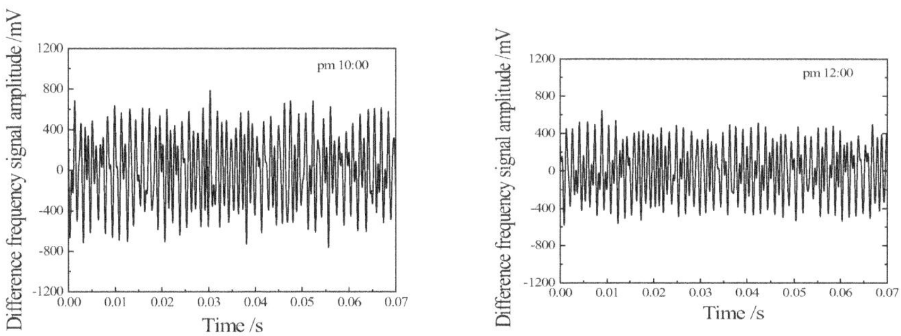

(a) 10:00 pm (b) Midnight

Fig. 9.22 Time-domain waveform of the NOHD difference-frequency signal in cloudy weather [4–6]

and midnight, respectively. The bit-error rate of the system at 10:00 pm is better than 10^{-4}, and the bit-error rate of the signal at midnight is better than 10^{-3}.

In the direct-detection system at 10:00 pm, the optical receiving antenna is coupled into the multimode fiber. The signal detected by the photoelectric detector is shown in Fig. 9.23. It can be clearly seen that the influence of atmospheric scintillation on the coupled optical power and the optical signal is very weak. The selection of the threshold has a great influence on the bit-error rate of the system.

In Fig. 9.23a, the equivalent power of the NOHD difference-frequency signal at 10:00 is 29 dBm, and the effective signal power of the direct-detection signal shown in Fig. 9.24 is 10.8 dBm. The NOHD gain obtained in the experiment is about 18.2 dB. In the same theoretical calculation, when $\alpha = 0.6$, $k_p = 0.7$, and $g = 10$, the NOHD gain $G = \alpha^2 k_p^4 g^4 P_l^2 R_L^2/2 = 18.39$ dB. It can be seen that the NOHD difference-frequency signal is nearly 20 dB stronger than the direct-detection signal.

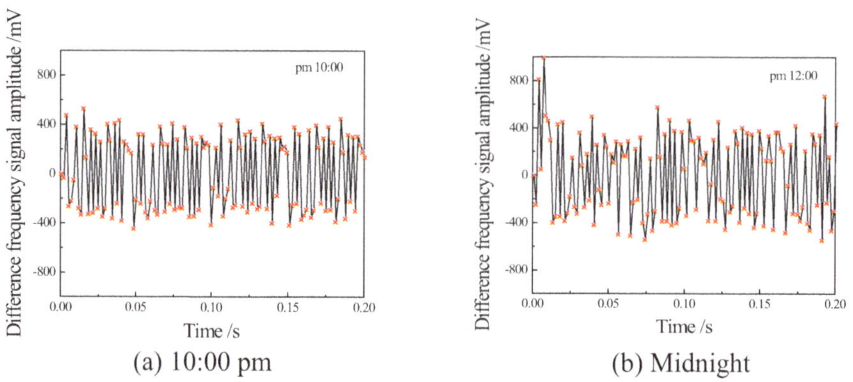

Fig. 9.23 Channel *I* signal after NOHD difference-frequency demodulation in cloudy weather [4–6]

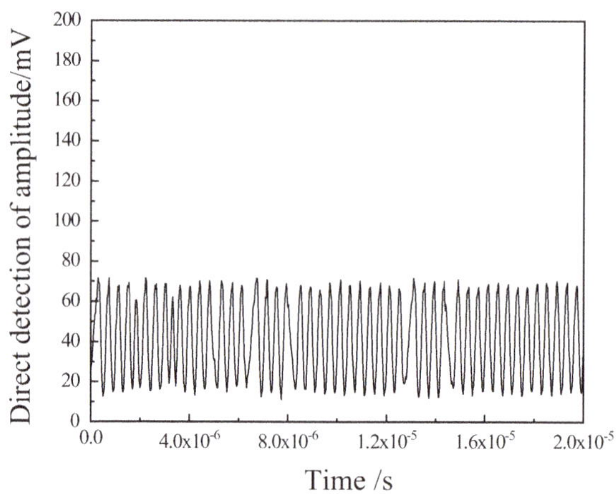

Fig. 9.24 Direct-detection waveform in cloudy weather

Figure 9.25 shows the NOHD and direct-detection power spectra of the measured signals at 10:00 pm. The energy of the SCM-QPSK modulated signal is mainly concentrated in the main lobe, the side-lobe power is significantly reduced, and the atmospheric-flicker noise is concentrated in the frequency band below 1 kHz [4]. Thus, the SCM-QPSK modulation is beneficial to the demodulation of the NOHD intermediate-frequency signal after the turbulent atmospheric channel.

When the measured received optical power is 11.7 dBm, the influence of atmospheric turbulence on the signal light is not obvious. The transmission eye diagram of the baseband signal after QPSK demodulation is shown in Fig. 9.26. The point divergence is ideal.

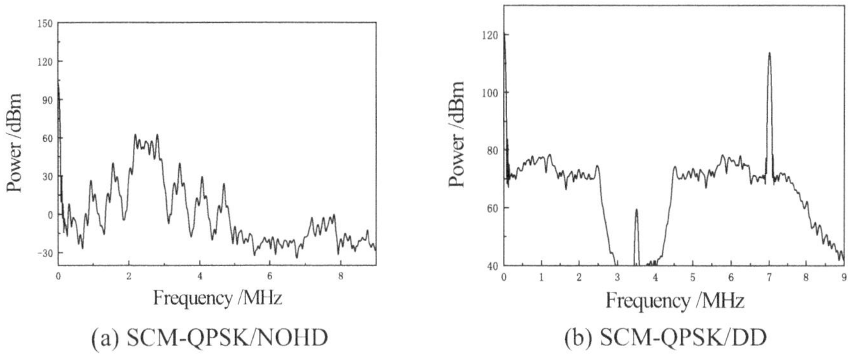

Fig. 9.25 Comparison of the measured difference-frequency signal power spectra in cloudy weather [5–7]

Fig. 9.26 Eye diagram of the baseband signal transmission in cloudy weather [5–7]

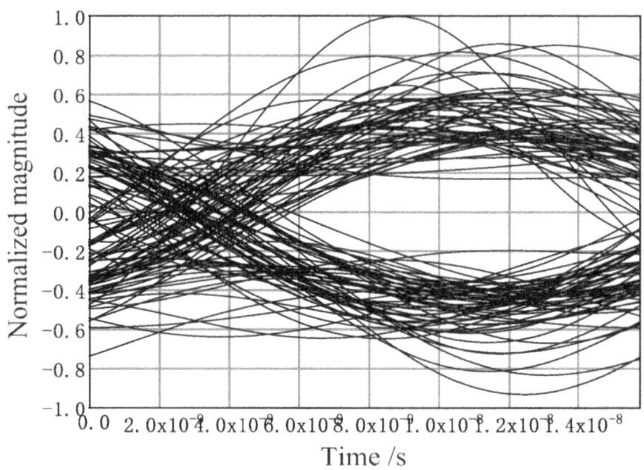

Figure 9.27 shows the demodulated signal constellation obtained by the carrier-phase estimation of the measured signal in cloudy weather. The constellation point rotates counterclockwise by $\Delta\theta$, indicating that atmospheric turbulence will cause phase noise; however, the distribution of each point is relatively concentrated and they are far from the judgment boundary. After filtering and carrier recovery processing, the system bit-error rate is 10^{-3}.

9.4.2.2 Measured Data and Analysis in Hazy Weather

On January 23, 2013, at 7:00 pm in hazy weather, the PM2.5 concentration was 299 μg/m^3, the PM10 concentration was 199 μg/m^3, the southwest wind was less than level 3, and the temperature was −3 to 10 °C. SCM-QPSK/NOHD and SCM-QPSK/direct-detection experiments were carried out. The communication channel was a 1.3 km turbulent atmospheric channel.

Fig. 9.27 Constellation diagram after carrier recovery in cloudy weather

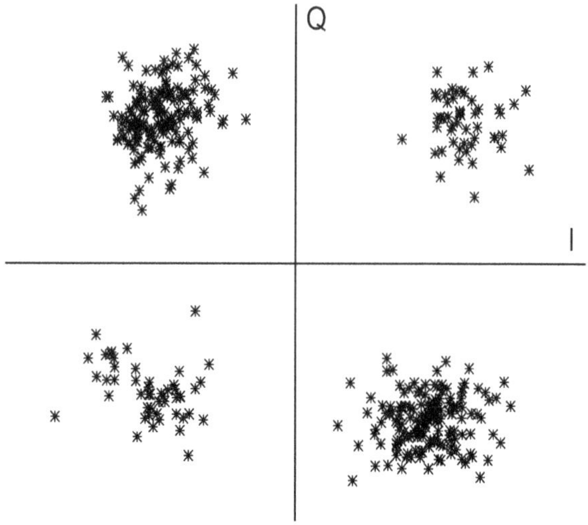

Figure 9.28 shows the output signal of the 3 dB coupler, and Fig. 9.29 shows the time-domain waveform of the direct-detection signal. Comparing the direct-detection data of hazy weather in Fig. 9.29 and the direct-detection data of cloudy weather in Fig. 9.25, it can be seen that the hazy weather has a significant influence on the signal; the attenuation increases and the flicker intensity increases.

Figure 9.30 shows a comparison of the power spectra of the NOHD difference-frequency signal and the direct-detection signal measured in hazy weather. In the power spectrum of the direct-detection signal in Fig. 9.30, the peak value of the subcarrier signal at 2.5 MHz is not obvious, indicating that the atmospheric environment in hazy weather greatly attenuates the signal light, and the atmospheric-channel noise is very obvious.

After NOHD detection, the 1 MHz frequency component of the difference-frequency signal is significantly amplified, indicating that NOHD can effectively suppress the influence of background and flicker noise on the signal light. For the NOHD method, because the photodetector can detect both the difference-frequency

Fig. 9.28 Output signal of the 3 dB coupler in hazy weather

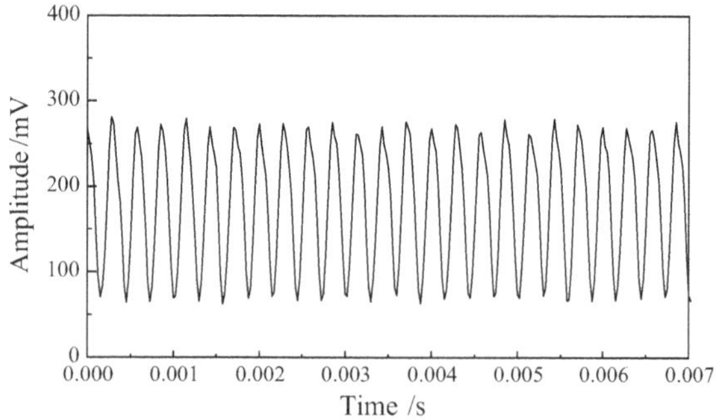

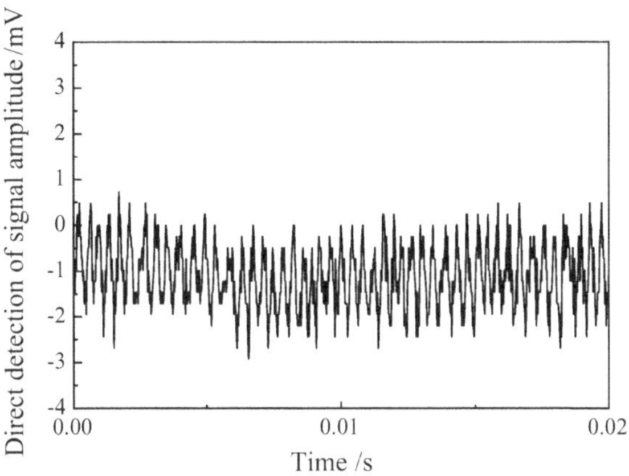

Fig. 9.29 Direct-detection signal in hazy weather

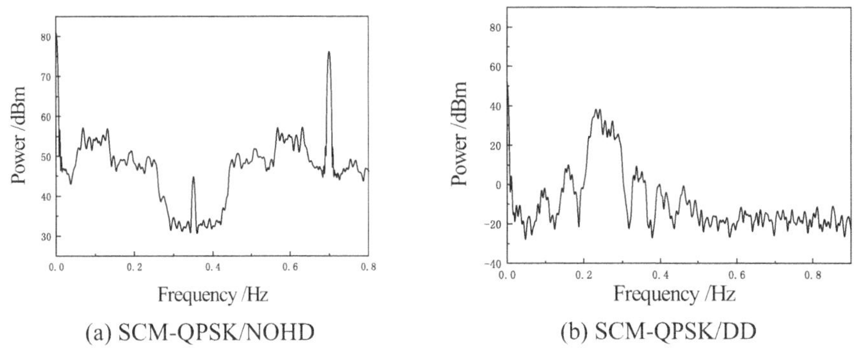

Fig. 9.30 Comparison of measured-signal power spectra in hazy weather [5–7]

term and the sum-frequency term, and the flicker noise is mainly concentrated in the low-frequency band, the sum-frequency term can also be selected during signal demodulation, thereby reducing the effect of flicker noise on the signal.

Figure 9.31 shows the I-channel signal obtained by demodulation after the carrier recovery of the sub-carrier difference frequency signal in haze weather. The time domain waveform of the signal before the judgment is filtered by the low-pass filter. It can be seen that the signal amplitude increases and the noise affects Not serious, easy to judge, the system bit error rate is about 10.3. Figure 9.32 shows the I-channel signal after demodulation of the direct detection signal. It can be seen from Fig. 9.32 that the signal amplitude is very small, it is difficult to demodulate the baseband signal correctly, and the bit error rate is higher than that of NOHD.

Figure 9.33 shows the eye diagram of channel *I* when the measured received optical power is 7 dBm. The "eye" of the eye diagram gradually becomes smaller

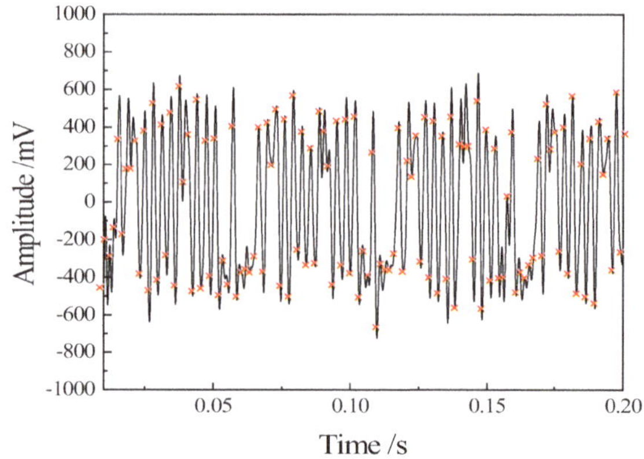

Fig. 9.31 Channel *I* signal after NOHD difference-frequency demodulation in hazy weather

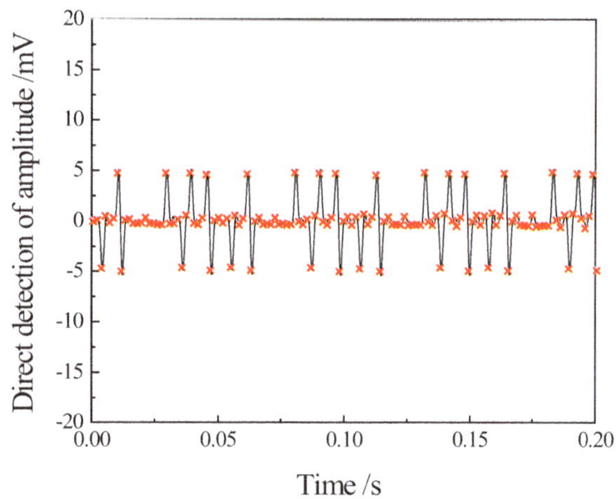

Fig. 9.32 Channel *I* signal after demodulation of the direct-detection signal in hazy weather

with the increase of the turbulence intensity. The intersymbol interference was large once, and the signal-to-noise ratio decreased.

The visibility of the atmospheric environment is determined by fog and haze in the atmospheric channel. The lower the visibility, the more serious the scattering and attenuation [1]. Figure 9.34 shows the constellation diagram of the actual measured signal after the carrier phase was estimated in hazy weather. The phase noise caused by atmospheric turbulence is relatively serious. Experiments showed that in hazy weather, when the SCM-QPSK modulated signal was transmitted for about 1.3 km, the bit-error rate of the direct-detection method was 10^{-1}, which is one to two orders of magnitude higher than that of the SCM-QPSK/NOHD method.

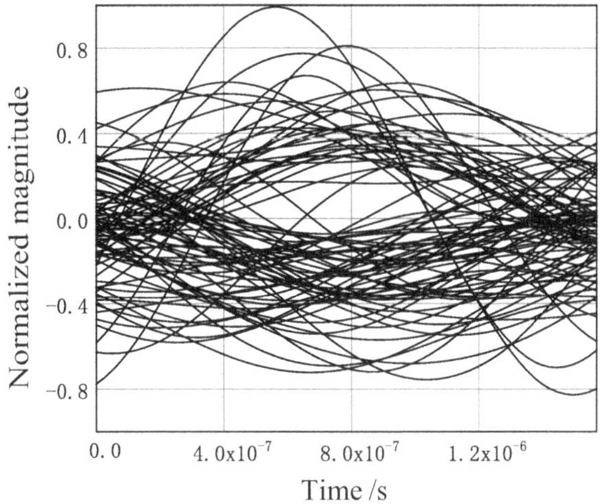

Fig. 9.33 Eye diagram of the baseband-signal transmission in hazy weather [5–7]

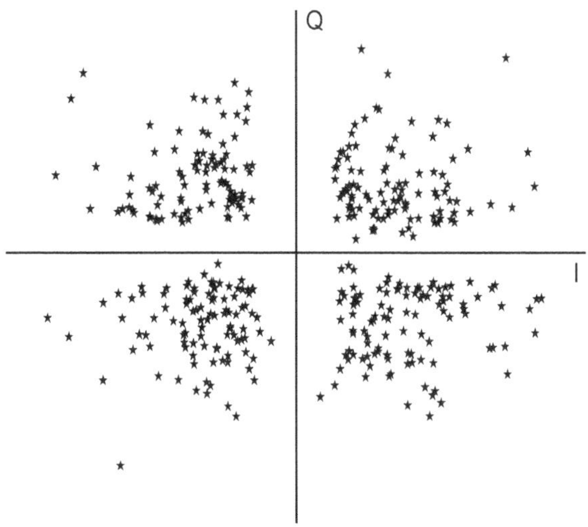

Fig. 9.34 Constellation diagram after carrier-phase recovery in hazy weather

9.4.3 Analysis of Experimental Results of Non-heterodyne Detection with Different Modulation Methods

To investigate the performance of the NOHD system under different subcarrier-modulation methods, experimental data were collected from a turbulent atmospheric channel at 2:00 pm and 6:00 pm on January 22, 2016, and the experimental weather was sunny and breezy.

Figure 9.35 shows the power spectrum of the NOHD received difference-frequency signal under different subcarrier-modulation modes. Figure 9.35a, b show the NOHD beat-frequency signal power spectrum and the direct-detection signal power spectrum, respectively, during SCM-amplitude-shift keying (ASK) modulation. It can be seen that the signal power spectrum curve is relatively stable when the beat-frequency signal frequency is 1.0424 MHz. The main lobe has a width of 0.5 MHz. The spectral line at the carrier frequency is not affected by atmospheric turbulence, and the power of the difference-frequency signal is 18.1 dB higher than that of the direct-detection signal.

Figure 9.35c, d show the NOHD difference-frequency signal power spectrum and the direct-detection signal power spectrum, respectively, during SCM-phase-shift keying PSK modulation. The subcarrier frequency in the direct-detection signal power spectrum is 2.5 MHz, and the main lobe width is 0.5 MHz. Noise spectral lines appear in the power spectrum of the direct-detection signal; however, the noise has little effect on the signal on sunny days, and the power of the difference-frequency signal is 13.4 dB higher than that of the direct-detection signal.

Figure 9.35e, f show the NOHD beat-frequency signal power spectrum and the direct-detection signal power spectrum, respectively, during SCM-QPSK modulation. It can be seen that the signal power-spectrum curve is relatively stable, the beat-frequency signal frequency is 1.0424 MHz, the main-lobe width is 0.425 MHz, and the difference-frequency signal power is 15.6 dB larger than the direct-detection signal power.

Figure 9.35g, h show the power spectra of the NOHD beat-frequency signal and the direct-detection signal when SCM-QPSK is modulated, respectively. The frequency of the beat-frequency signal is 1.0424 MHz, the main-lobe width is 0.45 MHz, and the power of the beat-frequency signal is 16.4dB larger than that of the direct-detection signal.

When SCM-ASK is modulated, the eye diagrams of the NOHD receiving IF signal and the direct-detection signal are shown in Fig. 9.36a, b, respectively. It can be seen that on sunny days, atmospheric turbulence has little effect on the received signal when NOHD receives it. When amplified, the signal decision threshold is 10 mV, and that of the direct detection is only 2 mV. Thus, the decision threshold is too low, and it can easily be affected and cause bit errors. When NOHD is used for reception, the error-reduction performance of SCM-ASK and SCM-PSK modulation is better than that of SCM-QPSK modulation and SCM-16PSK modulation.

Figure 9.37 shows the constellation diagrams of the actual measured signal after carrier phase estimation. Figure 9.37a, b show the demodulated signal-constellation diagrams of NOHD and direct detection, respectively, during SCM-QPSK modulation. The NOHD demodulated-signal constellation points have good convergence, and the bit-error rate is 10^{-5}. The direct-detection signal constellation points diverge, which means they are severely affected by the atmospheric channel, and the bit-error rate is on the order of 10^{-3}.

Figure 9.37c, d show the constellation diagrams of the SCM-16PSK/NOHD mode and SCM-16PSK/direct-detection demodulation signals, respectively. The bit-error rate of NOHD mode is 7×10^{-3}, and the receiving performance is worse than that of SCM-QPSK. The transmit power can be increased to reduce the bit-error rate.

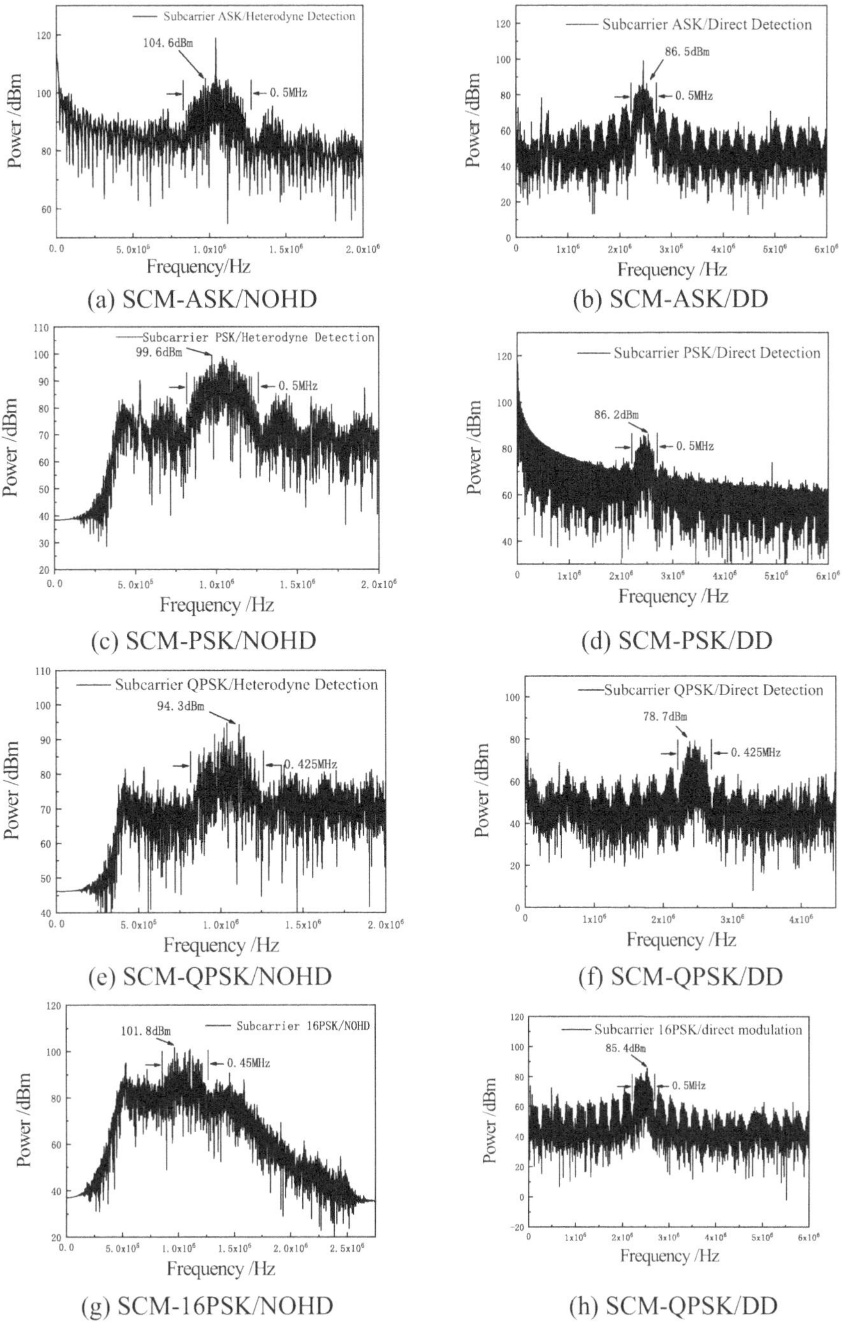

(a) SCM-ASK/NOHD (b) SCM-ASK/DD

(c) SCM-PSK/NOHD (d) SCM-PSK/DD

(e) SCM-QPSK/NOHD (f) SCM-QPSK/DD

(g) SCM-16PSK/NOHD (h) SCM-QPSK/DD

Fig. 9.35 Power spectra of received signals with different modulation methods [5–7]

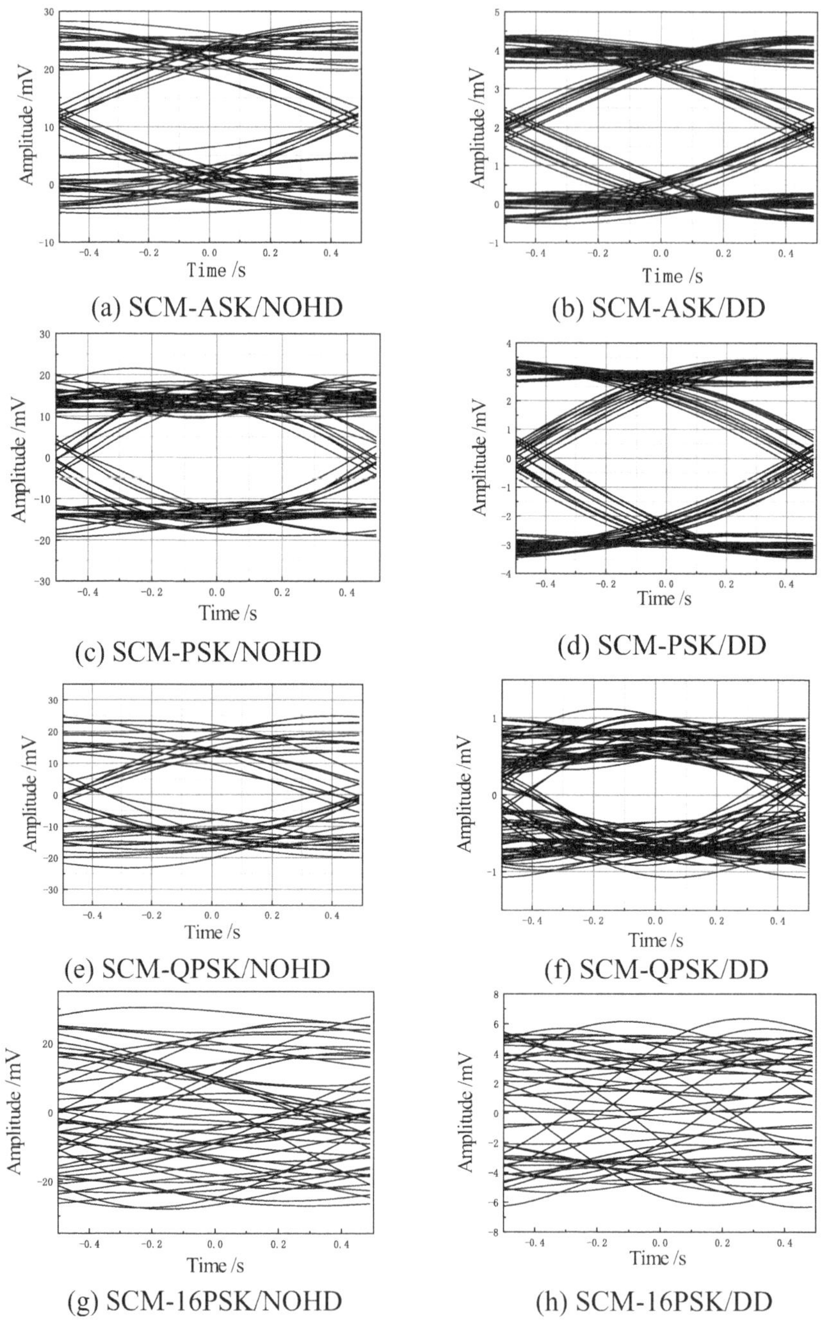

Fig. 9.36 Eye diagram of the NOHD-receiving and direct-detection signals in a turbulent channel [5–7]

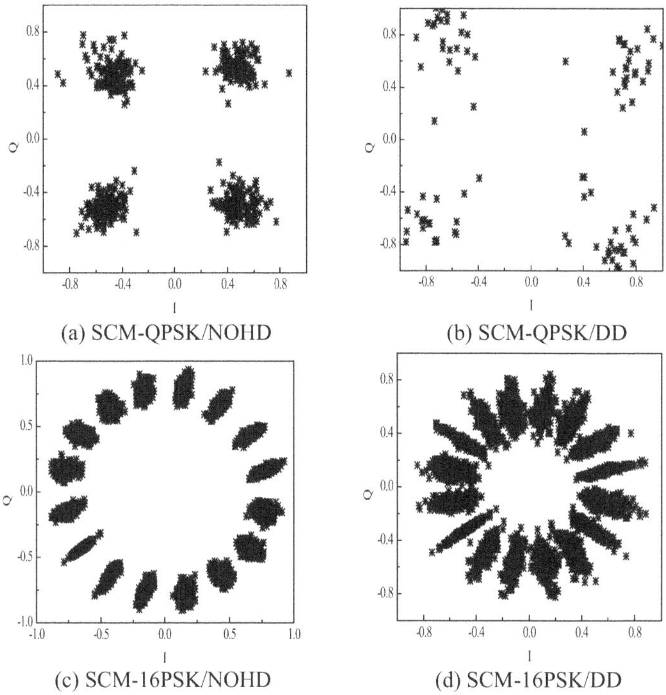

Fig. 9.37 Turbulent channel NOHD and direct-detection demodulation signal constellations [5–7]

9.4.4 SCM-OFDM Experimental Results and Analysis

At the receiving end of the experiment, the optical receiving antenna focuses the signal light onto the fiber end face and couples it into the multimode fiber. The signal generator generates a sinusoidal local-oscillator subcarrier signal to drive and modulate the local-oscillator analog semiconductor laser. The 3 dB coupler combines the signal light in the multimode fiber with the multimode fiber. The optical coupling of the local oscillator is modulated, and the photodetector detects the coupled light spot and generates a photocurrent. The signal is collected by the A/D acquisition card and processed offline by a PC.

The offline processing of the PC includes signal-square processing, synchronous demodulation of the difference-frequency signal after passing through the band-pass filter, carrier-frequency tracking, phase synchronization of the intermediate-frequency signal, and OFDM signal demodulation. In the experiment, a local-oscillator signal generator was used to generate a local-oscillator sinusoidal subcarrier to modulate the local-oscillator laser, and the frequency of the local-oscillator subcarrier was selected as 3.5 MHz.

Figure 9.38 compares the power spectra of the SCM-OFDM NOHD received beat-frequency signal and the direct-detection signal. Figure 9.38a, b show the NOHD

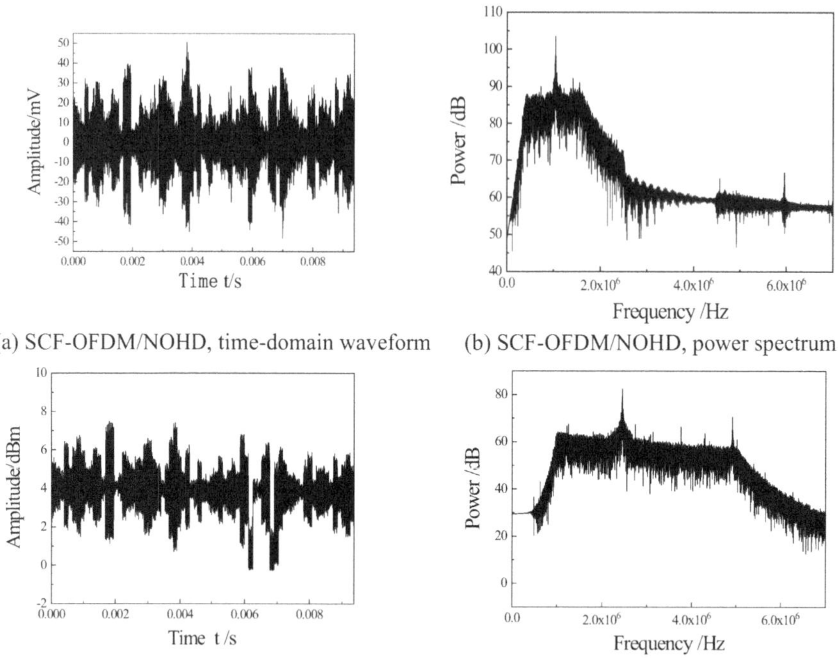

(a) SCF-OFDM/NOHD, time-domain waveform (b) SCF-OFDM/NOHD, power spectrum

(c) SCF-OFDM/DD, time-domain waveform (d) SCF-OFDM/DD, power spectrum

Fig. 9.38 Comparison of the time-domain waveforms and power spectra of the subcarrier difference-frequency and direct-detection signals [5–7]

difference-frequency signal and its power spectrum, respectively, and Fig. 9.38c, d show the direct-detection signal and its power spectrum, respectively. Comparing Fig. 9.38a, c, we can see that the time-domain waveform of the NOHD difference-frequency signal is less affected by noise. Comparing Fig. 9.38b, d, it can be seen that the heterodyne gain of NOHD mode is 17.2 dB higher than that of direct detection.

Figure 9.39 shows the constellation diagrams of the measured NOHD difference-frequency signal after carrier phase estimation. Figure 9.39a, b show the NOHD demodulated signal constellations of the SCM-OFDM optical signal at 10:00 pm and midnight, respectively. It can be seen from Fig. 9.39a that the NOHD-demodulation signal constellation points have good convergence at 10:00 pm, and the bit-error rate is 2×10^{-4}. It can be seen from Fig. 9.39b that the constellation points diverged, owing to the influence of the atmospheric channel, at midnight, and the bit-error rate increased by an order of magnitude.

The SCM/NOHD field-transmission experiment was carried out over a 1 km turbulent atmospheric link, and the maximum transmission rate of this experimental system reached 20 Mbit/s. The experiment tested the NOHD received data in cloudy weather and hazy weather. The experimental results showed that the heterodyne gain

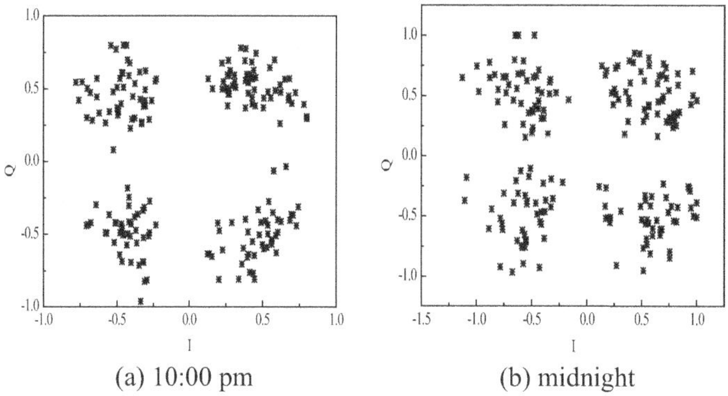

Fig. 9.39 Comparison of constellation diagrams of NOHD demodulated signals at different times [5–7]

of the NOHD experimental system was about 20 dB higher than that of direct detection. Without channel coding, after carrier recovery and synchronous demodulation, the bit-error rate was one to two orders of magnitude lower than direct detection.

When NOHD was used to receive optical signals of different subcarrier-modulation modes, the SCM-QPSK/NOHD mode demodulated the signal-constellation point convergence better, and the bit-error rate was 10^{-5}. When NOHD was used for reception, the error-reduction performance of SCM-ASK and SCM-PSK modulation was better than that of SCM-QPSK modulation and SCM-16PSK modulation. The SCM-16PSK/NOHD mode had a bit-error rate of 7×10^{-3}, and the receiving performance was worse than that of the SCM-QPSK/NOHD mode. The communication quality can be improved by increasing the transmit optical power.

The NOHD receiving system of the SCM-OFDM signal had good performance under weak turbulence. The NOHD demodulated signal constellation points had good convergence at 10:00 pm, and the bit-error rate was 2×10^{-4}. At midnight, the bit-error rate increased by an order of magnitude.

Chapter Summary

The non-optical heterodyne-detection (NOHD) method was used to detect SCM-OFDM optical signals, the basic principle of the SCM-OFDM/NOHD system was introduced, and its mathematical model was given. The closed expressions of the average carrier-to-noise ratio and the average bit-error rate of the subcarriers were derived. The numerical analysis showed that when the subcarrier-modulation coefficient was >0.2%, the average carrier-to-noise ratio obtained by subcarrier heterodyning was obviously better than that of direct detection. The highest was increased by about 20 dB.

When the received optical power was on the order of microwatts, the optimal average carrier-to-noise ratio could be obtained by selecting the local-oscillator optical power of about −11 dBm, and the average carrier-to-noise ratio gradually decreased as the local-oscillator optical power increased.

The subcarrier difference-frequency signal and its power spectrum were obtained by experimental simulation, and the constellation diagram and bit-error rate under different turbulence intensities were measured. The experimental results were in good agreement with the theoretical results, which fully shows that the NOHD method is feasible for detecting SCM-OFDM optical signals. It can improve the bit-error rate and detection sensitivity of the system.

Through theoretical analysis and simulation experiments, it was concluded that nonlinear modulation reduced the signal-to-noise ratio, increased the bit-error rate, and scattered and rotated the constellation diagram at a certain angle. This should be compensated for, if necessary. The orthogonality of the subcarriers caused by the frequency offset was reduced.

When the normalized frequency offset was 0.002, the BER curve was similar to that without a frequency offset, and the frequency offset could be ignored during demodulation at the receiving end. However, when the normalized frequency offset increased to 0.004, the system bit-error rate increased significantly, and it was necessary to compensate the frequency offset to reduce the system bit-error rate.

References

1. Xu G, Chen Z, Guo R (2013) Photoelectric detection technology and system design. National Defense Industry Press, Beijing, pp 111–120
2. Ishimaru A (1999) Wave propagation and scattering in random media. Wiley-IEEE Press
3. Buske I, Riede W (2009) A mobile adaptive optics system for compensation of atmospheric turbulence-induced phase distortions. In: CLEO Europe—EQEC 2009. European, p 1
4. Chen C, Yang H, Jiang H et al (2009) Research progress of atmospheric turbulence influence suppression technology in atmospheric optical communication. J Military Eng 30(6):779–791
5. Chen J. Key technology and experimental research of subcarrier modulation non-optical heterodyne detection optical wireless communication. Xi'an University of Technology, Xi'an
6. Chen J, Ke X (2016) Research on bit error performance of subcarrier-orthogonal frequency division multiplexing system based on subcarrier heterodyne detection. J Opt 02:32–40
7. Ke X, Chen J (2014) Experimental study on non-optical heterodyne detection of 1 km atmospheric laser communication system. Chin J Appl Sci 04:379–384

Chapter 10
Multi-band Carrier-Free Amplitude-Phase Modulation

In order to further improve the communication capacity of a system, the multi-band carrier-free amplitude-phase modulation method, based on carrier-free amplitude-phase (CAP) modulation technology, has been proposed. As an extension of orthogonal frequency-division multiplexing (OFDM) (although they are not necessarily related), this chapter will mainly study the application of CAP modulation and multi-band CAP modulation in optical wireless communication systems.

10.1 Introduction

10.1.1 Research Background and Significance

At present, three commonly used modulation methods are pulse-amplitude modulation (PAM), direct-detection orthogonal frequency-division multiplexing (DD-OFDM), and carrier-free amplitude-phase (CAP) modulation [1, 2]. CAP modulation is based on quadrature amplitude modulation (QAM), although it is carrier-free. Compared with PAM, DD-OFDM and CAP modulation can greatly improve the system flexibility and spectrum-utilization rate without increasing the speed requirement of the digital signal processor (DSP) [3].

Compared with DD-OFDM, CAP modulation does not require fast Fourier transforms (FFTs) or inverse fast Fourier transforms (IFFTs) in the modulation process; hence, its computational complexity and power loss are low [4]. Therefore, CAP modulation is a good choice for optical wireless communication systems. This chapter will introduce the application of CAP and multi-band CAP modulation to atmospheric laser-communication systems.

X. Ke, *Principles and Applications of Optical Wireless Orthogonal Frequency-Division Multiplexing*, Optical Wireless Communication Theory and Technology,
https://doi.org/10.1007/978-981-97-7973-4_10

10.1.2 Research Progress at Home and Abroad

10.1.2.1 Research on CAP Modulation Technology at Home and Abroad

AT&T Bell Laboratories proposed CAP modulation [5], based on QAM, in the mid-1970s, as a multi-level and multi-dimensional modulation technology. In 1997, Shalash et al. [6] simulated three-dimensional (3D)-CAP modulation technology using unshielded twisted-pair (UTP) copper-cable transmissions, and compared a 2D-CAP modulation system with the 3D-CAP modulation system in terms of communication capacity and system complexity. They concluded that the 3D-CAP modulation technology increased the system communication capacity.

In 1997, Zhou et al. [7] compared CAP modulation technology with QAM technology, and applied 16CAP modulation technology in a local area network (LAN) system composed of twisted-pair copper wires. The transmission rate was 51.84 Mbit/s. In 1998, Saltzberg explained the relationship between CAP modulation and QAM—that is, CAP modulation is carrier-free QAM [8]—and compared the CAP modulation system with a discrete multi-tone (DMT) modulation system in terms of system performance and cost.

In 1999, Shalash et al. [9] introduced the principle of K-dimensional CAP modulation and demodulation, and simulated digital subscriber-line (DSL) systems with 3D-CAP and 2D-CAP modulation. They found that the communication capacity of the 3D-CAP modulation system increased by 50%, compared with the 2D-CAP modulation system; however, the corresponding output signal-to-noise ratio decreased.

In 1999, Gao et al. [10] used the least mean squares (LMS) equalization algorithm in a CAP modulation system for equalizing the signal at the receiving end, and improved the algorithm on the premise of LMS equalization. The simulation results of the improved LMS algorithm showed that it had a faster convergence speed and could recover the original sequence at the receiving end of the system without distortion.

In 2000, Gao et al. introduced the principle of minimum mean-squared error (MMSE), and designed a shaping/matching filter in a CAP modulation system using this method [11]. In 2001, Gao et al. [12] explained the principle of CAP modulation and demodulation, and simulated and analyzed several methods of designing CAP modulation and demodulation digital filters, such as the window function, MMSE, weighted least squares (WLS), etc. It was concluded that when the taps of the designed digital filter were small, the MMSE algorithm was better.

In 2002, Luo et al. [13] compared CAP modulation technology with DMT modulation technology, and concluded that although DMT modulation has the advantages of better bandwidth utilization and a stronger ability to resist narrow-band interference, the CAP modulation system has lower power consumption and a simpler implementation; hence, CAP modulation is also good for asymmetric digital subscriber-line (ADSL) systems.

In 2010, Ingham et al. adopted CAP modulation technology in an optical digital communication-link system with a low cost and high spectrum utilization. They simulated and analyzed the signal eye diagram and the power loss of the system. A transmission rate of 10 Gbit/s was realized in a 15 km optical-fiber transmission [14].

In 2011, Wieckowski et al. [15] studied the application of multilevel CAP (16CAP and 64CAP) modulation technology in commercial fast Ethernet, and the transmission rate reached 300 Mbit/s. In 2011, Ingham et al. [16] studied an optical data-communication system with CAP modulation. Under different optical-fiber transmission distances, the power losses of a CAP modulation system and a non-return to zero (NRZ) modulation system were simulated and compared, and the transmission rate reached 40 Gbit/s.

In 2012, Wei et al. [17] discussed PAM, CAP, OFDM, and other modulation modes, using different optical-fiber transmission modes (single-mode optical fiber (SMF) and multimode optical fiber (MMF)) and different communication distances. The power losses of various modulation systems were analyzed, and the transmission rate was 28 Gbit/s.

In 2012, Geng et al. conducted an experimental study on a 64CAP modulated home-network system, and analyzed the influence of the received system power on the transmission performance. The transmission sequence was generated based on a field-programmable gate array (FPGA), and the transmission medium was 50 m of plastic fiber, with a transmission rate of 1.5 Gbit/s [18].

In 2013, Stepniak et al. [19] conducted experimental research on indoor optical network systems with different levels of CAP modulation (4CAP, 16CAP, and 64CAP). At the receiving end of the system, two equalization methods were adopted, namely, a feed-forward equalizer (FFE) and a decision-feedback equalizer (DFE), and the effects of these equalization methods on the system bit-error rate and signal constellation were compared and analyzed. The results indicated that the bit-error rate of the system was lower after DFE equalization, in which the transmission medium was polymeric fiber, and the transmission rate was 2.1 Gbit/s.

In 2013, Tao et al. [20] used a cascaded multi-modular algorithm (CMMA) to equalize the signals in an optical-fiber communication system modulated by CAP. In 2013, Wang et al. [21] adopted the decision-directed least mean squares (DD-LMS) algorithm for subsequent equalization in a short-distance 32CAP modulated single-mode optical-fiber communication system. They simulated and analyzed the influence of equalizer taps on the bit-error rate of the system under different equalizer step sizes. The transmission rate was 40 Gbit/s.

In 2013, Wei et al. [22] simulated the link power budget and power loss of a 16/64CAP modulated Ethernet system and a 16/64QAM-OFDM system. The transmission medium was single-mode fiber with a transmission rate of 100 Gbit/s. In 2013, Penty et al. [23] reduced the sensitivity to time jitter of a traditional CAP modulation system by using CAP/QAM transmission/reception technology in an Ethernet system, and analyzed the signal eye diagram after CAP and QAM demodulation. The simulation showed that the signal eye diagram after QAM receiver demodulation was better, and the communication rate was 100 Gbit/s.

In 2014, Zhang et al. simulated and experimentally verified a CAP-64QAM fiber radio-frequency system, in which the communication distance was 40 km and the communication rate was 24 Gbit/s [24]. In 2014, Estaran et al. conducted an experimental study on a CAP-modulated optical network communication system, and analyzed the influence of the optical transmission power on the bit-error rate of the system. The transmission distance was 451 km and the transmission rate was 300 Gbit/s [25].

In 2014, Tao et al. conducted an experimental study on a short-distance optical communication system with 128/256CAP modulation, and simulated and analyzed the effects of the multi-modulus algorithm (MMA) and modified MMA on the system performance. The transmission distance was 40 km and the communication rate was 10 Gbit/s [26].

In 2014, Tao et al. [27] introduced three modulation modes commonly used in short-distance optical-communication systems, namely, CAP, PAM, and OFDM, and compared and analyzed their advantages and disadvantages. In 2014, Stepniak [28] compared the performance of CAP modulation systems with different dimensions, and introduced the design of a modulation filter and equalizer in 3D-CAP modulation.

In 2015, Kelly et al. [29] conducted an experimental study on a full-duplex indoor optical-communication system with 16CAP modulation, in which the uplink communication rate of the system was 2 Gbit/s and the downlink communication rate was 10 Gbit/s. In 2015, Lin et al. [30] studied a multi-dimensional CAP-modulated visible-light communication system, and simulated and analyzed the influence of a sampling-clock offset at the receiving end on the system bit-error rate. The results showed that the larger the clock offset, the higher the bit-error rate.

In 2016, Ochi et al. [31] studied an Ethernet system with 64CAP modulation, and simulated and analyzed the influence of the sampling frequency and resolution of the analog-to-digital (A/D) conversion in the modulation system on the power loss of the system. The results showed that the higher the sampling frequency and resolution of the A/D and D/A conversions, the smaller the power loss of the system. In 2016, Jaafar et al. [32] studied high-dimensional CAP modulation technology, and simulated and analyzed the time-domain characteristics of shaping and matched filters in a 3D-/4D-CAP modulation system.

10.1.2.2 Domestic and Foreign Research on Multi-Band CAP Modulation Technology

In 2013, Olmedo et al. [33] conducted experimental research on a multi-band CAP-modulation data-communication system. The experimental results showed that all signals at the receiving end had a certain phase distortion, and the phase distortion of different sub-channels was different. The influence of the optical receiving power on the performance of the communication system was analyzed using simulation. They found that the higher the receiving power was, the lower the system bit-error rate was. The optical-fiber transmission distance was 15 km and the total transmission rate was 102 Gbit/s.

In 2013, Zhang et al. [34] proposed and experimentally verified a wavelength-division multiplexing passive optical network (WDM-PON) system that supported 11 channels. The channel spacing was 28 GHz, and the downlink rate of a single channel was 10 Gbit/s. Multi-band CAP-modulation technology (five bands) was applied. The optical-fiber transmission distance was 40 km, and the total downlink rate was 102Gb/s.

In 2014, Olmedo et al. [35] introduced the principle of multi-band CAP modulation and demodulation, and experimentally studied an optical data-communication system with multi-band CAP modulation. They simulated and analyzed the effects of the received optical power, transmission bandwidth, dispersion, etc., on the performance of different band CAP modulation systems. The optical-fiber transmission distance was 15 km and the transmission rate was 100 Gbit/s.

In 2015, Wang et al. [36] applied multi-band CAP modulation technology in a polarization-division multiplexing optical communication system. They simulated and analyzed the influence of the received optical power on the system performance, and experimentally verified it. The transmission distance was 15 km and the transmission rate was 56 Gbit/s.

In 2015, Wei et al. [37] conducted a simulation study on a multi-band CAP modulation/ QAM demodulation Ethernet system. The results showed that in the CAP modulation and demodulation system, the larger the sampling-clock offset at the receiving end, the more serious the phase distortion of the demodulated signal. In the CAP modulation/ QAM demodulation system, the sampling-clock jitter did not distort the phase of the signal; that is, using QAM demodulation technology could reduce the sensitivity of the CAP modulation system to time jitter; however, it correspondingly increased the system complexity. The optical-fiber transmission distance was 15 km and the transmission rate was 100 Gbit/s.

In 2015, Haigh et al. [38] of the University of London simulated and experimentally verified a multi-band CAP modulation visible-light communication system, and compared the performance of different frequency-band CAP modulation systems. It was concluded that the larger the frequency band, the lower the system bit-error rate, and the better the communication performance. The signal light source was a light-emitting diode (LED) and the transmission distance was 1 m.

In 2015, Wei et al. [39] studied a four-band CAP modulation/ QAM demodulation Ethernet system, simulated and analyzed the changes of the signal eye diagram before and after equalization at the receiving end, and compared and analyzed the power loss of the system under different transmission distances. The transmission rate was 400 Gbit/s.

In 2015, Son et al. [40] of Aston University in England studied the application of multi-band CAP modulation technology in a visible-light transmission system with limited bandwidth. Under different band CAP modulation conditions, the influence of the signal-to-noise ratio on the bit-error rate of each sub-band signal at the receiving end was simulated and analyzed.

In 2015, Werfli et al. [41] of Northumbria University simulated a visible-light communication system with different frequency bands (band number $n = 1, 2, 5,$

and 10), in which a decision-feedback equalizer was used at the receiver to compensate the signal, and the equalization effect was good. In 2015, Dong et al. [42] of Huaqiao University studied a passive optical-network system with multi-band CAP modulation, and discussed the influence of the optical received power on the system performance. The system communication rate was 40 Gbit/s.

In 2016, Puerta et al. [43] of the Danish University of Science and Technology simulated and experimentally verified a multi-band CAP-modulated ultra-wideband radio-communication system. The transmission distances were 2 m and 3.5 m and the transmission rate was 10 Gbit/s.

10.2 Research on an Atmospheric Laser-Communication System Based on CAP Modulation

Because CAP modulation has many advantages, such as high spectrum utilization, low cost, and simple operation, it has great application value for optical wireless communication systems. This section discusses atmospheric laser-communication systems modulated by CAP.

10.2.1 CAP Modulation and Demodulation Principle

10.2.1.1 Origin of CAP Modulation

AT&T Bell Laboratories proposed CAP modulation, that is, a variation of CAP modulation in QAM [16], in the mid-1970s. CAP modulation and QAM are both single-carrier modulation methods. CAP modulation is a multi-dimensional and multi-level modulation technology.

In the CAP-modulation process, the bandpass pulse is directly formed using a shaping filter instead of a multiplier. The transmitted data stream is reflected by changing the in-phase and quadrature waveforms; that is, the signal is modulated and demodulated using a shaping filter at the transmitter and a matched filter at the receiver. Therefore, it is called carrier-free amplitude-phase modulation.

Compared with QAM, CAP modulation is simpler and less expensive because it does not require a multiplier to modulate the signal at the transmitter or a carrier-recovery mechanism to demodulate the signal at the receiver. However, there is no difference between CAP modulation and QAM in terms of spectrum utilization and transmission performance, except that the carrier component is missing [14]. This carrier-free feature also makes CAP modulation more flexible when changing the spectrum [7]. Figure 10.1 shows the non-normalized power spectrum of the same input signal when modulated by 16QAM and 16CAP.

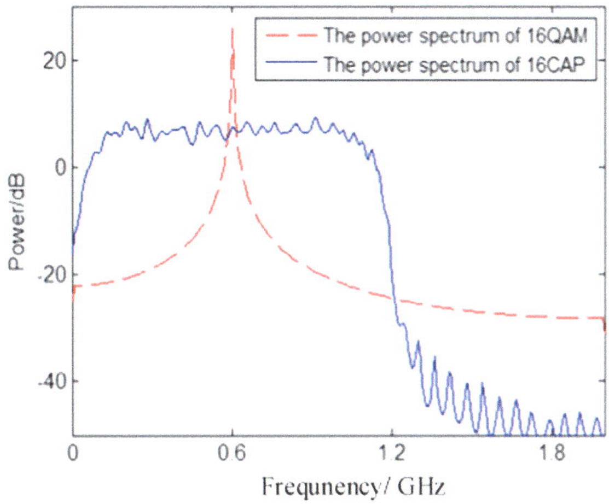

Fig. 10.1 Power spectra of 16QAM and 16CAP signals

The power spectrum of the modulated signal in Fig. 10.1 is estimated using the Burg method, in which the carrier frequency in 16QAM is 0.6 GHz, and the center frequency of the digital filter in 16CAP modulation is 0.6 GHz. CAP modulation is a carrier-free QAM. Because the carrier part does not carry useful information and consumes a certain amount of power, the power-utilization rate of CAP modulation is higher than that of QAM.

As can be seen from Fig. 10.1, the power of the CAP modulation signal is evenly distributed over the entire signal bandwidth, while the power of the QAM signal is mainly concentrated at the carrier frequency. For example, at the frequency of 0.6 GHz, the power of the QAM signal is about 15 dB higher than that of the CAP modulation signal, while at the non-carrier frequencies, the power of the QAM signal is about 30 dB lower than that of the CAP modulation signal.

10.2.1.2 CAP Modulation and Demodulation Principle

As can be seen from Fig. 10.2, at the sender, the input random sequence is first mapped, that is, encoded, by a high order, and then a multilevel symbol sequence is formed. To match the high sampling frequency of the shaping filter (the sampling frequency of the filter is M/T, where the sampling order is M, and the symbol period T), it is necessary to upsample the generated complex multilevel symbol sequence. Then, after up-sampling, zero values are inserted between two consecutive symbols.

Then, the signal is I/Q separated by taking the real part of the complex sequence up-sampled as the in-phase component, that is, the I component, and taking the imaginary part as the orthogonal component, that is, the Q component. Then, the I and Q components of the signal are modulated by an I-shaping filter and a Q-shaping filter, respectively (see Fig. 10.2a, b) for the time-domain waveforms, respectively). Finally, the outputs of these two filters are subtracted, and the analog signal output

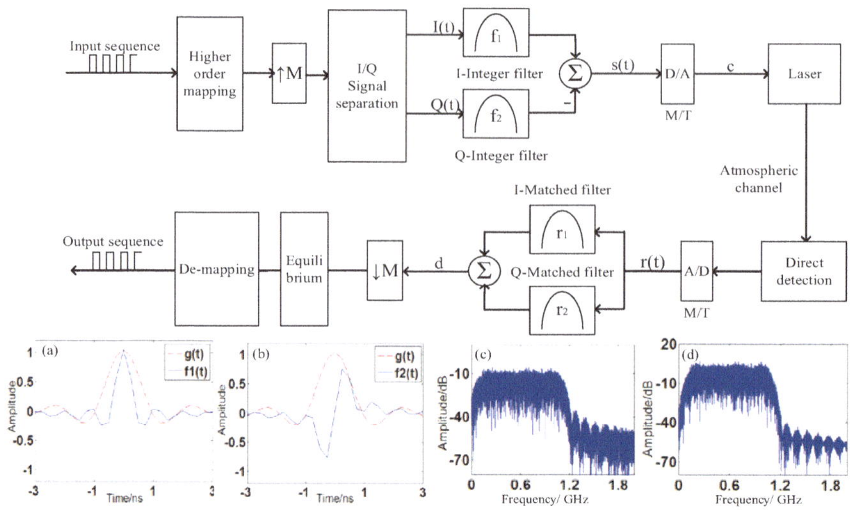

Fig. 10.2 Principle block diagram of CAP modulation. **a** time-domain diagram of the I-shaping filter; **b** time-domain diagram of the Q-shaping filter; **c** spectrum of the CAP signal; **d** spectrum of the signal at the receiving end

through a digital/analog (D/A) conversion is the generated CAP modulation signal (see Fig. 10.2c for the frequency spectrum of this signal).

The expressions of the I-shaping filter, Q-shaping filter, and generated CAP modulation signal $s(t)$ are [20] respectively,

$$f_1(t) = g(t)\cos(2\pi f_c t) \tag{10.1}$$

$$f_2(t) = g(t)\sin(2\pi f_c t) \tag{10.2}$$

$$g(t) = \frac{\sin\left[\frac{\pi t}{T}(1-\alpha)\right] + \frac{4\alpha t}{T}\cos\left[\frac{\pi(1+\alpha)t}{T}\right]}{\frac{\pi t}{T}\left[1-\left(\frac{4\alpha t}{T}\right)^2\right]} \tag{10.3}$$

$$s(t) = I(t)\otimes f_1(t) - Q(t)\otimes f_2(t). \tag{10.4}$$

Equation (10.3) shows the root-mean-square raised cosine roll-off function, where T is the symbol period and α is the roll-off coefficient. In Eqs. (10.1) and (10.2), f_c is the center frequency of the filter, and the shaping filters form a Hilbert transform pair. In Eq. (10.4), I and Q are the real and imaginary parts of the complex sequence after up-sampling, respectively, and ⊗ is the convolution symbol.

At the receiving end, the received signal is directly detected, which can reduce the complexity of the system [24]. The received analog signal is converted into a digital signal through an A/D conversion and output. The output digital signal is demodulated through the matched filter of the I/Q channel (see Fig. 10.2d for the frequency spectrum of the demodulated signal). The output signals of the I/Q matched filters are added and then downsampled, which is the inverse process of upsampling. The signal is then equalized to compensate for the signal distortion caused in the transmission process. The original input sequence can be obtained by demapping the equalized signal; that is, by decoding. The expressions of the I-matched filter and Q-matched filter at the receiving end are [20] respectively,

$$r_1(t) = f_1(-t) = f_1(t) \tag{10.5}$$

$$r_2(t) = f_2(-t) = -f_2(t). \tag{10.6}$$

10.2.2 Simulation Research

10.2.2.1 Simulation Conditions

Figures 10.7, 10.8, 10.9, 10.10, 10.11, 10.12, 10.13, 10.14 and 10.15 shows an atmospheric-channel model diagram. A gamma–gamma light-intensity fluctuation distribution model is adopted in the channel simulation. When the transmitted signal $s(t)$ is transmitted through a turbulent atmospheric channel, it is mainly affected by two types of noise: background noise $n(t)$ and multiplicative noise $k(t)$. When the background noise is large [44], the noise superimposed on the signal can be regarded as additive white Gaussian noise (AWGN).

In this simulation, the background light noise adopts an AGWN model with a non-zero mean value, and the multiplicative noise is gamma–gamma noise; hence, the signal received by the receiving antenna is $y(t) = k(t)s(t) + n(t)$. The simulated signal-to-noise ratio is $SNR = 16$ dB, and the fluctuation variance of the atmospheric light intensity is $\sigma_R^2 = 0.05$.

In a CAP modulation system, the main parameters that affect the system performance include the length of the filter l, the roll-off coefficient of the filter α, and the sampling-clock offset of the receiver k. See Table 10.1 for specific simulation parameters.

10.2.2.2 Simulation Analysis of the Roll-Off Coefficient

The shaping filter is composed of the product of the root-mean-square raised-cosine roll-off function $g(t)$ (see Eq. 10.3) and a cosine function. Roll-off coefficient α represents the filter's ability to retain the high-frequency information of the signal [14]. The closer α is to 0 ($0 \le \alpha \le 1$), the higher the spectrum-utilization rate of the system will be; however, the less edge information of the signal will be retained. Thus,

Table 10.1 Simulation-parameter values

Parameter	Value
Input sequence	213 random integers between 0 and 15
High-order mapping	16QAM
Upsampling	$M = 4$
Roll-off coefficient	$\alpha = 0.2$
Filter length	$l = 61$
Symbol rate	$R_s = 1$GBaud
Channel model	Gamma–gamma turbulent channel model
Channel noise	Mixed noise ($n(t)$, $k(t)$)
16CAP modulation signal bandwidth	$B = (1 + \alpha)/T = 1.2$ GHz

the size of α should be reasonably selected according to the system requirements. Figure 10.3 shows the relationship between the roll-off coefficient α and the bit-error rate (BER), as well as the constellation and eye diagrams of the corresponding signal.

Figure 10.3a shows the relationship between the roll-off coefficient α and the bit-error rate. As can be seen from the figure, with the increase of α, the filter retains more high-frequency information of the transmitted signal, which greatly decreases

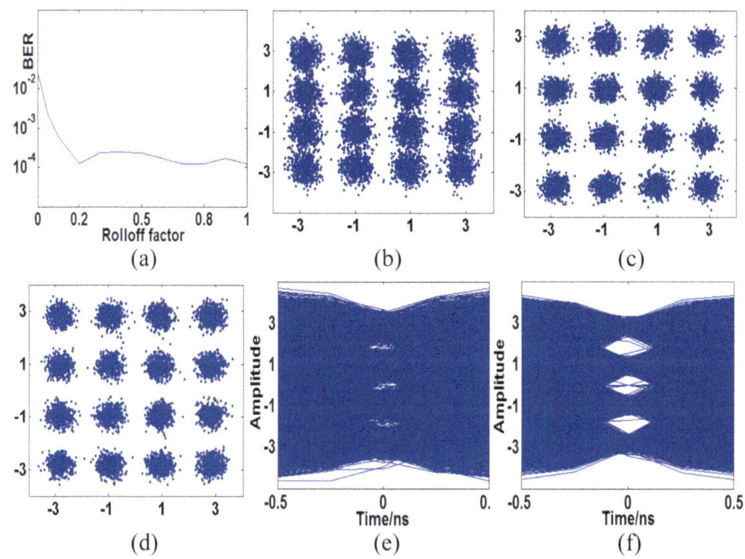

Fig. 10.3 **a** Relationship between the roll-off coefficient α and *BER*; **b–d** are the constellation diagrams of the signal at the receiving end when $\alpha = 0$, 0.2, and 0.5, respectively; **e** and **f** show the eye diagrams of the in-phase components of the demodulated signals when $\alpha = 0$ and 0.2, respectively

the value of the system at first and then stabilizes. When the roll-off coefficient value is <0.1, the bit-error rate value is large, and the system performance is poor.

For example, when $\alpha = 0$, the high-frequency information of the transmitted signal is severely missing; hence, the constellation points of the received signal are extremely scattered and the value is large ($BER = 0.0244$). When α is ≥ 0.1, the value is small (close to 10^{-4}) and the system performance is good. Therefore, to strike a balance between a high spectrum utilization and a low *BER*, α is usually set between 0.1 and 0.2.

Figure 10.3b–d are the corresponding signal constellations of the receiving end when $\alpha = 0, 0.2$, and 0.5. As shown in the figure, when the roll-off coefficient $\alpha = 0$, the convergence of the signal constellation points is poor. With the increase of the α value, comparing the three figures, it can be seen that the divergence degree of the signal constellation points decreases in turn. When the roll-off coefficient $\alpha > 0.2$, the bit-error rate of the system does not decrease with the increase of α; hence, the convergence of the constellation points of the corresponding signals basically remains unchanged.

Comparing Figure 10.3c, d, it can be seen that the simulation results are consistent with the above analysis. When $\alpha = 0.5$, the roll-off coefficient is large. For example, when the bandwidth required by the signal is large and the spectrum-utilization rate of the system is low, the roll-off coefficient should be reduced to improve the transmission performance of the system.

The spectrum-utilization rate and bit-error rate of the system are weighed, and $\alpha = 0.2$ is chosen. Figure 10.3e, f are the eye diagrams of the corresponding signals at the receiving end when α equals 0 and 0.2, respectively. Comparing the two figures, it can be seen that when $\alpha = 0.2$, the "eyes" of the signal eye diagram are more open, the eye diagram is more correct, and the intersymbol interference (ISI) is less; thus, the recovered signal has less distortion and a lower bit-error rate.

10.2.2.3 Simulation Analysis of the Filter Length

The length of the filter l is the degree of reservation of the designed infinite impulse-response (IIR) filter [37]. The bigger the filter l, the more the signal sidelobe information is reserved. This helps the receiver to recover the original signal, but increases the computational complexity. Therefore, it is very important to choose a reasonable filter length. Figure 10.4 shows the relationship between the filter length l and bit-error rate *BER*, and the constellation and eye diagrams of the corresponding signal.

Figure 10.4a shows the relationship between the filter length and the bit-error rate *BER*. Figure 10.4b–e show the signal constellations at the receiving end corresponding to $l = 11, 31, 61$, and 91, respectively Fig. 10.4f shows the eye diagram when $l = 61$. It can be seen from Fig. 10.4a that the bit-error rate of the system decreases rapidly with the increase of l, and then basically tends to stabilize.

When $l < 31$, the filter retains less sidelobe information from the transmitted signal, resulting in a higher bit-error rate ($BER > 10^{-3}$), a serious divergence of

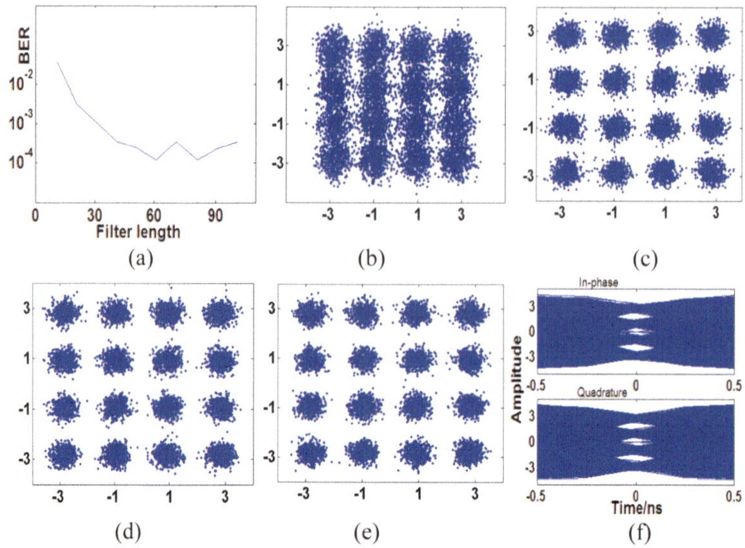

Fig. 10.4 **a** Relationship diagram of the filter length l and BER; **b**–**e** are the signal constellations of the receiving end when $l = 11, 31, 61,$ and 91, respectively; **f** signal eye diagram after down-sampling at the receiving end at $l = 61$

the signal constellation points, and poor system performance; for example, see the signal constellation in Fig. 10.4b when $l = 11$. With the increase of l, more sidelobe information of the transmitted signal is retained by the filter; thus, the bit-error rate will gradually decrease, and the convergence of the constellation points of the corresponding signal will improve.

For example, when comparing Fig. 10.4b–d, the divergence degree of the constellation points of the signal decreases in turn. When $l > 61$, the farther the sidelobe is from the main lobe in the signal spectrum, the smaller its amplitude, and the smaller its impact on the system performance. Thus, the bit-error rate basically does not change with l, and the divergence degree of the constellation points of the corresponding signals is similar. For example, by comparing Fig. 10.4d, e, it can be seen that the divergence degree of the constellation points of the corresponding signals is basically the same.

Based on the simulation results and theoretical analysis, $l = 61$ was selected, because the system error rate is small ($BER = 1.2207 \times 10^{-4}$) and the computational complexity is moderate. Figure 10.4f shows the $l = 61$ eye diagram of the corresponding signal. It can be seen from the figure that the "eyes" of the signal eye diagram are well opened, the eye diagram is correct, and the system performance is good.

10.2.2.4 Simulation Analysis of the Sampling-Clock Offset

Figure 10.5 shows the constellation and eye diagrams before the signal at the receiving end is unbalanced, when the offset k is 0, 1, and 2, respectively, when $l = 61, \alpha = 0.2$, and upsampling $M = 10$.

In the eye diagram of the CAP modulated signal, the eye-opening degree is very small; hence, it is necessary to sample at the moment when the eye-opening degree is the largest to best recover the original signal. The larger the sampling-clock offset at the receiving end, the greater the intersymbol interference, and the greater the I/Q two-channel crosstalk of the signal; that is, the more serious the distortion of the recovered signal, the higher the bit-error rate, and the worse the system performance.

Figure 10.5a–c show the constellations of the corresponding signal at the receiving end before it is unbalanced when the offset $k = 0$, 1, and 2. It can be seen from the figure that when the sampling clock is offset, the phase of the signal at the receiving end will be offset. The larger the sampling offset, the greater the divergence of the constellation of the corresponding signal, the more serious the phase offset, and the higher the bit-error rate BER. For example, when $k = 0$, $BER = 1.2207 \times 10^{-4}$; when $k = 1$, $BER = 0.4499$. Therefore, to compensate the phase offset caused by the sampling-clock offset [21], it is necessary to equalize the signal at the later stage of the receiving end; that is, the phase must be compensated.

Figure 10.5d–f show the eye diagrams of the corresponding signal at the receiving end before it is equalized, when the sampling-clock offset $k = 0$, 1, and 2. The degree to which the "eyes" of the eye graph are opened reflects the strength of the ISI. The wider the "eyes" are opened and the more correct the eye graph is, the smaller the

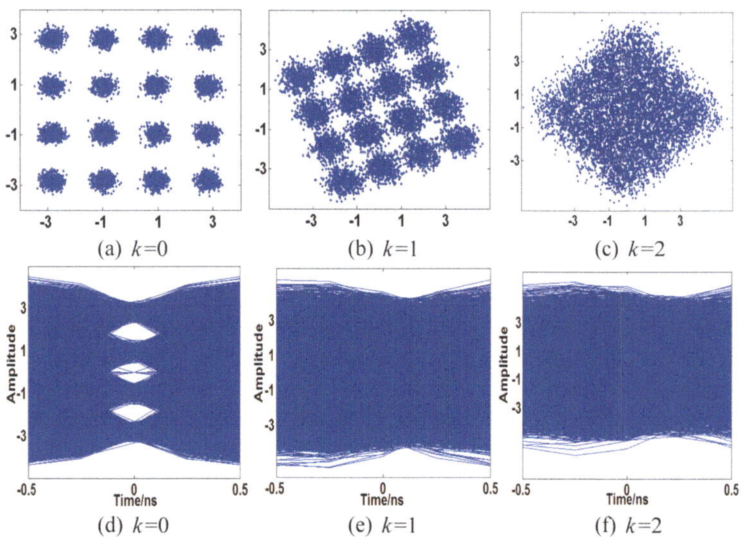

Fig. 10.5 **a–c** Are the constellation diagrams of the demodulated signals before equalization; **d–f** are the eye diagrams of the demodulated signals before equalization

ISI is, and vice versa. The blur degree of the eye trace reflects the noise. The bigger the noise, the wider and more blurred the eye trace. Comparing Fig. 10.5d–f, it can be seen that with the increase of k, the intersymbol interference and the two-branch crosstalk on the demodulated signal at the receiving end increases, resulting in the gradual closing of the "eyes" of the signal eye diagram. Finally, the "eyes" are completely closed, as shown in Fig. 10.5e, f.

From Fig. 10.5, we can see that the sampling-clock offset at the receiving end will have a great impact on the correct demodulation of the signal; that is, the error rate of the signal will increase rapidly and the system performance will decrease. However, in an actual communication system, a sampling-clock offset is inevitable; thus, it is usually necessary to add a phase-compensation algorithm at the receiver end of the communication system to reduce the signal distortion caused by the sampling-clock offset.

A commonly used equalization algorithm is the constant-modulus algorithm (CMA) [45], which is simple and easy to implement. However, for high-order CAP modulation, the equalization effect of CMA is not obvious, and in the equalization process, CMA only uses the amplitude information of the signal, not the phase information. Therefore, the CMA algorithm cannot compensate for the phase distortion of the signal, and the original signal can be correctly recovered using phase recovery after the decider.

Improved CMA algorithms, such as the cascaded multi-modulus algorithm (CMMA) [46] or multi-modulus algorithm (MMA), are typically used to balance the phase rotation. In addition to the above classical CMA algorithm and its derivatives, the decision-assisted least-mean-square algorithm (DD-LMS) is also a widely used blind-equalization algorithm. However, in practical applications, it is generally necessary to adopt other blind-equalization algorithms to pre-converge before using DD-LMS [47]; otherwise, the equalization effect will be poor.

10.2.3 MMA Blind-Equalization Algorithm

10.2.3.1 Basic Principle of the MMA Equalization Algorithm

The MMA algorithm is a blind-equalization algorithm. When the receiver uses this algorithm for equalization, it does not need to borrow from the training sequence; rather, it only needs to send the prior information of the sequence itself for equalization, so that the sequence output by the equalizer can be the same as the original sequence, to the greatest extent. Because the MMA algorithm needs to equalize both the real and imaginary parts of the signal, the amplitude and phase information of the signal should be used simultaneously in the equalization process, so that the phase distortion of the signal at the receiving end can be well compensated.

The formula for updating the tap coefficient of the MMA algorithm is [48]

$$W(n+1) = W(n) - \mu e^*(n)X(n), \tag{10.7}$$

where

$$e(n) = e_R(n) + je_I(n) \tag{10.8}$$

$$e_R(n) = y_R(n)\left[y_R^2(n) - r_R\right] \tag{10.9}$$

$$e_I(n) = y_I(n)\left[y_I^2(n) - r_I\right] \tag{10.10}$$

$$r_R = E\left[s_R^4(n)\right]/E\left[s_R^2(n)\right] \tag{10.11}$$

$$r_I = E\left[s_I^4(n)\right]/E\left[s_I^2(n)\right]. \tag{10.12}$$

$W(n)$ is the tap-weight vector of the equalizer at the moment; μ is the algorithm step size; $e(n)$ is the error function; $X(n)$ is the input vector of the equalizer at the moment; $^*(\cdot)$ is the complex conjugate; $e_R(n)$ and $e_I(n)$ are the real and imaginary parts of the error function, respectively; $y_R(n)$ and $y_I(n)$ are, respectively, the real and imaginary parts of the equalizer output sequence $y(n)$; r_R and r_I are the real and imaginary parts of the difference constant of the transmission sequence $s(n)$, in turn; $s_R(n)$ and $s_I(n)$ are the real and imaginary parts of the source symbol $s(n)$, respectively; and $E[\cdot]$ indicates the average.

10.2.3.2 MMA-Equalization Simulation Analysis

(1) **Constellation simulation analysis**

Figure 10.6 shows the constellations before and after signal equalization at the receiving end with offsets of 1 and 2, respectively, when $l = 61, \alpha = 0.2$, upsampling $M = 10$, and a gamma–gamma turbulent channel model (mixed noise) are used.

Figure 10.6a, b show the constellations before the signal equalization at the receiving end, corresponding to offsets $k = 1$ and 2, respectively, and Fig. 10.6c, d show the constellations after the signal equalization at the receiving end, corresponding to offsets $k = 1$ and 2, respectively.

From the previous simulation analysis of the sampling clock, it can be seen that the sampling-clock offset at the receiving end will cause the phase of the signal to rotate, which leads to a rapid increase in the bit-error rate of the signal. For example, the constellations in Fig. 10.6a, b are all offset by a certain angle, and the larger the offset k value, the more serious the phase distortion of the constellation and the greater the divergence of the constellation points.

Comparing Fig. 10.6a–d, it can be seen that after the demodulation signal at the receiving end is equalized by the MMA algorithm, the phase distortion is well compensated, the divergence of the constellation points is reduced, and the bit-error rate is greatly reduced. This shows that the MMA algorithm equalizes effectively.

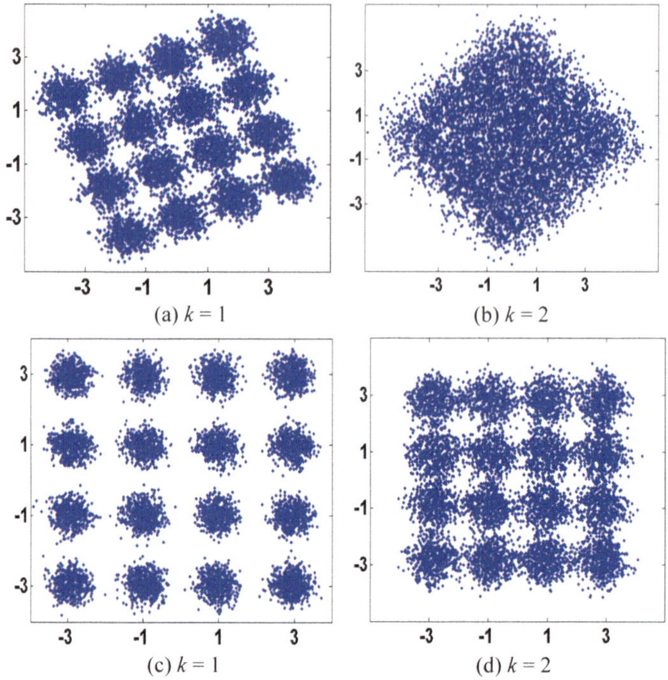

Fig. 10.6 Constellation diagrams before (**a**, **b**) and after (**c**, **d**) the demodulation signal equalization

(2) **Eye-diagram simulation analysis**

When l=61, $\alpha = 0.2$, and upsampling $M = 10$, with a gamma–gamma turbulent channel model (mixed noise), Fig. 10.7 shows the eye diagrams of the corresponding signal at the receiving end before and after equalization, with offsets $k = 1$ and 2, respectively.

Figure 10.7a, b show the eye diagrams of the corresponding signal at the receiving end before equalization, when the offset $k = 1$ and 2, respectively, and Fig. 10.7c, d show the eye diagrams of the corresponding signal at the receiving end after equalization, when the offset $k = 1$ and 2, respectively.

From the previous simulation analysis of the sampling clock, it can be seen that the larger the sampling-clock offset, the greater the intersymbol interference and I/Q two-channel crosstalk on the demodulated signal at the receiving end; that is, the higher the bit-error rate, and the smaller the opening degree of the signal eye diagram.

For example, the "eyes" in Fig. 10.7a, b are completely closed, indicating that the intersymbol interference and two-channel interference are very large at this time. Comparing Fig. 10.7a, c, it can be seen that when the sampling-clock offset is small, after the MMA-algorithm equalization, not only is the phase distortion of the signal compensated to some extent, but the intersymbol interference and two-channel crosstalk of the signal are restrained to some extent, and the system performance is improved significantly.

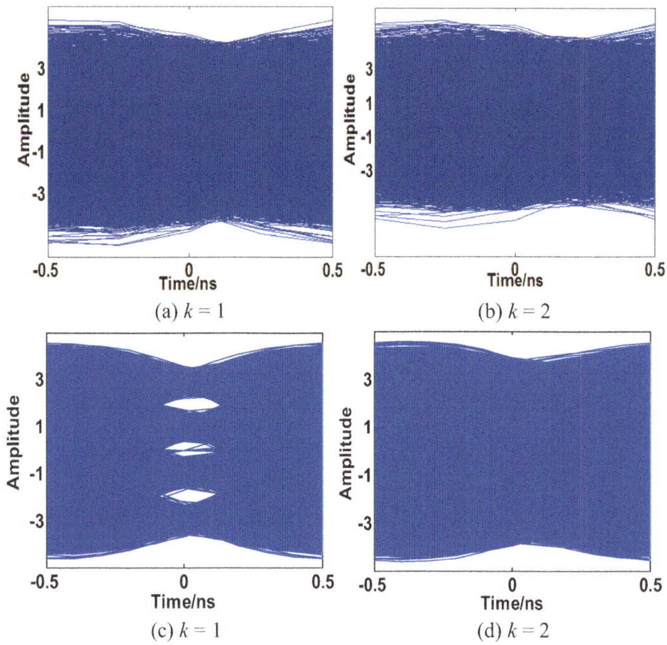

Fig. 10.7 Eye diagrams before (**a**, **b**) and after (**c**, **d**) demodulation signal equalization

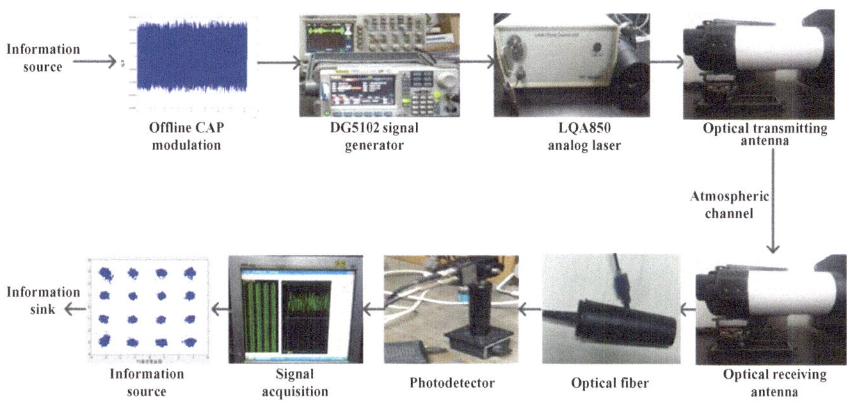

Fig. 10.8 Block diagram of the CAP modulation experiment

Comparing Fig. 10.7b, d, it can be seen that when the sampling-clock offset is large, the phase rotation of the signal can only be improved to some extent after equalization by the MMA algorithm; however, the suppression effect on the intersymbol interference and two-channel crosstalk is small. Moreover, as shown in Fig. 10.7d, the "eyes" are still closed.

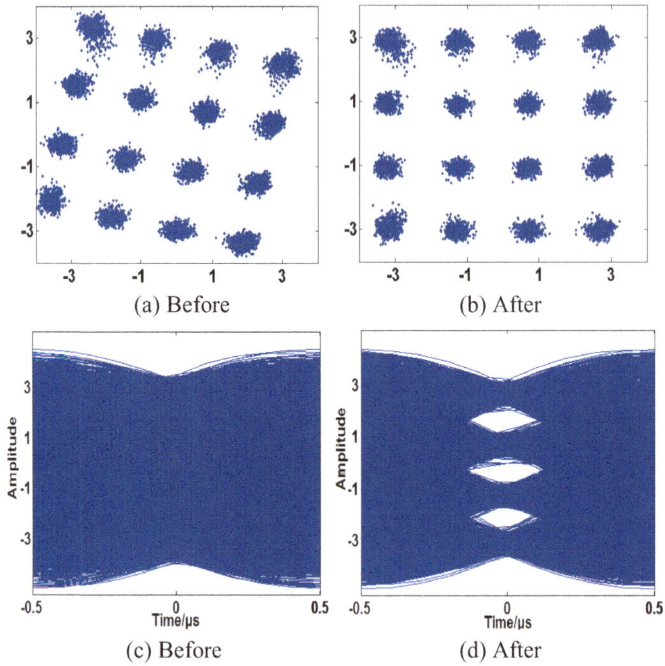

Fig. 10.9 **a**, **b** Show constellation diagrams before and after the demodulated signal is equalized, respectively; **c**, **d** show the eye diagrams before and after equalization of the in-phase component of the demodulated signal, respectively

Fig. 10.10 Demodulated signal power-spectrum diagram

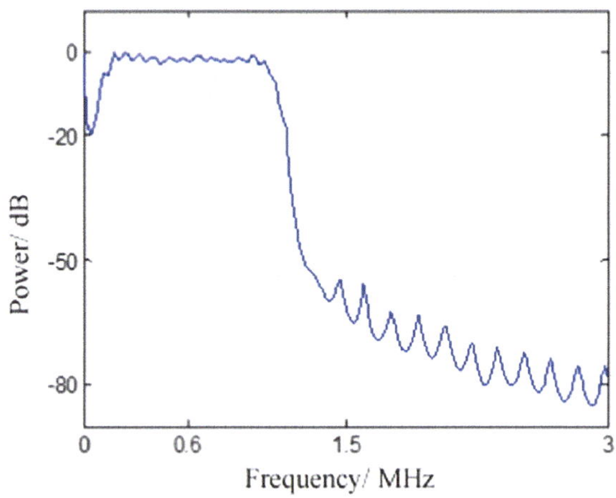

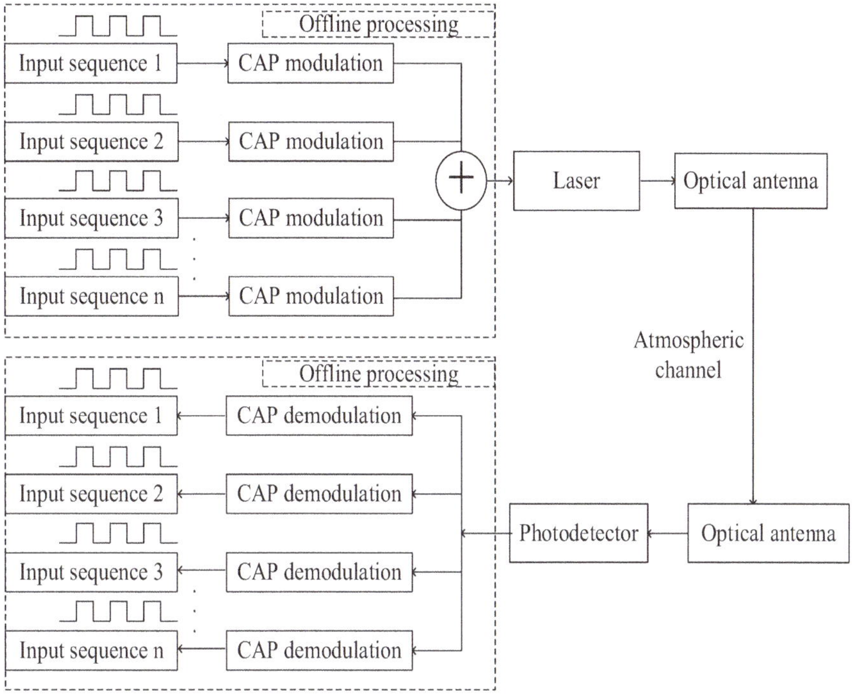

Fig. 10.11 System principle block diagram of multi-band CAP modulation

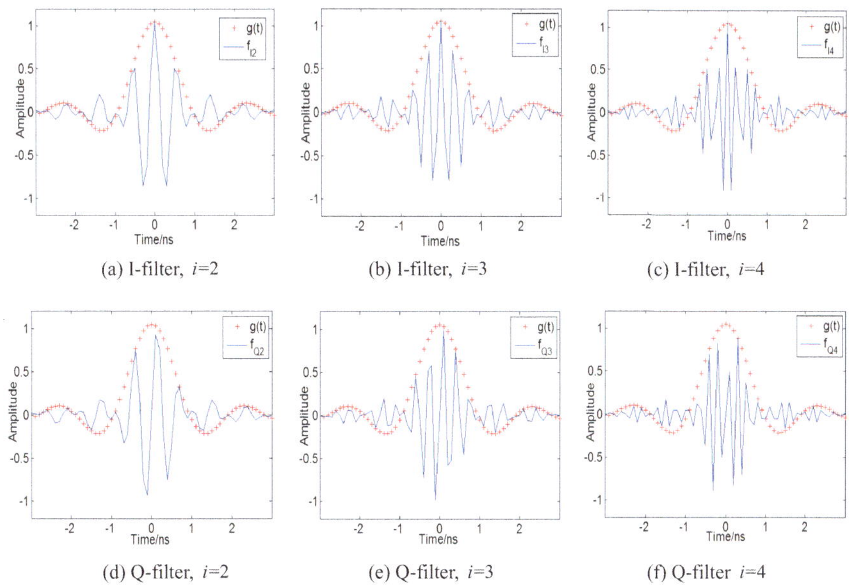

Fig. 10.12 Time-domain diagram of each CAP modulation filter [49, 50]

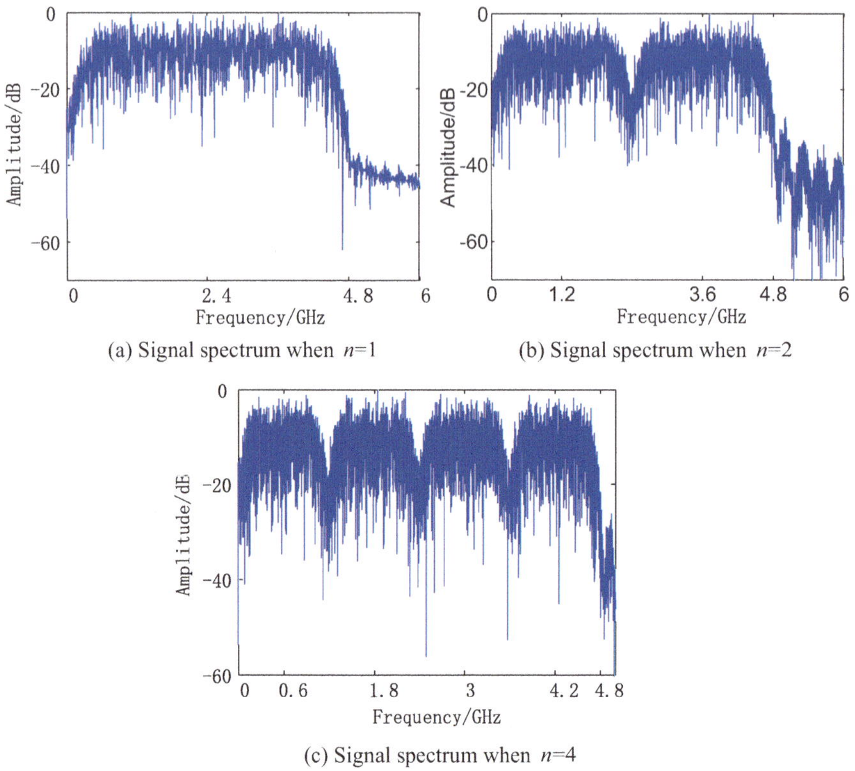

Fig. 10.13 Spectrum diagrams of the multi-band 16CAP signal generated by the sender [49, 50]

10.2.4 Experimental Study

10.2.4.1 Experimental Principle

Figure 10.8 shows a schematic diagram of the atmospheric laser-communication experiment with CAP modulation. At the sending end, 2^{13} random integers between 0 and 15 are generated offline using MATLAB, and modulated by CAP to generate 16CAP-modulated digital signals. Then, the modulated digital signals are converted into analog signals using Rigol's UltraStation software, and the amplitude and offset of the converted 16CAP analog signals are adjusted using a Rigol DG5102 signal generator to make them suitable for transmission in wireless laser channels.

Finally, the signal output by the signal generator is loaded into an 850-nm analog laser (LQA850, Newport), the 16CAP signal is loaded onto the optical carrier for transmission, and the light beam is transmitted to the receiving end through the atmosphere.

At the receiving end, the receiving optical antenna performs optical-fiber coupling on the received optical signal, detects the optical signal through a photodetector,

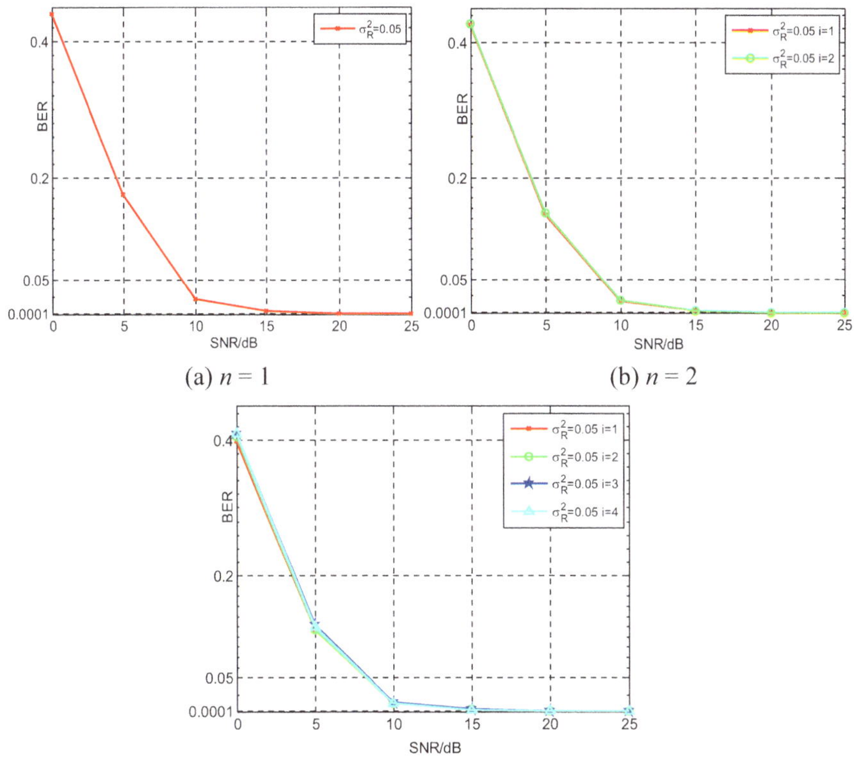

(a) $n = 1$ (b) $n = 2$ (c) $n=4$

Fig. 10.14 Relationship between the SNR and BER of a multi-band 16CAP signal [49, 50]

converts the detected optical signal into an electrical signal, and inputs it into a PC for signal collection. Finally, the collected signal is subjected to offline CAP demodulation and other processing to obtain the original input sequence.

This experiment is a short-distance atmospheric laser-communication experiment. See Table 10.1 for the specific parameter settings of the CAP modulation under the transmitter line, in which the transmission power P = 27.62 mW.

10.2.4.2 Analysis of Experimental Results

(1) **Analysis of the constellation and eye diagrams of the receiver signal**

Figure 10.9 shows the constellation and eye diagrams of the demodulated signal at the receiver, before and after equalization. Figure 10.9a, b show the constellation diagrams of the demodulated signal before and after the equalizer, respectively.

As can be seen from Fig. 10.9a, in the actual CAP modulation system, the sampling-clock offset at the receiving end is inevitable, which causes the signal

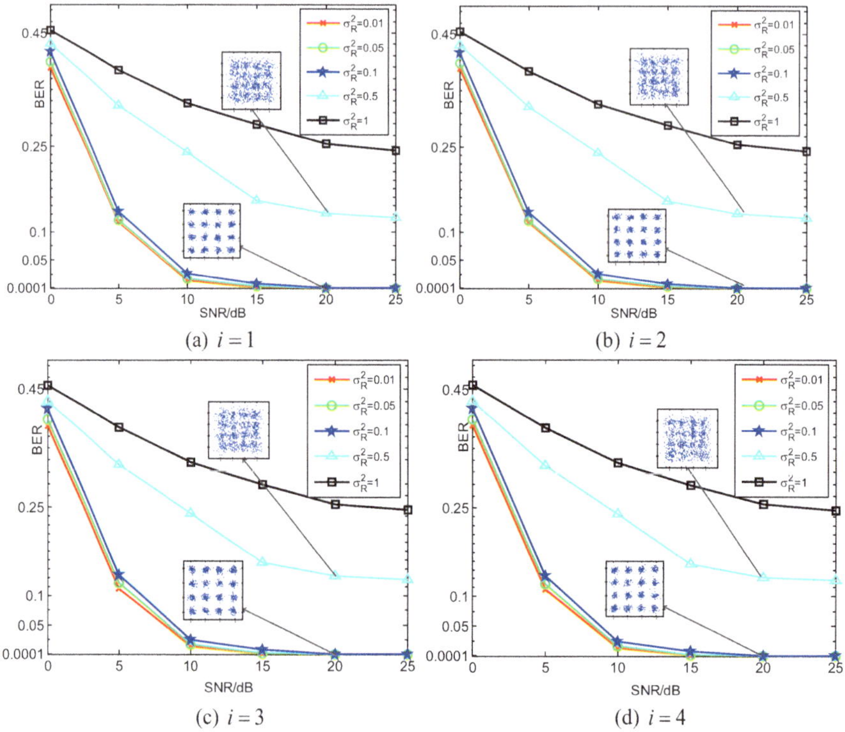

Fig. 10.15 Relationship between the BER and SNR of each signal at the receiving end [49, 50]. The constellations embedded in each figure are for signals with SNR = 20 dB and $\sigma_R^2 = 1, 0.5$, and 0.05, respectively

phase to distort and the signal constellation to rotate by a certain angle. Therefore, to correctly recover the original sequence, this phase distortion must be compensated, which is consistent with the above simulation analysis. Comparing Fig. 10.9a, b, the phase distortion caused by the sampling-clock offset is compensated after equalization by the MMA algorithm, and the equalization effect is good.

It can be seen from Fig. 10.9c that the sampling-clock offset at the receiving end increases the influence of the intersymbol interference and I/Q two-channel crosstalk on the signal, and makes the "eyes" of the signal eye diagram smaller or even completely closed. Comparing Fig. 10.9c, d, it can be seen that after equalization by the MMA algorithm, the eye diagram of the demodulated signal is compensated to a larger degree of eye opening, and the eye diagram is correct. This indicates that after equalization, the intersymbol interference and two-channel crosstalk are well suppressed, the distortion degree of the recovered signal at the receiving end is greatly reduced, and the system performance is good.

From the experimental results, it can be seen that in a short-distance wireless laser communication, the original sequence can be recovered well using the parameters obtained by simulation, the system error rate is low, and the transmission performance is good.

(2) **Analysis of the power spectrum of the receiver signal**

Figure 10.10 shows the normalized power spectrum of the demodulated signal at the receiver.

As shown in Fig. 10.10, the power spectrum of the signal is estimated using the Burg parametric-model spectrum-estimation method, where the model order is 80, the signal modulation frequency is $f_c = 0.6$ MHz, the sampling frequency is $f = 10$ MHz, and the signal bandwidth is $B = 1.2$ MHz. The figure only shows the power-spectrum distribution of the signal within 3 MHz. It can be seen from the figure that the signal power inside the bandwidth, $B = 1.2$ MHz, is about 80 dB higher than the signal power outside the bandwidth; that is, the signal energy is mainly concentrated in the bandwidth, and the signal-power distribution in the bandwidth is relatively stable.

10.3 Research on an Atmospheric Laser-Communication System with Multi-band CAP Modulation

To further improve the transmission performance of a communication system, multi-band CAP modulation, based on CAP modulation technology, has been proposed. Compared with single-band CAP modulation, multi-band CAP modulation can provide higher spectrum utilization and a lower system error rate; hence, it has also been favored in recent years. This section will analyze the influence of the frequency-band number, light-intensity fluctuation variance, etc., on the performance of a multi-band 16CAP modulation system.

10.3.1 Principle of Multi-band CAP Modulation and Demodulation

Figure 10.11 shows the schematic diagram of a multi-band CAP modulation system. At the transmitting end, independent random sequences from n channels are added into one channel signal after CAP modulation (the center frequency of the CAP modulation in each channel is different). Then, the generated multi-band CAP modulation signal is loaded into the optical signal through the semiconductor laser for transmission, and finally, the optical signal is transmitted through the transmitting antenna.

At the receiving end, the optical antenna transmits the received transmission signal to the photodetector, which converts it into the corresponding electrical signal for output. Finally, the collected electrical signal is processed offline by MATLAB, and the original input sequence is recovered by each CAP demodulator. The specific analysis is as follows.

According to Eqs. (10.1), (10.2), (10.5), and (10.6) in the single-band CAP modulation system, the expressions of the I/Q shaping filter and I/Q matched filter of each sub-path are

$$f_{i1}(t) = g(t)\cos(2\pi f_{ci}t) \tag{10.13}$$

$$f_{i2}(t) = g(t)\sin(2\pi f_{ci}t) \tag{10.14}$$

$$r_{i1}(t) = f_{i1}(-t) = g(-t)\cos(2\pi f_{ci}t) \tag{10.15}$$

$$r_{i2}(t) = f_{i2}(-t) = g(-t)\sin(2\pi f_{ci}t), \tag{10.16}$$

where i is the signal (i = 1, 2, 3, …n), f_{ci} is the modulation frequency of the signal, and $g(t)$ is the root-mean-square raised cosine roll-off function.

According to Fig. 10.11, the multi-band CAP modulation signal generated by the transmitter can be expressed as [40]

$$\begin{aligned} s(t) = \sum_{i=1}^{n}[s_{iI}(t) - s_{iQ}(t)] = \sum_{i=1}^{n}\sum_{k=1}[I_{i,k}g(t-kT)\cos 2\pi f_{ci}(t-kT) \\ - Q_{i,k}g(t-kT)\sin 2\pi f_{ci}(t-kT)], \end{aligned} \tag{10.17}$$

where n is the number of sub-bands, $s_{iI}(t)$ and $s_{iQ}(t)$ are the signals generated by the first channel signal modulated by the I/Q filters, $I_{i,k}$ and $Q_{i,k}$ are, respectively, the symbol values of the real and imaginary parts of the complex sequence generated by the first channel signal after 16QAM high-order mapping in the kth symbol period, and T is the symbol period.

At the receiving end, the received signal is demodulated by the I/Q matched filter of each sub-channel, and the output expressions of the I/Q matched filter of the ith channel are [40]

$$\begin{aligned} y_{iI}(t) &= \left\{\sum_{j=1}^{n}[s_{jI}(t) - s_{jQ}(t)]\right\} \otimes [g(-t)\cos 2\pi f_{ci}t] \\ &\cong \sum_{k}\left\{I_{i,k}\cos[2\pi f_{ci}(t-kT)]h(t-kT) - Q_{i,k}\sin[2\pi f_{ci}(t-kT)]h(t-kT)\right\} \end{aligned} \tag{10.18}$$

$$y_{iQ}(t) = \left\{\sum_{j=1}^{n}[s_{jI}(t) - s_{jQ}(t)]\right\} \otimes [g(-t)\sin 2\pi f_{ci}t]$$

$$\cong \sum_k \left\{ Q_{i,k} \cos[2\pi f_{ci}(t - kT)]h(t - kT) - I_{i,k} \sin[2\pi f_{ci}(t - kT)]h(t - kT) \right\}, \tag{10.19}$$

where $\otimes$ is the convolution symbol, k is the kth symbol period, and $h(t) = [g(t) \otimes g(-t)]/2$.

It can be known from Eq. (10.18) that the signal at the receiving end passes through the I-matched filter of the I-channel. There is interference from the Q-channel in the I-channel, but no interference from the other channels. Therefore, when sampling is performed at the best sampling time $t = kT$, the crosstalk from the Q-channel to the I-channel can be eliminated; that is, the I-channel signal can be recovered without distortion, and the Q-channel signal can be accurately recovered in the same manner.

10.3.2 Simulation Study

10.3.2.1 Simulation Channel Model

Consider Figs. 10.7, 10.8, 10.9, 10.10, 10.11, 10.12, 10.13, 10.14 and 10.15 as the atmospheric-channel simulation model diagram. A gamma–gamma turbulent channel model is selected in the simulation. When transmitting signal $s(t)$ in the atmosphere, it is mainly interfered with by background-light noise $n(t)$ and multiplicative noise $y(t)$. Because the background-noise field interferes strongly in atmospheric laser communication [44], in this simulation, an AWGN model with a non-zero mean value is selected, and gamma–gamma noise is selected for $y(t)$. It can be seen from Fig. 7.15 that the signal received by the optical antenna is $y(t) = k(t)s(t) + n(t)$. The simulated signal-to-noise ratio is $SNR = 16$ dB, and the fluctuation variance of the atmospheric light intensity is $\sigma_R^2 = 0.05$.

10.3.2.2 Determination of Simulation Parameters

In the simulation, a four-channel-input four-channel-output system is mainly studied; that is, $n = 4$ and 16CAP modulation is adopted. The total symbol rate of the system is $R_s = 4$ GBaud (that is, the total bit rate of the system $R_b = 16$ Gb/s). Because $n = 4$, the symbol rate of each channel is $R_{si} = 1$ GBaud (where $i = 1, 2, 3, 4$). In the modulation, the filter roll-off coefficient and key parameters are calculated as follows:

(1) Bandwidth required by the system B: $B = 4.8$ GHz can be obtained from $B = (1 + \alpha)/T$, where T is the symbol period; $1/T = R_s = 4$ GHz. In the same manner, the required bandwidth $B_i = B/n = 1.2$ GHz of the ith ($i = 1, 2, 3, 4$) channel and the symbol period $T_i = 1$ns of the ith channel.

Table 10.2 Simulation-parameter values

Parameter	Value
Input sequence	Four independent groups of 2^{13} random integers between 0 and 15
High-order mapping	16 QAM
Upsampling	$M = 10$
Roll-off coefficient	$\alpha = 0.2$
Filter length	$l = 61$
Symbol rate	$R_s = 4$ GBaud
Channel model	Gamma–gamma turbulent channel model
Channel noise	Mixed noise ($n(t)$ and $k(t)$)
Bandwidth of multi-band CAP signal	$B = (1 + \alpha)/T = 4.8$ GHz
Modulation frequency f_{ci}	$f_{c1} = 0.6$ GHz$f_{c2} = 1.8$ GHz $f_{c3} = 3$ GHz$f_{c4} = 4.2$ GHz

(2) Modulation center frequency f_{ci} of the ith ($i = 1, 2, 3, 4$) channel: $f_{c1} = 0.6$ GHz, $f_{c2} = 1.8$ GHz, $f_{c3} = 3.0$ GHz, and $f_{c4} = 4.2$ GHz can be obtained from $f_{ci} = [(1 + \alpha)(2i - 1)]/(2T_i)$.

(3) Up-sampling order M in CAP modulation: $f_{Nyquist}$ is defined as the minimum sampling frequency required to recover the original input sequence without distortion, and f_{tx} is the actual sampling frequency of the system transmitter. Equation (10.23) can be derived from Eqs. (10.20), (10.21), and (10.22). The specific formulas are as follows [35]:

$$f_{Nyquist} > 2R_s(1 + \alpha) \tag{10.20}$$

$$f_{tx} = R_s N_s / n \tag{10.21}$$

$$f_{tx} > f_{Nyquist} \tag{10.22}$$

$$N_s > 2n(1 + \alpha), \tag{10.23}$$

where N_s is the number of samples per symbol. $N_s > 9.6$ can be obtained from n = 4, $\alpha = 0.2$. In this simulation, $N_s = 10$; that is, M = 10 is selected. See Table 10.2 for specific simulation parameters.

10.3.2.3 Analysis of Simulation Results

(1) **Time-domain diagram of each sub-channel CAP modulation filter**

The time-domain diagram of each sub-channel CAP modulation filter is shown in Fig. 10.12.

It can be seen from Eqs. (10.1), (10.2), and (10.3) that the impulse response of the CAP modulation filter is obtained by multiplying the sine/cosine function and the root-mean-square cosine roll-off function, so the shaping filters of the I/Q channels form a pair of Hilbert transform pairs; that is, the impulse-response phase difference of the two filters is 90, as shown in Fig. 10.12a, d.

To prevent the signals in each frequency band from interfering with each other, the center-frequency value of each CAP modulation filter is different; that is, the carrier frequency of each channel is different (see Table 10.2 for specific values). Consequently, the time-domain waveforms of the I/Q shaping filters corresponding to each frequency band are also different, as shown in Fig. 10.12b, c, e, and f.

(2) **Spectrum diagram of a multi-band 16CAP modulation signal**

Figure 10.13 shows the spectrum diagrams of the generated multi-band 16CAP modulation signal, in which Fig. 10.1a–c respectively show the number of bands when $n = 1, 2$, and 4, and $B = 4.8$ GHz.

In a multi-band CAP modulation system, the modulation-frequency selection of each sub-channel ($i = 1, 2 \ldots n$) will ensure that the signals of adjacent channels do not interfere with each other; that is, the center frequencies of the modulation filters of each sub-channel are different. For example, the center frequencies of each sub-band in Fig. 10.13b, c are different, so that the sub-bands do not overlap with each other.

Comparing Fig. 10.13a–c, it can be seen that under a certain total bandwidth, the larger the frequency-band value, the smaller the corresponding sub-band bandwidth value, and the flatter the signal band. That is, the smaller the frequency-dependent attenuation of the signal [41], and the easier it is for the receiver to correctly demodulate the original sequence.

(3) **Analysis of the influence of the signal-to-noise ratio on the bit-error rate of a multi-band 16CAP modulation system**

Figure 10.14 shows the $SNR - BER$ relationship diagram of a multi-band 16CAP modulation signal, in which Figure 10.14a–c correspond respectively to frequency-band numbers $n = 1, 2$, and 4, with light-intensity fluctuation variance $\sigma_R^2 = 0.05$. The following can be obtained from Eq. (10.23): When $n = 1, N_s > 2.4$; when $n = 2$, $N_s > 4.8$; and when $n = 4$, $N_s > 9.6$; hence, in this simulation, the upsampling order is selected as $M = 3, 5$, and 10, in turn.

It can be seen from Fig. 10.14a–c) that when $SNR < 15$ dB, with the increase of the signal-to-noise ratio, the bit-error rate of the signal at the receiving end decreases greatly. When SNR > 15 dB, with the increase of the SNR, the bit-error rate of the corresponding signal tends to stabilize. Under a certain total bandwidth, with the increase of the band number n, the bandwidth value of the corresponding sub-band decreases, so that the signal band is flatter. Thus, the frequency-dependent attenuation of the signal is smaller, the power loss of the system is smaller, and the bit-error rate is lower [35].

Comparing Fig. 10.14a–c, it can be seen that under the same signal-to-noise ratio, when the number of frequency bands n increases, the corresponding bit-error rate decreases. For example, when SNR = 5 dB and the number of frequency bands are $n = 1, 2$, and 4, the corresponding bit-error rates are $BER_1 = 0.3384$, $BER_2 = 0.3002$, and $BER_4 = 0.2803$. It can be seen from Fig. 10.14b, c that with a certain number of frequency bands, the $SNR - BER$ relationship curves among different sub-bands i are similar. This indicates that there is no mutual interference among different sub-bands, which is consistent with the theoretical analysis.

(4) **Analysis of the influence of the light-intensity fluctuation variance on the bit-error rate of a four-band 16CAP modulation system**

From the above simulation analysis, it can be seen that in a multi-band CAP modulation system, with the increase of band number n, the bit-error rate of the signal at the receiving end can be reduced; however, the corresponding number of modulation filters ($2n$) will also increase, which means the computational complexity of the system will increase. Therefore, this simulation mainly studies the band number $n = 4$. Figure 10.15 shows the relationship between the system error rate and signal-to-noise ratio of each sub-channel signal ($i = 1, 2, 3, 4$) at the receiving end under different light-intensity fluctuation variances σ_R^2, where σ_R^2 is 0.01, 0.05, 0.1, 0.5, and 1.

Comparing Fig. 10.15a–d, it can be seen that when the value of σ_R^2 is constant, the BER value of each sub-channel signal at the receiving end is similar to the SNR and the corresponding BER is basically the same; this proves that there is almost no mutual interference among the channels.

It can be seen from Fig. 10.15 that when the variance σ_R^2 of the light-intensity fluctuation is small (such as 0.01, 0.05, and 0.1), the signal is mainly affected by additive noise; that is, the BER value mainly changes with the change of SNR. When SNR < 15 dB, the BER value of the corresponding signal decreases rapidly with the increase of SNR. When SNR > 15 dB continues to increase, the BER value of the corresponding signal decreases slowly and finally tends to stabilize.

For example, in Fig. 10.15a, when $\sigma_R^2 = 0.05\ and$ SNR $= 20dB$, the corresponding BER value reaches 10^4, and with the increase of SNR, the BER remains basically unchanged. With the increase of σ_R^2, the BER–SNR curve of the corresponding signal changes gradually. When the variance σ_R^2 of the light-intensity fluctuation is large (for example, when σ_R^2 is 0.5 and 1), the signal is mainly affected by multiplicative noise; that is, the system bit-error rate decreases with the decrease of σ_R^2. The larger the value of σ_R^2, the worse the system performance; that is, the more serious the signal distortion and the larger the corresponding BER.

For example, in Fig. 10.15b, when $\sigma_R^2 = 0.5$, the corresponding BER value decreases and finally stabilizes at about 0.13. From the constellation diagram of the corresponding signal when $\sigma_R^2 = 0.5$, it can be seen that the distribution of constellation points is chaotic and the recovered-signal distortion is severe. When $\sigma_R^2 = 0.1$, the BER decreases slowly with the increase of SNR, and finally stabilizes at about 0.26, which results in poor system performance.

10.3.3 Experimental Research

10.3.3.1 Experimental Principle

Figure 10.16 shows an experimental schematic diagram of a multi-band 16CAP modulation wireless laser-communication system. The experimental process is as follows.

The process at the sending end is as follows:

(1) Generation of a multi-band 16CAP modulation signal: Four independent channels of 212 random integers between 0 and 15 are generated offline in MATLAB; the information sources pass through their respective CAP modulators (each channel has a different modulation frequency) to generate four channels of 16CAP modulated digital signals, and the four channels of modulated signals are added.
(2) Digital-to-analog conversion: The generated four-band 16CAP modulated digital signal is converted to an analog signal, using the UltraStation software, and output.
(3) DG5102 signal generator: The signal generator adjusts the amplitude and offset of the converted four-band 16CAP modulated analog signal to make it suitable for transmission in the atmospheric laser channel.
(4) 650 nm red laser: This infrared laser is used to roughly align the transmitting/receiving ends.
(5) LQA850 analog laser: The four-band 16CAP modulated signal output by the signal generator is loaded into the 850 nm analog laser, that is, the optical carrier, for transmission. The laser is used to roughly align the antennas at both ends.
(6) Transmitting optical antenna: The optical-transmission signal is transmitted through the optical antenna, and the resulting light beam is transmitted to the receiving end through the atmosphere.

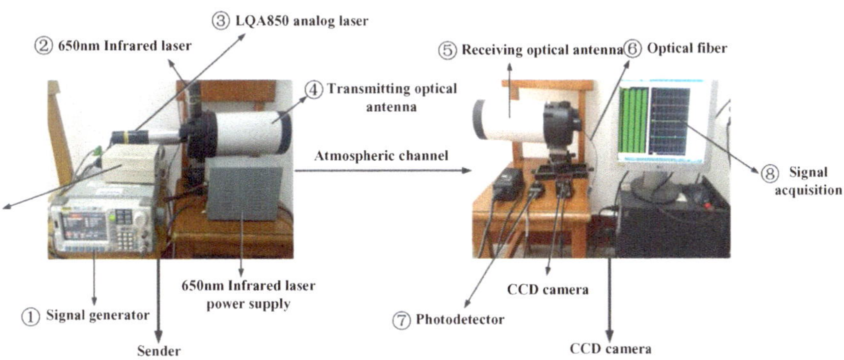

Fig. 10.16 Schematic diagram of an atmospheric laser-communication experiment with multi-band 16CAP modulation

The process at the receiving end is as follows:

(1) Receiving optical antenna: This antenna receives the optical signal emitted by the transmitting antenna.
(2) Optical fiber: An optical fiber is used to couple the received optical signal and fine align the antennas at both ends with a charge-coupled device (CCD) camera.
(3) Photoelectric detector: The wavelength of the detectable light wave is 200 nm $\leq \lambda \leq$ 1000 nm. The detector detects the signal coupled into the optical fiber and converts it into the corresponding electrical signal for output.
(4) PC signal acquisition: The electrical signal converted by the photoelectric detector is input into the PC for signal acquisition and processing.
(5) Offline processing: Finally, the collected signal is subjected to a series of processing, such as offline CAP demodulation, to obtain the original input sequence.

10.3.3.2 Analysis of Experimental Results

(1) **Constellation analysis of demodulated signals of each sub-band**

Figure 10.17 shows the constellation of each sub-channel signal at the receiving end, before and after equalization.

Figure 10.17a–d show the constellations of the demodulated signal of the *i*th channel ($i = 1, 2, 3, 4$) before equalization, respectively. Figure 10.17e–h show the

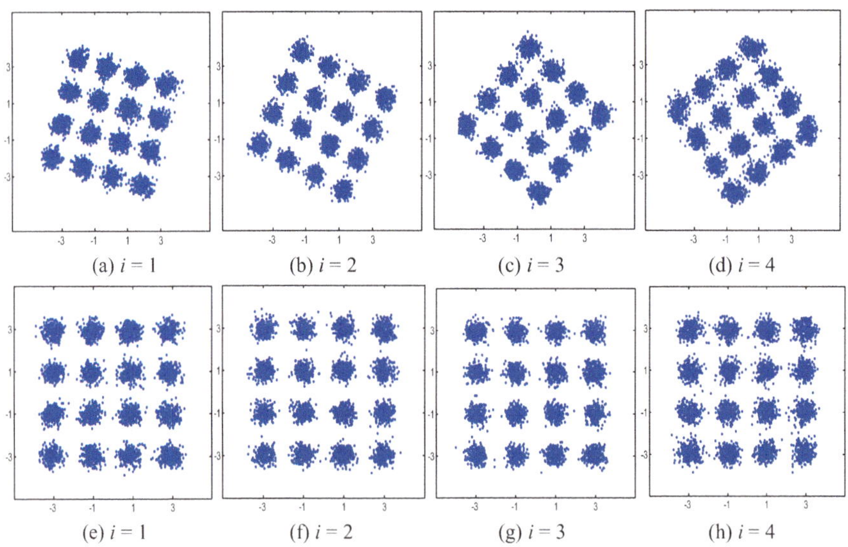

Fig. 10.17 Constellations before (**a–d**) and after (**e–h**) signal equalization at the receiving end [49, 50]

constellations of the equalized demodulated signal of channel i (i = 1, 2, 3, 4), respectively. See Table 10.1 for the specific parameter values in the experiment.

It can be seen from Fig. 10.17a–d that the sampling-clock offset at the receiving end will cause a phase offset of each sub-channel signal at the receiving end [18]. The phase distortion of each sub-channel signal is different; therefore, the phase offset of each sub-channel signal constellation is different. The larger i is, the more serious the phase shift of the signal is.

It can be seen from Fig. 10.17e–h that after equalization with the MMA algorithm, the phase distortion of the signal has been greatly improved, and the system error rate has been greatly reduced, indicating that the equalization effect is good. The experimental results show that for near-distance atmospheric laser communication, the four-band 16CAP modulated signal sent by the sender was well recovered at the receiver; that is, the bit-error rate of each sub-channel signal was very low ($BER = 2.4462x10^{-4}$), and the bit-error rate of each frequency-band signal at the receiver was basically the same. This proves that there is no crosstalk between the signals of each frequency band.

(2) **Eye-diagram analysis of the demodulated signals of each sub-band**

Figure 10.18 shows the eye diagram of each sub-channel signal at the receiving end before and after equalization.

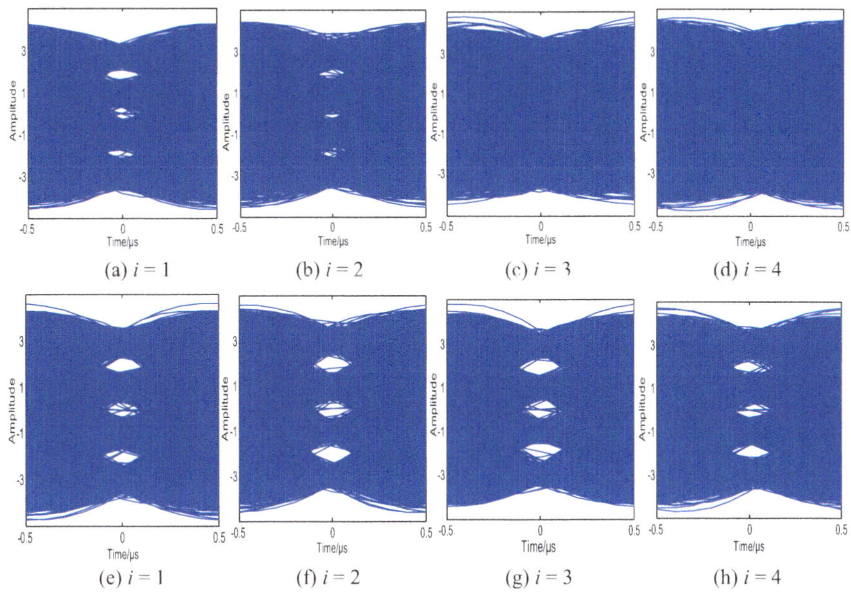

Fig. 10.18 Eye diagrams before (**a–d**) and after (**e–h**) signal equalization at the receiving end [49, 50]

Figure 10.18a–d show the eye diagrams of the demodulated signal of the *i*th channel ($i = 1, 2, 3, 4$) before equalization. Figure 10.18e–h show the eye diagrams of the demodulated signal of the *i*th channel ($i = 1, 2, 3, 4$) after equalization.

It can be seen from Fig. 10.18a–d that the intersymbol interference and I/Q two-channel crosstalk caused by the sampling-clock offset reduces the "eyes" of each sub-channel signal eye diagram, and the larger the i value, the smaller the corresponding eye diagram.

It can be seen from Fig. 10.18e–h that after the signal is equalized by the MMA algorithm, the intersymbol interference and I/Q two-channel crosstalk of each sub-channel signal are greatly suppressed, and the "eyes" of the corresponding signal eye diagrams are more open than before equalization. The BER of each sub-channel signal is also reduced from $BER > 10^3$ to $BER > 2.4462 \times 10^4$, and the transmission performance of the system is good.

Chapter Summary

This chapter analyzed the power-spectrum characteristics of CAP modulation and quadrature amplitude modulation (QAM) signals. Using a gamma–gamma channel model, it studied the effects of the roll-off coefficient α, number of filter taps l, and sampling-clock offset k at the receiver on the system bit-error rate, signal constellation, and eye diagram. It also discussed the changes in the signal constellation and eye diagrams before and after multi-modulus algorithm (MMA) equalization.

In the multi-band CAP modulation system, the time-domain characteristics of each sub-channel modulation filter were studied, and the spectrum characteristics of different band CAP modulation signals were analyzed. Under a gamma–gamma turbulence model, the influences of different light-intensity fluctuation variance values σ_R^2 and signal-to-noise ratios on the performance of a multi-band CAP modulation system were discussed, and the experimental results of an atmospheric laser transmission with four-band CAP modulation were analyzed.

The simulation results showed that in a CAP modulation system, the values of the roll-off coefficient α and filter-tap number l affected the transmission performance of the system; hence, they should be reasonably selected according to specific requirements. The larger the offset k was, the more serious the phase distortion of the signal was. After MMA equalization, the phase offset was well compensated.

In a multi-band CAP modulation system, the larger the number of frequency bands, the lower the bit-error rate of the system. When the variance σ_R^2 was small, the system bit-error rate mainly decreased with the increase of the signal-to-noise ratio, and when σ_R^2 was large, the system bit-error rate mainly decreased with the decrease of σ_R^2. The experimental results showed that all the signals at the receiving end had a certain phase distortion. After equalization by the MMA algorithm, the phase distortion was compensated, and the transmission performance of the system was good.

References

1. Ke X, Yin Z (2009) Coding theory in wireless laser communication system. Science Press, Beijing, pp 23–27
2. Wang Z, Tao L, Wang Y et al (2009) Performance study of 10G/40G carrier-free amplitude-phase modulation system. Opt Optoelectron Technol
3. Zhong K, Zhou X, Gui G et al (2015) Experimental study of PAM-4,CAP-16 and DMT for 100 Gb/s short reach optical transmission systems. OSA 23(2):227026
4. Wang K, Nirmalathas A, Lim C et al (2014) Free-space 120 Gb/s reconfigurable card-to-card optical wireless interconnects with 16-CAP modulation. In: 2014 conference on lasers and electro-optics (CLEO)-laser science to photonic applications, pp SM4J-7
5. Wang Y, Huang X, Tao L et al (2015) 1.8-Gb/s WDM visible light communication over 50 m outdoor free space transmission employing CAP modulation and receiver diversity technology. In: 2015 optical fiber communications conference and exhibition, pp M2F-2
6. Shalash A, Parhi KK (1997) Three-dimensional carrierless AM/PM line code for the unshielded twisted pair cables. In: IEEE international symposium on circuits and systems, vol 3, pp 2136–2139
7. Zhu Z, Ye N, Han S et al (1992) Principle and implementation of carrier-free amplitude-phase modulation system. Data Commun 2:63–67
8. Saltzberg BR (1998) Comparisons of single-carrier and multitoned digital modulation for ADSL applications. IEEE Commun Mag 36(11):114–121
9. Shalash AF, Parhi KK (1999) Multidimensional carrierless AM/PM systems for digital subscriber loops. IEEE Trans Commun 47(11):1655–1667
10. Gao J, Leung YH (1999) A new adaptive equalizer for carrierless amplitude and phase receivers. In: IEEE conference publications, vol 3, pp 90–93
11. Gao J, Leung YH, Cantoni A (2000) Minimum mean square error receive filters for carrierless amplitude and phase modulation. IEEE Int Symp Circ Syst 4:629–632
12. Gao J, Leung YH, Sreeram V (2001) Digital filters for carrierless amplitude and phase receivers. In: Proceedings of IEEE Region 10 international conference on electrical and electronic technology, vol 2, pp 575–579
13. Luo T, Shi B (2002) The core technology of broadband ADSL modulation and coding. Fujian Comput 10:7–9
14. Ingham JD, Penty RV, White IH (2010) Carrierless amplitude and phase modulation for low-cost high-spectral-efficiency optical data communication links. In: 2010 laser science to photonic applications, p CThC5
15. Wieckowski M, Jensen JB, Monroy IT et al (2011) 300 Mbps transmission with 4.6 bit/s/Hz spectral efficiency over 50 m PMMA POF link using RC-LED and multi-lever carrierless amplitude phase modulation. In: 2011 optical fiber communication conference and exposition and the national fiber optic engineers conference, p NTuB8
16. Ingham JD, Penty RV, White IH (2011) 40Gb/s carrierless amplitude and phase modulation for low-cost optical data communication links. In: 2011 optical fiber communication conference and exposition and the national fiber optic engineers conference, p OThZ3
17. Wei JL, Ingham JD, Cunningham DG et al (2012) Performance and power dissipation comparisons between 28 Gb/s NRZ, PAM, CAP and optical OFDM systems for data communication applications. J Lightwave Technol 30(20):3273–3280
18. Geng L, Penty RV, White IH et al (2012) FEC-free 50 m 1.5 Gb/s plastic optical fiber link using cap modulation for home networks. In: 2012 38th European conference and exhibition on optical communications, p Th1B4
19. Stepniak G, Siuzdak J (2013) Transmission beyond 2 Gbit/s in a 100m SI POF with multilevel cap modulation and digital equalization. In: 2013 optical fiber communication conference and exposition and the national fiber optic engineers conference, p NTu3J-5
20. Tao L, Ji Y, Liu J et al (2013) Advanced modulation formats for short reach optical communication systems. IEEE Netw 27(6):6–13

21. Tao L, Wang Y, Gao Y et al (2013) 40 Gb/s 32CAP system with DD-LMS equalizer for short reach optical transmissions. IEEE Photon Technol Lett 25(23):2346–2349
22. Wei JL, Cunningham DG, Penty RV et al (2013) Study of 100 Gigabit Ethernet using carrierless amplitude/phase modulation and optical OFDM. J Lightwave Technol 31(9):1367–1373
23. Wei JL, Geng L, Penty RV et al (2013) 100 Gigabit Ethernet transmission enabled by carrierless amplitude and phase modulation using QAM receivers. In: 2013 optical fiber communication conference and exposition and the national fiber optic engineers conference, p OW4A.5
24. Zhang J, Yu J, Li F et al (2014) Demonstration of 24-Gb/s carrierless amplitude and phase modulation (CAP) 64QAM radio-over-fiber system over 40-GHz Mm-wave fiber-wireless transmission. In: OFC 2014, p M2D-5
25. Estaran J, Olmedo MI, Zibar D et al (2014) First experimental demonstration of coherent CAP for 300-Gb/s metropolitan optical networks. In: OFC 2014, p Th3K-3
26. Tao L, Wang Y, Gao Y et al (2014) High order cap system using DML for short reach optical communications. IEEE Photon Technol Lett 26(13):1348–1351
27. Wang Z, Tao L, Huang X et al (2014) Research on advanced modulation format used in short-distance optical communication system. Opt Commun Res 3:11–15
28. Stepniak G (2014) Comparison of efficiency of N-dimensional CAP modulations. J Lightwave Technol 32(14):2516–2523
29. Wang K, Nirmalathas A, Lim C et al (2015) Experimental demonstration of full-duplex optical wireless personal area communication system with 16-CAP modulation. In: 2015 optical fiber communications conference and exhibition, p M2F-7
30. Lin S, Wang J-B, Wang J-Y et al (2015) Low-timing-sensitivity waveform design for carrierless amplitude and phase modulation in visible light communications. IET J Mag 9(6):317–324
31. Ochi H, Sasaki S (2016) Feasibility study of 100G Ethernet with carrierless amplitude and phase modulation. In: 2016 21st optoelectronics and communications conference held jointly with 2016 international conference on photonics in switching, pp WA2–44, 1–3
32. Jaafar MB, Othman MB, Mohd Ridzuan N et al (2016) Simulation of high dimensionality carrierless amplitude phase modulation technique. In: 2016 IEEE 6th international conference on photonics, pp 1–3
33. Olmedo MI, Tianjian Z, Jensen JB et al (2013) Towards 400GBASE 4-lane solution using direct detection of multiCAP signal in 14 GHz bandwidth per lane. In: 2013 optical fiber communication conference and exposition and the national fiber optic engineers conference, pp PDP5C-10
34. Zhang J, Yu J, Li F et al (2013) 11 × 5 × 10 Gb/s WDM-CAP-PON based on optical single side band multi-level multiband carrierless amplitude and phase modulation with direct detection. In: 39th European conference and exhibition on optical communication, pp 1–3
35. Olmedo MI, Zuo T, Jensen JB et al (2014) Multiband carrierless amplitude phase modulation for high capacity optical data links. J Lightwave Technol 32(4):798–804
36. Wang Z, Tao L, Wang Y et al (2015) 56 Gb/s direct-detection polarization multiplexing multiband CAP transmission. Chin Opt Lett 13(8):1–4
37. Wei J, Cheng Q, Cunningham DG et al (2015) 100 Gb/s hybrid multiband CAP/QAM signal transmission over a single wavelength. J Lightwave Technol 33(2):415–423
38. Haigh PA, Chvojka P, Zvanovec S et al (2015) Experimental verification of visible light communications based on multi-band CAP modulation, pp 1–3
39. Wei JL, Cheng Q, Cunningham DG et al (2015) High performance 400 Gigabit Ethernet links using hybrid multiband CAP/QAM scheme. In: 2015 optical fiber communication conference and exhibition, pp 1–3
40. Haigh PA, Le ST, Zvanovec S et al (2015) Multiband carrierless amplitude and phase modulation for bandlimited visible light communications systems. IEEE Wirel Commun 22(2):46–53
41. Werfli K, Haigh PA, Ghassemlooy Z et al (2015) Multiband carrierless amplitude and phase modulation with Decision feedback equalization for bandlimited VLC systems. In: 2015 4th international workshop on optical wireless communications, vol 7342255, pp 6–10

42. Dong Z, Yu J, Lu J (2015) Bandwidth-efficient WDM-CAP-PON using digital Hilbert single-sideband modulation. IEEE Photon J 7(5):1–8
43. Puerta R, Rommel S, Olmos JJV et al (2016) 10Gb/s ultra-wideband wireless transmission based on multiband carrierless amplitude phase modulation. In: 2016 IEEE 17th annual wireless and microwave technology conference, vol 7483847, pp 1–4
44. Ke X, Deng L (2016) Optical wireless communication. Science Press, Beijing, pp 92–102
45. Xin W, Ying G, Kong Y. Normal blind equalization algorithm: eliminating phase ambiguity caused by constant phase shift and Doppler shift. J Electron Inf
46. Chen L, Yu J, Li X (2016) PDM-16QAM vector signal generation and detection based on intensity modulation and direct detection. Opt Commun 371:15–18
47. Zhang J, Jacky G (2009) Joint CMA+DDLMS blind equalization algorithm. J Tsinghua Univ 49(10):1681–1684
48. Ni Y, Du X, Xiao R et al (2012) Multi-modulus blind equalization algorithm based on high-order QAM genetic optimization. In: 2012 8th international conference on natural computation, pp 679–682
49. Li M (2017) Research on optical wireless communication system with multi-band CAP modulation. Xi'an University of Technology, Xi'an
50. Ke X, Li M (2017) Research on optical wireless communication system without carrier amplitude and phase modulation. Infrared Laser Eng 46(11)

The manufacturer's authorised representative in the EU is Springer Nature Customer Service Centre GmbH, Europaplatz 3, 69115 Heidelberg, Germany. If you have any concerns regarding our products, please contact ProductSafety@springernature.com

Printed and bound by CPI Group (UK) Ltd, Croydon, CR0 4YY
30/09/2026
02225302-0017